AF692763

Heat Transfer Analysis: Recent Challenges and Applications

Heat Transfer Analysis: Recent Challenges and Applications

Guest Editors

Andrzej Frąckowiak
Bartosz Ciupek
Łukasz Brodzik

MDPI

Basel • Beijing • Wuhan • Barcelona • Belgrade • Novi Sad • Cluj • Manchester

Guest Editors
Andrzej Frąckowiak
Institute of Thermal Engineering
Poznan University of Technology
Poznan
Poland

Bartosz Ciupek
Department of Fuels and Renewable Energy
Poznan University of Technology
Poznan
Poland

Łukasz Brodzik
Institute of Thermal Engineering
Poznan University of Technology
Poznan
Poland

Editorial Office
MDPI AG
Grosspeteranlage 5
4052 Basel, Switzerland

This is a reprint of the Special Issue, published open access by the journal *Energies* (ISSN 1996-1073), freely accessible at: https://www.mdpi.com/journal/energies/special_issues/DA0UNX5PD8.

For citation purposes, cite each article independently as indicated on the article page online and as indicated below:

Lastname, A.A.; Lastname, B.B. Article Title. *Journal Name* **Year**, *Volume Number*, Page Range.

ISBN 978-3-7258-5987-0 (Hbk)
ISBN 978-3-7258-5988-7 (PDF)
https://doi.org/10.3390/books978-3-7258-5988-7

© 2025 by the authors. Articles in this book are Open Access and distributed under the Creative Commons Attribution (CC BY) license. The book as a whole is distributed by MDPI under the terms and conditions of the Creative Commons Attribution-NonCommercial-NoDerivs (CC BY-NC-ND) license (https://creativecommons.org/licenses/by-nc-nd/4.0/).

Contents

Plant Extracts as Functional Food Ingredients

Guest Editors

Jaroslawa Rutkowska
Antonella Pasqualone

MDPI

Basel • Beijing • Wuhan • Barcelona • Belgrade • Novi Sad • Cluj • Manchester

Guest Editors
Jaroslawa Rutkowska
Faculty of Human Nutrition
Warsaw University of Life
Sciences (WULS-SGGW)
Warsaw
Poland

Antonella Pasqualone
Department of Soil, Plant and
Food Science (DISSPA)
University of Bari Aldo Moro
Bari
Italy

Editorial Office
MDPI AG
Grosspeteranlage 5
4052 Basel, Switzerland

This is a reprint of the Special Issue, published open access by the journal *Foods* (ISSN 2304-8158), freely accessible at: https://www.mdpi.com/journal/foods/special_issues/Plant_Extracts_Functional.

For citation purposes, cite each article independently as indicated on the article page online and as indicated below:

Lastname, A.A.; Lastname, B.B. Article Title. *Journal Name* **Year**, *Volume Number*, Page Range.

ISBN 978-3-7258-7749-2 (Hbk)
ISBN 978-3-7258-7750-8 (PDF)
https://doi.org/10.3390/books978-3-7258-7750-8

© 2026 by the authors. Articles in this reprint are Open Access and distributed under the Creative Commons Attribution (CC BY) license. The reprint as a whole is distributed by MDPI under the terms and conditions of the Creative Commons Attribution-NonCommercial-NoDerivs (CC BY-NC-ND) license (https://creativecommons.org/licenses/by-nc-nd/4.0/).

Contents

About the Editors

Jaroslawa Rutkowska

Jaroslawa Rutkowska graduated with an M.Sc. in Human Nutrition and Food Technology (in 1995) from the Warsaw University of Life Sciences (WULS-SGGW), Poland. She earned a PhD in Food Technology and Human Nutrition in 2002 and a postdoctoral degree (habilitation) in 2013 (with specialization in Food Chemistry). She is a full professor at the Institute of Human Nutrition Sciences, Faculty of Human Nutrition, Warsaw University of Life Sciences. She supervised three doctoral theses. She is a member of the editorial board of the *Foods* and *Discover Food* journals. In the years 2016–2020, Professor Jaroslawa Rutkowska was appointed as the Head of PhD Studies in the Food Technology and Human Nutrition course at her alma mater. She also served as the Head of the Instrumental Analysis Chair from 2009 to 2023. She participated in 12 projects. Among them, heading a project that resulted in the design and implementation of chromatographic techniques to standardize radiation-sterilized cells and tissues for transplants at the National Centre of Cell and Tissue Banking in Poland is the most important. Her research focuses mainly on food chemistry and food quality, particularly food lipids. Her scholarly contributions are evidenced by nearly 90 publications, including 60 papers in high-ranking international journals, and over 1000 citations, yielding an H-index of 20.

Antonella Pasqualone

Antonella Pasqualone is an Associate Professor of Food Science and Technology at the University of Bari Aldo Moro (Bari, Italy), where she teaches "Cereal-based Food Technology," "Food Quality and Safety Certification," and "Food Science and Technology." She is a member of the quality assurance group for the master's degree course in Food Science and Technology at the University of Bari and a member of the Board of the Doctoral School in Soil and Food Sciences at the same university. Her main research interest is in the field of cereal- and legume-based foods to optimize nutritional properties, reduce or reuse waste, and reevaluate traditional but neglected ethnic foods. In this field, she has been the scientific responsible of numerous national and international projects. She has been invited as fellow professor by the Brussels Institute of Advanced Studies (BrIAS) in Brussels, Belgium, in 2022 and 2023. Her research output consists of 284 publications in international peer-reviewed scientific journals; h-index = 50 (Scopus database).

Preface

Plant derived bioactive compounds have gained considerable importance due to their potential health benefits. The extraction process for these compounds is determined by their chemical nature and can be modified by various factors to increase its effectiveness (and, if possible, improve environmental safety). This Reprint presents a collection of published articles that highlight current progress in advancing knowledge of extraction methods, composition, and potential health-related properties of plant extracts. The collected contributions can provide innovative approaches for the selective extraction of functional compounds, such as polysaccharides and phenolics, and help enhance their applications as fortified food ingredients. This Reprint also delves into the potential of utilizing agro-by-products to obtain rich bioactive extracts. By assembling these insights, this Reprint develops knowledge about obtaining functional ingredients with numerous applications for novel and future food products.

Jaroslawa Rutkowska and Antonella Pasqualone
Guest Editors

foods

Editorial

Plant Extracts as Functional Food Ingredients

Jaroslawa Rutkowska [1,*] and Antonella Pasqualone [2]

1 Institute of Human Nutrition Sciences, Faculty of Human Nutrition, Warsaw University of Life Sciences (WULS-SGGW), Nowoursynowska st.159c, 02-776 Warsaw, Poland
2 Department of Soil, Plant and Food Science (DISSPA), University of Bari 'Aldo Moro', Via Amendola 165/a, 70126 Bari, Italy; antonella.pasqualone@uniba.it
* Correspondence: jaroslawa_rutkowska@sggw.edu.pl

1. Introduction

Plant extracts are a source of phytochemicals with multiple biological activities, so they have always been the basis for the composition of pharmaceuticals, supplements, and functional foods. In fact, the recognition of the relationship between food and health has led to growing demand for functional foods that provide additional physiological benefits beyond their basic nutrition [1]. The key aims of this research area are the identification of new plant sources, the development of effective extraction methods to recover plants' bioactive compounds, and the study of new food applications for the obtained extracts. These steps are complemented by appropriate qualitative–quantitative analyses of both the extracts and the enriched foods to assess their biological properties (in vitro and in vivo) while evaluating the effect of the extracts on the physicochemical and sensory characteristics of the foods being functionalized.

The bioactive compounds typically present in plant extracts can be highly diverse and can be hydrophilic or lipophilic in nature; they include polyphenols, carotenoids, chlorophylls, tocopherols and tocotrienols, phytoestrogens, phytosterols, polyunsaturated fatty acids, peptides, and polysaccharides. The extraction of these compounds is conditioned by their chemical nature and can be modified by various factors (pH, temperature, pressure) to increase yield and, if possible, improve environmental safety. Several technologies have recently been developed for more environmentally friendly extraction of functional compounds from plant sources, such as ultrasound-assisted extraction (UAE), supercritical (SC) fluid extraction, microwave-assisted extraction (MAE), pulsed electric fields (PEF), and pressurized liquid extraction (PLE) [2–5], although their full industrial scalability is still under study. Also, the use of natural deep eutectic solvents (NADES) has been proposed as a green option compared to most organic solvents commonly used in extraction processes (e.g., methanol, acetone, hexane) [4,5].

Thermomechanical stress associated with the processing steps in food preparation can degrade bioactive compounds, necessitating the adjustment of processing parameters by choosing milder conditions where possible [6]. Once bioactive compounds have been added to food, the digestion process can affect their bioavailability and beneficial effects [7]. Effective encapsulation methodologies are thus needed to improve the stability of bioactive compounds and preserve their characteristics until they are delivered to the biological target [8].

Besides having a positive effect on health, functional foods may also possess improved technological features, such as longer storability, compared to conventional foods in the same category. So, another benefit of the addition of plant extracts is improved shelf-life, generally observed when the added compounds possess antioxidant properties [9].

However, we must carefully consider the effect of plant extracts on the physicochemical and especially sensory characteristics of foods. Negative sensory effects and color alterations may result from supplementation, which could be mitigated by balancing the amount of extract added or through encapsulation, which is useful in reducing undesirable flavors or tastes [10,11]. Future research addressing this problem is essential to significantly improve consumers' acceptance of these foods.

This Special Issue presents articles that discuss all of the abovementioned research directions, from optimizing extraction methods to evaluating the effects of extracts as new ingredients (including assessing the impact of digestion), and thus represents a valuable resource for researchers and food industry professionals.

2. Overview of the Contributions

Thirty-six articles were submitted to this Special Issue. Following a rigorous review process, 16 articles were published, including 14 research articles, 1 review, and 1 communication. The articles published in this Special Issue address the most recent advances in this field.

Plants are a valuable source of polysaccharides and have received much attention from researchers due to their strong biological activity, low cytotoxicity, safety, and beneficial roles in the modulation of gut microbiota composition. In this Special Issue, three papers present new knowledge about plant polysaccharides: Phomkaivon et al. [contribution 1], Wu et al. [contribution 6], and Fang et al. [contribution 16].

Green macroalgae *Ulva rigida* C. are rich in biologically active phytochemicals and possess high levels of polysaccharides (about 18% of their dry weight). Phomkaivon et al. [contribution 1] conducted a study on water extraction to obtain the maximum level of polysaccharides from oven-dried *Ulva rigida* (green macroalgae). The extraction factors analyzed included the biomass–solvent ratio, extraction temperature, extraction time, and extraction cycle. The results indicated that higher extraction temperatures led to a significant increase in polysaccharide yields. Polysaccharide extracts obtained from *Ulva rigida* exhibited antioxidative potential, probably because of the substantial presence of phenolics and flavonoids. The supplementation of polysaccharides extracted from *Ulva rigida* promoted *Lactobacillus rhamnosus* growth during cultivation. These findings demonstrated that polysaccharide extracts from *Ulva rigida* have the potential to serve as functional ingredients future food products, with numerous applications in the nutraceutical and pharmaceutical industries.

Tartary buckwheat green leaves, underappreciated by-products in the buckwheat industry, are rich in pectic polysaccharides. Wu et al. [contribution 6] aimed to optimize the extraction conditions of pectic polysaccharides from Tartary buckwheat green leaves using deep-eutectic-solvent-based techniques. The efficiency and selectiveness of these methods were compared with those of conventional hot water. The results revealed that both deep-eutectic-solvent extraction (DESE) and high-pressure-assisted deep-eutectic-solvent extraction (HPDEE) not only improved the extraction efficiency of pectic polysaccharides from Tartary buckwheat green leaves (TBP), but also regulated their structural properties and beneficial effects. In addition to high extraction efficiency, both methods enabled the extraction of valuable rhamnogalacturonan-I (RG-I) pectic fractions. High antioxidative potential and anti-glycosylation properties were noted in TBP extracts obtained using DESE and HPDEE. These findings demonstrated that developed deep-eutectic-solvent-based extraction methods, especially the HPDEE method, are promising techniques for obtaining selective RG-I from TBP.

Fang et al. [contribution 16] comprehensively reviewed the most recent literature about Konjac glucomannan (KGM), a major polysaccharide from the corm of *Amorphophallus* konjac plants that grow in Southeast Asia and Africa. KGM, a hydrocolloid dietary fiber, has attracted increasing attention as a bioactive polysaccharide due to its functional and nutraceutical features. Because of its water absorptivity, stability, film-forming, thickening, and emulsifying properties, KGM is used as a food additive. In the review by Fang et al. [contribution 16], the chemical structure, physicochemical properties, extraction, and purification process of KGM were discussed. On an industrial scale, KGM is obtained via dry-processing, which includes the washing, peeling, slicing, fixing, drying, grinding, and screening of KGM. In the wet-processing of KGM, a liquid medium is used that dissolves the KGM and removes soluble impurities. Using some solvents, ethanol prevents agglomeration. The damp method has many advantages, but is more expensive than dry methods. Regarding KGM's structure and properties, its high molecular weight and ability to form high viscosity and adsorb water were underlined. Its physical and chemical properties make KGM an excellent preservative and fat substitute to maintain the proper texture in foods for diabetic patients. Using KGM is beneficial in food products with attributes important for diabetes, including a low-glycemic index, reduced fat, and the ability to maintain weight loss and fullness. The review by Fang et al. [contribution 16] also discusses the biomedical function of KGM in type 2 diabetes mellitus, specifically regulating blood lipids, glucose homeostasis, oxidative stress and inflammation, and the gut microflora. KGM was shown to have the most significant potential for lowering LDL and total cholesterol, promoting weight loss, and aiding in diabetic control. The administration of KGM to diabetic rats had an anti-diabetic effect by increasing gene up-regulation and insulin pathway expression. This resulted in normalized insulin secretion and lowered blood sugar levels. KGM positively regulates oxidative stress and inflammation, indicating that it may effectively prevent or treat diabetes. In conclusion, Fang et al. [contribution 16] indicated that KGM can be used as a food additive to improve food products' taste, flavor, and appearance. However, the authors indicated that high doses of KGM may cause minor side effects such as hiccups, bloating, and diarrhea, and there is a risk of choking because of its high water absorption capacity.

In recent years, there has been significant interest in utilizing by-products and waste as sources of valuable functional ingredients [contributions 3 and 4]. In the study by Meng et al. [contribution 3], walnut meal was used as a raw material to prepare polypeptide, and its angiotensin-converting enzyme (ACE)-inhibitory activity was investigated. The fraction of ACE exhibited strong antioxidant capacity and total reducing power in vitro. The ACE inhibition retention rate was more than 90% after in vitro simulated gastrointestinal digestion, which provided a reference for the further development of ACE-inhibiting peptides in walnuts. The main finding of Meng et al.'s study provides a reference for applying low-molecular-weight walnut peptide as an antioxidant and ACE inhibitor. Glišic et al. [contribution 4] utilized post-harvest sunflower and maize stalk residues to prepare ethanolic extract, which can be used as a functional ingredient in pork liver pâtés. A substantial level of phenolic compounds was noted in the maize stalk residue extract and sunflower stalk ethanolic extract. When incorporated into pork liver pâtés, these extracts exhibited a noticeable antimicrobial effect. The crop ethanolic extracts increased the n-6 and total PUFA contents in the pâtés and improved the PUFA/SFA ratio. However, they proved ineffective in preventing lipid oxidation. Adding crop residue extracts influenced the instrumental color and intensified the flavor and aroma of the pork liver pâtés, resulting in lower product acceptability. The results of Glišic' et al. [contribution 4] indicated the potential of these extracts obtained from agricultural residues as a source of natural preservatives for meat product applications. These findings could be valuable in developing functional pork liver pâtés without added synthetic preservatives.

Flaxseed and rapeseed are emerging as key sources of phytochemicals in the functional food arena. In addition to being one of the richest sources of PUFAs, especially α-linolenic acid, flaxseed, and rapeseed, they are considered an essential source of high-quality protein. Jarošová et al. [contribution 2] evaluated the protein profiles of flaxseed products with different protein contents (seed cake, flour, and protein concentrate). The proteomic profile of flaxseed products included 2560 protein groups; among them, 33 had a relative abundance within the protein pool of 69–95%. Alkaline solubilization increased the relative abundance of 11S globulin-type proteins in the resulting protein concentrate, which have high potential for food applications due to their properties and amino acid composition. Selecting a suitable cultivar could be an important factor in obtaining a protein concentrate with the highest abundance of 11S globulins.

Amarowicz et al.'s study [contribution 7] thoroughly examined rapeseed's antioxidative potential and phenolic compounds. Rapeseed was derived from three varieties (Castilla, California, and Nelson F1) and was cultivated using different nitrogen and sulfur fertilization techniques. Diversifying fertilization significantly influenced the content of phenolic compounds in extracts of rapeseed. Intensive fertilization reduced the content of phenolic compounds in the Nelson F1 seeds, while economic fertilization decreased the phenolic content in the California variety. Antioxidant assessments through ABTS and FRAP assays revealed cultivar-specific responses. The California variety exhibited higher antioxidant activity in intensively cultivated seeds, whereas the Nelson F1 variety showed stronger antiradical activity in spare-cultivated seeds. These findings highlight the complex relationship between fertilization practices, phenolic compound accumulation, and the antioxidant activity of rapeseed.

In the human diet, fruits are a source of functional ingredients such as polyphenols (flavonoids, flavonols, and anthocyanins), phenolic acids, carotenoids, and vitamins. In this Special Issue, three papers analyzed the composition and properties of fruit extracts: Meremäe et al. [contribution 8], Dominguez-Valencia et al. [contribution 9], and Stockton and Al-Dujaili [contribution 15].

Meremäe et al. [contribution 8] studied the antibacterial effect of aqueous extract and 30% ethanolic extracts of chokeberry, blackcurrant, and rowan berries and their pomaces on the growth of *Listeria monocytogenes*, *Staphylococcus aureus*, *Escherichia coli*, and *Campylobacter jejuni*. Also, the polyphenolic profile and antioxidant activity of the extracts of the analyzed berries and their pomaces were investigated. The ethanolic extracts of chokeberry and blackcurrant berries and their pomaces were found to contain the highest levels of polyphenols and antioxidative potential compared with rowan berries and their pomace. The antibacterial activity of the extracts depended on the extraction solvent and mainly occurred for the ethanolic extracts. The ethanolic chokeberry extract had the most potent antibacterial activity against *Staphylococcus aureus*, which could be explain by it possessing the highest content of total polyphenols. Also, the antibacterial effect of ethanolic and aqueous extracts of blackcurrant and its pomace against Staphylococcus aureus and *Listeria monocytogenes* should be highlighted. This is probably due to the more diverse polyphenolic composition and higher content of anthocyanins (e.g., delphinidin rutinoside, cyaniding hexoside 2, and cyanidin rutinoside) in blackcurrant and its pomace compared with other berries and their pomaces. The findings obtained by Meremäe et al. [contribution 8] suggest that chokeberry and blackcurrant and their pomaces are a good source of polyphenols with antioxidative properties, and they also exhibit antibacterial activity against some foodborne pathogenic bacteria.

Dominguez-Valencia et al. [Contribution 9] characterized the hydrophilic compounds extracted from elderberries and the main lipophilic bioactive compounds retained in their solid residues. Also, their study investigated the effect of encapsulation on the concentration of hydrophilic bioactive compounds and their antioxidant capacity. The polyphenol profile of hydrophilic elderberry extract was rich with cyanidin-derived anthocyanins, characterized by high amounts of cyanidin-3-O-glucoside and cyanidin-3-O-sambubioside and non-anthocyanins compounds, such as hydroxycinnamic acids (5-O-caffeoylquinic acids) and flavonols (rutin). Encapsulation led to a significant reduction in the extract's phenolic and anthocyanin contents and antioxidant capacity, but it also made the extract more stable. Solid residues of elderberry extracts were rich in macronutrient lipids (about 17 g/100 g) and proteins (about 15.7 g/100 g). A high content of essential fatty acids and high proportions of tocopherols were noted in the lipophilic fraction of the solid residue. The critical finding of Dominguez-Valencia et al. [contribution 9] was the development of a process for obtaining a novel, stable, and versatile food ingredient (encapsulated elderberry extract). This process includes extracting and stabilizing (encapsulating) hydrophilic compounds (mainly phenolics and anthocyanins), utilizing solid residue to extract lipophilic bioactive compounds, and possibly producing protein isolates.

Polyphenols in fruits have received much attention because of their health-promoting functionalities. Stockton and Al-Dujaili [contribution 15] evaluated the effect of pomegranate extract and juice intake on satiety parameters in healthy volunteers. The experimental group included twenty-eight subjects (men and women aged 18–65 years) with a body mass index (BMI) between 18 and 34.9 kg/m^2. Over 3 weeks, the participants consumed pomegranate capsules (210 mg punicalgin) served with pomegranate juice. After this period, the participants recorded their satiety using visual analog scales. The results of the study were promising. Participants who consumed pomegranate capsules had higher satisfaction and feelings of fullness than those in the placebo group. The authors declared that further trials with a larger number of volunteers and an assessment of the effect of the pomegranate extract and juice on satiety parameters are required.

While plants are valuable sources of bioactive phytochemicals that can be used as functional ingredients and nutraceuticals in food products, it is also essential to evaluate their medicinal properties. Kelebek et al. [contribution 5] focused on studying the effects of infusion time, temperature, and solvent conditions on the color, antioxidant capacity, total phenolic content, phenolic profile, and antimicrobial activity of Guayusa (*Ilex guayusa* Loes.) ethanolic and water extracts. The phenolic profile of the extracts included 29 compounds with a dominant share of chlorogenic acid and its derivatives. The highest phenolic content in water and ethanolic extract was obtained at 70 °C-8 h. Antimicrobial activity analysis showed that Guayusa tea extracts generally have high levels of activity against Gram-positive pathogenic bacteria. Depending on the digestion stage, the total levels of phenolic compounds in the samples prepared with both solvents increased after oral intake. These findings provide valuable insights into functional foods, with implications for the tea industry and public health.

Nunes-Silva et al. [contribution 12] conducted a comprehensive analysis of decoctions, infusions, and hydroethanolic extracts of three plants, lemon balm (*Melissa officinalis* L.), sage (*Salvia officinalis* L.), and spearmint (*Mentha spicata* L.), to assess their polyphenolic profiles, bioactive properties, and anti-inflammatory and cytotoxic properties. The extracts had a higher content of total phenolic acids than total flavonoids. Some phenolic acids, rosmarinic, salvianolic, and lithospermic acid A, were assayed in all extracts. The extracts revealed antimicrobial activity against primary foodborne pathogens. Among the studied extracts, spearmint extract showed anti-inflammatory potential. It should be noted that all analyzed extracts showed antibacterial, antifungal, and antioxidant potential. The findings of this investigation emphasize the potential value of sage, spearmint, and lemon balm extracts as natural food ingredients to prevent spoilage, provide health benefits, and replace artificial additives in food products.

Kim et al. [contribution 10] demonstrated the preventive effect of *Allium macrostemon* Bunge extract against dysfunction in adipose tissue and the liver under co-exposure to bisphenol A (BPA) and a high-fat diet using a mouse model. *Allium macrostemon* supplementation improved adipose tissue dysfunction and reduced hepatic stress levels in the high-fat-diet- and BPA-treated mice. The authors underlined that further research investigating the specific impact of *Allium macrostemon* Bunge extract is necessary to determine its suitability as a potential food component to alleviate the harmful effect of these inflammation factors.

Calonico and De La Rosa-Millan [contribution 11] analyzed the role of bioactive compounds of hot water extracts of the Mexican medicinal plants *Ludwigia octovalvis*, *Cnidoscolus aconitifolius*, and *Crotalaria longirostrata* against some digestive enzymes. All three plant extracts exhibited enzymatic inhibition. The highest antioxidative potential showed extracts obtained from Cnidoscolus aconitifolius. The study stated that flavonoids can inhibit the activity of α-amylase and α-glucosidase, pivotal enzymes involved in the breakdown of carbohydrates.

Cuminum cyminum L. (cumin) is a herbaceous plant from the *Apiaceae* family. Cumin seeds are rich in bioactive constituents, such as terpenes, phenols, and flavonoids, and are appreciated for their many biological functions. Ishida et al. [contribution 14] investigated the effect of the oral administration of cumin seed aqueous extract in ovalbumin-induced allergic rhinitis. The study was conducted using mice as animal models. The in vivo experiments showed that the oral administration of cumin seed aqueous extract reduced the sneezing frequency in mice with allergic rhinitis induced by ovalbumin. Their findings suggest that ingesting aqueous extract of cumin seeds improves T-cell balance and alleviates the allergy symptoms of allergic rhinitis in vivo. These findings may provide insights into the biological functions of cumin seed ingredients in functional foods and contribute to the effective utilization of cumin seed residue after oil extraction.

Pagliari et al. [contribution 13] evaluated cinnamon *Cinnamomum verum* J. Presl bark extract with respect to its antioxidant and anti-inflammatory bioactivity after simulated digestion. They analyzed the chemical profiles of the pure and digested extracts, as well as their cellular effects, in in vitro models such as Caco2 and the intestinal barrier. The results showed that the digestive process reduces the total content of polyphenols, especially tannins, while preserving other bioactive compounds, such as cinnamic acid. These findings revealed that aqueous solvents can maintain the beneficial properties of cinnamon bark. Also, their antioxidant and anti-inflammatory properties were maintained after digestion.

3. Conclusions

This Special Issue showcases the diverse approaches to using eco-sustainable processes to obtain plant extracts. The findings of the studies indicated that using aqueous solvents can maintain the properties of plant extracts that are beneficial for human health. The studies also underscored innovations in utilizing agro-by-products to obtain rich bioactive extracts. These studies align with the goal of developing sustainable systems, reducing waste, and promoting healthier diets and could be of interest to the producers of functional food to encourage the effective utilization of plant extracts in the development of new products.

Author Contributions: Conceptualization, J.R. and A.P.; writing—original draft preparation, J.R. and A.P.; writing—review and editing, J.R. and A.P. All authors have read and agreed to the published version of the manuscript.

Funding: This research received no external funding.

Institutional Review Board Statement: Not applicable.

Informed Consent Statement: Not applicable.

Data Availability Statement: Not applicable.

Acknowledgments: As Guest Editors, we would like to express our appreciation to all authors whose valuable work was published in this Special Issue and thus contributed to its success.

Conflicts of Interest: The authors declare no conflicts of interest.

List of Contributions:

1. Phomkaivon, N.; Pongponpai, P.; Kosawatpat, P.; Thongdang, B.; Panutai, W. Extraction, characterisation and evaluation of antioxidant and probiotic growth potential of water-soluble polysaccharides from *Ulva rigida* macroalgae. *Foods* **2024**, *13*, 1630. https://doi.org/10.3390/foods13111630.
2. Jarošová, M.; Roudnický, P.; Bárta, J.; Zdráhal, Z.; Bártová, V.; Stupková, A.; Lorenc, F.; Bjelková, M.; Kyselka, J.; Jarošová E.; et al. Proteomic profile of flaxseed (*Linum usitatissimum* L.) products as influenced by protein concentration method and cultivar. *Foods* **2024**, *13*, 1288. https://doi.org/10.3390/foods13091288.
3. Meng, M.; She, Z.; Feng, Y.; Zhang, J.; Han, R.; Qi Y.; Sun L.; Sun, H. Optimization of extraction process and activity of angiotensin-converting enzyme (ACE) inhibitory peptide from walnut meal. *Foods* **2024**, *13*, 1067. https://doi.org/10.3390/foods13071067.
4. Glišić, M.; Bošković Cabrol, M.; Čobanović, N.; Starčević, M.; Samardžić, S.; Veličković, I.; Maksimović, Z. The effects of sunflower and maize crop residue extracts as a new ingredient on the quality properties of pork liver pâtés. *Foods* **2024**, *13*, 788. https://doi.org/10.3390/foods13050788.
5. Kelebek, H.; Sasmaz H.K.; Aksay O.; Selli S.; Kahraman O.; Fields C. Exploring the impact of infusion parameters and in vitro digestion on the phenolic profile and antioxidant Capacity of Guayusa (*Ilex guayusa* Loes.) Tea Using Liquid Chromatography, Diode Array Detection, and Electrospray Ionization Tandem Mass Spectrometry. *Foods* **2024**, *13*, 694. https://doi.org/10.3390/foods13050694.
6. Wu, D.T.; Lei, J.; Li, J.; Qu Mo, M.M.; Li, W.B.; Huang, Y.J.; Hu, Y.C.; Wang, A.L.; Zou, L. Efficient and selective extraction of rhamnogalacturonan-I-enriched pectic polysaccharides from tartary buckwheat leaves using deep-eutectic-solvent-based techniques. *Foods* **2024**, *13*, 625. https://doi.org/10.3390/foods13040625.
7. Amarowicz, R.; Cwalina-Ambroziak, B.; Janiak, M.A.; Damszel, M.; Stępień, A.; Sulewska, K.; Karamać, M.; Penkacik, K. Effect of fertilization on phenolics of rapeseeds and their antioxidant potential. *Foods* **2024**, *13*, 561. https://doi.org/10.3390/foods13040561.
8. Meremäe, K.; Raudsepp, P.; Rusalepp, L.; Anton, D.; Bleive, U.; Roasto, M. In vitro antibacterial and antioxidative activity and polyphenolic profile of the extracts of chokeberry, blackcurrant, and Rowan Berries and their pomaces. *Foods* **2024**, *13*, 421. https://doi.org/10.3390/foods13030421.
9. Domínguez-Valencia, R.; Cittadini, A.; Pateiro, M.; Munekata, P.E.S.; Lorenzo, J.M. Elderberry lipophilic and hydrophilic bioactive compounds: characterization and extract encapsulation. *Foods* **2023**, *12*, 4233; https://doi.org/10.3390/foods12234233.
10. Kim, J.; Kim, N.H.; Youn, I.; Kyoung Seo, E.; Kim, C.Y. Effects of allium macrostemon bunge extract on adipose tissue inflammation and hepatic endoplasmic reticulum stress in high-fat diet-fed and Bisphenol A-treated C57BL/6N mice. *Foods* **2023**, *12*, 3777. https://doi.org/10.3390/foods12203777.
11. Calonico, K.; De La Rosa-Millan, J. Digestion-related enzyme inhibition potential of selected Mexican medicinal plants (*Ludwigia octovalvis* (Jacq.) P.H. Raven, *Cnidoscolus aconitifolius* and *Crotalaria longirostrata*). *Foods* **2023**, *12*, 3529. https://doi.org/10.3390/foods12193529.
12. Silva, B.N.; Cadavez, V.; Caleja, C.; Pereira, E.; Calhelha, R.C.; Añibarro-Ortega, M.; Finimundy, T.; Kostić, M.; Soković, M.; Teixeira, J.A.; et al. Phytochemical composition and bioactive potential of *Melissa officinalis* L., *Salvia officinalis* L. and *Mentha spicata* L. extracts. *Foods* **2023**, *12*, 947. https://doi.org/10.3390/foods12050947.

13. Pagliari, S.; Forcella, M.; Lonati, E.; Sacco, G.; Romaniello, F.; Rovellini, P.; Fusi, P.; Palestini, P.; Campone L.; Labra M.; et al. Antioxidant and anti-inflammatory effect of cinnamon (*Cinnamomum verum* J. Presl) bark extract after in vitro digestion simulation. *Foods* **2023**, *12*, 452. https://doi.org/10.3390/foods12030452.
14. Ishida, M.; Miyagawa, F.; Nishi, K.; Sugahara, T. Aqueous extract from *Cuminum cyminum* L. seed alleviates ovalbumin-induced allergic rhinitis in mouse via balancing of helper T cells. *Foods* **2022**, *11*, 3224. https://doi.org/10.3390/foods11203224.
15. Stockton, A.; Al-Dujaili, E.A.S. Effect of pomegranate extract consumption on satiety parameters in healthy volunteers: A preliminary randomized study. *Foods* **2022**, *11*, 2639. https://doi.org/10.3390/foods11172639.
16. Fang, Y.; Ma, J.; Lei, P.; Wang, L.; Qu, J.; Zhao, J.; Liu, F.; Yan, X.; Wu, W.; Jin, L.; Ji. H.; Sun, D. Konjac glucomannan: An emerging specialty medical food to aid in the treatment of type 2 diabetes mellitus. *Foods* **2023**, *12*, 363. https://doi.org/10.3390/foods12020363.

References

1. ILSI–International Life Sciences Institute, North America Technical Committee on Food Components for Health Promotion. Safety Assessment and Potential Health Benefits of Food Components Based on Selected Scientific Criteria. *Crit. Rev. Food Sci. Nutr.* **1999**, *39*, 203–206. [CrossRef]
2. Soquetta, M.B.; Terra, L.D.M.; Bastos, C.P. Green technologies for the extraction of bioactive compounds in fruits and vegetables. *CyTA-J. Food* **2018**, *16*, 400–412. [CrossRef]
3. Ranjha, M.M.A.N.; Kanwal, R.; Shafique, B.; Arshad, R.N.; Irfan, S.; Kieliszek, M.; Kowalczewski, P.Ł.; Irfan, M.; Khalid, M.Z.; Roobab, U.; et al. A Critical Review on Pulsed Electric Field: A Novel Technology for the Extraction of Phytoconstituents. *Molecules* **2021**, *26*, 4893. [CrossRef] [PubMed]
4. Tapia-Quirós, P.; Granados, M.; Sentellas, S.; Saurina, J. Microwave-assisted extraction with natural deep eutectic solvents for polyphenol recovery from agrifood waste: Mature for scaling-up? *Sci. Tot. Environ.* **2024**, *912*, 168716. [CrossRef] [PubMed]
5. Martins, R.; Barbosa, A.; Advinha, B.; Sales, H.; Pontes, R.; Nunes, J. Green extraction techniques of bioactive compounds: A state-of-the-art review. *Processes* **2023**, *11*, 2255. [CrossRef]
6. Pasqualone, A.; Delvecchio, L.N.; Gambacorta, G.; Laddomada, B.; Urso, V.; Mazzaglia, A.; Ruisi, P.; Di Miceli, G. Effect of supplementation with wheat bran aqueous extracts obtained by ultrasound-assisted technologies on the sensory properties and the antioxidant activity of dry pasta. *Nat. Prod. Comm.* **2015**, *10*, 1739–1742. [CrossRef]
7. Carbonell-Capella, J.M.; Buniowska, M.; Barba, F.J.; Esteve, M.J.; Frígola, A. Analytical methods for determining bioavailability and bioaccessibility of bioactive compounds from fruits and vegetables: A review. *Compr. Rev. Food Sci. Food Saf.* **2014**, *13*, 155–171. [CrossRef] [PubMed]
8. da Silva, B.V.; Barreira, J.C.; Oliveira, M.B.P. Natural phytochemicals and probiotics as bioactive ingredients for functional foods: Extraction, biochemistry and protected-delivery technologies. *Trends Food Sci. Technol.* **2016**, *50*, 144–158. [CrossRef]
9. Leyva-Porras, C.; Román-Aguirre, M.; Cruz-Alcantar, P.; Pérez-Urizar, J.T.; Saavedra-Leos, M.Z. Application of Antioxidants as an Alternative Improving of Shelf Life in Foods. *Polysaccharides* **2021**, *2*, 594–607. [CrossRef]
10. Kovsari, E.; Gilani, P.S.; Shokri, S.; Borazgh, A.M.; Rezagholizade-Shirvan, A.; Nia, A.P. Influence of green pepper extract on the physicochemical, antioxidant, and sensory properties of stirred yogurt. *Food Chem. X* **2024**, *21*, 101070. [CrossRef] [PubMed]
11. Abtahi, S.A.; Khoshkhoo, Z.; Khorshidian, N.; Mohammadi, M. Alginate Microcapsules Loaded with Eryngium billardieri Extract and Its Application in Apple Juice. *J. Food Qual.* **2024**, *2024*, 6858230. [CrossRef]

Disclaimer/Publisher's Note: The statements, opinions and data contained in all publications are solely those of the individual author(s) and contributor(s) and not of MDPI and/or the editor(s). MDPI and/or the editor(s) disclaim responsibility for any injury to people or property resulting from any ideas, methods, instructions or products referred to in the content.

foods

MDPI

Article

Extraction, Characterisation and Evaluation of Antioxidant and Probiotic Growth Potential of Water-Soluble Polysaccharides from *Ulva rigida* Macroalgae

Naraporn Phomkaivon [1], Preeyanut Pongponpai [2], Prapat Kosawatpat [3], Bussaba Thongdang [3] and Wanida Pan-utai [4,*]

1 Department of Food Chemistry and Physics, Institute of Food Research and Product Development, Kasetsart University, Bangkok 10900, Thailand; ifrnpph@ku.ac.th
2 Faculty of Science, Prince of Songkla University, Songkhla 90110, Thailand; preeyanut1415@gmail.com
3 Phetchaburi Coastal Aquaculture Research and Development Center, Coastal Aquaculture Research and Development Division, Department of Fisheries, Phetchaburi 76100, Thailand; prapat1120@gmail.com (P.K.); nok.annann@gmail.com (B.T.)
4 Department of Applied Microbiology, Institute of Food Research and Product Development, Kasetsart University, Bangkok 10900, Thailand
* Correspondence: ifrwdp@ku.ac.th

Abstract: *Ulva rigida* green macroalgae contain a variety of polysaccharides. A recent study investigated the optimum concentration and yield of polysaccharide extraction from oven-dried *U. rigida* biomass using a water-soluble polysaccharide extraction method that adhered to safety standards. This study utilised complete factorial experiments to examine the effects of varying factors on polysaccharide extraction. Results showed a positive correlation between increased levels of all factors and higher polysaccharide extraction yield. This study also found that the main factors and their interaction had a significant impact on the extracted polysaccharides from *U. rigida*. The highest polysaccharide content and yield were 9.5 mg/mL and 189 mg/g, respectively. Water-soluble polysaccharides demonstrated the presence of reducing sugar (8 mg/g), phenolics (0.69 mg/g) and flavonoids (1.42 mg/g) and exhibited antioxidant properties. Results revealed that freeze-dried polysaccharide powders were primarily composed of the monosaccharide rhamnose. Preliminary results on the effect of these powders on probiotics demonstrated that supplementation of polysaccharides from *U. rigida* promoted viable *Lactobacillus rhamnosus* ATCC 53103 growth during cultivation. This discovery has the potential to revolutionise the human food industry and promote the development of functional ingredients for novel and future food products, with numerous applications in the nutraceutical and pharmaceutical industries.

Keywords: *Ulva*; macroalgae; water extraction; polysaccharide; antioxidant; monosaccharide; probiotic

1. Introduction

Modern consumers are more aware of their well-being, with increasing interest in functional foods. There has been growing awareness of the importance of promoting healthy and sustainable diets as one of the critical strategies to safeguard human and environmental health [1]. Functional foods can be produced by enriching traditional foods with bioactive compounds. These compounds can be sourced from various materials including plants, macroalgae and microalgae and have demonstrated significant health benefits. However, the incorporation of these compounds into functional food formulations requires the extraction of the desired compounds. In the food industry, there is a growing preference for natural products, emphasising product safety, quality and functionality, as well as utilising environmentally friendly techniques that avoid harmful solvents. To achieve these goals, researchers and industry practitioners are exploring various methods of extraction that eliminate the need for toxic solvents [2]. This approach ensures that the production process is environmentally sustainable while maintaining the final product's safety and quality [3].

Seaweeds, also referred to as macroalgae, are categorised based on their chlorophyll nature, cell wall chemistry and flagella presence [2]. These marine plants are a rich and abundant source of biologically active phytochemicals, which include carotenoids, phycobilins, fatty acids, polysaccharides, vitamins, sterols, tocopherols and phycocyanins [4]. Phytochemicals show potential for use in healthcare due to their remarkable biological activity. However, the nutritional and chemical composition of seaweeds can vary based on factors such as species, geographical origin, seasonal and environmental fluctuations, harvest time, water temperature and processing techniques [5]. These factors must be considered when evaluating the potential use of seaweeds as a source of beneficial phytochemicals. Macroalgae are known to possess a significant quantity of carbohydrates accumulated within their cells. These carbohydrates, including monosaccharide, oligosaccharide and polysaccharide, provide the human body with ample energy and support various physiological functions and physical activities. Essential carbohydrates cannot be substituted with alternative energy sources [6]. Green macroalgae are rich in polysaccharides, which are of considerable interest [7].

The *Ulva* species of macroalgae exhibit a cell wall composition primarily constituted by polysaccharides, accounting for approximately 18% of dry weight [8]. Methods for polysaccharide extraction from cells include hot water, acidic and alkaline conditions and mechanical-assisted techniques through microwave, ultrasonic and enzymatic hydrolysis [9]. Due to its high polarity, non-toxicity, safety in production and ability to effectively penetrate plant tissues, water is the favoured solvent for extracting polysaccharides. The conventional approach for extracting polysaccharides involves hot-water extraction [10]. Recent investigations have revealed that *Ulva* polysaccharides have diverse physiological properties such as antioxidant, anticoagulant, antitumour, anti-ageing and immune regulatory activities [11]. These findings have been substantiated in various scientific studies, further highlighting the immense potential of polysaccharides from *Ulva* spp. as chemical agents and medicines in the medical and agricultural industries [12,13]. Polysaccharides derived from macroalgae are considered functional prebiotics. These polysaccharides cannot be easily digested by the human gastrointestinal tract but they can be utilised by the gut microbiota to generate metabolites and produce several beneficial effects [14]. Polysaccharides from macroalgae exhibit the potential to support probiotic growth in the host gut for balanced health and can be applied in healthy food and ingredients [15].

The objective of this study was to investigate the maximum polysaccharides extracted from oven-dried *U. rigida* green macroalgae using full factorial experiments. The water-soluble extract under various extraction conditions was evaluated for polysaccharides, reducing sugar, phenolics and flavonoid contents and antioxidant capacities. Freeze-dried polysaccharides extracted from *U. rigida* were determined for monosaccharide composition and their potential to promote bacteria as a prebiotic substance.

2. Materials and Methods

2.1. Macroalgae Preparation

Green macroalgae *Ulva rigida* C. Agardh or Sea Lettuce were collected from the Phetchaburi Coastal Aquacultural Research and Development Center, Coastal Aquaculture Research and Development Division, Department of Fisheries, Thailand [16]. *U. rigida* algae were cultured in seawater, with salinity controlled at 30–32 ppm. The maximum *U. rigida* biomass was harvested after 21 days of cultivation and washed with fresh water to remove the residual seawater [17]. Fresh macroalgae were dried in a hot air oven, with the temperature controlled at 60 °C for 6 h. The oven-dried biomass preparation was milled to 0.5 mm size and stored in a polyethylene bag in the dark at room temperature.

2.2. Proximate Composition

The chemical composition of *U. rigida* oven-dried biomass was analysed according to AOAC standard methods [18]. Crude protein content was determined using the Kjeldahl method and crude lipid was extracted via Soxhlet extraction using petroleum ether. The resulting lipid was dried until constant weight. Crude fibre content was assessed using both acid and alkaline digestion methods, with the fibre residue also dried until constant weight. Ash content was analysed by igniting dried samples in an electric furnace at 550 °C. Finally, carbohydrate content was calculated by subtracting the sum of moisture, protein, lipid, fibre and ash contents from 100 g of dry matter.

2.3. Polysaccharide Extraction Experiments

Water-soluble polysaccharide extraction from *U. rigida* was performed using a two-level full factorial design with four factors to optimise the significant conditions for maximum polysaccharide extraction. These factors included biomass–solvent ratio, extraction temperature, extraction time and cycle of extraction (Table 1). The *U. rigida* biomass was mixed with deionised (DI) water, with the extraction temperature controlled in a water bath (WNB10, Memmert, Schwabach, Germany). The mixture was hand-mixed at 10 min intervals and the water-soluble polysaccharides extracted were separated using a centrifuge at 3660× *g* for 20 min (Frontier^{TM} 2000 Multi Centrifuges, Ohaus, Parsippany, NJ, USA). DI water was added to the biomass pellet residue for the second extraction cycle, and the water-soluble polysaccharides extracted from *U. rigida* were subjected to further analysis. Moreover, the schematic diagram clearly illustrates the workflow in Figure 1.

Table 1. Two-level full factorial design experiments for water-soluble polysaccharide extraction from *U. rigida* macroalgae.

Factor	Levels	
Biomass-solvent ratio (w/v)	1:20	1:40
Extraction temperature (°C)	70	90
Extraction time (h)	1	2
Number of extraction cycles	1	2

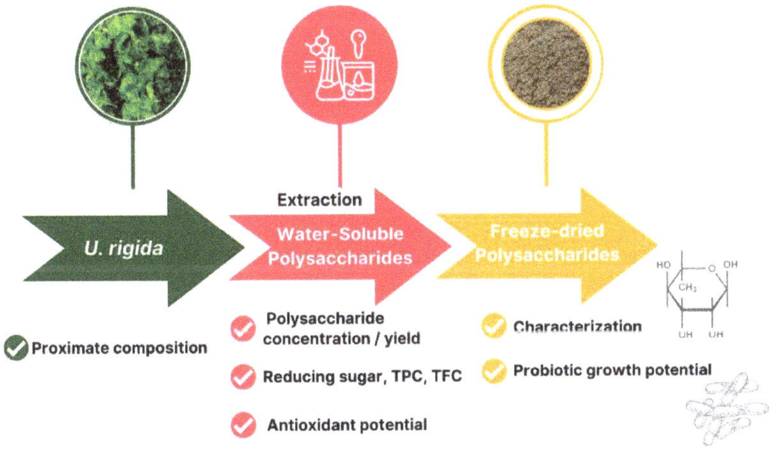

Figure 1. Schematic diagram outlining the process of extracting water-soluble polysaccharides from *U. rigida* macroalgae, followed by characterisation and evaluation of freeze-dried polysaccharides for their probiotic growth potential.

2.4. Analysis

2.4.1. Determination of Polysaccharides

Polysaccharide content was determined via the phenol–sulphuric method according to Wang, et al. [19] with minor modification. First, 0.5 mL of water-soluble polysaccharides extracted from *U. rigida* was placed in a test tube. Then, 5% phenol was added, and the mixture was thoroughly mixed using a vortex mixer. After that, 2.5 mL of sulphuric acid was added, and the mixture was incubated at room temperature for 20 min. Finally, the samples were measured at 490 nm using a spectrophotometer (SP-8001, UV-Vis Spectrophotometer, Metertech, Taipei, Taiwan) with glucose used as the standard. The polysaccharide content was calculated and expressed as glucose in mg/mL (C_{PS}). Polysaccharide extraction yield (*yield*) was calculated with Equation (1) and expressed as milligrams per gram of dried biomass (mg/g).

$$Yield\ (\mathrm{mg/g}) = \frac{C_{PS} \times V}{Biomass} \tag{1}$$

where C_{PS} is the concentration of polysaccharide (mg/mL), V is the DI solvent extraction (mL) and *Biomass* is the dried biomass (g).

2.4.2. Determination of Reducing Sugar

The reducing sugars present were analysed using the dinitrosalicylic acid (DNS) method following Miller (1959) [20] with some modifications. A sample of 0.25 mL was mixed with 0.5 mL of DI water and 0.25 mL of the DNS reagent. This mixture was then boiled in a water bath for 15 min and left to cool in cold water to stop the reaction. After that, 2 mL of DI water was added and the absorbance of the solution was measured at 540 nm using a spectrophotometer (SP-8001, UV-Vis Spectrophotometer, Metertech, Taiwan). Glucose was used as the standard for reducing sugar and expressed as glucose in milligrams per gram of dried biomass (mg/g).

2.4.3. Total Phenolic Content

Total phenolic content (TPC) was measured with the Folin–Ciocalteu method, as described by Pan-utai et al. [17]. Briefly, 20 µL of the sample or standard was mixed with 10% Folin–Ciocalteu reagent (100 µL) in a 96-well plate and the mixture was incubated in the dark at room temperature for 8 min. After that, 80 µL of 7.5% sodium carbonate and 50 µL of DI water were added and the resulting mixture was thoroughly mixed and incubated at 40 °C for 30 min. The absorbance was measured at 750 nm using a microplate reader (M965+, Microplate Reader, Metertech, Taiwan), and gallic acid was used as the standard. The TPC was expressed as milligram gallic acid equivalent per gram of dried biomass (mg GAE/g).

2.4.4. Total Flavonoid Content

Total flavonoid content (TFC) was determined according to the method of Liu et al. [21] with some modifications. A 100 µL aliquot of sample or standard was mixed with an equal amount of aluminium chloride solution. The mixture was then incubated at room temperature for 10 min and the absorbance was measured at 405 nm using a microplate reader (M965+, Microplate Reader, Metertech, Taiwan). Quercetin was used as the standard. The TFC was expressed as milligram quercetin equivalent per gram of dried biomass (mg QE/g).

2.4.5. Antioxidant Activity via DPPH Assay

The DPPH radical-scavenging activity assay was performed following Hung, et al. [22] with minor modifications. In brief, a 100 µL sample or standard was combined with 100 µL of 200 µM DPPH solution (2,2-diphenyl-1-picrylhydrazyl, Sigma-Aldrich, Singapore). After 30 min incubation in the absence of light at room temperature, the absorbance was measured at 750 nm using a microplate reader (M965+, Microplate Reader, Metertech,

Taiwan). Ascorbic acid was used as the standard. The antioxidant capacity was expressed as mg ascorbic acid equivalent per gram of dried biomass (mg AAE/g).

2.4.6. Antioxidant Activity via ABTS Assay

The ABTS radical-scavenging activity assay was conducted following the method described by Pan-utai et al. [17]. In brief, 245 mM ammonium persulphate at 505.05 and 5.05 μL, respectively, and 7 mM ABTS (22′-azino-bis (3-ethylbenzothiazoline-6-sulphonic acid) diammonium salt were mixed to create the ABTS radical solution, and the combination was left to incubate for 16 h at ambient temperature in the dark. The prepared solution was diluted with DI water to provide 0.7 absorbance at 750 nm. Afterwards, 190 μL of ABTS solution was combined with 10 μL of either the sample or the standard on a 96-well plate and incubated for 5 min at room temperature in the dark. A microplate reader (M965+, Microplate Reader, Metertech, Taiwan) was used to measure the absorbance at 750 nm, with ascorbic acid used as a standard. The amount of antioxidant capacity per gram of dried biomass was reported as mg ascorbic acid equivalent (mg AAE/g).

2.4.7. Antioxidant Activity via FRAP Assay

The ferric-reducing antioxidant power (FRAP) assay was conducted following the method described by Pan-utai et al. [17]. In summary, the reagent was made by mixing 10 mM TPTZ (2,4,6-tris(2-pyridyl)-s-triazine) with 300 mM sodium acetate (pH 3.6) in 40 mM HCl and 20 mM ferric chloride at 25, 2.5 and 2.5 mL, respectively. After that, 190 μL of FRAP reagent was combined with 10 μL of either the sample or the standard in a 96-well plate and incubated at room temperature for 30 min in the dark. The absorbance was measured using a microplate reader at 593 nm (M965+, Microplate Reader, Metertech, Taiwan), with ascorbic acid used as a standard. The antioxidant capacity was expressed as mg AAE/g or milligrams of ascorbic acid equivalent.

2.5. Monosaccharide Determination

Water-soluble polysaccharides from *U. rigida* were prepared under the optimum extraction condition using a biomass–solvent ratio of 1:20 (*w*/*v*) and an extraction temperature of 90 °C for 2 h. The supernatant was separated via centrifugation at 3660× *g* for 20 min. The extracted polysaccharides were then stored in a freezer at −30 °C for 18–24 h, followed by freeze-drying 30–60 Pa for 38 h. The resulting freeze-dried polysaccharides from *U. rigida* were used to determine monosaccharides as the substrate to evaluate their effect on probiotics.

Monosaccharide analysis was performed using HPAEC-PAD. The freeze-dried polysaccharides extracted from *U. rigida* were dissolved in ultrapure water (18 MΩ-cm, Merck Millipore, Darmstadt, Germany) at 2 mg/mL concentration. For acid hydrolysis, 200 μL of sample was mixed with 200 μL of 0.2 N trifluoroacetic acid (TFA) for 2 h at 100 °C following the method described by Weitzhandler, et al. [23] with some modifications. The mixture was evaporated under pressure until dryness and then redissolved with deionised water. Both samples were measured via high-performance anion exchange chromatography (HPAEC) using a pulsed amperometric detector (PAD). The system (Dionex™ ICS 5000, Thermo Scientific, Vacaville, CA, USA) was equipped with a CarboPac™ PA20 column (6 μm particle size, 3 × 150 mm) and CarboPac™ PA20 guard column (6 μm particle size, 3 × 30 mm). The isocratic of 16 mM sodium hydroxide at a flow rate of 0.4 mL/min was conducted for 20 min. The injection volume was 5 μl and the column temperature was set at 30 °C. The waveform "standard carbohydrate quad" and a pH-Ag/AgCl reference electrode were used. Six monosaccharides including arabinose, galactose, glucose, fructose, rhamnose and xylose (Sigma-Aldrich, Singapore) were used as standards, with data processed using Chromeleon™ 6.8 Chromatography Data System software.

2.6. Effect of Probiotic

The procedure was modified from the method presented by Song, et al. [24] to analyse the impact of polysaccharides from *U. rigida* on probiotic growth. *Lactobacillus rhamnosus* ATCC 53103 was used as the reference bacterial strain for the probiotics. First, *L. rhamnosus* ATCC 53103 was prepared in MRS broth and incubated at 37 °C for 18 h. Then, 10% (v/v) of *L. rhamnosus* ATCC 53103 was transferred to MRS broth and incubated at 37 °C for another 12 h. The cells were centrifuged at 3660× g for 10 min, washed twice with 0.1% peptone and resuspended in 0.1% peptone. The cells were then diluted to an absorbance of 0.2 at 600 nm to prepare the probiotic bacterial suspension. Next, a *L. rhamnosus* ATCC 53103 bacterial suspension of 10% (v/v) was inoculated into minimum medium broth supplemented with 2% (w/v) freeze-dried polysaccharides from *U. rigida*, while glucose and inulin were utilised as the controls. The bacterial mixture was cultured at 37 °C for 48 h and samples were collected at 0, 6, 12, 24 and 48 h to evaluate the viable cells of *L. rhamnosus* ATCC 53103 using the plate-count method on MRS agar.

2.7. Statistical Analysis

Results were shown as mean values ± standard deviation (SD) for each experiment run in triplicate, with statistical analysis carried out utilising SPSS version 25.0 (IBM Corp., Armonk, NY, USA). The effects of the four parameters—biomass–solvent ratio, extraction temperature, extraction duration and extraction cycle—on polysaccharide content and extraction yield were assessed using two levels of full factorial design. All experiment parameters were compared using Duncan's multiple range test (DMRT) at a significance level of 0.05.

3. Results

Ulva rigida macroalgae are a significant source of polysaccharides which have various applications. Water is a safe solvent that shows promise for extracting bioactive compounds. To achieve the maximum concentration and yield of polysaccharides, oven-dried *U. rigida* biomass was extracted using water. The optimum extraction condition was determined using a two-level complete factorial design of essential factors. Various assays were conducted to determine the reducing sugar, total phenolic and flavonoid contents, as well as antioxidant capacities. The water-soluble polysaccharides extracted were also found to support bacterial growth.

3.1. Proximate Composition

Table 2 presents the chemical composition of oven-dried *U. rigida* macroalgae. The proximate composition analysis revealed that green *Ulva* species primarily comprised polysaccharides with high carbohydrate content.

Table 2. Proximate composition of *U. rigida* oven-dried macroalgae.

Composition	% Dry Weight
Crude protein	24.29 ± 0.29
Crude lipid	0.62 ± 0.09
Crude fibre	6.12 ± 0.01
Ash	20.84 ± 0.26
Carbohydrate	48.14 ± 0.48

3.2. Water-Soluble Polysaccharide Extraction

This study analysed polysaccharide concentration (C_{PS}) and extraction yield from *U. rigida*. Various factors were taken into account including biomass–solvent ratio (ratio), extraction temperature (temp) and time (time) and cycles of extraction (cycle). Analyses of the main effects and interactions between polysaccharide concentration and extraction yield at various conditions are shown in Table 3. Results indicated that all dependent variables significantly differed between the main factors and interaction ($p < 0.05$), except for the yield based on the biomass–solvent ratio. C_{PS} and yield showed no significant interactions between extraction temperature and time, extraction time and cycle, biomass–solvent ratio and temperature and time, biomass–solvent ratio and extraction time and cycle and also biomass–solvent ratio and extraction temperature and time and cycle. Therefore, biomass–solvent ratio, extraction temperature and time and extraction cycle affected C_{PS} and yield.

Table 3. Main effects and interactions of polysaccharide concentration and extraction yield from *U. rigida* macroalgae under various conditions.

Factor	Dependent Variable	*SS*	*df*	*MS*	F-Value	*p*-Value
Ratio	C_{PS} Yield	53.464 68.032	1 1	53.464 68.032	1385.158 2.064	0.000 [a] 0.157
Temp	C_{PS} Yield	19.465 15,363.200	1 1	19.465 15,363.200	504.299 466.001	0.000 [a] 0.000 [a]
Time	C_{PS} Yield	5.706 3357.666	1 1	5.706 3357.666	147.822 101.846	0.000 [a] 0.000 [a]
Cycle	C_{PS} Yield	174.699 134,727.813	1 1	174.699 134,727.813	4526.183 4086.606	0.000 [a] 0.000 [a]
Ratio × Temp	C_{PS} Yield	0.707 356.780	1 1	0.707 356.780	18.319 10.822	0.000 [a] 0.002 [a]
Ratio × Time	C_{PS} Yield	1.881 297.472	1 1	1.881 297.472	48.743 9.023	0.000 [a] 0.004 [a]
Ratio × Cycle	C_{PS} Yield	8.690 1912.674	1 1	8.690 1912.674	225.144 58.016	0.000 [a] 0.000 [a]
Temp × Time	C_{PS} Yield	0.108 91.530	1 1	0.108 91.530	2.806 2.776	0.100 0.102
Temp × Cycle	C_{PS} Yield	7.273 5805.468	1 1	7.273 5805.468	188.436 176.093	0.000 [a] 0.000 [a]
Time × Cycle	C_{PS} Yield	0.092 87.173	1 1	0.092 87.173	2.373 2.644	0.130 0.110
Ratio × Temp × Time	C_{PS} Yield	0.001 5.664	1 1	0.001 5.664	0.024 0.172	0.877 0.680
Ratio × Temp × Cycle	C_{PS} Yield	0.223 164.259	1 1	0.223 164.259	5.769 4.982	0.020 [a] 0.030 [a]
Ratio × Time × Cycle	C_{PS} Yield	0.001 14.367	1 1	0.001 14.367	0.017 0.436	0.898 0.512
Temp × Time × Cycle	C_{PS} Yield	0.333 222.513	1 1	0.333 222.513	8.629 6.749	0.005 [a] 0.012 [a]
Ratio × Temp × Time × Cycle	C_{PS} Yield	0.057 1.946	1 1	0.057 1.946	1.475 0.059	0.230 0.809

SS: sum of squares; *df*: degree of freedom; *MS*: mean square; C_{PS}: polysaccharide concentration (mg/mL), Yield: extraction yield (mg/g). [a] Significant factors ($p < 0.05$).

Table 4 shows multiple comparisons between the various extraction conditions. C_{PS} and yield under various conditions ranged from 0.18 to 7.28 mg/mL and from 7.27 to 159.23 mg/g, respectively. A biomass–solvent ratio of 1:20 (*w*/*v*), extraction temperature at 90 °C for 2 h and the first extraction cycle gave the highest water-soluble polysaccharide concentration of 7.28 mg/mL, whereas the highest extraction yield was over 159 mg/g from a biomass–solvent ratio 1:40 (*w*/*v*) at 90 °C for 2 h in the first extraction cycle. Among the different conditions, polysaccharide content and yield in the first cycle were higher than in the second cycle.

Table 4. Polysaccharide concentration and extraction yield of *U. rigida* macroalgae under various conditions.

Ratio (*w*/*v*)	Temperature (°C)	Time (h)	Cycle	C_{PS} (mg/mL)	Yield (mg/g)
1:20	70	1	1	4.16 ± 0.38 [d]	83.26 ± 7.58 [f]
			2	0.78 ± 0.17 [i]	15.58 ± 3.39 [jk]
		2	1	4.88 ± 0.16 [c]	97.57 ± 3.11 [d]
			2	1.76 ± 0.21 [g]	35.28 ± 4.13 [gh]
	90	1	1	5.98 ± 0.11 [b]	119.49 ± 2.11 [c]
			2	1.41 ± 0.21 [h]	28.29 ± 4.13 [hi]
		2	1	7.28 ± 0.27 [a]	145.56 ± 5.29 [b]
			2	2.17 ± 0.15 [f]	43.42 ± 2.89 [g]
1:40	70	1	1	2.20 ± 0.20 [f]	87.82 ± 8.10 [ef]
			2	0.18 ± 0.03 [k]	7.27 ± 1.30 [k]
		2	1	2.37 ± 0.19 [f]	94.91 ± 7.39 [de]
			2	0.36 ± 0.07 [jk]	14.56 ± 2.79 [jk]
	90	1	1	3.49 ± 0.32 [e]	139.41 ± 12.89 [b]
			2	0.53 ± 0.08 [ij]	21.12 ± 3.30 [ij]
		2	1	3.98 ± 0.05 [d]	159.23 ± 1.85 [a]
			2	0.69 ± 0.20 [i]	27.62 ± 7.98 [hi]

Data in the same column with different superscripts are significantly different ($p < 0.05$). Data were calculated from triplicate experimental values ± standard deviation (SD).

Figures S1–S3 display the concentrations and yields of water-soluble polysaccharides extracted from *U. rigida* under various conditions. Results showed that longer extraction time led to higher polysaccharide concentration and extraction yield compared to different biomass–solvent ratios (Figure S1). Therefore, polysaccharide extraction was most efficient at higher extraction temperatures. Figure S2 illustrates the correlation between polysaccharide concentration and extraction yield at different temperatures. Higher biomass concentration led to higher polysaccharide concentrations at various temperature extractions (Figure S2a,c). Extraction yield at 70 °C showed no significant difference between different biomass concentrations, while the highest yield was obtained from the lower biomass concentration at 90 °C (Figure S2b,d). The correlation between extraction times is demonstrated in Figure S3. At higher biomass concentrations for different extraction times, higher concentrations of polysaccharides were observed. Conversely, when lower biomass concentrations were utilised during different extraction times, higher extraction yields were noted.

Table 5 presents polysaccharide concentrations and extraction yields obtained under different conditions. The highest concentration of polysaccharides (9.45 mg/mL) was achieved after two extraction cycles with a biomass–solvent ratio of 1:20 (*w*/*v*) at 9 °C for 2 h. The highest extraction yields obtained at biomass–solvent ratios of 1:20 and 1:40 (*w*/*v*) were not significantly different (189 mg/g and 187 mg/g, respectively). The optimum extraction temperature was 90 °C for 2 h, with higher polysaccharide concentrations and yields obtained at longer extraction times and higher temperatures. The highest polysaccharide concentration and extraction yield were obtained from the highest biomass–solvent ratio (1:20% *w*/*v*), at an extraction temperature of 90 °C, with an extraction time of 2 h and only one cycle. Water-soluble polysaccharides extracted from *U. rigida* macroalgae at optimum conditions were prepared and used as the freeze-dried polysaccharide preparation for monosaccharide composition and probiotic growth determination.

Table 5. Total water-soluble polysaccharide extraction from *U. rigida* macroalgae.

Ratio (*w/v*)	Temperature (°C)	Time (h)	C_{PS} (mg/mL)	Yield (mg/g)
1:20	70	1	4.942 ± 0.46 [d]	98.843 ± 9.12 [f]
		2	6.643 ± 0.11 [c]	132.846 ± 2.12 [d]
	90	1	7.389 ± 0.16 [b]	147.780 ± 3.18 [c]
		2	9.449 ± 0.36 [a]	188.970 ± 7.20 [a]
1:40	70	1	2.377 ± 0.18 [g]	95.089 ± 7.16 [f]
		2	2.737 ± 0.18 [f]	109.464 ± 7.21 [e]
	90	1	4.014 ± 0.40 [e]	160.534 ± 16.16 [b]
		2	4.671 ± 0.17 [d]	186.856 ± 6.63 [a]

Data in the same column with different superscripts are significantly different ($p < 0.05$). Data were calculated from triplicate experimental values ± standard deviation (SD).

3.3. Reducing Sugar, Total Phenolic and Flavonoid Contents

U. rigida water-soluble extracts contain non-reducing polysaccharides and reducing mono or oligosaccharides. Table 6 presents the results of testing different polysaccharide conditions using soluble extraction to determine reducing sugar contents. The first extraction cycle at a biomass concentration ratio of 1:20 (*w/v*) yielded the highest amount of reducing sugar, while only the first extraction cycle resulted in reducing sugar when the biomass–solvent ratio was 1:40 (*w/v*). Highest reducing sugar yields of 7.95 mg/g and 6.58 mg/g at ratios of 1:20 and 1:40 (*w/v*), respectively, were obtained at an extraction temperature of 90 °C for 2 h.

Table 6. Reducing sugar, total phenolic and flavonoid contents of polysaccharides extracted from *U. rigida* macroalgae under various conditions.

Ratio (*w/v*)	Temperature °C	Time (h)	Cycle	Reducing Sugar (mg/g)	TPC (mg GAE/g)	TFC (mg QE/g)
1:20	70	1	1	8.02 ± 3.32 [a]	0.54 ± 0.10 [bc]	0.22 ± 0.04 [fgh]
			2	0.24 ± 0.15 [bc]	ND	0.11 ± 0.05 [hijk]
		2	1	6.50 ± 2.11 [a]	0.60 ± 0.06 [b]	0.21 ± 0.03 [fghi]
			2	0.63 ± 0.30 [bc]	ND	0.15 ± 0.03 [ghij]
	90	1	1	5.76 ± 4.36 [a]	0.52 ± 0.03 [c]	0.34 ± 0.09 [e]
			2	0.13 ± 0.13 [bc]	ND	0.23 ± 0.04 [fg]
		2	1	7.95 ± 2.36 [a]	0.69 ± 0.02 [a]	1.42 ± 0.16 [a]
			2	0.37 ± 0.41 [bc]	ND	0.28 ± 0.05 [ef]
1:40	70	1	1	0.54 ± 0.26 [bc]	0.18 ± 0.04 [e]	0.64 ± 0.16 [d]
			2	ND	ND	0.02 ± 0.01 [k]
		2	1	0.82 ± 2.24 [bc]	0.42 ± 0.03 [d]	0.53 ± 0.03 [d]
			2	ND	ND	0.03 ± 0.03 [k]
	90	1	1	1.89 ± 2.24 [b]	0.40 ± 0.12 [d]	0.83 ± 0.12 [c]
			2	ND	ND	0.08 ± 0.03 [jk]
		2	1	6.58 ± 1.06 [a]	0.60 ± 0.03 [b]	0.99 ± 0.06 [b]
			2	ND	ND	0.10 ± 0.03 [ijk]

ND: Not detected; Data in the same column with different superscripts are significantly different ($p < 0.05$). Data were calculated from triplicate experimental values ± standard deviation (SD).

Water-soluble polysaccharides extracted from *U. rigida* have various compositions. To determine the TPC and TFC, the extraction was carried out using different conditions, as shown in Table 6. TPC was obtained from the first cycle of extraction using two levels of biomass–solvent ratio, extraction temperature and time. TPC was not detected in all the experiments from the second extraction cycle. TPC ranged between 0.18 and 0.69 mg GAE/g, while TFC ranged from 0.02 to 1.42 mg QE/g under various conditions. The highest TPC and TFC values of 0.69 mg GAE/g and 1.42 mg QE/g were obtained from the highest biomass concentration, extraction temperature and time of 1:20 (*w/v*) at 90 °C and 2 h, respectively.

3.4. Antioxidant

Antioxidant properties of polysaccharides extracted from *U. rigida* were evaluated using DPPH, ABTS and FRAP assays under different conditions, as shown in Table 7. Water-soluble polysaccharides extracted at various conditions exhibited a DPPH radical-scavenging capacity ranging from 0.02 to 10.49 mg AAE/g. The highest DPPH was achieved at 70 °C for 1 h from a lower biomass concentration of 1:40 (*w/v*). ABTS radical-scavenging activity at different conditions ranged from 0.11 to 0.94 mg AAE/g. Polysaccharides extracted under various conditions were determined using the ferric reducing antioxidant power (FRAP) assay, with results ranging from 0.17 to 0.52 mg AAE/g. The antioxidant properties showed a variety of value trends from various antioxidant assays. The water-soluble polysaccharides extracted from *U. rigida* were a greenish colour visible to the naked eye, indicating high total phenolic and flavonoid contents and antioxidant capacities.

Table 7. Antioxidant activity of polysaccharides extracted from *U. rigida* macroalgae under various conditions.

Ratio (*w/v*)	Temperature °C	Time (h)	Cycle	DPPH (mg AAE/g)	ABTS (mg AAE/g)	FRAP (mg AAE/g)
1:20	70	1	1	6.47 ± 2.49 [abcd]	0.87 ± 0.07 [a]	0.26 ± 0.01 [c]
			2	6.60 ± 2.42 [abcd]	0.29 ± 0.04 [c]	0.08 ± 0.01 [e]
		2	1	8.76 ± 2.62 [ab]	0.94 ± 0.13 [a]	0.27 ± 0.00 [c]
			2	6.84 ± 2.29 [abcd]	0.27 ± 0.10 [c]	0.17 ± 0.01 [d]
	90	1	1	7.84 ± 1.86 [abc]	0.78 ± 0.03 [a]	0.24 ± 0.02 [c]
			2	5.27 ± 2.86 [bcde]	0.27 ± 0.08 [c]	0.17 ± 0.01 [d]
		2	1	4.10 ± 2.01 [cdef]	0.79 ± 0.05 [a]	0.27 ± 0.02 [c]
			2	5.17 ± 0.86 [bcde]	0.22 ± 0.09 [c]	0.18 ± 0.01 [d]
1:40	70	1	1	10.49 ± 3.56 [a]	0.57 ± 0.19 [b]	0.40 ± 0.04 [b]
			2	0.05 ± 0.04 [fg]	0.27 ± 0.09 [c]	0.25 ± 0.01 [c]
		2	1	1.37 ± 0.72 [efg]	0.78 ± 0.13 [a]	0.48 ± 0.03 [a]
			2	0.02 ± 0.02 [fg]	0.11 ± 0.08 [c]	0.25 ± 0.01 [c]
	90	1	1	ND	0.56 ± 0.13 [b]	0.51 ± 0.10 [a]
			2	0.03 ± 0.02 [fg]	0.11 ± 0.22 [c]	0.26 ± 0.01 [c]
		2	1	2.39 ± 7.21 [defg]	0.60 ± 0.10 [b]	0.52 ± 0.02 [a]
			2	0.09 ± 0.02 [fg]	0.29 ± 0.16 [c]	0.29 ± 0.01 [c]

ND: Not detected; Data in the same column with different superscripts are significantly different ($p < 0.05$). Data were calculated from triplicate experimental values ± standard deviation (SD).

3.5. Monosaccharide Composition

Freeze-dried polysaccharides extracted from *U. rigida* included numerous monosaccharides. Six monosaccharides were separated using HPAEC-PAD analysis between 6.507 and 12.217 min. Each calibration curve was plotted using linear regression calculations. The correlation coefficient (R^2) values ranged between 0.992 and 0.999 (Table S1). Six standards were used to identify the analyte peak area of the sample with acceptable accuracy.

Table 8 shows the monosaccharide content of water-soluble polysaccharides obtained from *U. rigida* acid hydrolysis of polysaccharides. The acid hydrolysed sample consisted of rhamnose as the significant component at 78.97 mg/g followed by xylose, glucose and galactose, with arabinose as the minor component at 0.33 mg/g sample. By contrast, only a small amount of glucose was detected in the polysaccharide extract. The low level of monosaccharides in the polysaccharide extract suggested that the main composition was oligosaccharides or long-chain polysaccharides composed of rhamnose, xylose and glucose.

Table 8. Monosaccharide contents (in mg/g) from freeze-dried polysaccharides extracted and acid hydrolysis of freeze-dried polysaccharides from *U. rigida*.

Composition	Polysaccharide	Acid Hydrolysis of Polysaccharide
Rhamnose	ND	78.97 ± 2.71
Arabinose	ND	0.33 ± 0.02
Galactose	ND	2.81 ± 0.40
Glucose	0.11 ± 0.10	5.24 ± 0.28
Xylose	ND	12.11 ± 0.07
Fructose	ND	ND

ND: Not detected; Data were calculated from triplicate experimental values ± standard deviation (SD).

4. Discussion

Green macroalgae, also known as seaweed, are a promising source of biochemicals with various nutritional benefits. They contain polysaccharides ranging from 15 to 76%, proteins ranging from 5 to 47%, minerals ranging from 7 to 36% and lipids ranging from 1 to 5% [25]. Seaweed has been used in traditional medicine for centuries and is considered a sustainable and eco-friendly source of nutrition. The potential health benefits of seaweed make it an attractive option for the food and pharmaceutical industries. Macroalgae are an excellent biomass source rich in various bioactive polysaccharides, peptides, polyphenols, pigments and other bioactive compounds, which have potential applications in the food industry and biomedical sectors as functional ingredients [26,27]. A significant proportion of carbohydrates was obtained from *U. rigida*, accounting for 48% of the dry biomass. Our results were consistent with the high carbohydrate content typically found in macroalgae, with one study reporting the carbohydrate content found in different samples of *U. rigida* ranging from 16.63 to 65.93% of dry biomass [28]. Another study found that the carbohydrate content in *U. fenestrate* varied between 25 and 38% of dry biomass, with the highest amount obtained under elevated irradiance and temperature conditions [29]. Environmental factors such as temperature, irradiance, nutrients, CO_2 levels and season significantly impact carbohydrate content and other biochemical compositions in microalgae [30,31]. The carbohydrates present in macroalgae mainly consist of polysaccharides, along with a small quantity of oligosaccharides and monosaccharides [32]. Macroalgae represent a cost-effective source of valuable and beneficial compounds [33].

Various methods have been applied for the extraction of bioactive compounds [7]. Hot-water extraction is the most reliable and effective technique for extracting bioactive compounds from algae, particularly polysaccharides [34]. Water, a safe and non-toxic solvent, can easily penetrate plant tissue, making hot-water extraction the conventional method for polysaccharide extraction [10]. Despite the challenges posed by algal cell walls, our results identified crucial factors for optimising extraction conditions to enhance the extraction of valuable macroalgal polysaccharides. The extraction of water-soluble polysaccharides from *U. rigida* was influenced by various extraction factors such as ratio, temperature, time and cycle. To achieve the highest extraction yield and bioactive compound extraction, both individual and combined extraction conditions were studied [35]. The solvent-to-biomass ratio is one of the factors that impact polysaccharide extraction. The evaluation of significant factors showed that increasing the biomass–solvent ratio led to high polysaccharide solubility in water extraction. Extraction temperature and time also impacted polysaccharide extraction. Higher extraction temperatures led to a greater recovery of bioactive compounds from *U. lactuca* [36]. Polysaccharide extraction from algae has been performed using several methods. Table 9 compares polysaccharide extraction yields for various algal species and solvents, highlighting the different methods employed. Among multiple species and extraction methods, our results showed a polysaccharide yield of 18.9% as the optimum result. Extraction methods affect algal yield which also depends on algal species, cultivation methods, season and environmental conditions [7,37]. The data obtained indicate that higher extraction temperatures lead to a significant increase in polysaccharide yields, as a result of enhanced dissolution [10]. One of the main advantages of this extraction method is that it is simple and cost-effective [38]. According to the findings in Table 9, hot-water extraction at 90 °C for 2 h provides a significantly higher yield compared to using an 80% ethanol solution at 73 °C for 2 h and 6 min. Additionally, it also provides a superior yield compared to other hot-water extraction techniques (ranging from 80 °C to 97 °C) with a shorter duration. This method of extraction is not only environmentally friendly but also energy-saving and solvent-free. It is considered safe for food production and significantly reduces the need for solvent and acid separation from the polysaccharides. Furthermore, polysaccharide extracted using hot water tends to have an original profile compared to acidic solvents, which can cause acid hydrolysis during the extraction process.

Table 9. Polysaccharide extraction yield for various algal species and conditions.

Macroalgae	Condition	Yield (%)	Reference
Sargassum polycystum	Acidic water, 60 °C, 1 h	59.80	[39]
Chlorella sp.	Water, 80 °C, 3 h	9.62	[40]
Rhodosorus sp.	Water, 84 °C, 2 h	9.29	[9]
Sargassum thunbergii	Water, 97 °C, 3.50 h	7.53	[41]
Grifolafrondosa	Water, 210 °C, 43.65 min	25.10	[42]
U. intestinalis	Acidic solvent, Ultrasonic-assisted 200 W, 86 °C	26.77	[43]
U. rigida	80% ethanol, 73 °C, 2 h 6 min	13.22	[44]
U. pertusa	80% ethanol, Ultrasonic-assisted 600 W, 43.63 min	41.91	[45]
U. rigida C. Agardh	Water, 90 °C, 2 h	18.90	This study

Water-soluble extracts from macroalgae release polysaccharides and also other bioactive substances. Monosaccharides are reducing sugars containing hemiacetal or hemiketal groups, while polysaccharides are non-reducing. During water-soluble extraction, the reducing sugar in the supernatant extracted from *U. rigida* was maximised at 8 mg/g. A previous study reported maximum reducing sugar extracted from *U. prolifera* biomass of 0.156 g/g using a thermochemical method at 0.9 M H_2SO_4 and 121 °C for 50 min [46]. Enzymatic hydrolysis of *U. fasciata* macroalgae yielded maximum reducing sugar levels of 1.61 g/L at 6 h hydrolysis time [47]. Reducing sugar production in *Gracilaria verrucosa* macroalgae was achieved via ultrasonication with an acid catalyst followed by enzymatic hydrolysis [48]. Maximum reducing sugar content was achieved with extreme extraction or hydrolyses at high temperatures using acidic solutions and mechanical-assisted techniques. Our results found that glucose was expressed as a reducing sugar during polysaccharide extraction under non-extreme conditions with a mild solvent such as water. Macroalgae contain secondary metabolites that form phenolic and flavonoid compounds and are recognised for their antioxidant properties and essential functions in numerous biological processes [49,50]. Our results determined similar total phenolic and flavonoid contents as found in previous studies. In *U. lactuca* green macroalgae and *Laurencia obtuse* red macroalgae, increasing the biomass–solvent ratio increased the total phenolic compound content [36,51]. A methanolic extraction of *U. clathrate* and *U. intestinalis* from the Persian Gulf found that flavonoid contents varied from 8 to 33 mg RE/g [52]. Chemical constituents and phenolic and flavonoid contents from macroalgae vary depending on the season, environment and extraction conditions. The macroalga *U. rigida* demonstrated excellent antioxidant properties with DPPH, ABTS and FRAP assays. Green macroalgae possess noteworthy antioxidant potential, which could be applied in several industries including medicine, cosmetics, dietary supplements and food [52]. This discovery has significant implications for various applications in these sectors. Our results suggested that green macroalgae are a valuable source of antioxidants for multiple applications and warrant further investigation. These findings expand the current knowledge base and can be shared with external stakeholders. Polysaccharides extracted from green macroalgae exhibited antioxidant properties. They can be used as antioxidant ingredients in the food industry. Furthermore, their potential as a hepatoprotective agent in the pharmaceutical industry has also been demonstrated [12].

The carbohydrates present in macroalgae are composed of different sugar monomers [53], with extraction methods and conditions significantly impacting the yield, composition and structure of the resultant product. Changes in the bioactivity of polysaccharides impact the quality and efficacy of the final product [10]. Various monosaccharides were reported in polysaccharides from *Ulva* spp., with macroalgae including glucose, rhamnose, arabinose, xylose, mannose, galactose, fucose, glucosamine and glucuronic acid [8,54,55]. The monosaccharide composition and ratio in polysaccharides from *Ulva* spp. varied depending on species, growth conditions and harvesting season [4,56]. Among sixblade species and one filamentous species (Table 10), rhamnose (10.90–257.00 mg/g) was the significant component in *Ulva*, followed by glucose (3.60–183.10 mg/g) and xylose (4.40–99.00 mg/g) [57–60]. Galactose was detected at 0.42–28.27 mg/g [57–60]. Mannose was found in *U. meridionalis* (filamentous) and *U. ohnoi* as a minor component [60]. *U. rigida* from Chile contained arabinose, fucose and fructose in small amounts, while these monosaccharides were not detected in *U. rigida* from New Zealand. Our results indicated that the pattern of monosaccharides in *U. rigida* from different sources was mainly composed of rhamnose, xylose and glucose. The complexity of monosaccharide components, glycosidic linkage and sulphated group substitution of *Ulva* polysaccharides contributed to their biological antioxidant activity and immunomodulatory and hypolipidemic activity, as well as increasing the proliferation of probiotics in the intestinal microbiome [8,58,61]. Furthermore, the antioxidant activity of *Ulva* polysaccharides is significantly correlated with molecular weight. Notably, the degraded polysaccharide from *U. pertusa* Kjellm with a low molecular weight of 28.2 kDa demonstrated substantially higher scavenging activity for superoxide and hydroxyl radicals, as well as the strongest reducing power and metal chelating activity, in comparison to higher molecular weight variants (58.0–151.7 kDa) [62]. *Ulva* polysaccharides are endowed with unique attributes including hydrogels, nanofibers, 2D structures and 3D porous structures, rendering them an ideal option for drug delivery and tissue engineering applications. Their capacity to effectively stabilise and release drugs or selectively target bioactive compounds in hydrophilic matrix systems makes them particularly promising for a range of biotechnological applications [63,64]. In addition, rhamnose-rich oligosaccharides and polysaccharides are potential bioactive compounds that can serve as functional food ingredients and also in the pharmacological and medical fields [4,8,65].

Table 10. Monosaccharide compositions from various *Ulva* spp. species.

Ulva Species	Source	Ara	Fuc	Fru	Gal	Glc	Man	Rha	Xyl	Ref
U. australis	New Zealand	ND	ND	ND	3.00	ND	ND	251.00	99.00	[57]
U. fasciata	Philippines	ND	ND	ND	0.42	16.59	ND.	33.62	4.89	[58]
U. meridionalis	Japan	ND	ND	ND	28.27	144.43	5.40	66.56	12.34	[60]
U. ohnoi	Japan	ND	ND	ND	1.20	89.43	0.17	58.48	21.72	[60]
U. pertusa Kjellm	China	ND	ND	ND	2.20	3.60	ND	10.90	4.40	[66]
U. rigida	Chile	0.7	0.1	1.4	11.70	183.10	ND	81.20	38.50	[59]
U. rigida	New Zealand	ND	ND	ND	<1.00	ND	ND	257.00	27.00	[57]
Ulva spp.	New Zealand	ND	ND	ND	8.00	ND	ND	201.00	73.00	[57]

Ara = arabinose, Fuc = fucose, Fru = fructose, Gal = galactose, Glc = glucose, Man = mannose, Rha = rhamnose and Xyl = Xylose; ND: Not detected.

The consumption of probiotics can improve the health of the host. However, ensuring the survival of these probiotics in challenging environments is difficult. One promising solution is using polysaccharides, which are non-toxic, highly compatible with biological systems and biodegradable. These polysaccharides can form a protective layer around the probiotics, providing a physical barrier and aiding in their delivery, thereby making them a potential solution for probiotic delivery [67]. During the growth of *Lactobacillus rhamnosus* ATCC 53103 probiotic bacteria, freeze-dried polysaccharides extracted from *U. rigida* were replaced as a substrate in the culture medium. Results showed that viable *L. rhamnosus* ATCC 53103 cells could grow for up to 48 h, with a maximum of 6.85 log CFU/mL (Figure 2). The control medium contained glucose with maximum viable cells of 10.86 log CFU/mL. Inulin, a commercial prebiotic, showed similar cell growth compared to polysaccharides from *U. rigida*, with maximum cell growth of 7.2 log CFU/mL. Polysaccharides possess the ability to act as prebiotics, meaning they can influence the composition of the gut microbiome, promote gut barrier health and regulate the production of metabolites by the gut microbiota [68]. Prebiotics improve immunity, resist pathogens, influence metabolism, increase mineral absorption and enhance health [69]. Our results are in concordance with a previous study that used ulvan polysaccharide as a prebiotic in yoghurt products [70]. Our findings suggest that *L. rhamnosus* probiotic, when promoted with polysaccharides from *U. rigida*, has the potential to be effective in a similar way to previous reports. Polysaccharides such as laminaran, porphyran and ulvan extracted from brown, red and green macroalgae were found to promote the growth of *L. rhanmosus* ATCC 53103, *L. plantarum* ATCC 10241 and *Bifidobacterium breve* ATCC 15700 [71]. Our research findings suggest that the polysaccharides present in *U. rigida* can act as carbon sources to promote cell growth. These polysaccharides also possess hydrolysis enzymes that are not present in the human genome [68]. Polysaccharides found in green macroalgae consist of sulphated L-rhamnose and D-glucoronic acid. The whole genome sequence of *L. rhamnosus* ATCC 53103 suggests that certain enzymes are involved in the hydrolysis process, with gene GCF_000011045.1_00051 being responsible for hydrolysis enzymes that play a role in β-glucosidase, α-L-rhamnosidase and β-glucan 1,4-α-glucosidase [71,72]. This study investigated the feasibility of utilising polysaccharides derived from *U. rigida* to support bacterial growth. Results suggested that these polysaccharides could enhance the viability and functionality of probiotic microorganisms, thus contributing to the quality and safety of food products. These findings underscore the potential value of this novel approach for improving the quality and safety of food products and advancing our understanding of the interplay between polysaccharides, beneficial bacteria and human health. Polysaccharides extracted from *U. lactuca*, incorporated with a yoghurt formulation at 1–2%, were reported to have good physicochemical properties, as well as stimulating the growth and activity of probiotic bacteria comprising *Lactobacillus acidophilus*, *Streptococcus thermophilus* and *Bifidobacterium* sp. [70]. Moreover, the combination of macroalgae-derived polysaccharides with probiotics can be used as dietary supplements to enhance the health of aquatic animals [73].

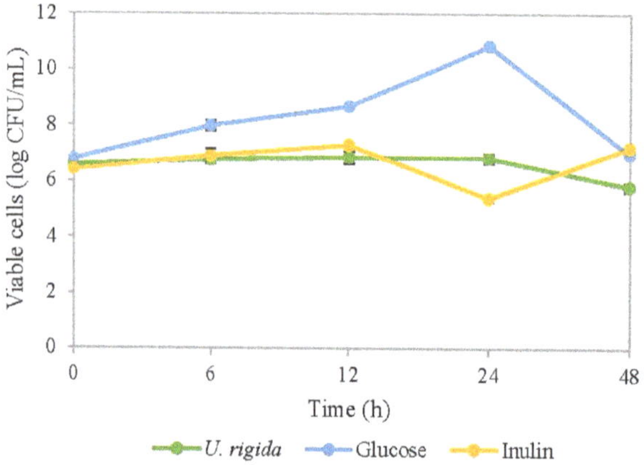

Figure 2. Viable cells of *Lactobacillus rhamnosus* ATCC 53103 under different supplemented ingredients in culture broth.

5. Conclusions

U. rigida C. Agardh green macroalgae contain significant amounts of polysaccharides. This study evaluated the maximum concentration and yield of a safe water-soluble polysaccharide extraction from oven-dried *U. rigida* biomass using full factorial experiments. Increasing all factors resulted in higher levels of polysaccharides extracted. The main factors and their interaction had a significant effect on the extracted polysaccharides from *U. rigida*. The water-soluble polysaccharides contained reducing sugar, phenolics and flavonoids and exhibited antioxidant activities. Freeze-dried polysaccharide powders mainly consist of the monosaccharide rhamnose. Preliminary results on the effect of these powders on probiotics suggested that when supplemented with polysaccharides from *U. rigida*, viable *L. rhamnosus* ATCC 53103 growth was promoted during cultivation. This discovery has the potential to impact the human food industry through the development of functional ingredients for novel and future food products, with various other applications in the nutraceutical and pharmaceutical industries.

Supplementary Materials: The following supporting information can be downloaded at https://www.mdpi.com/article/10.3390/foods13111630/s1, Figure S1: Polysaccharide extraction from *U. rigida* macroalgae under various conditions. (a,b) C_{PS} and extraction yield at 1:20 (*w*/*v*) biomass–solvent ratio, respectively; (c,d) C_{PS} and extraction yield at 1:40 (*w*/*v*) biomass–solvent ratio, respectively; Figure S2: Polysaccharide extraction from *U. rigida* macroalgae under various conditions. (a,b) C_{PS} and extraction yield at 70 °C, respectively; (c,d) C_{PS} and extraction yield at 90 °C, respectively; Figure S3: Polysaccharide extraction from *U. rigida* macroalgae under various conditions. (a,b) C_{PS} and extraction yield at an extraction time of 1 h, respectively; (c,d) C_{PS} and extraction yield at an extraction time of 2 h, respectively; Table S1: Standard curves of six monosaccharide reference standards.

Author Contributions: Conceptualisation, N.P. and W.P.-u.; methodology, N.P. and W.P.-u.; formal analysis, N.P., P.P. and W.P.-u.; resources, P.K. and B.T.; writing—original draft preparation, N.P. and W.P.-u.; writing—review and editing, N.P. and W.P.-u.; visualisation, N.P., P.K., B.T., P.P. and W.P.-u.; project administration, N.P.; funding acquisition, N.P. All authors have read and agreed to the published version of the manuscript.

Funding: This research was financially supported by Kasetsart University Research and Development Institute (KURDI), FF(KU)18.67.

Institutional Review Board Statement: Not applicable.

Informed Consent Statement: Not applicable.

Data Availability Statement: The original contributions presented in the study are included in the article and Supplementary Materials, further inquiries can be directed to the corresponding author.

Conflicts of Interest: The authors declare no conflicts of interest. The funders had no role in the design of the study, the collection, analyses or interpretation of data, the writing of the manuscript or in the decision to publish the results.

References

1. Donn, P.; Prieto, M.A.; Mejuto, J.C.; Cao, H.; Simal-Gandara, J. Functional foods based on the recovery of bioactive ingredients from food and algae by-products by emerging extraction technologies and 3D printing. *Food Biosci.* **2022**, *49*, 101853. [CrossRef]
2. Lordan, S.; Ross, R.P.; Stanton, C. Marine bioactives as functional food ingredients: Potential to reduce the incidence of chronic diseases. *Mar. Drugs* **2011**, *9*, 1056–1100. [CrossRef] [PubMed]
3. Biesalski, H.-K.; Dragsted, L.O.; Elmadfa, I.; Grossklaus, R.; Müller, M.; Schrenk, D.; Walter, P.; Weber, P. Bioactive compounds: Safety and efficacy. *Nutrition* **2009**, *25*, 1206–1211. [CrossRef] [PubMed]
4. Xu, J.; Liao, W.; Liu, Y.; Guo, Y.; Jiang, S.; Zhao, C. An overview on the nutritional and bioactive components of green seaweeds. *Food Prod. Process. Nutr.* **2023**, *5*, 18. [CrossRef]
5. Lafeuille, B.; Tamigneaux, É.; Berger, K.; Provencher, V.; Beaulieu, L. Impact of Harvest Month and Drying Process on the Nutritional and Bioactive Properties of Wild Palmaria palmata from Atlantic Canada. *Mar. Drugs* **2023**, *21*, 392. [CrossRef] [PubMed]
6. Biris-Dorhoi, E.S.; Michiu, D.; Pop, C.R.; Rotar, A.M.; Tofana, M.; Pop, O.L.; Socaci, S.A.; Farcas, A.C. Macroalgae—A Sustainable Source of Chemical Compounds with Biological Activities. *Nutrients* **2020**, *12*, 3085. [CrossRef] [PubMed]

7. Jönsson, M.; Allahgholi, L.; Sardari, R.R.R.; Hreggviðsson, G.O.; Nordberg Karlsson, E. Extraction and Modification of Macroalgal Polysaccharides for Current and Next-Generation Applications. *Molecules* **2020**, *25*, 930. [CrossRef] [PubMed]
8. Ning, L.; Yao, Z.; Zhu, B. Ulva (Enteromorpha) Polysaccharides and Oligosaccharides: A Potential Functional Food Source from Green-Tide-Forming Macroalgae. *Mar. Drugs* **2022**, *20*, 202. [CrossRef] [PubMed]
9. Wang, N.; Dai, L.; Chen, Z.; Li, T.; Wu, J.; Wu, H.; Wu, H.; Xiang, W. Extraction optimization, physicochemical characterization, and antioxidant activity of polysaccharides from *Rhodosorus* sp. SCSIO-45730. *J. Appl. Phycol.* **2022**, *34*, 285–299. [CrossRef]
10. Liu, F.; Chen, H.; Qin, L.; Al-Haimi, A.A.N.M.; Xu, J.; Zhou, W.; Zhu, S.; Wang, Z. Effect and characterization of polysaccharides extracted from *Chlorella* sp. by hot-water and alkali extraction methods. *Algal Res.* **2023**, *70*, 102970. [CrossRef]
11. Tang, Z.; Gao, H.; Wang, S.; Wen, S.; Qin, S. Hypolipidemic and antioxidant properties of a polysaccharide fraction from Enteromorpha prolifera. *Int. J. Biol. Macromol.* **2013**, *58*, 186–189. [CrossRef]
12. Zhong, Q.; Wei, B.; Wang, S.; Ke, S.; Chen, J.; Zhang, H.; Wang, H. The Antioxidant Activity of Polysaccharides Derived from Marine Organisms: An Overview. *Mar. Drugs* **2019**, *17*, 674. [CrossRef]
13. Jiang, F.; Chi, Z.; Ding, Y.; Quan, M.; Tian, Y.; Shi, J.; Song, F.; Liu, C. Wound dressing hydrogel of enteromorpha prolifera polysaccharide–polyacrylamide composite: A facile transformation of marine blooming into biomedical material. *ACS Appl. Mater. Interfaces* **2021**, *13*, 14530–14542. [CrossRef]
14. Wang, S.; Zhang, B.; Chang, X.; Zhao, H.; Zhang, H.; Zhao, T.; Qi, H. Potential use of seaweed polysaccharides as prebiotics for management of metabolic syndrome: A review. *Crit. Rev. Food Sci. Nutr.* **2023**, 1–21. [CrossRef]
15. Patel, A.K.; Singhania, R.R.; Awasthi, M.K.; Varjani, S.; Bhatia, S.K.; Tsai, M.-L.; Hsieh, S.-L.; Chen, C.-W.; Dong, C.-D. Emerging prospects of macro- and microalgae as prebiotic. *Microb. Cell Factories* **2021**, *20*, 112. [CrossRef]
16. Paopun, Y.; Thanomchat, P.; Roopkham, C.; Umroong, P.; Pan-utai, W.; Satmalee, P.; Kosawatpat, P.; Thongdang, B.; Tamtin, M. Structural Development of Marine Green Alga (*Ulva rigida* C. Agardh, 1823) during Cultivation. *Trends Sci.* **2023**, *20*, 6747. [CrossRef]
17. Pan-utai, W.; Pantoa, T.; Roytrakul, S.; Praiboon, J.; Kosawatpat, P.; Tamtin, M.; Thongdang, B. Ultrasonic-Assisted Extraction and Antioxidant Potential of Valuable Protein from *Ulva rigida* Macroalgae. *Life* **2023**, *13*, 86. [CrossRef]
18. Horwitz, W.; Latimer, G.W. *Official Methods of Analysis of AOAC International*; AOAC International Gaithersburg: Rockville, MD, USA, 2000; Volume 1.
19. Wang, C.; Li, J.; Cao, Y.; Huang, J.; Lin, H.; Zhao, T.; Liu, L.; Shen, P.; Julian McClements, D.; Chen, J.; et al. Extraction and characterization of pectic polysaccharides from Choerospondias axillaris peels: Comparison of hot water and ultrasound-assisted extraction methods. *Food Chem.* **2023**, *401*, 134156. [CrossRef]
20. Miller, G.L. Use of Dinitrosalicylic Acid Reagent for Determination of Reducing Sugar. *Anal. Chem.* **1959**, *31*, 426–428. [CrossRef]
21. Liu, X.-m.; Liu, Y.; Shan, C.-h.; Yang, X.-q.; Zhang, Q.; Xu, N.; Xu, L.-y.; Song, W. Effects of five extraction methods on total content, composition, and stability of flavonoids in jujube. *Food Chem. X* **2022**, *14*, 100287. [CrossRef]
22. Hung, Y.-H.R.; Chen, G.-W.; Pan, C.-L.; Lin, H.-T.V. Production of Ulvan Oligosaccharides with Antioxidant and Angiotensin-Converting Enzyme-Inhibitory Activities by Microbial Enzymatic Hydrolysis. *Fermentation* **2021**, *7*, 160. [CrossRef]
23. Weitzhandler, M.; Barreto, V.; Pohl, C.; Jandik, P.; Cheng, J.; Avdalovic, N. CarboPac™ PA20: A new monosaccharide separator column with electrochemical detection with disposable gold electrodes. *J. Biochem. Biophys. Methods* **2004**, *60*, 309–317. [CrossRef] [PubMed]
24. Song, Z.; Li, M.; Du, J.; Zhang, K. Effects of non-starch polysaccharides from pure wheat malt beer on beer quality, in vitro antioxidant, prebiotics, hypoglycemic and hypolipidemic properties. *Food Biosci.* **2022**, *47*, 101780. [CrossRef]
25. Silva, A.; Silva, S.A.; Carpena, M.; Garcia-Oliveira, P.; Gullón, P.; Barroso, M.F.; Prieto, M.; Simal-Gandara, J. Macroalgae as a source of valuable antimicrobial compounds: Extraction and applications. *Antibiotics* **2020**, *9*, 642. [CrossRef] [PubMed]
26. Berri, M.; Olivier, M.; Holbert, S.; Dupont, J.; Demais, H.; Le Goff, M.; Collen, P.N. Ulvan from *Ulva armoricana* (Chlorophyta) activates the PI3K/Akt signalling pathway via TLR4 to induce intestinal cytokine production. *Algal Res.* **2017**, *28*, 39–47. [CrossRef]
27. Anisha, G.S.; Augustianath, T.; Padmakumari, S.; Singhania, R.R.; Pandey, A.; Patel, A.K. Ulvan from green macroalgae: Bioactive properties advancing tissue engineering, drug delivery systems, food industry, agriculture and water treatment. *Bioresour. Technol. Rep.* **2023**, *22*, 101457. [CrossRef]
28. Balar, N.; Sharnagat, P.; Kumari, P.; Mantri, V.A. Variation in the proximate composition of edible marine macroalga *Ulva rigida* collected from different coastal zones of India. *J. Food Sci. Technol.* **2019**, *56*, 4749–4755. [CrossRef]
29. Olsson, J.; Toth, G.B.; Oerbekke, A.; Cvijetinovic, S.; Wahlström, N.; Harrysson, H.; Steinhagen, S.; Kinnby, A.; White, J.; Edlund, U.; et al. Cultivation conditions affect the monosaccharide composition in *Ulva fenestrata*. *J. Appl. Phycol.* **2020**, *32*, 3255–3263. [CrossRef]
30. Sarker, M.Y.; Bartsch, I.; Olischläger, M.; Gutow, L.; Wiencke, C. Combined effects of CO_2, temperature, irradiance and time on the physiological performance of *Chondrus crispus* (Rhodophyta). *Bot. Mar.* **2013**, *56*, 63–74. [CrossRef]
31. Queirós, A.S.; Circuncisão, A.R.; Pereira, E.; Válega, M.; Abreu, M.H.; Silva, A.M.S.; Cardoso, S.M. Valuable Nutrients from *Ulva rigida*: Modulation by Seasonal and Cultivation Factors. *Appl. Sci.* **2021**, *11*, 6137. [CrossRef]
32. Fu, Y.; Jiao, H.; Sun, J.; Okoye, C.O.; Zhang, H.; Li, Y.; Lu, X.; Wang, Q.; Liu, J. Structure-activity relationships of bioactive polysaccharides extracted from macroalgae towards biomedical application: A review. *Carbohydr. Polym.* **2024**, *324*, 121533. [CrossRef] [PubMed]

33. Salehi, B.; Sharifi-Rad, J.; Seca, A.M.; Pinto, D.C.; Michalak, I.; Trincone, A.; Mishra, A.P.; Nigam, M.; Zam, W.; Martins, N. Current trends on seaweeds: Looking at chemical composition, phytopharmacology, and cosmetic applications. *Molecules* **2019**, *24*, 4182. [CrossRef] [PubMed]
34. Kronusová, O.; Kaštánek, P.; Koyun, G.; Kaštánek, F.; Brányik, T. Factors Influencing the Production of Extracellular Polysaccharides by the Green Algae Dictyosphaerium chlorelloides and Their Isolation, Purification, and Composition. *Microorganisms* **2022**, *10*, 1473. [CrossRef]
35. Domínguez, R.; Zhang, L.; Rocchetti, G.; Lucini, L.; Pateiro, M.; Munekata, P.E.S.; Lorenzo, J.M. Elderberry (*Sambucus nigra* L.) as potential source of antioxidants. Characterization, optimization of extraction parameters and bioactive properties. *Food Chem.* **2020**, *330*, 127266. [CrossRef]
36. Pappou, S.; Dardavila, M.M.; Savvidou, M.G.; Louli, V.; Magoulas, K.; Voutsas, E. Extraction of Bioactive Compounds from *Ulva lactuca*. *Appl. Sci.* **2022**, *12*, 2117. [CrossRef]
37. Flores-Contreras, E.A.; Araújo, R.G.; Rodríguez-Aguayo, A.A.; Guzmán-Román, M.; García-Venegas, J.C.; Nájera-Martínez, E.F.; Sosa-Hernández, J.E.; Iqbal, H.M.N.; Melchor-Martínez, E.M.; Parra-Saldivar, R. Polysaccharides from the Sargassum and Brown Algae Genus: Extraction, Purification, and Their Potential Therapeutic Applications. *Plants* **2023**, *12*, 2445. [CrossRef]
38. Xue, J.; Su, J.; Wang, X.; Zhang, R.; Li, X.; Li, Y.; Ding, Y.; Chu, X. Eco-Friendly and Efficient Extraction of Polysaccharides from *Acanthopanax senticosus* by Ultrasound-Assisted Deep Eutectic Solvent. *Molecules* **2024**, *29*, 942. [CrossRef] [PubMed]
39. Digala, P.; Saravanan, M.; Dhanraj, M.; Pamarthi, J.; Muralidharan, S.; Narikimelli, A.; Dinakaran, K.P.; Arokiyaraj, S.; Vincent, S. Optimized extraction of sulfated polysaccharide from brown seaweed *Sargassum polycystum* and its evaluation for anti-cancer and wound healing potential. *S. Afr. J. Bot.* **2022**, *151*, 345–359. [CrossRef]
40. Song, H.; He, M.; Gu, C.; Wei, D.; Liang, Y.; Yan, J.; Wang, C. Extraction Optimization, Purification, Antioxidant Activity, and Preliminary Structural Characterization of Crude Polysaccharide from an Arctic *Chlorella* sp. *Polymers* **2018**, *10*, 292. [CrossRef]
41. Yuan, X.; Zeng, Y.; Nie, K.; Luo, D.; Wang, Z. Extraction Optimization, Characterization and Bioactivities of a Major Polysaccharide from *Sargassum thunbergii*. *PLoS ONE* **2015**, *10*, e0144773. [CrossRef]
42. Yang, L.; Qu, H.; Mao, G.; Zhao, T.; Li, F.; Zhu, B.; Zhang, B.; Wu, X. Optimization of subcritical water extraction of polysaccharides from *Grifola frondosa* using response surface methodology. *Pharmacogn. Mag.* **2013**, *9*, 120–129. [CrossRef] [PubMed]
43. Kazemi, M.; Fathi, M.; Jahanbin, K.; Taghdir, M.; Abbaszadeh, S. Optimization of ultrasonic-assisted hot acidic solvent extraction of ulvan from Ulva intestinalis of the Persian Gulf: Evaluation of structural, techno-functional, and bioactivity properties. *Food Hydrocoll.* **2023**, *142*, 108837. [CrossRef]
44. Akbal, A.; Şahin, S.; Güroy, B. Optimization of ultrasonic-assisted extraction of polysaccharides from *Ulva rigida* and evaluation of their antioxidant activity. *Algal Res.* **2024**, *77*, 103356. [CrossRef]
45. Le, B.; Golokhvast, K.S.; Yang, S.H.; Sun, S. Optimization of Microwave-Assisted Extraction of Polysaccharides from *Ulva pertusa* and Evaluation of Their Antioxidant Activity. *Antioxidants* **2019**, *8*, 129. [CrossRef] [PubMed]
46. Dave, N.; Varadavenkatesan, T.; Singh, R.S.; Giri, B.S.; Selvaraj, R.; Vinayagam, R. Evaluation of seasonal variation and the optimization of reducing sugar extraction from *Ulva prolifera* biomass using thermochemical method. *Environ. Sci. Pollut. Res.* **2021**, *28*, 58857–58871. [CrossRef]
47. Ami, J.; Mensah, M.; Asiedu, N.Y.; Thygesen, A. Optimization of Reducing Sugar Concentration from *Ulva fasciata* Using Cellulase via Response Surface Methodology Techniques. *Ind. Biotechnol.* **2023**, *19*, 23–32. [CrossRef]
48. Park, M.-R.; Jeong, G.-T. Production of reducing sugar in *Gracilaria verrucosa* using physio-chemical pretreatment and subsequent enzymatic hydrolysis. *Algal Res.* **2021**, *60*, 102531. [CrossRef]
49. Prabakaran, G.; Moovendhan, M.; Arumugam, A.; Matharasi, A.; Dineshkumar, R.; Sampathkumar, P. Quantitative analysis of phytochemical profile in marine microalgae *Chlorella vulgaris*. *Int. J. Pharm. Biol. Sci.* **2018**, *8*, 562–565.
50. Putra, N.R.; Fajriah, S.; Qomariyah, L.; Dewi, A.S.; Rizkiyah, D.N.; Irianto, I.; Rusmin, D.; Melati, M.; Trisnawati, N.W.; Darwati, I.; et al. Exploring the potential of *Ulva lactuca*: Emerging extraction methods, bioactive compounds, and health applications—A perspective review. *S. Afr. J. Chem. Eng.* **2024**, *47*, 233–245. [CrossRef]
51. Topuz, O.K.; Gokoglu, N.; Yerlikaya, P.; Ucak, I.; Gumus, B. Optimization of Antioxidant Activity and Phenolic Compound Extraction Conditions from Red Seaweed (*Laurencia obtuse*). *J. Aquat. Food Prod. Technol.* **2016**, *25*, 414–422. [CrossRef]
52. Farasat, M.; Khavari-Nejad, R.A.; Nabavi, S.M.; Namjooyan, F. Antioxidant Activity, Total Phenolics and Flavonoid Contents of some Edible Green Seaweeds from Northern Coasts of the Persian Gulf. *Iran. J. Pharm. Res.* **2014**, *13*, 163–170. [PubMed]
53. Bayu, A.; Warsito, M.F.; Putra, M.Y.; Karnjanakom, S.; Guan, G. Macroalgae-derived rare sugars: Applications and catalytic synthesis. *Carbon Resour. Convers.* **2021**, *4*, 150–163. [CrossRef]
54. Cho, M.; Yang, C.; Kim, S.M.; You, S. Molecular characterization and biological activities of watersoluble sulfated polysaccharides from *Enteromorpha prolifera*. *Food Sci. Biotechnol.* **2010**, *19*, 525–533. [CrossRef]
55. Qi, X.; Mao, W.; Gao, Y.; Chen, Y.; Chen, Y.; Zhao, C.; Li, N.; Wang, C.; Yan, M.; Lin, C.; et al. Chemical characteristic of an anticoagulant-active sulfated polysaccharide from *Enteromorpha clathrata*. *Carbohydr. Polym.* **2012**, *90*, 1804–1810. [CrossRef] [PubMed]
56. Samarasinghe, M.B.; van der Heide, M.E.; Weisbjerg, M.R.; Sehested, J.; Sloth, J.J.; Bruhn, A.; Vestergaard, M.; Nørgaard, J.V.; Hernández-Castellano, L.E. A descriptive chemical analysis of seaweeds, *Ulva* sp., *Saccharina latissima* and *Ascophyllum nodosum* harvested from Danish and Icelandic waters. *Anim. Feed Sci. Technol.* **2021**, *278*, 115005. [CrossRef]

57. Kidgell, J.T.; Carnachan, S.M.; Magnusson, M.; Lawton, R.J.; Sims, I.M.; Hinkley, S.F.R.; de Nys, R.; Glasson, C.R.K. Are all ulvans equal? A comparative assessment of the chemical and gelling properties of ulvan from blade and filamentous Ulva. *Carbohydr. Polym.* **2021**, *264*, 118010. [CrossRef] [PubMed]
58. Magdugo, R.P.; Terme, N.; Lang, M.; Pliego-Cortés, H.; Marty, C.; Hurtado, A.Q.; Bedoux, G.; Bourgougnon, N. An Analysis of the Nutritional and Health Values of *Caulerpa racemosa* (Forsskål) and *Ulva fasciata* (Delile)—Two Chlorophyta Collected from the Philippines. *Molecules* **2020**, *25*, 2901. [CrossRef] [PubMed]
59. Pezoa-Conte, R.; Leyton, A.; Anugwom, I.; von Schoultz, S.; Paranko, J.; Mäki-Arvela, P.; Willför, S.; Muszyński, M.; Nowicki, J.; Lienqueo, M.E.; et al. Deconstruction of the green alga *Ulva rigida* in ionic liquids: Closing the mass balance. *Algal Res.* **2015**, *12*, 262–273. [CrossRef]
60. Tsubaki, S.; Oono, K.; Hiraoka, M.; Ueda, T.; Onda, A.; Yanagisawa, K.; Azuma, J.-i. Hydrolysis of green-tide forming *Ulva* spp. by microwave irradiation with polyoxometalate clusters. *Green Chem.* **2014**, *16*, 2227–2233. [CrossRef]
61. Chen, R.; Xu, J.; Wu, W.; Wen, Y.; Lu, S.; El-Seedi, H.R.; Zhao, C. Structure–immunomodulatory activity relationships of dietary polysaccharides. *Curr. Res. Food Sci.* **2022**, *5*, 1330–1341. [CrossRef]
62. Qi, H.; Zhao, T.; Zhang, Q.; Li, Z.; Zhao, Z.; Xing, R. Antioxidant activity of different molecular weight sulfated polysaccharides from *Ulva pertusa* Kjellm (Chlorophyta). *J. Appl. Phycol.* **2005**, *17*, 527–534. [CrossRef]
63. Gupta, V.K.; Hariharan, M.; Wheatley, T.A.; Price, J.C. Controlled-release tablets from carrageenans: Effect of formulation, storage and dissolution factors. *Eur. J. Pharm. Biopharm.* **2001**, *51*, 241–248. [CrossRef]
64. Tziveleka, L.-A.; Ioannou, E.; Roussis, V. Ulvan, a bioactive marine sulphated polysaccharide as a key constituent of hybrid biomaterials: A review. *Carbohydr. Polym.* **2019**, *218*, 355–370. [CrossRef]
65. Kidgell, J.T.; Glasson, C.R.K.; Magnusson, M.; Sims, I.M.; Hinkley, S.F.R.; de Nys, R.; Carnachan, S.M. Ulvans are not equal—Linkage and substitution patterns in ulvan polysaccharides differ with Ulva morphology. *Carbohydr. Polym.* **2024**, *333*, 121962. [CrossRef]
66. Chen, J.; Zeng, W.; Gan, J.; Li, Y.; Pan, Y.; Li, J.; Chen, H. Physicochemical properties and anti-oxidation activities of ulvan from *Ulva pertusa* Kjellm. *Algal Res.* **2021**, *55*, 102269. [CrossRef]
67. Lin, Q.; Si, Y.; Zhou, F.; Hao, W.; Zhang, P.; Jiang, P.; Cha, R. Advances in polysaccharides for probiotic delivery: Properties, methods, and applications. *Carbohydr. Polym.* **2024**, *323*, 121414. [CrossRef]
68. Guo, Q.; Li, Y.; Dai, X.; Wang, B.; Zhang, J.; Cao, H. Polysaccharides: The Potential Prebiotics for Metabolic Associated Fatty Liver Disease (MAFLD). *Nutrients* **2023**, *15*, 3722. [CrossRef]
69. Peredo-Lovillo, A.; Romero-Luna, H.E.; Jiménez-Fernández, M. Health promoting microbial metabolites produced by gut microbiota after prebiotics metabolism. *Food Res. Int.* **2020**, *136*, 109473. [CrossRef]
70. Shalaby, S.; Amin, H. Potential Using of Ulvan Polysaccharide from *Ulva lactuca* as a Prebiotic in Synbiotic Yogurt Production. *J. Probiotics Health* **2019**, *7*, 208. [CrossRef]
71. Seong, H.; Bae, J.-H.; Seo, J.S.; Kim, S.-A.; Kim, T.-J.; Han, N.S. Comparative analysis of prebiotic effects of seaweed polysaccharides laminaran, porphyran, and ulvan using in vitro human fecal fermentation. *J. Funct. Foods* **2019**, *57*, 408–416. [CrossRef]
72. Xu, J.; Bjursell, M.K.; Himrod, J.; Deng, S.; Carmichael, L.K.; Chiang, H.C.; Hooper, L.V.; Gordon, J.I. A genomic view of the human-*Bacteroides thetaiotaomicron* symbiosis. *Science* **2003**, *299*, 2074–2076. [CrossRef] [PubMed]
73. Vidhya Hindu, S.; Chandrasekaran, N.; Mukherjee, A.; Thomas, J. A review on the impact of seaweed polysaccharide on the growth of probiotic bacteria and its application in aquaculture. *Aquac. Int.* **2019**, *27*, 227–238. [CrossRef]

Disclaimer/Publisher's Note: The statements, opinions and data contained in all publications are solely those of the individual author(s) and contributor(s) and not of MDPI and/or the editor(s). MDPI and/or the editor(s) disclaim responsibility for any injury to people or property resulting from any ideas, methods, instructions or products referred to in the content.

foods

MDPI

Article

Proteomic Profile of Flaxseed (*Linum usitatissimum* L.) Products as Influenced by Protein Concentration Method and Cultivar

Markéta Jarošová [1], Pavel Roudnický [2], Jan Bárta [1,*], Zbyněk Zdráhal [2], Veronika Bártová [1], Adéla Stupková [1], František Lorenc [3], Marie Bjelková [4], Jan Kyselka [5], Eva Jarošová [1], Jan Bedrníček [3] and Andrea Bohatá [1]

1 Department of Plant Production, Faculty of Agriculture and Technology, University of South Bohemia, Na Sádkách 1780, 370 05 České Budějovice, Czech Republic; jarosovam@fzt.jcu.cz (M.J.); vbartova@fzt.jcu.cz (V.B.); stupkova@fzt.jcu.cz (A.S.); jarose02@fzt.jcu.cz (E.J.); bohata@fzt.jcu.cz (A.B.)
2 Mendel Centre of Plant Genomics and Proteomics, Central European Institute of Technology, Masaryk University, Kamenice 753/5, 625 00 Brno, Czech Republic; pavel.roudnicky@ceitec.muni.cz (P.R.); zbynek.zdrahal@ceitec.muni.cz (Z.Z.)
3 Department of Food Biotechnology and Agricultural Products Quality, Faculty of Agriculture and Technology, University of South Bohemia, Studentská 1668, 370 05 České Budějovice, Czech Republic; lorencf@fzt.jcu.cz (F.L.); bedrnicek@fzt.jcu.cz (J.B.)
4 Department of Legumes and Technical Crops, Agritec Plant Research Ltd., Zemědělská 2520/16, 787 01 Šumperk, Czech Republic; bjelkova@agritec.cz
5 Department of Dairy, Fat and Cosmetics, University of Chemistry and Technology, Technická 5, 166 28 Prague, Czech Republic; jan.kyselka@vscht.cz
* Correspondence: barta@fzt.jcu.cz

Abstract: The research is focused on the quantitative evaluation of the flaxseed (*Linum usitatissimum* L.) proteome at the level of seed cake (SC), fine flour—sieved a fraction below 250 μm (FF)—and protein concentrate (PC). The evaluation was performed on three oilseed flax cultivars (Agriol, Raciol, and Libra) with different levels of α-linolenic acid content using LC-MS/MS (shotgun proteomics) analysis, which was finalized by database searching using the NCBI protein database for *Linum usitatissimum* and related species. A total of 2560 protein groups (PGs) were identified, and their relative abundance was calculated. A set of 33 quantitatively most significant PGs was selected for further characterization. The selected PGs were divided into four classes—seed storage proteins (11S globulins and conlinins), oleosins, defense- and stress-related proteins, and other major proteins (mainly including enzymes). Seed storage proteins were found to be the most abundant proteins. Specifically, 11S globulins accounted for 41–44% of SC proteins, 40–46% of FF proteins, and 72–84% of PC proteins, depending on the cultivar. Conlinins (2S albumins) were the most abundant in FF, ranging from 10 to 13% (depending on cultivar). The second most important class from the point of relative abundance was oleosins, which were represented in SC and FF in the range of 2.1–3.8%, but only 0.36–1.20% in PC. Surprisingly, a relatively high abundance of chitinase was found in flax products as a protein related to defence and stress reactions.

Keywords: flaxseed; cultivar; flour; protein concentrates; proteomic profile; 11S globulin

1. Introduction

Flaxseeds (*Linum usitatissimum* L.) have been used as human food or animal feed since historical times [1]. Flaxseed contains 30–46% oil, 18–30% protein, 20–35% fibre (about 10% is soluble fibre), 3–4% ash, and 4–8% moisture [1–5]. Minor constituents include cyanogenic glycosides, phytic acid, polyphenols, linatine, lignans (phytoestrogens), cyclolinopeptides, vitamins, selenium, and cadmium or trypsin inhibitors [2]. Due to their unique composition, flax seeds are classified as a functional food often referred to as a superfood [6,7]. Flaxseed oil is rich in essential polyunsaturated fatty acids, with n-3 alpha-linolenic acid (ALA; C18:3) 47–55% and n-6 linoleic acid (C18:2) represented by ≈15% in the oil of traditional cultivars [8]. Flaxseed oil with predominance of n-3 linolenic acid is not only used for human consumption but also, due to its drying properties, in industrial applications such as production of paints and varnishes, linoleum, oleo-chemicals, etc. [1,4,9,10]. The iodine value, which indicates the unsaturation level of the oils, is as high as 205 g I_2/g oil [11]. There are few vegetable oil plants that provide oil with a similar level of iodine value, e.g., Dragon's head oil (*Lallemantia iberica*) [12].

After mechanical extraction of the oil by pressing, flaxseed cake is a valuable material not only for farm animal feed, but also for the production of flaxseed flour, water-soluble fibre and protein concentrates [5,13–16], which, due to their functional properties, can be applied in a number of food applications in the production of quality foods that provide considerable health benefits to consumers [7,16]. High-protein flaxseed products can be used to enrich food products with nutritionally valuable components, improving their properties, including consistency, taste, and aroma. Flaxseed flours can be used in bakery products (bread, cookies, bars, biscuits, etc.), in pasta, but also, for example, in salad dressings or meat products [7]. Flaxseed protein flours have great potential in producing gluten-free products [17]. On the other hand, the food processing of flaxseeds and flaxseed products (e.g., roasting) can lead to several positive and negative changes, which also affect the protein and should be further monitored [18].

Flaxseed proteins in the form of concentrates or isolates are most commonly produced from defatted flaxseed meal stripped of water-soluble mucilage (which complicates protein concentration). First, alkaline solubilisation of the proteins is performed and then the proteins are precipitated by isoelectric precipitation [5,14,15]. The disadvantage of this most widely used method can be the varying extent of denaturation of flaxseed proteins (decreasing their solubility and functional properties), which can be caused by a high pH value (over 9) during the solubilisation step and may be improved by optimising the process conditions or partially hydrolysing the obtained proteins, which also yields valuable biologically active peptides [19–21], or by using an alternative method of protein concentration, e.g., protein extraction using ionic strength or membrane techniques [3,15].

Flaxseed protein is rich in glutamic and aspartic acid (including their amides) and the basic amino acid arginine. The total content of essential amino acids is 34%; leucine and lysine are the limiting amino acids in flaxseed protein, especially in children's diets [4,22]. Flaxseed proteins are divided into globulins (linin), which account for up to 80% of flaxseed proteins, and albumins (conlinin), which represent a minor component compared to globulins. Diverse information is given on the structure of linin or 11–12S protein. While Madhusudhan and Singh [23] present flaxseed major protein as a 12S protein with 5 non-identical subunits with a total molecular weight (MW) of 294 kDa, later papers describe an acidic (MW = 30.0–35.2 kDa) and basic (MW = 24.6 kDa) subunit in 11S flaxseed globulin [24] or present it as a protein with a total MW of 365 kDa, which, after separation on SDS-PAGE under reducing conditions, separates into bands with MWs of 20, 23, and 31 kDa [25]. Barvkar et al. [1], in their unique study of the proteomic profile of flaxseed, report 11S globulin (with the names legumin, glutenin type A or cupin) as the major protein. The second major protein group is the 1.6–2S albumin referred to as conlinin, which is composed of a single polypeptide chain of MW 16–18 kDa. Other important flaxseed proteins are reported to be the 7S storage protein, oleosins, late embryogenesis abundant (LEA) proteins and a number of metabolically important enzymes [1,26]. The importance of flaxseed proteins, especially 11S globulins and 2S albumins, in relation to food applications, human nutrition and medical applications has been sufficiently discussed in recent works [6,15,26–28].

Information on the flaxseed proteome is scarce, e.g., the work of Barvkar et al. [1], which deals with the evaluation of changes in the flaxseed proteome during its development, or the study of Klubicová et al. [29], which deals with changes in the flaxseed proteome in response to cultivation in the Chernobyl region under phytoremediation.

The evaluation of protein profiles of flaxseed products with different protein contents (seed cake, flour, protein concentrate), using a proteomic approach has not yet been performed or is not available. Similarly, information on the cultivar differences of these flaxseed products is not available. For these reasons, the results presented in this article can be considered to be new. The knowledge of the protein profile is not only important from a nutritional and technological point of view, but it is also important to better understand the physiology of mature seed, the influence genotype and method of processing on proteome (protein composition), and it enables more efficient verification of the addition of flaxseed products in food products or for monitoring the presence of potentially risky proteins.

The objectives of the present work were (a) to describe the protein profiles of flaxseed products using shotgun proteomics at the level of quantitative estimation (determination of relative abundance) of identified protein groups, (b) to evaluate the effects of the protein concentration approach used (sieving versus alkaline solubilisation/isoelectric precipitation) and the cultivar effect on the changes in the relative abundance of food-relevant flaxseed proteins such as 11S globulins, 2S albumins (conlinin), and oleosins.

2. Materials and Methods

2.1. Flaxseeds

Flaxseeds of three cultivars of oilseed flax (*Linum usitatissimum* L.) were grown at the experimental station of the Faculty of Agriculture and Technology of the University of South Bohemia in České Budějovice (GPS 48.9742 N, 14.4477 E; 380 m above sea level). These cultivars were Agriol (cultivar number LNU25748), yellow seed colour, low ALA content (Agritec Plant Research, Ltd., Šumperk, Czech Republic); Raciol (cultivar number LNU11435), yellow seed colour, medium ALA content (Agritec Plant Research, Ltd., Šumperk, Czech Republic); and Libra (cultivar number LNU19160), brown seed colour, high ALA content (Limagrain Nederland B.V., Rilland, The Netherlands). Flaxseed cultivars used were grown in 2020. The sowing of the crop took place on 7 April 2020. The soil was satisfactorily supplied with the main nutrients, and no nitrogen in mineral fertilisers was applied. The crop was treated against weeds with the herbicide Glean 75 WG (chlorsulfuron) at a 15 g/ha dose. The seeds were harvested on 27 August 2020 using Wintersteiger Nursery Master Elite small plot research combine harvester (Wintersteiger Seedmech, GmbH, Ried im Innkreis, Austria). After harvesting, the flaxseeds were cleaned of impurities, including moisture-increasing contaminants. Until analyses were carried out, seeds were stored in the dark in paper bags inside polypropylene bags at 4 °C and 42% relative humidity.

2.2. Preparation of Flaxseed Cake, Flour, and Protein Concentrate

Preparation of analysed flaxseed products is shown schematically in Figure 1. The seeds were de-mucified by water soaking in two steps. In the first step, 200 g of seeds was mixed with 800 mL of water preheated to 50 °C. The mucilage was extracted for 1 h and removed from the mixture. In the second step, the seeds were soaked in 600 mL of water for 1 h at 50 °C, while the second step was performed twice in total. Subsequently, the wet seeds were dried on a sieve for 2 days at 50 °C. Afterwards, the seeds were subjected to oil pressing using a Yoda YD-ZY-02A domestic press (Yoda, Jinjiang, China) using the "flaxseed" program. The obtained cake was subjected to repeated pressing. Subsequently, the defatted cake was ground using a Grindomix GM 200 knife mill (Retsch, Haan, Germany) at 10,000 rpm for 1 min to obtain seed cake (SC). The flour was further sieved on a stainless-steel sieve (Preciselect, Tišnov-Malhostovice, Czech Republic) with a mesh size of 250 µm to produce fine flaxseed flour (FF). FF was used as a starting material for the preparation of protein concentrate (PC) through alkaline solubilisation followed by isoelectric precipitation (AE/IEP). Protein solubilization (extraction) was carried out as follows: deionized water was added to FF in a centrifuge plastic tube at a ratio of 1:10 (*w*/*v*), the mixture was mixed properly and then the reaction was adjusted to pH 8.5 using 1M NaOH, the total extraction was carried out at room temperature for 2 h with continuous control and pH adjustment. Subsequently, centrifugation was performed at RPM 4500 for 10 min and at RT on a Rotina 420 R centrifuge (Hettich, Tuttlingen, Germany). The proteins in the obtained supernatant were then subjected to precipitation by decreasing the pH to a value of 4.5 using 2M HCl. After centrifugation (under the same conditions as above), the obtained pellets were dissolved in deionized water, the pH was adjusted to 7.0 using 1M NaOH, the sample was frozen (−20 °C) and then freeze-dried using an Alpha 1–4 LSC lyophilizer (Martin Christ, Osterode am Harz, Germany) at a pressure of 0.520 mBar, and a temperature of −50 °C for 72 h. The obtained lyophilizate was homogenized to a powder representing the protein concentrate (PC).

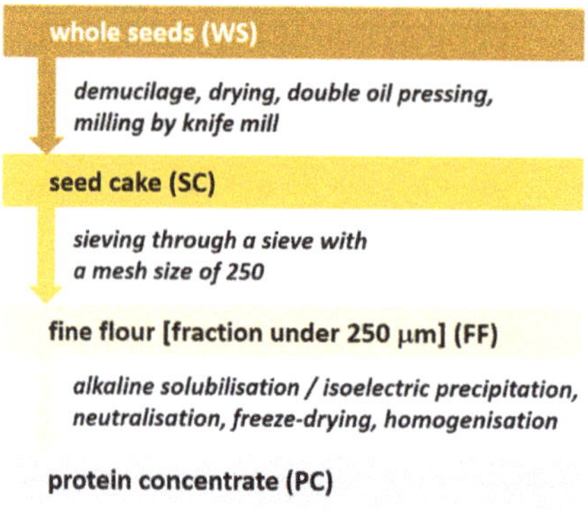

Figure 1. Preparation of analysed flaxseed products from whole seeds.

2.3. Proximate Composition Analysis

All the proximate analyses were determined according to Bárta et al. [30]. Moisture was determined gravimetrically by drying the samples in an oven to constant weight at 105 °C for 3 h. The protein content was determined by the modified Dumas combustion method using a rapid N cube (N/Protein analysis) instrument (Elementar Analysen System, Langenselbold, Germany). The protein content was calculated as the nitrogen content multiplied by a factor of 6.25. The fat content was measured using the Soxhlet extraction method using an ANKOM XT 10 Extractor (ANKOM Technology, Macedon, NY, USA) according to the manufacturer's manual. Petroleum ether was used as an extraction reagent. The fat content was calculated from the weight differences of the sample before and after extraction. The ash content was measured as the total amount of inorganic residues remaining after the high heat incineration of analysed samples at 550 °C for 5 h in a muffle furnace (muffle furnace LE 09/11, LAC s.r.o., Židlochovice, Czech Republic). The content of carbohydrates was determined as the remaining residue of the total fresh matter (FM) after subtracting the sum of moisture, protein, fat, and ash. Each sample was analysed in triplicate for all analyses.

2.4. One-Dimensional Electrophoresis (SDS-PAGE)

The flaxseed product samples were extracted with SDS extraction buffer [0.065 M Tris-HCl, pH 6.8; 2% (w/v) SDS; 5% (v/v) 2-sulphanylethanol] in a ratio of 1:10 (w/v) at 4 °C for 4 h. Cooled dual vertical slab units SE 600 (Hoefer Scientific Instruments, Holliston, MA, USA) were used for protein separation. Protein samples were separated with a discontinuous gel system (4% stacking and 11% resolving gel) in reduction conditions (modified according to Laemmli [31] and Bárta et al. [32]. Each research variant was analysed in triplicate. Protein detection was performed using Coomassie Brilliant Blue R-250.

2.5. LC-MS/MS Analysis and Proteomic Data Processing

Flaxseed product samples were processed in triplicate in a similar way as hemp samples in Bárta et al. [32]. Proteins from flaxseed product samples were isolated using SDT buffer (4% SDS, 0.1M DTT, 0.1M Tris/HCl, pH 7.6) in an Eppendorf ThermoMixer C (120 min, 95 °C, 750 rpm). Subsequently, samples were centrifuged (15 min, 20,000× g). Then, the supernatants (containing ca 100 μg of total protein) were processed using filter-aided sample preparation (FASP) as described elsewhere [33] using 0.5 μg of trypsin (sequencing grade; Promega). Resulting peptides were analysed using LC-MS/MS.

Obtained peptide mixtures were analysed using LC-MS/MS using the UltiMate 3000 RSLCnano system coupled to Orbitrap Exploris 480 mass spectrometer (Thermo Fisher Scientific, Waltham, MA, USA). Tryptic digests were online concentrated and desalted on trapping column (Acclaim PepMap 100 C18, dimensions 300 μm ID, 5 mm long, 5 μm particles, Thermo Fisher Scientific, heated to 40 °C). Trapping column with peptides was washed with 0.1% FA; the peptides were eluted in backflush mode (350 nL·min^{-1}) from the trapping column. Gradient elution (90 min, flow rate 250 nL·min^{-1}, 3–37% of mobile phase B; mobile phase A: 0.1% FA in water; mobile phase B: 0.1% FA in 80% ACN) was applied for separation on an analytical column (EASY-Spray column, 75 μm ID, 500 mm long, 2 μm particles, Thermo Fisher Scientific, heated to 50 °C).

A data-independent acquisition mode (DIA) was selected for data acquisition. The m/z range of 350–1400 at a resolution of 60,000 (at m/z 200) and maximum injection time of 55 ms was used for the survey scan. The m/z range of 200–2000 at 30,000 resolution (maximum injection time 55 ms, 27% relative fragmentation energy) was selected for HCD MS/MS spectra collection. Overlapping windows scheme in m/z range from 400 to 1000 were used as isolation window placements.

DIA-NN software (version 1.8.1) [34] was employed for the processing of acquired DIA data. The library free mode was used for searches against modified cRAP database (based on http://www.thegpm.org/crap/ (accessed on 22 November 2018); 111 sequences in total) and whole *Linum* genus database (https://www.ncbi.nlm.nih.gov/ipg/?term=txid4005[Organism:exp] (accessed on 30 September 2023, number of protein sequences: 125,854). Carbamidomethylation as fixed modification, with no optional modification and trypsin/P enzyme with 1 allowed missed cleavage and peptide length 7–30, were set for preparation of the library. The false discovery rate (FDR) control was set to 1% FDR. The initial test searches were performed to set MS1 and MS2 accuracies and scan window parameters (median value from all samples ascertained parameter values). MBR was switched on.

The main report file generated using the DIA-NN was further processed using the software container environment (https://github.com/OmicsWorkflows/KNIME_docker_vnc (accessed on 4 October 2023)), version 4.6.3a. The processing workflow is available upon request. Briefly, it covered (a) removal of low-quality precursors and contaminant protein groups, (b) estimation of intensity-based absolute quantities (iBAQ values) using DIA-TPA [35] algorithm using normalized precursor intensities, (c) relative iBAQ (riBAQ) values calculation for individual protein groups within individual sample replicates. DIA-TPA algorithm implementation calculated iBAQ (and following riBAQ) values for all protein groups having at least one proteotypic peptide and or protein groups without any proteotypic peptide, but where all proteins of the given protein group were reported only together.

The description of the identified protein groups presented in the results section of the article (Section 3.2) was based on information available in the NCBI or UniProt KB databases.

2.6. Statistical Analysis

Obtained data were analysed using software Statistica 12 (StatSoft Power Solutions Inc., Palo Alto, CA, USA). Two-way ANOVA (analyses of variance) and the variance components method was used for data evaluation. The means were compared using the Tukey HSD test. Differences between the variants were considered significant at $p < 0.05$ unless stated otherwise.

3. Results and Discussion

3.1. Proximate Composition of Flaxseed Products

The proximate chemical composition of the analysed flaxseeds (WS) and the flax products derived from them is given in Table 1. The individual processing steps from seed to flaxseed protein concentrate (WS, SC, FF, PC), as shown in Figure 1, lead to a gradual relative increase in the average protein content, from 18% (WS), over 30% (SC), and 36% (FF) to 70% or more (PC) of the three oilseed flax cultivars evaluated. The differences between the cultivars are noticeable, but the trend and the ratio of the gradual increase are similar for all the cultivars evaluated. The WS fat content of all three cultivars has decreased from the original 38–40% to below 10% in SC due to cold pressing (up to 50 °C). During sieving SC using a sieve with a mesh size of 250 μm, a fractionation effect occurred resulting in increased protein and fat content but less carbohydrates in flaxseed flour (FF), compared to the original SC.

The production of oil, flour, or protein from flaxseed is more difficult than for other oilseeds because of water-soluble mucilages (gums), which complicate processing, especially 'wet' processing. The presence of these mucilages and unsophisticated isolation methods have been the main reasons that have hindered the wider dissemination of flax protein [25,36]. Various methods have been tried to remove (or reduce) the mucilage content in flaxseeds, including the use of degradative enzymes and extraction using water at different temperatures [14,37]. Kaushik et al. [14] tested the effect of temperature during water extraction of the mucilage, prior to the actual processing of the seed. The authors concluded that water extraction at 60 °C can yield flaxseed protein isolate with purity of over 90%. Sieving is also important in changing the proportions of the content of the moulded fibre. The finer fractions of the flour (under a sieve with a size of about 250 μm) have a higher protein (and possibly fat) content and a lower carbohydrate (especially fibre) content, as sieving removes a greater proportion of the coarse fragments that represent the surface layers of the seed or hull [30].

Table 1. Proximate composition of whole flaxseeds (WS) and derived flaxseed products—seed cake (SC), fine flour (FF), and protein concentrate (PC).

Component	Cultivar	WS	SC	FF	PC
Protein (% FM)	Agriol	18.02 ± 0.58 Ad	30.19 ± 1.24 Ac	35.81 ± 0.50 Ab	78.34 ± 0.52 Aa
	Raciol	17.06 ± 0.66 Ad	29.27 ± 0.59 Ac	33.87 ± 0.89 Ab	70.09 ± 1.63 Ba
	Libra	17.84 ± 1.04 Ad	30.42 ± 1.04 Ac	36.02 ± 0.22 Ab	70.87 ± 1.28 Ba
Fat (% FM)	Agriol	37.76 ± 0.61 Ba	8.53 ± 0.47 Ac	10.94 ± 0.04 Ab	2.41 ± 0.77 Bd
	Raciol	38.24 ± 0.08 Ba	9.46 ± 0.16 Ac	12.04 ± 1.09 Ab	2.42 ± 0.36 Bd
	Libra	40.76 ± 0.54 Aa	9.22 ±1.20 Ac	12.21 ± 1.22 Ab	5.04 ± 1.27 Ad
Carbohydrates (% FM)	Agriol	34.27 ± 0.47 Ac	47.28 ± 1.23 Aa	38.75 ± 0.66 Ab	9.78 ± 0.54 Cd
	Raciol	34.86 ± 0.67 Ac	48.01 ± 0.31 Aa	39.68 ± 0.70 Ab	19.41 ± 2.15 Ad
	Libra	31.90 ± 1.69 Ac	46.29 ± 0.91 Aa	36.87 ± 0.82 Ab	15.99 ± 0.32 Bd
Ash (% FM)	Agriol	3.43 ± 0.12 Ac	5.36 ± 0.12 Bb	6.28 ± 0.03 Ba	2.71 ± 0.12 Ad
	Raciol	3.52 ± 0.09 Ac	5.27 ± 0.08 Bb	6.28 ± 0.02 Ba	2.72 ± 0.08 Ad
	Libra	3.52 ± 0.09 Ac	5.78 ± 0.13 Ab	7.01 ± 0.04 Aa	2.82 ± 0.15 Ad
Moisture (% FM)	Agriol	6.52 ± 0.04 Ab	8.63 ± 0.24 Aa	8.22 ± 0.21 Aa	6.76 ± 0.86 Ab
	Raciol	6.32 ± 0.05 Ab	7.99 ± 0.32 Aa	8.12 ± 0.13 Aa	5.36 ± 1.03 Bb
	Libra	5.98 ± 0.08 Ab	8.29 ± 0.14 Aa	7.89 ± 0.24 Aa	5.27 ± 0.15 Bb

Note: FM—fresh matter; different capital letters in the columns for individual component values indicate a statistically significant cultivar difference at $p < 0.05$ (Tukey HSD test); different lower-case letters for the component values in the rows indicate a statistically significant difference between the flaxseed products of the respective cultivar at $p < 0.05$ (Tukey HSD test).

3.2. Description of Main Identified Protein Groups in Flaxseed Products

The protein profiles of flaxseed products derived from seeds after oil pressing of the three cultivars separated using SDS-PAGE are shown in Figure 2.

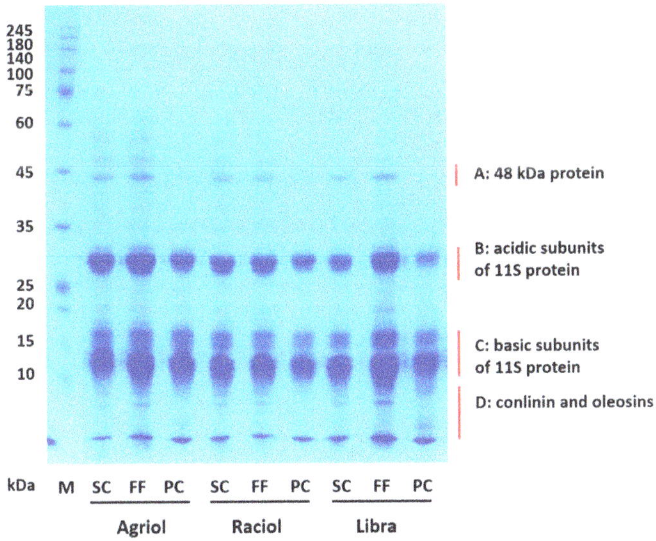

Figure 2. 1D SDS-PAGE profiles of flaxseed products (seed cake—SC, fine flour—FF, protein concentrate—PC) from three oilseed flax cultivars (Agriol, Raciol, Libra). Note: M—molecular mass standard (RotiMark Tricolor [8271.1], Carl Roth GmbH + Co. KG, Karlsruhe, Germany).

A typical electrophoretic profile of oilseed proteins separated under denaturing and reducing conditions [27,30] is evident in all samples and it agrees with reported studies dealing with flaxseed proteins [1,5,25,38,39]. The proteins can be divided into four MW regions: the protein with an MW of 48 kD (region A), the acidic subunits of the 11S globulin monomer (region B), the basic subunits of the 11S globulin monomer (region C), and the region of 2S albumin conlinin and oleosins (region D). There are minor differences between the seed protein profiles of the evaluated cultivars. The PC profiles of all three cultivars are slightly simpler with a predominance of protein bands in the regions of the acidic and basic subunits of the 11S protein monomer (globulin). This suggests that the chosen alkaline solubilisation (extraction) procedure at pH 8.5 and the subsequent precipitation of the extracted proteins caused by decreasing the pH to 4.5 leads to a higher relative abundance of 11S globulins (discussed more in Section 3.3). However, this suggestion can be taken with some reserve because the absolute amount of protein in the samples loaded on the electrophoretic gel differed.

After LC-MS/MS analysis of the peptide mixtures of the samples, the obtained data were subjected to qualitative and quantitative evaluation using available databases of primary protein sequences. Considering the unavailability of reference proteome for flax, the NCBI database (LINUM taxonomy) was primarily used, with the help of which 2560 protein groups (PGs) were identified (see Supplementary Table S1). For the assessment of relative abundance (determined from measured intensities, riBAQ values), the set of 2560 identified PGs was assumed to be 100%. For detailed evaluation, a set of 33 PGs with high relative abundance in the total protein profile (PGs above 0.5%) or belonging to important flaxseed proteins was selected. The PGs items in this set were divided into four classes (Table 2) into seed storage proteins, oleosins, proteins associated with defence and stress, and other selected important proteins.

Table 2. Description of selected protein groups found by proteomic processing of flaxseed products.

Accession	Name of Protein Groups	Description from UniProt KB/(Available Information from NCBI)
Seed storage proteins		
CAI0558954.1	unnamed protein product; LINTE	(cupin 11S legumin N/11S legumin seed storage globulin/N-terminal cupin)
CAI0432378.1	unnamed protein product; LINTE	(11S globulin-like/11S globulin seed storage protein)
CAI0558846.1	unnamed protein product; LINTE	(cupin_RmlC-like protein)
CAI0434421.1	unnamed protein product; LINTE	(cupin_RmlC-like protein) 11S globulin
CAI0418748.1	unnamed protein product; LINTE	(cupin_RmlC-like protein)
CAC94011.1	conlinin; LINUS	2S seed storage albumins family; nutrient reservoir activity
CAC94010.1	conlinin; LINUS	2S seed storage albumins family; nutrient reservoir activity
Oleosins		
ABB01617.1	oleosin low molecular weight isoform; LINUS	oleosin family; membrane; monolayer-surrounded lipid storage body
ABB01620.1	oleosin low molecular weight isoform, partial; LINUS	oleosin family; membrane; monolayer-surrounded lipid storage body
ABB01616.1	oleosin high molecular weight isoform; LINUS	oleosin family; membrane; monolayer-surrounded lipid storage body
ABB01624.1	oleosin high molecular weight isoform; LINUS	oleosin family; membrane; monolayer-surrounded lipid storage body
ABB01618.1	oleosin low molecular weight isoform; LINUS	oleosin family; membrane; monolayer-surrounded lipid storage body
ABB01619.1	oleosin low molecular weight isoform, partial; LINUS	oleosin family; membrane; monolayer-surrounded lipid storage body
Defence and stress-related proteins		
WMZ41542.1	chitinase; LINUS	(chitinase; involved in somatic embryogenesis in flax)
AAW31878.1	chitinase, partial; LINUS	chitinase activity; chitin binding; cell wall macromolecule catabolic process; defense response
AMY26620.1	late embryogenesis abundant (lea) group 1-embryo development; LINUS	LEA type 1 family; embryo development ending in seed dormancy
AVY09180.1	Sequence 3353 from patent US 9878004; UNKNOWN	(17.3 kDa class I heat shock protein)
CAI0408810.1	unnamed protein product; LINTE	(luminal-binding protein 5; heat shock 70 kDa protein)
1DWM_A	Chain A, Linum usitatissimum trypsin inhibitor; LINUS	(Serine proteinase inhibitor class)
Other selected important proteins		
AVY09175.1	Sequence 3348 from patent US 9878004; UNKNOWN	(11-beta-hydroxysteroid dehydrogenase)
CAI0407981.1	unnamed protein product; LINTE	(glucose and ribitol dehydrogenase/NADPH-dependent aldehyde reductase 1)
CAF22093.1	glyceraldehyde 3-phosphate dehydrogenase, partial; LINUS	glyceraldehyde-3-phosphate dehydrogenase family; oxidoreductase
BAL41455.1	fructose-bisphosphate aldolase 1, partial; LINGR	class I fructose-bisphosphate aldolase family; fructose-bisphosphate aldolase activity; glycolytic process
P48417.1	Allene oxide synthase, chloroplastic; LINUS	allene oxide synthase; activity; fatty acid biosynthetic process
ABM64783.1	cell wall glycosidase, partial; LINUS	hydrolase activity, hydrolyzing O-glycosyl compounds; carbohydrate metabolic process
CAI0384355.1	unnamed protein product; LINTE	(copper/zinc superoxide dismutase)
CAI0431450.1	unnamed protein product; LINTE	(manganese superoxide dismutase)
CAI0412769.1	unnamed protein product; LINTE	(linoleate 13S-lipoxygenase 2-1, chloroplastic; Lipoxygenase)
AVY09204.1	Sequence 3377 from patent US 9878004; UNKNOWN	(hypothetical protein DKX38_015183; malate dehydrogenase)
CAI0440995.1	unnamed protein product; LINTE	(putative mitochondrial malate dehydrogenase; malate dehydrogenase, mitochondrial)
CAI0411675.1	unnamed protein product; LINTE	(elongation factor 1-alpha; translation elongation factor)
AGN56421.1	non-specific lipid transfer protein 1; LINUS	lipid binding; lipid transport; plant LTP family
CAI0414382.1	unnamed protein product; LINTE	(thaumatin like protein)

Note: LINUS—*Linum usitatissimum*; LINTE—*Linus tenue*; LINGR—*Linus grandiflorum*; UNKNOWN.

3.2.1. Seed Storage Proteins

The seed storage proteins (SSPs) in flaxseed are mainly classified as 11S (12S) globulins and 2S albumins [3,16,25,27]. The globulin fraction is often referred as linin [3]; however, PGs are not listed under this name in the NCBI and UniProt KB databases. In addition to 11S globulins, we were able to find five items in the set of identified PGs (Table 2). In terms of relative abundance, the most significant items are CAI0558954.1, CAI0432378.1, and CAI0558846.1, which are designated as "cupin 11S legumins" or "11S globulin/legumin seed storage proteins" or "11S globulin-like proteins".

According to other authors [3,25,28], these are polymeric proteins with MW ≈ 300 kDa. The PGs we have found have primary polypeptide sequences of 367–536 amino acids in length, from which two sub-polypeptide chains, an acidic and a basic one, are cut during post-translational processing, similarly to proteins of other oilseeds [40], which then give rise to the final monomer by their association via a disulfide bond. Due to their high relative abundance, 11S globulins have a major impact on the amino acid composition of flaxseed proteins. The nutritional quality of flaxseed proteins is close to that of soya and canola proteins. Their essential amino acid index is reported at 69, whereas values of 79 and 75 are reported for soya and canola proteins, respectively. The significantly represented amino acids of flaxseed proteins include arginine, aspartic, and glutamic acids (including their amides) [3]. Flaxseed protein has a lysine to arginine ratio of about 0.25 and, therefore, has less lipidemic and atherogenic effects than, for example, soya protein (0.71), milk caseinate (2.15), or whey protein (5.38), so it is referred to as a "better" protein for the health of cardiac patients [14].

A second important part of flax SSPs are the 2S albumins, commonly referred to as conlinins [1,26,41]. In the set of identified proteins, two PGs (CAC94011.1, CAC94010.1) were found with the name conlinin and the described activity of the storage protein (Table 2). According to the accession number and the assigned GI number, these items are identical to the two proteins found by Liu et al. [41] in their work. These authors studied conlinin for its involvement in the emulsification activity of the flaxseed soluble gum (FSG) fraction, to which conlinins were co-transferred during the hot water isolation of FSG. The conlinins were found to be significantly involved in increasing the emulsification activity and stability of the emulsions formed. These are proteins that have been described in databases based on the cDNA sequence of two conlinin genes, cnl1 and cnl2 [42]. Earlier work [23] describes conlinins as proteins with a sedimentation coefficient of 1.6S and a MW of 15–18 kDa, while Liu et al. [41] found a MW by 2D-PAGE of around 11 kDa; however, the UniPro KB database lists a predicted MW of 19.063 kDa (169 amino-acids) for Q8LPD4 (CAC94011.1) and a predicted MW of 19.012 kDa (168 amino-acids) for Q8LPD3 (CAC94011.1).

Diverse physiological functions are generally reported for seminal 2S albumins. Souza [43] in his work states that they are not only proteins with a storage function as a source of indispensable amino acids (especially cysteine), but due to their alpha-helical structure and positively charged amino acid residues, they are often characterized by antimicrobial and antifungal activity. The more significant use of 2S albumins in food and biotechnological applications is hampered by their relatively high allergenicity [44]. Additionally, flaxseed polypeptides with MW from 20 to 38 kDa (or more), e.g., malate dehydrogenase, are suspected of allergenicity, too [3]. Allergies to flaxseed are rarely reported. However, Bueno-Díaz et al. [6] described recently the allergy of five patients to flaxseed and 2S albumin was the main allergen, and 11S globulin was a potential allergen.

In addition to 11S globulins and 2S albumins, other proteins of similar function have been reported in flaxseed. For example, Barvkar et al. [1] described glutelins, a 48 kDa proteins, and 7S globulins in flaxseeds. Globulins 7S were detected in seeds of other oilseed crops including hemp [32]. No proteins of this type were found in our study.

3.2.2. Oleosins

Oleosins (OLs) are proteins with MW ranging from 15 to 26 kDa [45,46] that participate in the structure of oil bodies (oleosomes). In oilseeds, oleosomes are mainly localized in the cotyledon's cells, where they form structures with diameters between 0.2 and 2 μm [47]. Oleosomes are composed of a monolayer membrane of phospholipids reinforced not only by OLs but also by other types of proteins such as caleosins or steroleocins [47–49]. Typically, the oleosome isolated from oil bodies contains about 94–98% neutral lipids, 0.6–2% phospholipids and 0.6% protein [49].

The OLs are among the most quantitatively important proteins in oilseeds. We found a total of six oleosin PGs, whose relative abundance in the seed extracts of the three cultivars evaluated ranged from 2.12 to 2.86% in analysed flaxseed products. Four PGs were the "oleosin low molecular weight isoforms" (MW in the range 14.991–16.039 kDa) and two the "oleosin high molecular weight isoforms" (MW 18.643 kDa or 18.712 kDa). OLs relative abundance in flaxseeds is lower compared to hempseeds (8%) based on the results of the recent study by Bárta et al. [32].

3.2.3. Defence- and Stress-Related Proteins

Within the PGs significantly related to flaxseed defence and its response to stress, we detected the enzyme chitinase, late embryogenesis abundant protein, heat shock proteins, and trypsin inhibitor. We observed a surprisingly high relative abundance of chitinase (represented by two PGs WMZ41542.1 and AAW31878.1) ranging from 6 to 10% in the flaxseed protein pool. Klubicová et al. [29] also found chitinase in mature flaxseeds harvested from remediated plots in the Chernobyl region, but at a significantly lower abundance of 0.3% (calculated as a percentage of spot volume in the evaluation of the information obtained on the gel after two-dimensional electrophoresis).

Among the proteins associated with the stress responses of the maturing seed, a PG with the designation "Late embryogenesis abundant (LEA) group 1—embryo development" (AMY26620.1) can be included. LEA proteins have been found in various organisms, including plants. In seed tissues, they act as protein and membrane protectors against drought stress and water deprivation [50].

The PG with the designation "Sequence 3353 from patent US 9878004" (accession number AVY09180.1) was found to be a quantitatively significant item (relative abundance of about 0.9%). Polypeptide of this PG consists of 158 amino acids, and the predicted MW is 17.99 kDa. The function of this PG in flaxseeds is unclear; however, a detailed search of the NCBI and UniProt KB databases revealed a high degree of sequence similarity (around 80%) to small heat shock protein (SHSP) type proteins. SHSPs are found in both prokaryotic and eukaryotic cells and are among the chaperones, and are generally part of the quality control mechanism of cellular proteins with a line of defence function against conditions that threaten the cellular proteome [51]. In plant seeds, the action of SHSPs is not only related to protection against heat stress. The expression of SHSPs in seeds occurs during the ripening process, leading to their accumulation in dry seeds, but their presence disappears after seed germination [52].

A significant finding is the presence of an item (1DWM_A) that forms part of a trypsin inhibitor. On the one hand, protease inhibitors are part of the natural defence of plants against harmful agents, which can be used biotechnologically or in plant protection [2,53], on the other hand, they can have adverse effects on human digestion and utilization of consumed proteins [54].

3.2.4. Other Selected Important Proteins

Among other proteins significant in their relative abundance are enzymes metabolically relevant to mature seed physiology, e.g., glyceraldehyde-3-phosphate dehydrogenase, fructose-bisphosphate aldolase and malate dehydrogenase, whose presence in flaxseeds is also confirmed in the work of Barvkar et al. [1]. Among the significant PGs found in the flax products we analysed is AVY09175.1, which is probably related to hydroxysteroid dehydrogenase activity and oleosome (oil body) membrane proteins [55]. PGs with significant relative abundance in flax seeds also include allene oxide synthase (P48417.1) with a relative abundance of about 0.6%, which is likely involved in fatty acid biosynthesis, and translational elongation factor (CAI0411675.1) with a relative abundance of about 1%.

3.3. Influence of Cultivar and Concentration Method on Relative Abundance of Main Protein Groups

The relative abundance of individual proteins contained in flaxseed products can be influenced by several factors. Our study evaluated the influence of protein concentration methods producing different flaxseed products (FF and PC) in combination with the influence of cultivars. In terms of evaluating the influence of the concentration method factor, two methods used were chosen as examples of dry and wet processes to increase the protein concentration in products [16,30,56]. The dry process is represented by sieving milled SC through a sieve with a mesh size of 250 μm, while the wet process is represented by the AE/IEP combination. The 250 μm mesh size sieve was chosen, as in our work of Bárta et al. [30], based on our previous experience when we tested sieves with mesh sizes ranging from 180 to 710 μm. The 250 μm mesh sieve provided the best result regarding the protein concentration effect and practical advantages (the sieve did not clog). Analysis of the data set of selected PGs revealed that for 30 PGs out of 33 in total, the concentration method factor had a significantly greater influence than the cultivar factor.

The estimation of the relative abundance of the PGs for the flaxseed products derived from the three cultivars is given in Table 3. From the results, it is evident that SSPs have the highest relative abundance in the flaxseed protein pool. The PGs (especially items CAI0558954.1, CAI0432378.1, CAI0558846.1; CAI0434421.1), which represent 11S globulin or linin, respectively, occupy a relative representation in SC for all three cultivars in the range of 41–44%. Most authors report higher values, e.g., Bekhit et al. [4] report a relative representation of 64–66% for 11S globulin, while another study [7] reports a relative abundance of 70–85% for globulins. The difference may be explained by the fact that in most papers the abundance of globulins is reported as the abundance of the protein fraction soluble in salt solution, which may also contain other proteins. On the other hand, the 7S protein fraction, which is assigned to globulins [1], was not found in our samples.

The wet process of protein concentration using AE/IEP led to a significant increase in the relative abundance of 11S proteins in PC samples (Table 3, Figure 3). This method increased the representation of 11S globulins from a level of 40–46% to a level of 72–84% for the materials of all three cultivars. The effect of cultivar on the concentration of 11S globulins was less significant for the most represented items and was estimated by statistical analysis (variance component) to be 3–11%, while the effect of the concentration method was in the range of 79–95%. The effect of sieving the SC on the concentration of 11S globulins was ambiguous depending on the cultivar (see Figure 3). Alkaline solubilisation (extraction) allows the transition of water-insoluble globulins to the soluble state due to the prevailing isoelectric point (pI) values of globulin proteins. The pI values for 11S flaxseed globulins are reported to be around pH 4.5 [14]. Considering the known fact that proteins are insoluble in environments equal to or close to their pI, alkaline extraction in the pH range of 7.5–9.0 (12.0) allows sufficient "distance" of proteins from their pI value and conversion to a soluble state [56,57]. Increasing the concentration of the alkaline medium, and thus higher pH during extraction, generally increases the yield of proteins [57,58], but also increases the risk of protein damage—denaturation, protein hydrolysis, amino acid racemization, and loss of essential amino acids, including the formation of lysinoalanine with nephrotoxic activity [58,59]. The low residual fat content of the original flour or its defatting with organic solvent is also important, as pH adjustment through hydroxides leads to soap formation when reacting with fat.

The fraction of conlinins (2S albumin) was found in the SC of the evaluated cultivars at the level of 7–9% (Table 3), which is lower than have been reported by other authors. Often, the albumin fraction is reported at 15–30% [5,15]. As can be seen from Figure 3, for conlinins, mechanical sieving (increase from the original 7–9% to 10–13%) has a higher concentration effect than AE/IEP (decrease to 7.5–11%). Sieving resulted in an increase in relative abundance for two cultivars (Agriol, Libra), but AE/IEP resulted in a very low abundance of OLs in PC for the class OLs.

Table 3. Relative abundance of selected protein groups in flaxseed products (seed cake—SC, fine flour—FF, protein concentrate—PC) of three oilseed flax cultivars expressed in relative % (mean ± std. deviation).

Accession	Protein Name	Agriol			Raciol			Libra		
		SC	FF	PC	SC	FF	PC	SC	FF	PC
Seed storage proteins										
CAI0558954.1	unnamed protein product; LINTE	29.13 c ± 1.30	30.44 c ± 0.32	53.37 a ± 0.45	31.09 bc ± 6.62	29.37 c ± 0.43	51.71 a ± 1.63	26.34 c ± 1.10	23.34 c ± 0.28	38.83 b ± 4.24
CAI0432378.1	unnamed protein product; LINTE	7.12 cd ± 0.14	8.06 cd ± 0.70	16.08 ab ± 0.83	6.39 d ± 0.77	7.94 cd ± 0.44	14.57 b ± 0.30	8.19 cd ± 0.93	8.89 c ± 0.19	16.88 a ± 0.95
CAI0558846.1	unnamed protein product; LINTE	6.69 b ± 0.15	6.42 b ± 0.29	11.86 a ± 0.55	5.21 b ± 0.43	5.69 b ± 0.26	10.91 a ± 0.28	6.37 b ± 0.68	6.30 b ± 0.10	13.23 a ± 0.20
CAI0434421.1	unnamed protein product; LINTE	0.82 bc ± 0.02	0.97 bc ± 0.05	2.44 a ± 0.10	0.65 c ± 0.15	0.92 bc ± 0.02	2.41 a ± 0.09	0.68 c ± 0.03	1.06 b ± 0.07	2.59 a ± 0.29
CAI0418748.1	unnamed protein product; LINTE	0.01 a ± 0.01	0.21 a ± 0.02	0.27 a ± 0.01	0.17 a ± 0.02	0.19 a ± 0.03	0.17 a ± 0.15	0.23 a ± 0.01	0.13 a ± 0.11	0.20 a ± 0.21
CAC94011.1	conlinin; LINUS	7.01 ab ± 1.09	9.62 ab ± 1.13	7.43 ab ± 1.41	5.90 b ± 1.29	9.01 ab ± 1.55	8.56 ab ± 0.93	7.72 ab ± 2.31	11.32 a ± 1.01	9.72 ab ± 2.51
CAC94010.1	conlinin; LINUS	0.24 c ± 0.07	0.58 bc ± 0.30	0.26 c ± 0.04	0.43 c ± 0.16	0.31 c ± 0.16	0.46 c ± 0.27	1.35 ab ± 0.60	2.05 a ± 0.25	1.46 a ± 0.28
Oleosins										
ABB01617.1	oleosin low molecular weight isoform; LINUS	0.12 b ± 0.00	0.24 ab ± 0.23	0.03 b ± 0.02	0.23 ab ± 0.15	0.15 ab ± 0.01	0.05 b ± 0.03	0.17 ab ± 0.02	0.51 a ± 0.29	0.11 b ± 0.07
ABB01620.1	oleosin low molecular weight isoform, partial; LINUS	0.17 a ± 0.01	0.29 a ± 0.20	0.03 a ± 0.01	0.30 a ± 0.12	0.17 a ± 0.01	0.06 a ± 0.03	0.26 a ± 0.11	0.38 a ± 0.26	0.12 a ± 0.07
ABB01616.1	oleosin high molecular weight isoform; LINUS	0.98 b ± 0.60	1.11 ab ± 0.46	0.12 c ± 0.03	1.77 a ± 0.11	0.60 bc ± 0.01	0.16 c ± 0.03	0.95 b ± 0.19	1.09 ab ± 0.11	0.41 bc ± 0.10
ABB01624.1	oleosin high molecular weight isoform; LINUS	0.64 abc ± 0.45	0.73 abc ± 0.55	0.14 c ± 0.01	0.18 c ± 0.01	1.11 ab ± 0.16	0.19 c ± 0.02	1.17 ab ± 0.25	1.39 a ± 0.32	0.44 bc ± 0.18
ABB01618.1	oleosin low molecular weight isoform; LINUS	0.14 d ± 0.003	0.18 bc ± 0.01	0.03 f ± 0.01	0.15 cd ± 0.01	0.19 b ± 0.01	0.04 ef ± 0.01	0.17 bc ± 0.01	0.22 a ± 0.01	0.06 e ± 0.004
ABB01619.1	oleosin low molecular weight isoform, partial; LINUS	0.07 ab ± 0.01	0.13 ab ± 0.09	0.01 b ± 0.005	0.14 ab ± 0.05	0.09 ab ± 0.01	0.02 a ± 0.01	0.14 ab ± 0.06	0.19 a ± 0.12	0.06 ab ± 0.04
Defence and stress-related proteins										
WMZ41542.1	chitinase; LINUS	9.38 a ± 0.15	7.63 b ± 0.06	1.28 e ± 0.05	10.39 e ± 0.58	9.94 a ± 0.25	3.14 d ± 0.59	6.18 c ± 0.56	5.25 c ± 0.12	3.21 d ± 0.31
AAW31878.1	chitinase, partial; LINUS	0.06 b ± 0.01	0.02 de ± 0.002	0.004 e ± 0.00004	0.10 a ± 0.0004	0.05 c ± 0.0003	0.01 e ± 0.002	0.07 b ± 0.01	0.03 d ± 0.003	0.01 e ± 0.0003
AMY26620.1	late embryogenesis abundant (lea) group 1-embryo development; LINUS	0.39 a ± 0.03	0.22 bc ± 0.01	0.01 d ± 0.003	0.38 a ± 0.07	0.16 c ± 0.03	0.03 d ± 0.005	0.32 ab ± 0.10	0.21 bc ± 0.02	0.02 d ± 0.01
AVY09180.1	Sequence 3353 from patent US 9878004; UNKNOWN	0.86 a ± 0.01	0.91 a ± 0.05	0.29 c ± 0.01	0.89 a ± 0.08	0.95 a ± 0.04	0.35 bc ± 0.05	0.85 a ± 0.03	0.96 a ± 0.02	0.45 b ± 0.06
CAI0408810.1	unnamed protein product; LINTE	0.46 a ± 0.01	0.22 c ± 0.01	0.04 d ± 0.01	0.35 b ± 0.03	0.20 c ± 0.01	0.04 d ± 0.004	0.38 b ± 0.03	0.18 c ± 0.003	0.06 d ± 0.01
1DWM_A	Chain A, Linum usitatissimum trypsin inhibitor; LINUS	1.28 a ± 0.21	0.60 bc ± 0.07	0.21 cd ± 0.01	1.12 a ± 0.29	0.62 b ± 0.09	0.19 d ± 0.04	0.97 ab ± 0.16	0.69 b ± 0.05	0.21 cd ± 0.05
Other selected important proteins										
AVY09175.1	Sequence 3348 from patent US 9878004; UNKNOWN	1.32 b ± 0.04	1.32 b ± 0.04	0.24 d ± 0.05	1.29 b ± 0.11	1.25 b ± 0.03	0.22 d ± 0.02	1.37 ab ± 0.05	1.48 a ± 0.01	0.55 c ± 0.04
CAI0407981.1	unnamed protein product; LINTE	0.88 c ± 0.02	0.85 c ± 0.01	0.25 e ± 0.08	0.97 bc ± 0.11	0.94 bc ± 0.05	0.29 e ± 0.04	1.11 ab ± 0.07	1.18 a ± 0.12	0.57 d ± 0.05
CAF22093.1	glyceraldehyde 3-phosphate dehydrogenase, partial; LINUS	0.38 ab ± 0.03	0.34 b ± 0.003	0.05 c ± 0.02	0.36 b ± 0.03	0.35 b ± 0.02	0.05 c ± 0.004	0.43 a ± 0.02	0.39 ab ± 0.01	0.10 c ± 0.02
BAL41455.1	fructose-bisphosphate aldolase 1, partial; LINGR	0.33 b ± 0.02	0.33 b ± 0.01	0.06 c ± 0.01	0.33 b ± 0.02	0.32 b ± 0.02	0.08 c ± 0.005	0.39 a ± 0.03	0.40 a ± 0.01	0.10 c ± 0.02
P48417.1	Allene oxide synthase, chloroplastic; LINUS	0.67 b ± 0.03	0.58 cd ± 0.01	0.02 e ± 0.01	0.58 cd ± 0.05	0.56 d ± 0.02	0.02 e ± 0.004	0.66 bc ± 0.04	0.75 a ± 0.03	0.07 e ± 0.01
ABM64783.1	cell wall glycosidase, partial; LINUS	0.04 bc ± 0.003	0.04 cd ± 0.003	0.03 d ± 0.004	0.05 b ± 0.01	0.06 a ± 0.002	0.04 c ± 0.001	0.01 e ± 0.0004	0.01 e ± 0.001	0.01 e ± 0.001
CAI0384355.1	unnamed protein product; LINTE	0.04 a ± 0.002	0.03 bc ± 0.002	0.01 d ± 0.002	0.04 a ± 0.01	0.03 c ± 0.001	0.01 d ± 0.001	0.04 a ± 0.003	0.04 ab ± 0.002	0.01 d ± 0.002
CAI0431450.1	unnamed protein product; LINTE	0.10 ab ± 0.002	0.09 b ± 0.002	0.02 d ± 0.004	0.09 ab ± 0.012	0.09 b ± 0.003	0.03 d ± 0.003	0.11 a ± 0.006	0.11 a ± 0.003	0.06 c ± 0.005
CAI0412769.1	unnamed protein product; LINTE	0.06 ab ± 0.005	0.05 b ± 0.003	0.003 c ± 0.001	0.07 a ± 0.012	0.06 ab ± 0.003	0.004 c ± 0.002	0.05 b ± 0.003	0.06 ab ± 0.004	0.01 c ± 0.003
AVY09204.1	Sequence 3377 from patent US 9878004; UNKNOWN	0.08 b ± 0.004	0.08 b ± 0.003	0.01 d ± 0.002	0.08 b ± 0.009	0.07 b ± 0.002	0.01 d ± 0.001	0.09 a ± 0.003	0.09 ab ± 0.003	0.03 c ± 0.001

Table 3. *Cont.*

Accession	Protein Name	Agriol			Raciol			Libra		
		SC	FF	PC	SC	FF	PC	SC	FF	PC
CAI0440995.1	unnamed protein product; LINTE	0.12 ab ± 0.007	0.10 b ± 0.006	0.01 c ± 0.003	0.11 ab ± 0.012	0.10 b ± 0.007	0.01 c ± 0.002	0.13 a ± 0.009	0.11 ab ± 0.010	0.02 c ± 0.002
CAI0411675.1	unnamed protein product; LINTE	1.01 ab ± 0.09	1.10 a ± 0.03	0.11 c ± 0.03	0.95 b ± 0.05	1.06 ab ± 0.04	0.10 c ± 0.01	1.09 a ± 0.05	1.12 a ± 0.04	0.22 c ± 0.02
AGN56421.1	non-specific lipid transfer protein 1; LINUS	0.54 a ± 0.03	0.49 ab ± 0.01	0.05 d ± 0.003	0.47 abc ± 0.03	0.43 bc ± 0.04	0.06 d ± 0.01	0.49 abc ± 0.05	0.42 c ± 0.02	0.05 d ± 0.01
CAI0414382.1	unnamed protein product; LINTE	0.26 a ± 0.02	0.25 ab ± 0.07	0.03 c ± 0.002	0.32 a ± 0.06	0.21 ab ± 0.03	0.04 c ± 0.01	0.24 ab ± 0.03	0.30 a ± 0.07	0.13 bc ± 0.04

Note: Different lower-case letters in rows indicate the statistically significant difference at $p < 0.05$ (Tukey HSD test).

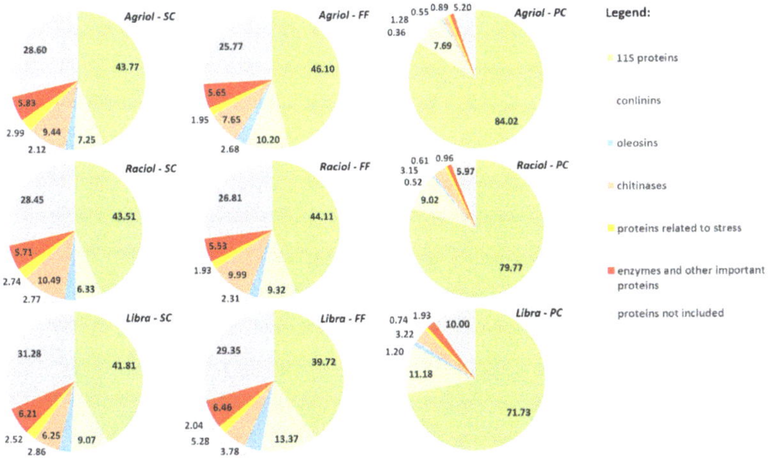

Figure 3. Relative proportions of protein classes depending on the flaxseed product (seed cake—SC, fine flour—FF, protein concentrate—PC) and cultivar.

In the case of chitinase (item WMZ41542.1) and most other PGs of the defence- and stress-related proteins class (items AMY26620.1, CAI0408810.1, 1DWM_A), sieving and AE/IEP leads to a decrease in their abundance in FF and PC. This may be explained by the fact that proteins related to defence and stress reactions are present in higher concentrations in the seed coat and other surface layers of the seed tissues, where they are "closer" to factors attacking or affecting the seed. Sieving with a 250 μm mesh size sieve leads to the separation of coarser particles from the FF. These particles are generated during milling from the surface layers of the seed coat.

For other selected important proteins (mainly quantitatively important enzymes), no significant difference in their relative abundance in SC and FF was found, but a significant decrease in their abundance in PC was observed. Thus, these types of proteins may be found in higher concentrations in the residual flour after the solubilisation phase of the AE/IEP procedure (in the pellet after centrifugation of the extraction mixture) and this material may be an interesting source for their recovery and possible use in biotechnological processes.

From the point of view of food, nutritional and medical applications of flaxseed proteins, the major fractions of proteins such as 11S globulins, 2S albumins and oleosins, are particularly important. From the results presented above, it is evident that mechanical sieving leads to an increase in the relative abundance of 2S albumins in the product. On the contrary, the AE/IEP method significantly increases the relative abundance of 11S globulins, so that the obtained PC can have up to twice higher abundance than SC or FF. This is important information when practical use of their functional properties is required. Kaushik et al. [14] reported that flaxseed protein isolate (also obtained using the AE/IEP method) excels in water binding, has a stabilizing effect on emulsions formed with sunflower oil and, as discussed in Section 3.2, it may be a "better" protein for cardiac patients due to its low lysine to arginine ratio (only 0.25). Flaxseed protein concentrate may be a good source of bioactive peptides having antibacterial, antioxidant, and antidiabetic effects as well as an angiotensin-converting enzyme inhibition effect due to the amino acid sequence of 11S globulin [15]. Merker et al. [28] highlighted the potential of flaxseed proteins including 11S globulins to produce peptides with an anticarcinogenic effect.

As already mentioned, the effect of cultivar on the different abundances of the PGs evaluated in flaxseed products was lower than the effect of the concentration method. However, for PCs, significant differences were found between cultivars in the relative abundance of 11S globulins. While the high ALA cultivar Libra had a relative abundance of 11S globulins (in total) of 71.73%, the low ALA cultivar Agriol had a relative abundance of 84.02%. These findings suggest that cultivar selection may significantly influence the representation of these proteins in PC, but this idea needs to be further developed in follow-up research by obtaining data from a larger set of cultivars.

4. Conclusions

A total of 2560 protein groups were found within the evaluated flaxseed products (seed cake, fine flour, protein concentrate) prepared from the three oilseed flax cultivars despite the incomplete sequence information for *Linum usitatissimum*. Subsequently, incomplete annotations made creating a functional description rather difficult for most of the identified protein groups.

The only 33 protein groups that were selected for detailed description were identified to have a relative abundance within the protein pool of 69–95%. The conditions combining alkaline solubilisation (at pH 8.5) and subsequent isoelectric precipitation of the proteins (at pH 4.5) significantly increased the relative abundance of 11S globulin-type proteins (from 41–44% to 72–84%) in the resulting protein concentrate, which have high potential for food applications due to their properties and amino acid composition.

Mechanical sieving of milled cake increased the relative amount of 2S albumin in the fine flour (up to the value of 9–13%, depending on the cultivar). Another important protein class, oleosins, was more present in seed cake or fine flour, depending on the cultivar, ranging from 2–4%. Both protein classes (2S albumin and oleosins) have been shown to decrease their abundance in protein concentrate compared to 11S globulins. A surprising and interesting finding is the high relative abundance of chitinase (up to 10% in seed cake and fine flour, respectively), which is part of the plant defence system against microorganisms. The majority of the other protein groups evaluated were enzymes (e.g., glyceraldehyde 3-phosphate dehydrogenase, fructose-bisphosphate aldolase, or allene oxide synthase), which crossed over to protein concentrate in only a small proportion.

The effect of cultivar on the relative abundance of the proteins evaluated is not negligible, but compared to the method of the protein concentration is significantly lower for the vast majority of identified protein groups. The low ALA cultivar Agriol had the highest relative abundance of 11S globulins (84%), whereas the high ALA cultivar Libra had the lowest relative abundance of 11S globulins (almost 72%). The selection of a suitable cultivar could be an important factor to obtain a protein concentrate with the highest abundance of 11S globulins.

Supplementary Materials: The following supporting information can be downloaded at: https://www.mdpi.com/article/10.3390/foods13091288/s1.

Author Contributions: Conceptualization, M.J. and J.B. (Jan Bárta); formal analysis, M.J., P.R., J.B. (Jan Bárta), Z.Z., V.B., A.S., F.L., M.B., J.K., E.J., J.B. (Jan Bedrníček) and A.B.; funding acquisition, J.B. (Jan Bárta), Z.Z., V.B., M.B. and J.K.; investigation, M.J., P.R., J.B. (Jan Bárta), Z.Z., V.B., A.S., F.L., M.B., J.K., E.J., J.B. (Jan Bedrníček) and A.B.; methodology, M.J., P.R., J.B. (Jan Bárta) and Z.Z.; project administration, J.B. (Jan Bárta) and V.B.; resources, M.B.; supervision, J.B. (Jan Bárta) and Z.Z.; writing—original draft, M.J., P.R., J.B. (Jan Bárta), Z.Z., V.B. and F.L.; writing—review and editing, M.J., P.R., J.B. (Jan Bárta), Z.Z., V.B. and F.L. All authors have read and agreed to the published version of the manuscript.

Funding: This study was supported by the research project QK1910302 of the Ministry of Agriculture of the Czech Republic and by the project GAJU 080/2022/Z of the Grant Agency of the University of South Bohemia. We acknowledge CEITEC Proteomics Core Facility of CIISB, Instruct-CZ Centre, supported by MEYS CR (LM2023042, e-INFRA CZ (ID:90254)).

Institutional Review Board Statement: Not applicable.

Informed Consent Statement: Not applicable.

Data Availability Statement: The original contributions presented in the study are included in the article/Supplementary Materials, further inquiries can be directed to the corresponding author.

Conflicts of Interest: Author Marie Bjelková is employed by Agritec Plant Research Ltd. She was involved in the prepa-ration of the experiment by selecting the cultivars of oilseed flax, supplying the seeds, preparing the cultivation technology for the experimental flax cultivars, and commenting obtained results for manuscript preparation. Agritec Plant Research did not contribute funds to the project and the preparation of the manuscript. The remaining authors declare that the research was conducted in the absence of any commercial or financial relationships that could be construed as a potential conflict of interest.

References

1. Barvkar, V.T.; Pardeshi, V.C.; Kale, S.M.; Kadoo, N.Y.; Giri, A.P.; Gupta, V.S. Proteome Profiling of Flax (*Linum usitatissimum*) Seed: Characterization of Functional Metabolic Pathways Operating during Seed Development. *J. Proteome Res.* **2012**, *11*, 6264–6276. [CrossRef] [PubMed]
2. Shim, Y.Y.; Gui, B.; Arnison, P.G.; Wang, Y.; Reaney, M.J.T. Flaxseed (*Linum usitatissimum* L.) Bioactive Compounds and Peptide Nomenclature: A Review. *Trends Food Sci. Technol.* **2014**, *38*, 5–20. [CrossRef]
3. Marambe, H.K.; Wanasundara, J.P.D. Protein from Flaxseed (*Linum usitatissimum* L.). In *Sustainable Protein Sources;* Nadathur, S.R., Wanasundara, J.P.D., Scanlin, L., Eds.; Academic Press: Cambridge, MA, USA, 2017; Chapter 8; pp. 133–144.
4. Bekhit, A.E.-D.A.; Shavandi, A.; Jodjaja, T.; Birch, J.; Teh, S.; Ahmed, I.A.M.; Al-Juhaimi, F.A.; Saeedi, P.; Bekhit, A.A. Flaxseed: Composition, Detoxification, Utilization, and Opportunities. *Biocatal. Agric. Biotechnol.* **2018**, *13*, 129–152. [CrossRef]
5. Lan, Y.; Ohm, J.-B.; Chen, B.; Rao, J. Physicochemical Properties and Aroma Profiles of Flaxseed Proteins Extracted from Whole Flaxseed and Flaxseed Meal. *Food Hydrocoll.* **2020**, *104*, 105731. [CrossRef]
6. Bueno-Díaz, C.; Biserni, C.; Martín-Pedraza, L.; de las Heras, M.; Blanco, C.; Vázquez-Cortés, S.; Fernández-Rivas, M.; Ba-tanero, E.; Cuesta-Herranz, J.; Villalba, M. Association Between the Seed Storage Proteins 2S Albumin and 11S Globulin and Severe Allergic Reaction after Flaxseed Intake. *J. Investig. Allergol. Clin. Immunol.* **2022**, *32*, 375–382. [CrossRef] [PubMed]
7. Mueed, A.; Shibli, S.; Korma, S.A.; Madjirebaye, P.; Esatbeyoglu, T.; Deng, Z. Flaxseed Bioactive Compounds: Chemical Composition, Functional Properties, Food Applications and Health Benefits-Related Gut Microbes. *Foods* **2022**, *11*, 3307. [CrossRef] [PubMed]
8. Čeh, B.; Štraus, S.; Hladnik, A.; Kušar, A. Impact of Linseed Variety, Location and Production Year on Seed Yield, Oil Content and Its Composition. *Agronomy* **2020**, *10*, 1770. [CrossRef]
9. Nykter, M.; Kymäläinen, H.-R.; Gates, F. Quality Characteristics of Edible Linseed Oil. *Agric. Food Sci.* **2006**, *15*, 402–413. [CrossRef]
10. Rabetafika, H.N.; Van Remoortel, V.; Danthine, S.; Paquot, M.; Blecker, C. Flaxseed Proteins: Food Uses and Health Benefits. *Int. J. Food Sci. Technol.* **2011**, *46*, 221–228. [CrossRef]
11. Anastasiu, A.-E.; Chira, N.-A.; Banu, I.; Ionescu, N.; Stan, R.; Rosca, S.-I. Oil productivity of seven Romanian linseed varieties as affected by weather conditions. *Ind. Crop. Prod.* **2016**, *86*, 219–230. [CrossRef]
12. Komartin, R.S.; Stroescu, M.; Chira, N.; Stan, R.; Stoica-Guzun, A. Optimization of oil extraction from *Lallemantia iberica* seeds using ultrasound-assisted extraction. *J. Food Meas. Charact.* **2021**, *15*, 2010–2020. [CrossRef]
13. Gutiérrez, C.; Rubilar, M.; Jara, C.; Verdugo, M.; Sineiro, J.; Shene, C. Flaxseed and Flaxseed Cake as a Source of Compounds for Food Industry. *J. Soil Sci. Plant Nutr.* **2010**, *10*, 454–463. [CrossRef]
14. Kaushik, P.; Dowling, K.; McKnight, S.; Barrow, C.J.; Wang, B.; Adhikari, B. Preparation, Characterization and Functional Properties of Flax Seed Protein Isolate. *Food Chem.* **2016**, *197 Pt A*, 212–220. [CrossRef]
15. Wu, S.; Wang, X.; Qi, W.; Guo, Q. Bioactive Protein/Peptides of Flaxseed: A Review. *Trends Food Sci. Technol.* **2019**, *92*, 184–193. [CrossRef]
16. Bárta, J.; Bártová, V.; Jarošová, M.; Švajner, J. Proteins of Oilseed Cakes, Their Isolation and Usage Possibilities. *Chem. Listy* **2021**, *115*, 472–480. (In Czech)
17. Maghaydah, S.; Alkahlout, A.; Abughoush, M.; Al Khalaileh, N.I.; Olaimat, A.N.; Al-Holy, M.A.; Ajo, R.; Choudhury, I.; Hayajneh, W. Novel Gluten-Free Cinnamon Rolls by Substituting Wheat Flour with Resistant Starch, Lupine and Flaxseed Flour. *Foods* **2022**, *11*, 1022. [CrossRef] [PubMed]
18. Waszkowiak, K.; Mikołajczak, B. The Effect of Roasting on the Protein Profile and Antiradical Capacity of Flaxseed Meal. *Foods* **2020**, *9*, 1383. [CrossRef] [PubMed]

19. Perreault, V.; Hénaux, L.; Bazinet, L.; Doyen, A. Pretreatment of flaxseed protein isolate by high hydrostatic pressure: Impacts on Protein Structure, Enzymatic Hydrolysis and Final Hydrolysate Antioxidant Capacities. *Food Chem.* **2017**, *221*, 1805–1812. [CrossRef] [PubMed]
20. Logarušić, M.; Radošević, K.; Bis, A.; Panić, M.; Slivac, I.; Srček, V.G. Biological Potential of Flaxseed Protein Hydrolysates Obtained by Different Proteases. *Plant Foods Hum. Nutr.* **2020**, *75*, 518–524. [CrossRef] [PubMed]
21. Ye, X.-P.; Xu, M.-F.; Tang, Z.-X.; Chen, H.-J.; Wu, D.-T.; Wang, Z.-Y.; Songzhen, Y.-X.; Hao, J.; Wu, L.-M.; Shi, L.-E. Flaxseed Protein: Extraction, Functionalities and Applications. *Food Sci. Technol.* **2022**, *42*, e22021. [CrossRef]
22. Qin, X.; Li, L.; Yu, X.; Deng, Q.; Xiang, Q.; Zhu, Y. Comparative Composition Structure and Selected Techno-Functional Elucidation of Flaxseed Protein Fractions. *Foods* **2022**, *11*, 1820. [CrossRef] [PubMed]
23. Madhusudhan, K.T.; Singh, N. Isolation and Characterization of a Small Molecular Weight Protein of Linseed Meal. *Phytochemistry* **1985**, *24*, 2507–2509. [CrossRef]
24. Marcone, M.F.; Kakuda, Y.; Yada, R.Y. Salt-soluble Seed Globulins of Various Dicotyledonous and Monocotyledonous Plants—I. Isolation/Purification and Characterization. *Food Chem.* **1998**, *62*, 27–47. [CrossRef]
25. Chung, M.W.Y.; Lei, B.; Li-Chan, E.C.Y. Isolation and Structural Characterization of the Major Protein Fraction from NorMan Flaxseed (*Linum usitatissimum* L.). *Food Chem.* **2005**, *90*, 271–279. [CrossRef]
26. Lorenc, F.; Jarošová, M.; Bedrníček, J.; Smetana, P.; Bárta, J. Structural Characterization and Functional Properties of Flaxseed Hydrocolloids and Their Application. *Foods* **2022**, *11*, 2304. [CrossRef] [PubMed]
27. Kotecka-Majchrzak, K.; Sumara, A.; Fornal, E.; Montowska, M. Proteomic Analysis of Oilseed Cake: A Comparative Study of Species-Specific Proteins and Peptides Extracted from Ten Seed Species. *J. Sci. Food Agric.* **2021**, *101*, 297–306. [CrossRef] [PubMed]
28. Merkher, Y.; Kontareva, E.; Alexandrova, A.; Javaraiah, R.; Pustovalova, M.; Leonov, S. Anti-Cancer Properties of Flaxseed Proteome. *Proteomes* **2023**, *11*, 37. [CrossRef] [PubMed]
29. Klubicová, K.; Berčák, M.; Danchenko, M.; Skultety, L.; Rashydov, N.M.; Berezhna, V.V.; Miernyk, J.A.; Hajduch, M. Agricultural Recovery of a Formerly Radioactive Area: I. Establishment of High-resolution Quantitative Protein Map of Mature Flax Seeds Harvested from the Remediated Chernobyl Area. *Phytochemistry* **2011**, *72*, 1308–1315. [CrossRef] [PubMed]
30. Bárta, J.; Bártová, V.; Jarošová, M.; Švajner, J.; Smetana, P.; Kadlec, J.; Filip, V.; Kyselka, J.; Berčíková, M.; Zdráhal, Z.; et al. Oilseed Cake Flour Composition, Functional Properties and Antioxidant Potential as Effects of Sieving and Species Differences. *Foods* **2021**, *10*, 2766. [CrossRef] [PubMed]
31. Laemmli, U.K. Cleavage of Structural Proteins during the Assembly of the Head of Bacteriophage T4. *Nature* **1970**, *227*, 680–685. [CrossRef] [PubMed]
32. Bárta, J.; Roudnický, P.; Jarošová, M.; Zdráhal, Z.; Stupková, A.; Bártová, V.; Krejčová, Z.; Kyselka, J.; Filip, V.; Říha, V.; et al. Proteomic Profiles of Whole Seeds, Hulls, and Dehulled Seeds of Two Industrial Hemp (*Cannabis sativa* L.) Cultivars. *Plants* **2024**, *13*, 111. [CrossRef]
33. Wiśniewski, J.R.; Zougman, A.; Nagaraj, N.; Mann, M. Universal Sample Preparation Method for Proteome Analysis. *Nat. Methods* **2009**, *6*, 359–362. [CrossRef] [PubMed]
34. Demichev, V.; Messner, C.B.; Vernardis, S.I.; Lilley, K.S.; Ralser, M. DIA-NN: Neural Networks and Interference Correction Enable Deep Proteome Coverage in High Throughput. *Nat. Methods* **2020**, *17*, 41–44. [CrossRef] [PubMed]
35. He, B.; Shi, J.; Wang, X.; Jiang, H.; Zhu, J.H. Label-free Absolute Protein Quantification with Data-independent Acquisition. *J. Proteom.* **2019**, *200*, 51–59. [CrossRef] [PubMed]
36. Oomah, B.D.; Mazza, G. Flaxseed Proteins—A Review. *Food Chem.* **1993**, *48*, 109–114. [CrossRef]
37. Wanasundara, P.K.J.P.D.; Shahidi, F. Removal of Flaxseed Mucilage by Chemical and Enzymatic Treatments. *Food Chem.* **1997**, *59*, 47–55. [CrossRef]
38. Waszkowiak, K.; Mikołajczak, B.; Kmiecik, D. Changes in Oxidative Stability and Protein Profile of Flaxseeds Resulting from Thermal Pre-treatment. *J. Sci. Food Agric.* **2018**, *98*, 5459–5469. [CrossRef] [PubMed]
39. Waszkowiak, K.; Mikołajczak, B.; Polanowska, K.; Wieruszewski, M.; Siejak, P.; Smułek, W.; Jarzębski, M. Protein Fractions from Flaxseed: The Effect of Subsequent Extractions on Composition and Antioxidant Capacity. *Antioxidants* **2023**, *12*, 675. [CrossRef] [PubMed]
40. Shewry, P.R. Seed proteins. In *Seed Technology and its Biological Basis*; Black, M., Bewley, J.D., Eds.; Sheffield Academic Press Ltd.: Sheffield, UK; CRC Press LLC: Boca Raton, FL, USA, 2000; pp. 42–84.
41. Liu, J.; Shim, Y.Y.; Poth, A.G.; Reaney, M.J.T. Conlinin in Flaxseed (*Linum usitatissimum* L.) Gum and Its Contribution to Emulsification Properties. *Food Hydrocoll.* **2016**, *52*, 963–971. [CrossRef]
42. Truksa, M.; MacKenzie, S.L.; Qiu, X. Molecular Analysis of Flax 2S Storage Protein Conlinin and Seed Specific Activity of Its Promoter. *Plant Physiol. Biochem.* **2003**, *41*, 141–147. [CrossRef]
43. Souza, P.F.N. The forgotten 2S albumin proteins: Importance, Structure, and Biotechnological Application in Agriculture and Human Health. *Int. J. Biol. Macromol.* **2020**, *164*, 4638–4649. [CrossRef] [PubMed]
44. Moreno, F.J.; Clemente, A. 2S Albumin Storage Proteins: What Makes them Food Allergens? *Open Biochem. J.* **2008**, *2*, 16–28. [CrossRef] [PubMed]
45. Bhatla, S.C.; Kaushik, V.; Yadav, M.K. Use of Oil bodies and Oleosins in Recombinant Protein Production and Other Biotechnological Applications. *Biotechnol. Adv.* **2010**, *28*, 293–300. [CrossRef] [PubMed]

46. Liu, C.; Wang, R.; He, S.; Cheng, C.; Ma, Y. The Stability and Gastro-intestinal Digestion of Curcumin Emulsion Stabilized with Soybean Oil Bodies. *LWT* **2020**, *131*, 109663. [CrossRef]
47. Nikiforidis, C.V. Structure and Functions of Oleosomes (Oil Bodies). *Adv. Colloid Interface Sci.* **2019**, *274*, 102039. [CrossRef] [PubMed]
48. Acevedo-Fani, A.; Dave, A.; Singh, H. Nature-Assembled Structures for Delivery of Bioactive Compounds and Their Potential in Functional Foods. *Front. Chem.* **2020**, *8*, 564021. [CrossRef] [PubMed]
49. Kara, H.H.; Araiza-Calahorra, A.; Rigby, N.M.; Sarkar, A. Flaxseed Oleosomes: Responsiveness to Physicochemical Stresses, Tribological Shear and Storage. *Food Chem.* **2024**, *431*, 137160. [CrossRef] [PubMed]
50. Li, Z.; Chi, H.; Liu, C.; Zhang, T.; Han, L.; Li, L.; Pei, X.; Long, Y. Genome-Wide Identification and Functional Characterization of LEA Genes during Seed Development Process in Linseed Flax (*Linum usitatissimum* L.). *BMC Plant Biol.* **2021**, *21*, 193. [CrossRef] [PubMed]
51. Carra, S.; Alberti, S.; Benesch, J.L.P.; Boelens, W.; Buchner, J.; Carver, J.A.; Cecconi, C.; Ecroyd, H.; Gusev, N.; Hightower, L.E.; et al. Small Heat Shock Proteins: Multifaceted Proteins with Important Implications for Life. *Cell Stress Chaperon.* **2019**, *24*, 295–308. [CrossRef] [PubMed]
52. Waters, E.R.; Vierling, E. Plant Small Heat Shock Proteins—Evolutionary and Functional Diversity. *New Phytol.* **2020**, *227*, 24–37. [CrossRef] [PubMed]
53. Clemente, M.; Corigliano, M.G.; Pariani, S.A.; Sánchez-López, E.F.; Sander, V.A.; Ramos-Duarte, V.A. Plant Serine Protease Inhibitors: Biotechnology Application in Agriculture and Molecular Farming. *Int. J. Mol. Sci.* **2019**, *20*, 1345. [CrossRef] [PubMed]
54. Anaya, K.; Cruz, A.C.B.; Cunha, D.C.S.; Monteiro, S.M.N.; dos Santos, E.A. Growth Impairment Caused by Raw Linseed Consumption: Can Trypsin Inhibitors Be Harmful for Health? *Plant Foods Hum. Nutr.* **2015**, *70*, 338–343. [CrossRef] [PubMed]
55. Shao, Q.; Liu, X.; Su, T.; Ma, C.; Wang, P. New Insights Into the Role of Seed Oil Body Proteins in Metabolism and Plant Development. *Front. Plant Sci.* **2019**, *10*, 1568. [CrossRef] [PubMed]
56. Contreras del Mar, M.; Lama-Muñoz, A.; Gutiérrez-Pérez, J.M.; Espínola, F.; Moya, M.; Castro, E. Protein Extraction from Agri-food Residues for Integration in Biorefinery: Potential Techniques and Current Status. *Bioresour. Technol.* **2019**, *280*, 459–477. [CrossRef]
57. Rodrigues, I.M.; Coelho, J.F.J.; Graça, M.; Carvalho, V.S. Isolation and Valorisation of Vegetable Proteins from Oilseed Plants: Methods, Limitations and Potential. *J. Food Eng.* **2012**, *109*, 337–346. [CrossRef]
58. Hadidi, M.; Aghababaei, F.; McClements, D.J. Enhanced Alkaline Extraction Techniques for Isolating and Modifying Plant-Based Proteins. *Food Hydrocoll.* **2023**, *145*, 109132. [CrossRef]
59. Hou, F.; Ding, W.; Qu, W.; Oladejo, A.O.; Xiong, F.; Zhang, W.; He, R.; Ma, H. Alkali Solution Extraction of Rice Residue Protein Isolates: Influence of Alkali Concentration on Protein Functional, Structural Properties and Lysinoalanine Formation. *Food Chem.* **2017**, *218*, 207–215. [CrossRef] [PubMed]

Disclaimer/Publisher's Note: The statements, opinions and data contained in all publications are solely those of the individual author(s) and contributor(s) and not of MDPI and/or the editor(s). MDPI and/or the editor(s) disclaim responsibility for any injury to people or property resulting from any ideas, methods, instructions or products referred to in the content.

foods MDPI

Article

Optimization of Extraction Process and Activity of Angiotensin-Converting Enzyme (ACE) Inhibitory Peptide from Walnut Meal

Meng Meng [1,2,3], Ziyi She [1,2,3], Yinyin Feng [1,2,3], Junhan Zhang [1,2,3], Ran Han [1,2,3], Yanlong Qi [4], Lina Sun [5,*] and Huiqing Sun [1,2,3,*]

1 State Key Laboratory of Food Nutrition and Safety, College of Food Science and Engineering, Tianjin University of Science and Technology, No. 29, 13th Avenue, Tianjin Economy Technological Development Area, Tianjin 300457, China; mengmeng@tust.edu.cn (M.M.); 15931795027@163.com (Z.S.)
2 Key Laboratory of Food Nutrition and Safety, Ministry of Education, Tianjin University of Science and Technology, No. 29, 13th Avenue, Tianjin Economy Technological Development Area, Tianjin 300457, China
3 College of Food Science and Engineering, Tianjin University of Science and Technology, No. 29, 13th Avenue, Tianjin Economy Technological Development Area, Tianjin 300457, China
4 Research Institute of Farm Products Storage and Processing, Xinjiang Academy of Agricultural Sciences, No. 403 Nanchang Road, Urumqi 830091, China; qiyanlong9@126.com
5 Institute of Agricultural Mechanization, Xinjiang Academy of Agricultural Sciences, No. 291 South Nanchang Road, Urumqi 830091, China
* Correspondence: lina_sun@xaas.ac.cn (L.S.); sunhuiqing@tust.edu.cn (H.S.); Tel./Fax: +86-0991-4500174 (L.S.); +86-022-60912421 (H.S.)

Abstract: In order to further realize the resource reuse of walnut meal after oil extraction, walnut meal was used as raw material to prepare polypeptide, and its angiotensin-converting enzyme (ACE) inhibitory activity was investigated. The ACE inhibitory peptides were prepared from walnut meal protein by alkaline solution and acid precipitation. The hydrolysis degree and ACE inhibition rate were used as indexes to optimize the preparation process by single-factor experiment and response surface method. The components with the highest ACE activity were screened by ultrafiltration, and their antioxidant activities were evaluated in vitro. The effect of gastrointestinal digestion on the stability of walnut peptide was analyzed by measuring molecular weight and ACE inhibition rate. The results showed that the optimal extraction conditions were pH 9.10, hydrolysis temperature 54.50 °C, and hydrolysis time 136 min. The ACE inhibition rate of walnut meal hydrolysate (WMH) prepared under these conditions was 63.93% ± 0.43%. Under the above conditions, the fraction less than 3 kDa showed the highest ACE inhibitory activity among the ACE inhibitory peptides separated by ultrafiltration. The IC_{50} value of scavenging ·OH free radical was 1.156 mg/mL, the IC_{50} value of scavenging DPPH free radical was 0.25 mg/mL, and the IC_{50} value of scavenging O_2^- was 3.026 mg/mL, showing a strong total reducing ability. After simulated gastrointestinal digestion in vitro, the ACE inhibitory rate of walnut peptide decreased significantly, but it still maintained over 90% ACE inhibitory activity. This study provides a reference for the application of low-molecular-weight walnut peptide as a potential antioxidant and ACE inhibitor.

Keywords: walnut meal; enzymatic hydrolysis; ACE inhibition rate; anti-oxidation; stability

1. Introduction

Walnut has a long planting history in China and a wide distribution range. Xinjiang walnut planting area belongs to the northwest of the four planting areas in our country. The planting area and yield of walnut have certain scale advantages compared with other provinces in China. Walnut is rich in high-quality protein, oil, cellulose, vitamins and other essential nutrients for the human body. Its use and medicinal value make walnut a food resource with great development prospects [1]. At present, the research is mainly focused on walnut oil. The protein content of walnut meal after oil extraction is more than 40%, but it has not been fully used, which not only wastes resources but also causes pollution. To enhance the depth of processing efficiency and overall utilization rate of walnut meal, further processing of peptides derived from readily accessible walnut meal was undertaken. Current research on the active ingredients in walnuts mainly focuses on the extraction of polyphenols, flavonoids, proteins, and walnut oil, as well as their antioxidant, antibacterial, anti-inflammatory, anti-cancer, and memory improvement effects [2]. Current research efforts center on identifying natural bioactive compounds that can effectively reduce blood pressure, with ACE inhibitory peptides gaining significant attention. In this context, researchers such as Dr. Kunlun Liu's team have successfully isolated and characterized four distinct peptides from maize embryo proteins, demonstrating ACE inhibitory properties. They delved deeper into understanding the unique inhibitory mechanisms of these peptides through kinetic studies and molecular docking analysis [3]. Duan et al. identified three novel ACE-inhibiting peptides by screening rapeseed protein sequences in a database using bioinformatics methods and synthesized and verified their biological activities [4].

Angiotensin-converting enzyme (ACE), a zinc dipeptide carboxypeptidase, is a key target in the pathogenesis of hypertension [5]. ACE plays a critical role in the human blood pressure regulatory system by catalyzing the cleavage of two terminal amino acids from angiotensin I to form angiotensin II, thereby inducing vasoconstriction and contributing to elevated blood pressure levels [6]. It can also cause blood pressure to rise through degradation of bradykinin, which deactivates it [7]. ACE inhibitors inhibit the activity of ACE to block the pathway that increases blood pressure to achieve the purpose of lowering blood pressure. At present, numerous synthetic antihypertensive medications, including captopril, lisinopril, and eplerenone, are employed in the clinical management of hypertension. However, effective long-term use of these medications can cause a variety of adverse side effects like coughing, rashes, diarrhea, high potassium levels, kidney issues, and taste alterations, which can gradually erode patient well-being [8]. ACE inhibitory peptides extracted from natural plants and animals not only have good antihypertensive efficacy, but also are non-toxic and have no side effects. Therefore, foodborne ACE inhibitory peptides have good application prospects.

The enzymatic hydrolysis product with high ACE inhibition activity was obtained from the walnut cake after oil extraction of Xinjiang walnut by alkaline protease enzymatic hydrolysis. Taking the enzymatic hydrolysis product as the research object, the ACE inhibition activity was used as the evaluation index, and the extraction conditions were optimized by the Box–Behnken test. Under optimal conditions, the peptide fraction was further separated by ultrafiltration to explore its antioxidant activity. In vitro digestion simulation was performed to study the gastrointestinal stability. This study can provide a reference for further development and utilization of walnut protein and other plant proteins.

2. Materials and Methods

2.1. Materials

The walnut cultivar was Xinjiang Wen185, which was produced in Wensu County, Aksu, Xinjiang.

Alkaline protease (20,000 U/g), angiotensin converting enzyme, Hippuryl-His-Leu-OH (HHL), pepsin, trypsin—Beijing Solaibao Technology Co., LTD., Beijing, China; N-hexane, sodium hydroxide, hydrochloric acid, boric acid, ethyl acetate, anhydrous ethanol, ferric chloride, potassium ferricyanide (all analytical pure)—Shanghai McLean Biochemical Technology Co., LTD., Shanghai, China; phenol, trichloroacetic acid (all analytically pure) —Shanghai Jizhi Biochemical Technology Co., LTD., Shanghai, China.

2.2. Preparation of Defatted Walnut Meal

Once the walnut meal has undergone crushing and subsequent sieving through an 80-mesh screen, it is mixed with n-hexane in a solid-to-liquid ratio of 1:5 (g/mL). The blend is then allowed to sit in a water bath maintained at a temperature of 45 °C for a duration of

1 h. Then, it is drained and filtered. The filter cake is collected and extracted again until the filtrate is colorless and transparent, and it is completely dried in an oven at 60 °C.

2.3. Preparation of Walnut Protein

Deionized water was incorporated into the nonfat walnut powder following a material-to-liquid ratio of 1:20 and thoroughly blended. Subsequently, the pH level of the mixture was adjusted up to 11, allowing for complete dissolution of the protein in the walnut powder under magnetic stirring at 53 °C for an interval of 1.5 h. After cooling, the mixture is centrifuged at 5000× *g* of centrifugal force for 15 min, and then the supernatant is carefully poured out.

2.4. Optimization of Preparation Process of Walnut Peptide by Enzymatic Hydrolysis

2.4.1. Single-Factor Test

The substrate concentration of 2% walnut protein was added into ultrapure water, and the substrate of alkaline protease 8000 U/g was added; the enzymatic hydrolysis temperature was 55 °C. During the enzymatic hydrolysis, the pH of the solution was kept constant at 9. Upon completion of a 2-h enzymatic hydrolysis period, the enzyme activity was terminated by boiling the mixture for 10 min. After the cooling phase, the supernatant was centrifuged at a centrifugal force of 7000× *g* for 15 min. The single-factor test factors and levels were presented in the Table 1.

Table 1. Single-factor test factors and levels.

Influencing Factors	Level Setting				
	1	2	3	4	5
Amount of enzyme added (U/g)	2000	4000	6000	8000	10,000
Temperature (°C)	40	45	50	55	60
pH	8.5	9	9.5	10	10.5
Time (min)	60	90	120	150	180

2.4.2. Response Surface Optimization Method

According to the initial findings from the one-factor-at-a-time tests, three significant variables were identified and selected for advanced optimization using response surface methodology. These factors included pH (Factor A), enzymatic temperature (Factor B), and enzymatic time (Factor C), utilizing the ACE inhibitory efficiency as the response parameter to gauge the outcomes. To systematically optimize these factors, a Box–Behnken experimental design was implemented. Table 2 presents the precise parameter settings and levels for each of these influencing factors in the designed experiment.

Table 2. Design factors and levels of Box–Behnken response surface test.

Influencing Factors		Level Setting		
		−1	0	1
A	pH	8.5	9	9.5
B	Temperature (°C)	50	55	60
C	Time (min)	60	120	150

2.5. Degree of Hydrolysis

The degree of hydrolysis (DH) was determined employing the pH-stat method, representing the proportion of peptide bonds that have been cleaved during the enzymatic hydrolysis process [9]. In the experiment, the pH of the enzymolysis system was kept at a stable level by adding NaOH solution continuously. DH is determined by calculating the amount of NaOH volume that needs to be added throughout the process, and the NaOH volume is proportional to the degree of cleavage of the peptide bond.

$$DH(\%) = \frac{B \times N_b}{Mp \times \alpha \times h_{tot}} \times 100$$

where: B is the volume of consumed NaOH standard solution, mL; N_b is the concentration of NaOH standard solution mol/L; α is the degree of dissociation of the amino group, where the value is 1; Mp is the total amount of protein in the substrate, g; h_{tot} is the millimolar number of substrate protein bonds per gram, mmol/g, and this value is taken to be 8.0 mmol/g.

2.6. Evaluation of ACE Inhibitory Activity

ACE (50 mU/mL), along with the substrate HHL (8 mmol/mL), were both solubilized in a borate buffer solution with a pH of 8.3. The components were then successively introduced into separate 1.5 mL centrifuge tubes as outlined in Table 3 and subsequently incubated in a water bath according to the prescribed conditions.

Table 3. In vitro detection methods of ACE inhibition rate.

Reagent	Amount Added (μL)		
	Experimental	Control	Blank
1 mol/L HCl	0	0	100
HHL	30	30	30
Peptide	10	0	10
Buffer	0	10	0
Operation 1		A water bath at 37 °C for 5 min	
ACE	10	10	10
Operation 2		A water bath at 37 °C for 30 min	
1 mol/L HCl	100	100	0

After the reaction is complete, 0.8 mL of ethyl acetate is introduced into the blend, followed by 20 s of agitation, and then precipitated for liquid separation. Then, carefully remove the 0.6 mL ethyl acetate layer and dry it in an oven heated to 100 °C. After cooling, 0.3 mL of deionized water was added to the dry tube, vigorously rotated for 30 s, and finally, the absorbance was measured at 228 nm wavelength.

$$\text{ACE inhibitory activity}(\%) = \frac{A_b - A_a}{A_b - A_c} \times 100$$

where: A_a is the absorbance of the experimental group; A_b is the absorbance of the control group; and A_c is the absorbance of the blank group.

2.7. Ultrafiltration Separation

The products were separated using centrifugal ultrafiltration tubes with molecular weight cutoffs of 10 kDa and 3 kDa. Initially, the membranes of these tubes were immersed in ultrapure water on ice for a period of 10 min. Subsequently, the water was discarded, and 10 mL of the hydrolyzed solution was loaded into each tube. The tubes were then centrifuged at a force of 5000× g under a temperature of 4 °C for a duration of 20 min.

2.8. Determination of Antioxidant Capacity

2.8.1. ·OH Free Radical Scavenging Ability

The ·OH radical scavenging capacity was assessed using the Fenton reaction system method. To carry out the process, it was necessary to introduce 0.1 mL of the peptide solution into a centrifuge tube with a capacity of 1.5 mL. Following this, 0.1 mL of a ferric sulfate solution with a concentration of 6 mmol/L and an equal amount of hydrogen peroxide solution were put into the tube. The last step involved topping off the mixture with 1 mL of a salicylic acid-ethanol solution with a concentration of 6 mmol/L. The solution was subjected to incubation in a water bath regulated at 37 °C for a period of one hour. Following this, we used a wavelength of 510 nm to determine the absorbance value A_1. To determine the absorbance values A_0 and A_2 we substituted the sample solution and hydrogen peroxide solution with distilled water, respectively. The experiment was conducted in three parallel trials, using Vc as our positive control.

2.8.2. DPPH Free Radical Scavenging Ability

To evaluate the capability of scavenging DPPH free radicals, the peptide solution (0.2 mL) was combined with DPPh-ethanol solution (0.2 mL, 0.2 mmol/L) and vortexed for thorough mixing. The mixture was then left to stand in the dark for 30 min, and the absorbance was subsequently measured at a wavelength of 517 nm, which was recorded as A_1. The absorbance values were recorded as A_0 and A_2 by anhydrous ethanol instead of polypeptide solution and DPPH-ethanol solution, respectively. With Vc as the positive control, the experiment was repeated in parallel 3 times.

2.8.3. O_2^- Scavenging Ability

The antioxidant capacity was evaluated by pyrogallol autoxidation method. Tris-HCl buffer 0.3 mL (pH 8.2, concentration 0.5 M) was prepared for the test. Then, 0.1 mL peptide solution was added, and after 20 min of water bath at 25 °C, 0.04 mL preheated pyrocatechol solution (25 mmol/L) was added. The solution underwent a rotation to ensure its accurate reaction for a duration of four minutes. Upon cessation of the reaction, a concentrated hydrochloric acid volume of 0.05 mL was promptly added. Following this, the resultant sample underwent centrifugation at 7000× g at a temperature of 4 °C for a period of three minutes. The absorbance was then measured at 325 nm and denoted as A_1. The absorbance values obtained by using distilled water instead of sample solution and pyrotriol solution were denoted as A_0 and A_2, respectively. The whole procedure was repeated three times with Vc as a positive control.

$$\text{free radical rate of clearance}/\% = \left(1 - \frac{A_1 - A_2}{A_0}\right) \times 100$$

2.8.4. Total Reducing Power

The Prussian Blue Assay method was utilized to evaluate the overall reducing capability. During this procedure, a 0.1 mL quantity of the peptide solution was dispensed into a 1.5 mL centrifuge tube, followed by the addition of 0.25 mL of phosphate buffer (adjusted to pH 6.6 and 0.2 Molar concentration). Next, 0.25 mL of a 1% potassium ferricyanide solution was mixed in. Subsequently, 0.25 mL of a 10% trichloroacetic acid solution was introduced. After thorough vortex mixing, the reading was taken at a wavelength of 700 nm. Parallel experiments were performed three times with Vc as a positive control. A higher OD value indicates a stronger total reducing power of the test sample.

2.9. Simulating Gastrointestinal Digestion In Vitro

The method of Zhang Li for in vitro simulation of gastrointestinal digestion was slightly modified [10]. The fractions smaller than 3 kDa were initially separated and then mixed with the simulated gastric fluid at a ratio of 1:20 (mass/volume). This simulated fluid comprised a combination of equal volumes of 0.02 mol/L NaCl, 0.02 mol/L KCl, and 0.02 mol/L $NaHCO_3$. The pH level of the mixture was modified to 1.5 through the inclusion of 0.5 mol/L HCl, which facilitated the dissolution of 4% pepsin. Prior to the introduction of the peptide solution, the solution was subjected to a pre-incubation period of 5 min at a temperature of 37 °C. The hydrolysis process was carried out at 37 °C for a duration of 2.5 h, following which enzyme activity was terminated by heating the samples in boiling water at 100 °C for 10 min.

The pH of the remaining solution was re-set to 7.5 using a 1 mol/L NaOH solution. After that, 4% trypsin was added, and the solution was subjected to another round of digestion for 2.5 h at a temperature of 37 °C. The activity of the enzyme was then stopped through heat inactivation by boiling the sample for 10 min after collection. After cooling down, centrifuge the mixture at 7000× *g* for 15 min at a temperature of 4 °C. A portion of the supernatant was freeze dried for measuring ACE inhibitory activity, while the remainder was stored in a −20 °C refrigerator for future use.

2.10. Determination of the Relative Molecular Mass of Peptides

The molecular mass distribution was analyzed by HPLC [11]. The enzyme solution of walnut meal was prepared into a 2 mg/mL solution and passed through a 0.45 μm filter membrane. The TSK gel 2000 SWXL chromatographic column with dimensions of 7.8 mm × 30 cm was utilized, where the mobile phase comprised a ratio of acetonitrile, water, and trifluoroacetic acid in volumes of 10:40:0.05, respectively. The detection wavelength was 214 nm, the temperature recorded by the column stood at 30 °C, while the flow rate was clocked at 0.5 mL/min. Moreover, the injection volume was noted to be 10 μL. LC solution Version 1.22 SP1 was used for automatic data processing and calculation.

To prepare a molecular weight standard curve, five different peptide standards with varying molecular weights were employed: Cytochrome C (Mw = 12,200 Da), Aprotinin (Mw = 6511.44 Da), Bacitracin (Mw = 1421.69 Da), Gly-Gly-Tyr-Arg (Mw = 451.48 Da), and Ala-Ala-Ala (Mw = 189.17 Da). The curve was constructed to calculate the distribution of peptide molecular weights in the sample.

3. Results

3.1. Single-Factor Test of Enzymatic Hydrolysis Process

3.1.1. Enzymatic Effect

As shown in Figure 1, with the increment of enzyme dosage, there was a corresponding increase in the binding of the enzyme to the substrate, consequently leading to a heightened rate of enzymatic reaction. When the enzyme concentration was 6000 U/g, the DH value reached 16.74%, the ACE inhibition rate attained a considerable 63.85% mark at an enzyme concentration of 6000 U/g., the binding between enzyme and substrate was approximately saturated, and the reaction rate tended to be flat. Therefore, 6000 U/g was selected as the condition of enzyme addition in subsequent experiments.

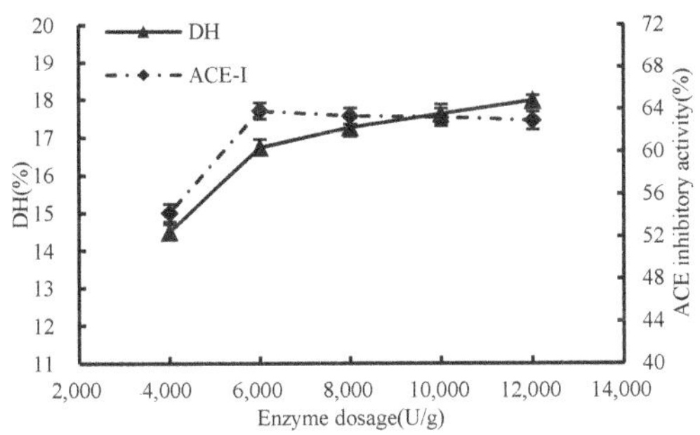

Figure 1. Effect of enzyme dosage on hydrolysis degree and ACE inhibition rate.

3.1.2. Effect of Temperature

As shown in Figure 2, the degree of hydrolysis first increased and then decreased during the temperature rise. The highest inhibition rates of DH and ACE were 16.73% and 63.65%, respectively, when the temperature was 55 °C. However, t as the temperature for enzymatic hydrolysis kept escalating, and a decrease in the efficiency of the process was observed; it is possible that the deleterious effects of high temperatures on the enzyme's molecular stability and the ability of its binding sites to attach to the substrate may be the root cause of the issue, thus causing a drop in the degree of hydrolysis. Consequently, 55 °C was determined as the optimal enzymatic hydrolysis temperature for the subsequent single-factor experiments.

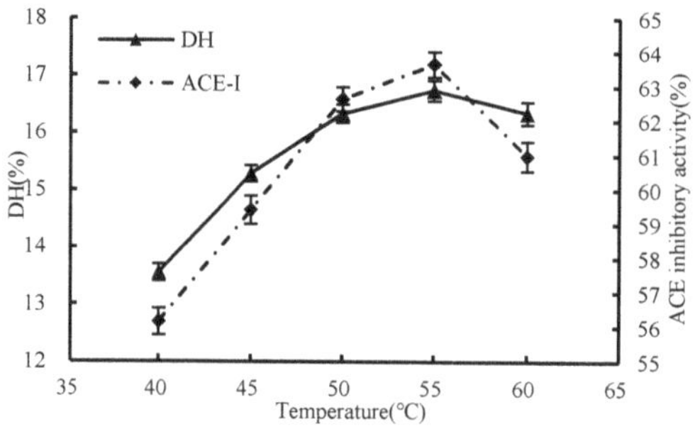

Figure 2. Influence of enzymatic hydrolysis temperature on hydrolysis degree and ACE inhibition rate.

3.1.3. Effect of pH

The inhibition activity of ACE demonstrated an upward trend initially, followed by a decline, with the increase in pH level (Figure 3). The peak of ACE inhibition rate was recorded at a pH value of 9. DH increased continuously from pH 8.5 to 10 and then decreased. Raising the pH level can impede the optimal performance of the enzyme by interrupting its association with the substrate, which consequently lessens the extent of hydrolysis. Nevertheless, the highest degree of ACE inhibition was observed at pH 9, suggesting that ACE inhibition was not directly related to the degree of hydrolysis. Therefore, in order to protect the potency of the peptide from any harmful effects of excess ·OH, and to guarantee a robust hydrolysis while not hindering the peptide's activity, it is vital to sustain a moderate pH level throughout the enzymatic process. Finally, the optimal pH of enzymatic hydrolysis in this experiment was determined to be 9 as the condition for subsequent single-factor experiments.

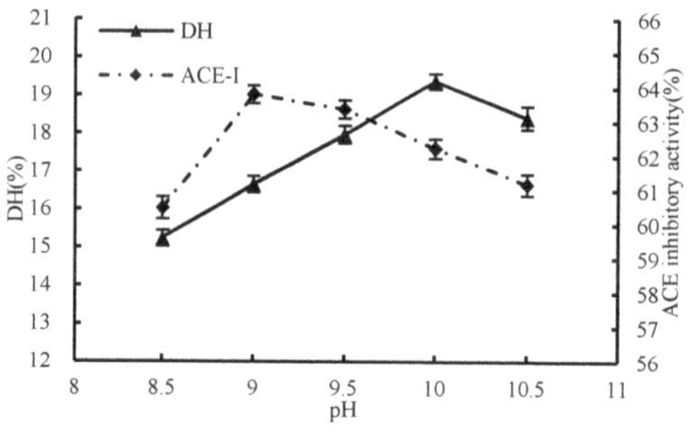

Figure 3. Effect of pH on hydrolysis degree and ACE inhibition rate.

3.1.4. Effect of Time

The results in Figure 4 showed that the DH and ACE inhibition rates increased with the prolongation of the enzymatic hydrolysis time. Within the initial 2 h, the reaction progressed swiftly, with the enzyme rapidly bonding to the substrate during the early stages of enzymatic hydrolysis. However, beyond the 2-h mark, the increase in both DH and the ACE inhibition rate began to slow down. At the 2-h mark of enzymatic hydrolysis, the ACE inhibitory activity had reached 63.17%. By the 3-h point, it further increased marginally to 63.88%. The enzymatic hydrolysis time of 2 h was selected as the optimal condition in terms of saving time and cost.

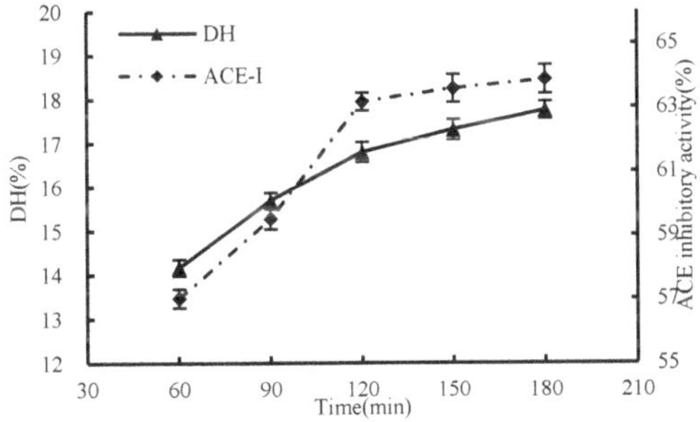

Figure 4. Effect of enzymatic hydrolysis time on hydrolysis degree and ACE inhibition rate.

3.2. Response Surface Optimization of Enzymatic Hydrolysis Process

3.2.1. Response Surface Experiments—Design and Results

Based on the findings from the single-factor tests where pH, hydrolysis temperature, and hydrolysis time were treated as independent variables, to enhance the preparation of ACE inhibitory peptides via the alkaline protease enzymatic hydrolysis of walnut protein, a response surface methodology experiment was undertaken, with the ACE inhibition rate serving as the dependent response variable. The primary objective was to refine and optimize the process parameters, resulting in an ideal outcome. The detailed experimental design and outcomes are presented in Table 4.

Table 4. Box–Behnken response surface test design and results.

Test No.	A pH	B Temperature	C Time	Y ACE Inhibition Rate (%)
1	0	0	0	64.64
2	−1	0	1	60.03
3	0	0	0	63.34
4	1	0	−1	58.32
5	−1	1	0	57.06
6	−1	−1	0	60.26
7	0	−1	1	62.18
8	0	0	0	63.53
9	0	−1	−1	59.63
10	1	1	0	60.74
11	1	−1	0	60.48
12	−1	0	−1	57.83
13	0	0	0	63.66
14	0	0	0	63.16
15	1	0	1	62.96
16	0	1	1	61.37
17	0	1	−1	57.64

3.2.2. Response Surface Regression Equation

The regression equation was fitted using Design Expert 12, wherein the three independent variables of pH, enzymatic hydrolysis temperature, and enzymatic hydrolysis time were represented by A, B, and C, respectively. The quadratic multinomial regression model was obtained through multiple fitting regression of the ACE inhibition rate: $Y = 63.67 + 0.915A - 0.7175B + 1.64C + 0.865AB + 0.61AC + 0.295BC - 2.23A^2 - 1.81B^2 - 1.66C^2$.

Table 5 illustrates the outcomes of the regression analysis of the rate model for ACE inhibition and the correlation regression coefficients. If the p-value of the regression model is less than 0.01, it is considered to be highly statistically significant. Moreover, the lack-of-fit term's p-value stands at 0.97, exceeding the 0.05 benchmark, hence indicating that the model provides a good fit for the data and can accurately predict the corresponding regression value from the equation with consistent reliability. The *p*-values of A (pH), B (hydrolysis temperature), and C (hydrolysis time) were all less than 0.01, indicating that they had significant effects on ACE inhibition rate. The joint impact of pH and enzymatic hydrolysis temperature on ACE inhibition rate was found to be considerable, as the interaction term AB's *p*-value was less than 0.01. Furthermore, the *p*-value of AC term (interaction between pH and enzymatic hydrolysis time) was less than 0.05, indicating that their interaction also significantly affected ACE inhibition rate. In contrast, B (enzymolysis temperature) and C (enzymolysis time) had no significant effect on ACE inhibition rate.

Table 5. Regression analysis of the ACE inhibition rate response surface regression model.

Source	Sum of Square	df	Mean Square	F-Value	*p*-Value
Model	88.58	9	9.84	49.21	<0.0001 **
A-pH	6.70	1	6.70	33.49	0.0007 **
B-temperature	4.12	1	4.12	20.59	0.0027 **
C-time	21.52	1	21.52	107.59	<0.0001 **
AB	2.99	1	2.99	14.96	0.0061 **
AC	1.49	1	1.49	7.44	0.0294 *
BC	0.3481	1	0.3481	1.74	0.2286
A^2	20.85	1	20.85	104.27	<0.0001 **
B^2	13.73	1	13.73	68.63	<0.0001 **
C^2	11.54	1	11.54	57.70	0.0001 **
Residual	1.40	7	0.20		
Lack of Fit	0.0704	3	0.0235	0.0707	0.9726
Pure Error	1.33	4	0.3324		
Cor Total	89.98	16			
R^2 = 0.9844 Adj R^2 = 0.9644 C.V.% = 0.73					

Note: $p < 0.01$ was considered extremely significant and indicated by **; $p < 0.05$ was considered significant and indicated by *.

At the same time, the regression coefficient R^2 = 0.9844 was basically consistent with the adjusted regression coefficient R^2 = 0.9644; the test results demonstrate a strong concurrence between the model fitted and the actual data, which could be used to explain 96.44% of the data. The equation had high credibility, the signal-to-noise ratio was small (C.V.% = 0.73), and the data-fitting degree was good with small differences. The F-value reflects the influence relationship of A single factor on ACE inhibition rate; the causal relationship between the factors can be ranked in order of enzyme hydrolysis time (C) having the greatest impact, followed by pH (A), and then enzyme hydrolysis temperature (B).

3.2.3. The Results of Interaction

As both the pH value and enzymatic temperature rise concurrently, the ACE inhibition rate of the enzymatic solution initially shows an increase, followed by a subsequent decline, as depicted in Figure 5, affirmatively demonstrating that the interaction between pH and enzymatic temperature exerts a highly significant impact on the ACE inhibition rate. Referring to Figure 6, the contour plot depicting the interaction between pH and digestion time reveals an elliptical shape, with a steep response surface gradient. This indicates that both pH and its interaction with the enzyme solution temperature have substantial effects on the ACE inhibition rate, which aligns with the findings from the variance analysis.

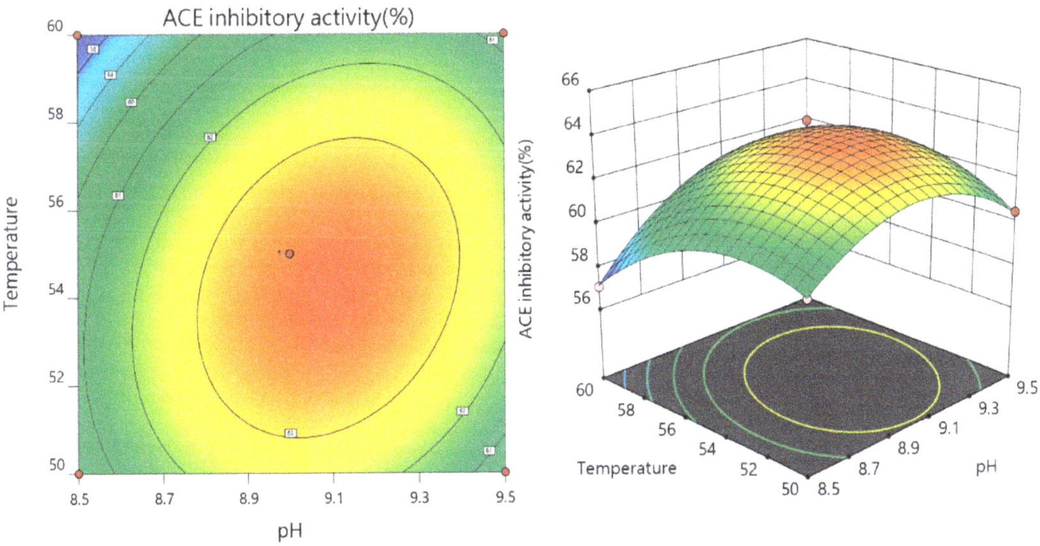

Figure 5. Effect of pH and temperature of enzymatic hydrolysis on the rate of ACE inhibition.

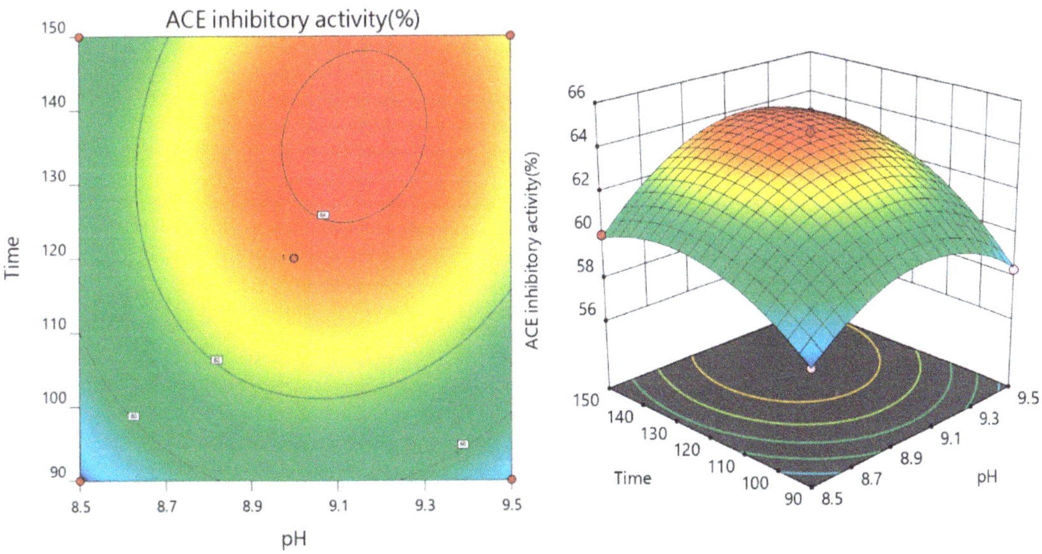

Figure 6. Interaction of pH and enzymatic hydrolysis time on ACE inhibition rate.

3.2.4. Verification of the Optimal Enzymatic Hydrolysis Process

Response surface analysis showed that the optimal process conditions for ACE inhibition of peptidase hydrolysis were pH 9.13, hydrolysis temperature 54.54 °C, and hydrolysis time 136.04 min, and the highest ACE inhibition rate was 64.26% under these conditions. Considering the actual situation, the extraction process parameters were adjusted. The optimal enzymatic hydrolysis process conditions were pH 9.10, enzymatic hydrolysis temperature 54.50 °C, and enzymatic hydrolysis time 136.00 min. The ACE inhibition rate of walnut protease hydrolysate obtained was 63.93% ± 0.43%, which was within 1% of the theoretical prediction value.

3.3. *Ultrafiltration Separation*

Ultrafiltration is a membrane separation technique that separates solutes based on their molecular weight. The enzymatic hydrolysates were separated with ultrafiltration centrifuge tubes of different molecular weight sizes to obtain the preliminarily isolated peptide fractions. The ultrafiltration results showed (Table 6) that the yield of WRH-II (3000~10,000 Da) was higher at 39.82%. However, it is noteworthy that the sample's ACE inhibition rate is markedly inferior in comparison to that of WMH-I, and its IC_{50} for ACE inhibition is 0.299. When the concentration is 1 mg/mL, the ACE inhibition rate obtained by the experiment is only 63.93%, which is significantly lower than the data obtained by Liu et al. [12]. The literature proved that the inhibition rate of ACE reached as high as 82.4% in components with molecular weight less than 3 kDa. Given this significant difference, we decided to proceed with WMH-I for our subsequent investigations.

Table 6. Fractional ultrafiltration yield and ACE inhibition activities. Different letters represent significant differences ($p < 0.05$).

Group	Molecular Weight (Da)	Rate of Yield (%)	IC_{50} (mg/mL)
WMH	Stoste	100	0.401 ^{b}
WMH-I	<3000	26.16	0.299 ^{a}
WMH-II	3000~10,000	39.82	0.652 ^{c}
WMH-III	>10,000	32.48	1.518 ^{d}

3.4. *In Vitro Antioxidant Activity*

3.4.1. ·OH Scavenging Ability

As shown in Figure 7, the capacity to remove ·OH gradually increased as the concentration of WMH-I rose from 1 mg/mL to 6 mg/mL, resulting in a sharp increase in the clearance rate when the concentration was between 1 and 3 mg/mL. At a concentration below 6 mg/mL, the ·OH clearance rate for Vc was 96.6%, while that for WMH-I was slightly lower at 82.6%. This variation suggests that WMH-I's performance, while lower than that of Vc, boasts an impressive capacity for ·OH elimination. The IC_{50} value of WRH-I for ·OH clearance stood at 1.156 mg/mL.

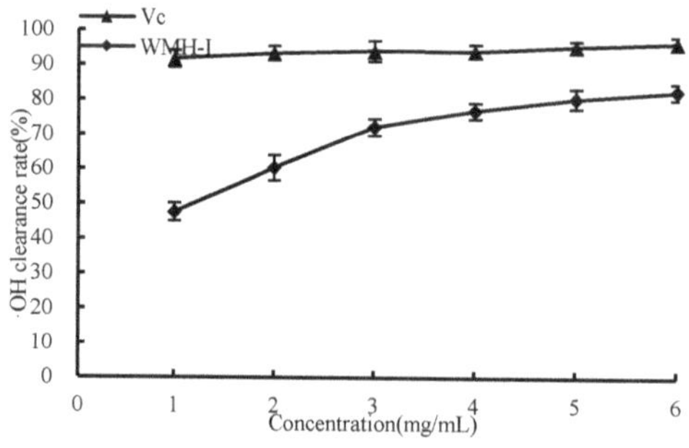

Figure 7. Clearance of ·OH by WMH-I and Vc.

3.4.2. DPPH Scavenging Ability

In Figure 8, it is evident that between the concentrations of 0.1 mg/mL and 0.6 mg/mL, a noticeable trend can be discerned. Both Vc and WMH-I demonstrate an escalating trend in their scavenging capabilities against DPPH free radicals as the sample concentration rises. Notably, between concentrations of 0.1 mg/mL and 0.4 mg/mL, WMH-I displays a marked increase in its ability to scavenge DPPH rapidly. At the concentration of 0.6 mg/mL, Vc displayed an exceptional DPPH clearance rate of 94.2%, while WMH-I displayed a clearance rate of 81.1% for the identical radicals, albeit being slightly less efficient compared to Vc in terms of its DPPH clearance ability. Despite this difference, these results still highlight a robust scavenging ability of WMH-I against DPPH radicals. The determined IC_{50} measurement for the DPPH radical scavenging capability of WMH-1 revealed a value of 0.25 mg/mL.

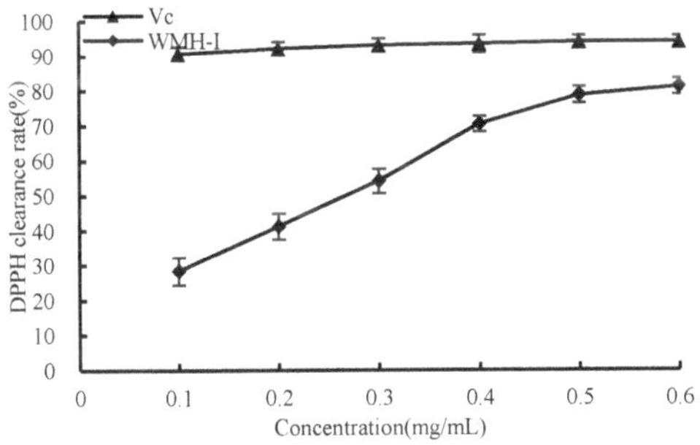

Figure 8. Scavenging rate of DPPH free radical by WMH-I and Vc.

3.4.3. O_2^- Scavenging Ability

O_2^- is a weak free radical that is produced by mitochondrial electron transport. The aforementioned is also regarded as an antecedent to the production of reactive oxygen species (ROS), which is accountable for the manifestation of a plethora of ailments [13]. As shown in Figure 9, the content of WMH-I demonstrates a fluctuation, ranging from 0.2 mg/mL to 1.2 mg/mL; there is a noticeable enhancement in its superoxide anion scavenging rate in correspondence with the increase in concentration. However, the rate of increase follows a relatively modest slope, showing a gradual and cumulative improvement that adheres to a dose-dependent pattern. Regarding the scavenging activity against O_2^-, the calculated IC_{50} value for WMH-I is 3.026 mg/mL, which reflects the concentration needed to achieve 50% scavenging efficiency for O_2^-.

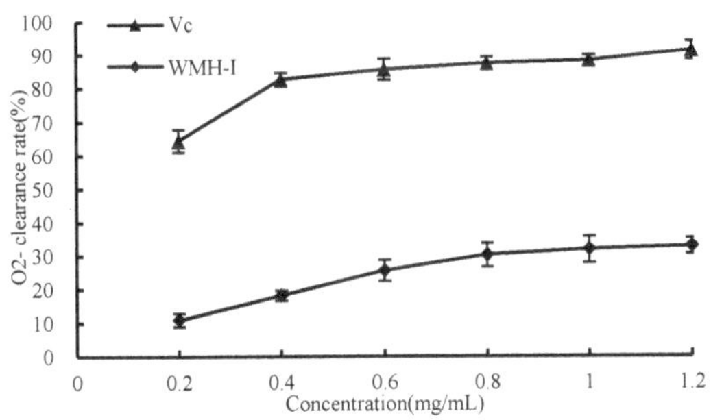

Figure 9. Clearance of superoxide anion by WMH-I and Vc.

3.4.4. Total Reducing Power

As shown in Figure 10, as the concentration increased from 1 mg/mL to 6 mg/mL, there was a corresponding increase in the overall reducing capacities of both Vc and WMH-I. Nonetheless, it is noteworthy that throughout this concentration range, WMH-I consistently demonstrated a significantly lesser reducing power compared to that of Vc. The results showed that WRH-I has a certain total reducing capacity.

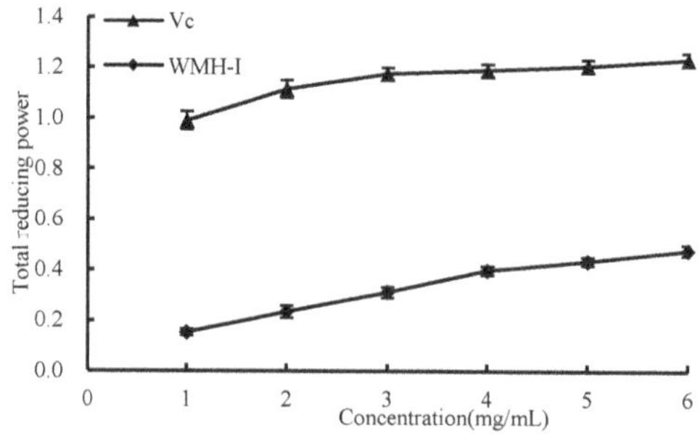

Figure 10. Total reducing power of WMH-I and Vc.

3.5. Changes in ACE Inhibition Rate during Digestion

The results of changes in ACE inhibition rate during digestion were illustrated in the Figure 11. The residence time of food in the stomach is generally 2–2.5 h, pepsin is present in gastric juice, and the pH value ranges from 1 to 2. After evaluating the rate of ACE inhibition at a concentration of 2 mg/mL during the process of digestion, it was observed that the inhibition rate experienced a noteworthy decrease ($p < 0.05$) following 2.5 h of simulated gastric digestion. However, even post-decrease, the ACE inhibition rate remained appreciable at 60.41% ± 1.27%. Further, following 2.5 h of intestinal digestion, there was another significant decline ($p < 0.05$) in the ACE inhibition rate, yet it still managed to reach 63.66% ± 0.74%. The inhibition rate of ACE was also significantly decreased when simulated gastrointestinal digestion was performed at the same time ($p < 0.05$), but the inhibitory activity of ACE was higher than that of single treatment. After digestion, the inhibitory activity of the peptide ACE decreased, which was similar to the change pattern studied by Zhenqiu Xu et al. [14]. However, in spite of that, Osama Magouz and his colleagues reported divergent findings, in which they found a significant increase in the ACE inhibitory activity of the fermented buttermilk-derived peptides post-gastrointestinal digestion [15]. During the gastrointestinal digestion process, enzymatic processes instigated by molecules like pepsin and trypsin are responsible for breaking down peptides into smaller peptide fragments or singular amino acids. This enzymatic breakdown can lead to the disruption of certain structural elements that possess ACE inhibitory properties, thereby contributing to a decrease in overall ACE inhibitory activity [16]. The short peptides produced in different digestion processes are inconsistent, which inevitably leads to different degrees of changes in their inhibition rates.

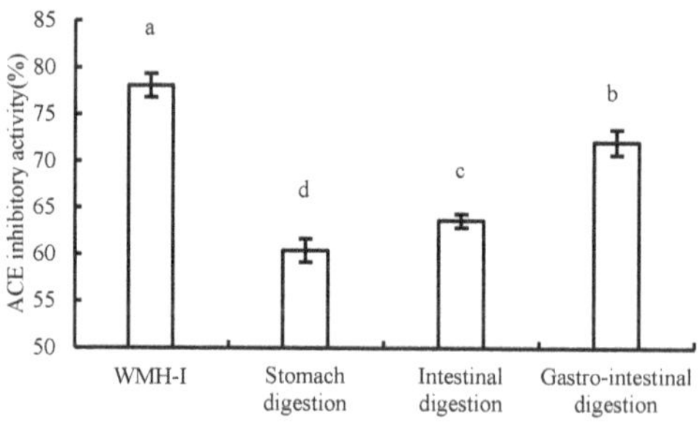

Figure 11. Comparison of ACE inhibitory activity during digestion. Data are expressed as means ± SD for three independent experiments; different letters marked significant differences as determined by one-way analysis of variation multiple test ($p < 0.05$).

3.6. Changes in Molecular Mass Distribution during Digestion

The distribution of molecular masses smaller than 3 kDa was analyzed by HPLC for different digestion processes. As shown in Figure 12, the percentage of peptides with a molecular weight below 1000 Da in the digested hydrolysate underwent a notable augmentation, mirroring the findings of Zhengli Lin's investigation [11]. The extent of reduction in the share of peptides exhibiting a molecular weight of over 2000 Daltons varied. During digestion, trypsin acts on the peptide bond linked by lysine and arginine, which further hydrolyze the peptide of 1000 to 2000 Da into smaller fragments. Wen Wenjun showed that the molecular weight of wheat was in the range of 70 to 3000 Da after gastrointestinal digestion, and it was a peptide with effective ACE inhibitory activity [17].

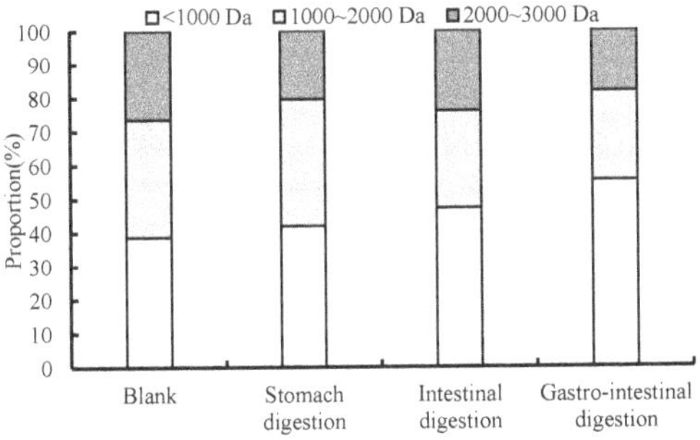

Figure 12. Molecular weight distribution of digestive products under different digestive processes.

4. Discussion

In recent years, there has been a growing interest in ACE-inhibiting peptides sourced from natural origins, including those extracted from both animals and plants. However, they are often ignored due to their low bioavailability in vivo. In the present study, walnut protein hydrolysates treated with alkaline protease showed higher ACE inhibitory activity after optimized extraction conditions. We have verified that the protein hydrolysate's ACE inhibitory function is unequivocally associated with the molecular weight of its constituents; hence, we proceeded to fractionate the walnut hydrolysate based on the varying sizes of its molecular weights.

The smallest molecular weight fraction (<3 kDa) of hydrolyzed walnut protein exhibited the greatest ACE inhibitory activity. Peptides with low molecular weight are more likely to bind to target molecules than peptides with high molecular weight and therefore exhibit higher ACE inhibitory activity [18]. Consistent with the earlier scientific literature, our investigation confirms that peptides of lower molecular weights generally display a higher potency in terms of inhibiting ACE activity [10]. In addition, the process of ultrafiltration aids in preserving hydrophobic amino acids at the terminal position of the peptide, promoting inhibition [19].

Antioxidants function to impede or mitigate the oxidative harm inflicted upon cells and tissues within the body. These free radicals, characterized by their heightened reactivity, can inflict damage on cellular components such as proteins and DNA, ultimately causing oxidative stress and injury [20]. O_2^- serves as the initial reactive oxygen species engendered in the human body, setting off the generation of a host of other ROS. ROS play integral roles in nearly all biological and disease-related processes occurring within living organisms, with O_2^- often recognized as a key responsive element in these mechanisms [21]. The superoxide anion, being highly reactive, readily transforms into various alternative forms of ROS, including peroxy nitrite, ·OH, and hydrogen peroxide, among others [22].

The state of oxidative stress results from the imbalance between oxidation and the body's antioxidant capacity, and ROS levels in the system are elevated [23]. Excess ROS can rapidly react with nitric oxide (NO) to produce peroxynitrites that are highly toxic to biofilms and intracellular structures [24]. This results in a loss of the effectiveness of NO, which disrupts the NO-mediated vasodilation. It further aggravates endothelial dysfunction and ultimately promotes the occurrence and development of hypertension [25].

Therefore, the antioxidant effect of ACE inhibitors contributes to the prevention and improvement of hypertension.

Bioactive compounds present in food play a vital role in the absorption process of the gastrointestinal tract and can only showcase their physiological effects once they are assimilated into the bloodstream [26]. Despite evidence showing that numerous food-originating peptides can indeed be fully absorbed, it remains true that a significant number of proteins continue to be susceptible to degradation by proteases located on the surface of intestinal microvilli [27]. It has been reported that dipeptides and tripeptides can be absorbed by the body entirely through intestinal endothelial cells [28]. Therefore, the assessment of the durability of ACE-inhibiting peptides during the process of digestion establishes a theoretical foundation for utilizing them as functional elements in food products. This approach aims to enhance the efficient absorption and availability of these active peptides in the human system.

5. Conclusions

In this study, the enzymatic preparation process of walnut peptidase was optimized by single-factor test and response surface test, and the optimal process conditions were pH 9.10, enzymatic temperature 54.50 °C, and enzymatic time 136 min. Under these conditions, the ACE inhibition rate of the enzymatic hydrolysate reached 63.93% ± 0.43%, and the IC_{50} value of ACE activity of fractions less than 3 kDa obtained after ultrafiltration was 0.299. The IC_{50} value of ·OH, DPPH radical, and O_2^- was 1.156 mg/mL, 0.25 mg/mL, and 3.026 mg/mL, respectively, and it has strong antioxidant capacity and total reducing power in vitro. ACE inhibition retention rate was more than 90% after simulated gastrointestinal digestion in vitro, which provided reference for further development of ACE-inhibiting peptides in walnut.

Author Contributions: Methodology, M.M., Z.S. and J.Z.; Software, M.M. and Z.S.; Formal analysis, R.H.; Investigation, J.Z., R.H., Y.Q. and L.S.; Resources, Y.Q.; Writing—original draft, M.M., Z.S. and H.S.; Writing—review & editing, L.S. and H.S.; Visualization, Y.F.; Supervision, Y.F.; Funding acquisition, L.S. All authors have read and agreed to the published version of the manuscript.

Funding: This research was funded by [Young scientific and technological backbone Innovation Ability Training Program of Xinjiang Academy of Agricultural Sciences] [xjnkq-2020004], [Tianjin Science and Technology Plan Project of China] [22JCQNJC01480], [Key Research and Development (R&D) Plan of Xinjiang Uygur Autonomous Region] [2022B02048; 2022B02048-2], [China Central Finance Forest and Grass Science and Technology Demonstration Project] [Xin [2023]TG No. 19] and Xinjiang Academy of Agricultural Sciences independent cultivation project (xjnkyzzp-2022003). The APC was funded by [Key Research and Development (R&D) Plan of Xinjiang Uygur Autonomous Region] [2022B02048; 2022B02048-2].

Institutional Review Board Statement: Not applicable.

Informed Consent Statement: Not applicable.

Data Availability Statement: The original contributions presented in the study are included in the article, further inquiries can be directed to the corresponding authors.

Conflicts of Interest: The authors declare no conflict of interest.

References

1. Liu, C.; Zhang, Z.; Shang, Y.; Li, S.; Xia, J.; Tian, Y.; Jia, Y.; Ma, A. Progress in the preparation, identification and biological activity of walnut peptides. *J. Future Foods* **2024**, *4*, 205–220. [CrossRef]
2. Xie, J.; Hong, Z.; Dai, J.; Li, T.; Bai, Y.; Zhang, L.; Hu, X.; Chen, J.; Sheng, J.; Tian, Y. Isolation and identification of anti-colorectal cancer peptides from walnut proteins and associated in silico analysis. *J. Funct. Foods* **2024**, *112*, 105952. [CrossRef]
3. Liu, K.L.; Gao, Z.R.; Li, Q.Y.; Zhang, H.Y. Identification and mechanistic study of four novel ACE inhibitory peptides from maize germ protein hydrolysates. *LWT* **2023**, *186*, 115254. [CrossRef]
4. Duan, X.J.; Dong, Y.F.; Zhang, M.; Li, Z.H.; Bu, G.H.; Chen, F.S. Identification and molecular interactions of novel ACE inhibitory peptides from rapeseed protein. *Food Chem.* **2023**, *422*, 136085. [CrossRef] [PubMed]

5. Asoodeh, A.; Yazdi, M.M.; Chamani, J. Purification and characterisation of angiotensin I converting enzyme inhibitory peptides from lysozyme hydrolysates. *Food Chem.* **2012**, *131*, 291–295. [CrossRef]
6. Xue, L.; Yin, R.; Howell, K.; Zhang, P. Activity and bioavailability of food protein-derived angiotensin-I-converting enzyme–inhibitory peptides. *Compr. Rev. Food Sci. Food Saf.* **2021**, *20*, 1150–1187. [CrossRef] [PubMed]
7. Lee, S.Y.; Hur, S.J. Antihypertensive peptides from animal products, marine organisms, and plants. *Food Chem.* **2017**, *228*, 506–517. [CrossRef] [PubMed]
8. Gallego, M.; Mora, L.; Toldrá, F. Health relevance of antihypertensive peptides in foods. *Curr. Opin. Food Sci.* **2018**, *19*, 8–14. [CrossRef]
9. Shao, B.; Huang, X.Y.; Xu, M.D.; Cheng, D.W.; Li, X.L.; Li, M.Q. Peptides isolated from black soybean synergistically inhibit the activity of angiotensin converting enzyme (ACE). *J. Funct. Foods* **2023**, *106*, 105604. [CrossRef]
10. Zhang, J.; Li, M.; Zhang, G.; Tian, Y.; Kong, F.; Xiong, S.; Zhao, S.; Jia, D.; Manyande, A.; Du, H. Identification of novel antioxidant peptides from snakehead (*Channa argus*) soup generated during gastrointestinal digestion and insights into the anti-oxidation mechanisms. *Food Chem.* **2021**, *337*, 127921. [CrossRef]
11. Lin, Z.L.; Wu, H.; Zhang, M.M. Isolation, identification, and structure-activity relationship of novel ACE inhibitory peptides from earthworm protein in vitro gastrointestinal digestion product. *Food Biosci.* **2023**, *55*, 103010. [CrossRef]
12. Liu, M.; Du, M.; Zhang, Y.; Xu, W.; Wang, C.; Wang, K.; Zhang, L. Purification and Identification of an ACE Inhibitory Peptide from Walnut Protein. *J. Agric. Food Chem.* **2013**, *61*, 4097–4100. [CrossRef] [PubMed]
13. Obeng, E.; Ding, F.; He, X.J.; Shen, J.L. Bioimaging of superoxide anions in living cells and Perfect visualization with fluorescence probes and their applications. *Dye. Pigment.* **2022**, *199*, 109964. [CrossRef]
14. Xu, Z.; Wu, C.; Sun-Waterhouse, D.; Zhao, T.; Waterhouse, G.I.N.; Zhao, M.; Su, G. Identification of post-digestion angiotensin-I converting enzyme (ACE) inhibitory peptides from soybean protein Isolate: Their production conditions and in silico molecular docking with ACE. *Food Chem.* **2021**, *345*, 128855. [CrossRef] [PubMed]
15. Magouz, O.; Mehanna, N.; Khalifa, M.; Sakr, H.; Gensberger-Reigl, S.; Dalabasmaz, S.; Pischetsrieder, M. Profiles, antioxidative and ACE inhibitory activity of peptides released from fermented buttermilk before and after simulated gastrointestinal digestion. *Innov. Food Sci. Emerg. Technol.* **2023**, *84*, 103266. [CrossRef]
16. Escudero, E.; Mora, L.; Toldrá, F. Stability of ACE inhibitory ham peptides against heat treatment and digestion. *Food Chem.* **2014**, *161*, 305–311. [CrossRef] [PubMed]
17. Wen, W.J.; Li, S.J.; Gu, Y.; Wang, S.; Wang, J.P. Effects of Starch on the Digestibility of Gluten under Different Thermal Processing Conditions. *J. Agric. Food Chem.* **2019**, *67*, 7120–7127. [CrossRef] [PubMed]
18. Garzón, A.G.; Veras, F.F.; Brandelli, A.; Drago, S.R. Purification, identification and in silico studies of antioxidant, antidiabetogenic and antibacterial peptides obtained from sorghum spent grain hydrolysate. *LWT* **2022**, *153*, 112414. [CrossRef]
19. Silvestre, M.P.C.; Silva, M.R.; Silva, V.D.M.; de Souza, M.W.S.; Lopes, C.D.; Afonso, W.D. Analysis of whey protein hydrolysates: Peptide profile and ACE inhibitory activity. *Braz. J. Pharm. Sci.* **2012**, *48*, 747–757. [CrossRef]
20. Hasan, M.R.; Haque, M.M.; Hoque, M.A.; Sultana, S.; Rahman, M.M.; Ali Shaikh, M.A.; Sarker, M.K.U. Antioxidant activity study and GC-MS profiling of Camellia sinensis Linn. *Heliyon* **2024**, *10*, e23514. [CrossRef]
21. Zhang, W.; Zhang, J.; Li, P.; Liu, J.; Su, D.; Tang, B. Two-photon fluorescence imaging reveals a Golgi apparatus superoxide anion-mediated hepatic ischaemia-reperfusion signalling pathway. *Chem. Sci.* **2019**, *10*, 879–883. [CrossRef] [PubMed]
22. Sawyer, D.T.; Valentine, J.S. How super is superoxide? *Acc. Chem. Res.* **1981**, *14*, 393–400. [CrossRef]
23. Lushchak, V.I. Free radicals, reactive oxygen species, oxidative stress and its classification. *Chem. Biol. Interact.* **2014**, *224*, 164–175. [CrossRef] [PubMed]
24. Laursen, J.B.; Somers, M.; Kurz, S.; McCann, L.; Warnholtz, A.; Freeman, B.A.; Tarpey, M.; Fukai, T.; Harrison, D.G.J.C. Endothelial regulation of vasomotion in apoE-deficient mice: Implications for interactions between peroxynitrite and tetrahydrobiopterin. *Circulation* **2001**, *103*, 1282–1288. [CrossRef]
25. Bitar, M.S.; Wahid, S.; Mustafa, S.; Al-Saleh, E.; Dhaunsi, G.S.; Al-Mulla, F. Nitric oxide dynamics and endothelial dysfunction in type II model of genetic diabetes. *Eur. J. Pharmacol.* **2005**, *511*, 53–64. [CrossRef]
26. Hu, X.T.; Xu, F.R.; Li, J.L.; Li, J.; Mo, C.; Zhao, M.; Wang, L.F. Ultrasonic-assisted extraction of polysaccharides from coix seeds: Optimization, purification, and digestibility. *Food Chem.* **2022**, *374*, 131636. [CrossRef] [PubMed]
27. Neha, K.; Haider, M.R.; Pathak, A.; Yar, M.S. Medicinal prospects of antioxidants: A review. *Eur. J. Med. Chem.* **2019**, *178*, 687–704. [CrossRef]
28. Li, X.; Wei, L.; Jia, L.; Li, M.; Zhu, L.; Liu, L.; Gao, Y.J.P. Identification and characterization of cow's milk proteins from the rat intestinal lymph using a proteomic strategy. *Proteomics* **2013**, *13*, 2649–2656. [CrossRef]

Disclaimer/Publisher's Note: The statements, opinions and data contained in all publications are solely those of the individual author(s) and contributor(s) and not of MDPI and/or the editor(s). MDPI and/or the editor(s) disclaim responsibility for any injury to people or property resulting from any ideas, methods, instructions or products referred to in the content.

foods

MDPI

Article

The Effects of Sunflower and Maize Crop Residue Extracts as a New Ingredient on the Quality Properties of Pork Liver Pâtés

Milica Glišić [1,*,†], Marija Bošković Cabrol [1,2,†], Nikola Čobanović [1], Marija Starčević [3], Stevan Samardžić [4], Ivona Veličković [5] and Zoran Maksimović [4]

1 Faculty of Veterinary Medicine, Department of Food Hygiene and Technology, University of Belgrade, 11000 Belgrade, Serbia; marija.boskoviccabrol@unipd.it (M.B.C.); cobanovic.nikola@vet.bg.ac.rs (N.Č.)

2 Department of Agronomy, Food, Natural Resources, Animal and Environment (DAFNAE), University of Padova, 35020 Padova, Italy

3 Serbian Armed Forces, 11000 Belgrade, Serbia; marijadok@gmail.com

4 Faculty of Pharmacy, Department of Pharmacognosy, University of Belgrade, 11000 Belgrade, Serbia; stevan.samardzic@pharmacy.bg.ac.rs (S.S.); zmaksim1@pharmacy.bg.ac.rs (Z.M.)

5 Faculty of Biology, Institute of Botany and Botanical Garden "Jevremovac", University of Belgrade, 11000 Belgrade, Serbia; ivona@bio.bg.ac.rs

* Correspondence: glisic.mica@gmail.com or milica.glisic@vet.bg.ac.rs; Tel.: +381-64-2593160

† These authors contributed equally to this work.

Abstract: The present study aimed to evaluate the antioxidant capacity of ethanolic extracts from post-harvest sunflower and maize stalk residues, and their impact on the chemical composition, physicochemical parameters, lipid oxidative stability, microbiological properties, and sensory characteristics of pork liver pâtés over a 90-day storage period. Four formulations were prepared: a control group (CON), a batch with butylated hydroxytoluene as a synthetic antioxidant (BHT), 1% ethanolic extract from sunflower residues (SSRE), and 1% ethanolic extract from maize residues (MSRE). The MSRE had a higher total phenol content and showed better antioxidant activity relative to the SSRE ($p < 0.01$). The addition of SSRE decreased the lightness and increased the redness in the pork liver pâtés, with these pâtés showing the highest total color difference compared to the control ($p < 0.01$). The crop extracts increased the n-6 and total PUFA contents in pâtés and improved the PUFA/SFA ratio ($p < 0.01$). Formulations containing crop residue extracts showed higher TBARs and POV values than the control and BHT group ($p < 0.01$), indicating a pro-oxidant effect and accelerated lipid oxidation in pâtés during storage. As far as microbiological quality, the presence of crop residue extracts decreased the total viable count, lactic acid bacteria, and psychotropic aerobic bacteria ($p < 0.01$). The incorporation of crop extracts in the pork pâtés impaired their sensory quality, particularly color, odor, aroma, and flavor, and decreased their overall acceptability. These results indicated that, while the crop residue extracts were not as effective as synthetic antioxidants in preserving the lipid stability of pâtés, they demonstrated potential for enhancing the microbial quality of this type of meat product.

Keywords: agricultural waste; sunflower and maize stalks; antioxidant activity; meat emulsions; lipid oxidation; microbial quality; sensory properties

1. Introduction

Liver pâtés are traditional cooked meat products mainly manufactured from pork and veal liver, pork backfat, and other characteristic ingredients, depending on consumer habits (nuts, fruits, spices, and herbs, etc.) [1–3]. This type of meat product is highly appreciated by consumers all over the world due to its nutrient content, added-value ingredients, and good sensory attributes [2,4–6]. Liver pâtés are an important source of high-quality proteins, vitamins (B1, B12, and folic acid), and heme iron, which makes this type of meat products even more attractive, especially for children and women, who may be more susceptible to iron deficiency anemia [3,7].

Despite the presence of nitrite, sodium ascorbate, and some ingredients with proven antioxidant activity in liver pâtés, lipid oxidation during refrigerated storage has been reported as the main cause of deterioration in product quality and the development of unpleasant aromas [1,6]. This lipid oxidation limits the quality and acceptability of liver pâtés as a consequence of the degradation of polyunsaturated fatty acids and heme pigments, followed by the release of iron and alterations in texture, flavor, color, and nutritional value [4,5,8]. The high oxidative instability of liver pâtés can be ascribed to their high amount of fat (around 35%) and non-heme iron (approximately 30 mg/g of product), with the latter being considered as the major pro-oxidant compound in meat products [6]. Additionally, the manufacturing process of liver pâtés involves the chopping, mixing, and cooking of meat, fat, and liver, which increases oxidative instability and favors reactions between free fatty acids and atmospheric oxygen in the presence of catalysts such as heat and metalloproteins [3,6], where a high temperature during the heat treatment of liver pâtés reduces their activation energy for the development of oxidative reactions and leads to the decomposition of previously formed hydroperoxides [6]. Liver pâté matrix has a low level of natural endogenous antioxidants (mainly tocopherol), justifying the wide use of exogenous antioxidants, such as sodium nitrite and/or synthetic exogenous antioxidants (butylated hydroxyanisole (BHA), butylated hydroxytoluene (BHT), and propyl, octyl, and dodecyl gallates) [3,5,6,8]. However, the use of synthetic preservatives in meat products has been associated with health risks related to the carcinogenic potential of residual nitrite, N-nitrosamines, and some lipid oxidation products, and the possible hazardous effects of synthetic antioxidants on consumer health [3,5,9]. In this regard, the meat industry has been making efforts not only to directly reduce the content of potentially hazardous preservatives, but also to look for alternatives to substitute synthetic additives with food ingredients derived from natural sources that could be used as safe antimicrobials and antioxidants [3,9].

In the last decade, several studies have examined the antioxidant properties of different plant extracts (containing mainly polyphenolic compounds) as natural additives in various meat products, including grape and tea extracts in meat patties and cooked products [9], palm and persimmon by-products, and sage and rosemary essential oils in liver pâtés [3,5,6,8]. It has been demonstrated that the utilization of natural antioxidants could successfully substitute synthetic ones in meat products, whereby some of the examined extracts have approximately seven times higher antioxidant potential than BHT and BHA [5,6]. Nevertheless, natural antioxidants are usually more expensive than their synthetic counterparts and may negatively affect the color and/or flavor of the meat product [9].

A considerable amount of various plant materials generated annually, including waste from different agro-industrial sectors, such as seeds, peels, husks, leaves, and stalks, etc., or poorly harvested and discarded residues on the land, have recently gained increased attention as abundantly available and cost-effective renewable raw materials for the production of value-added compounds [10]. In this context, the European Union generates an average annual agricultural biomass production of 956 megatons of dry matter, with 442 megatons accounting for residue production; cereals contribute 74.3% and oil-bearing crops contribute 16.5% to the total agricultural waste [11]. The majority of these residues originate from maize, an economically significant crop that stands out as one of the world's foremost staple cereals, with an annual production exceeding 1.2 billion metric tons. Simultaneously, sunflower ranks among the largest oilseed crops, surpassing 55 million metric tons per annum [12].

By-products from the agrifood industry are regarded as a good source of natural antioxidants, characterized by a substantial presence of polyphenols, which: (i) improve the technological, nutritional, and microbiological properties of meat products; (ii) provide physiological benefits for human health as they have antiviral, anti-inflammatory, anti-cancer, anti-tumor, and hepatoprotective properties; and (iii) valorize the by-product in question, which entails reductions in waste and environmental pollution, attaining the lower price of antioxidants and, consequently, increasing their profitability [10,13–16]. To the best of the authors' knowledge, the potential of sunflower and maize agro-industrial wastes as natural sources of functional antioxidants for incorporation into the model meat-based matrix has not yet been evaluated. Considering all mentioned above, the present study aimed to determine the effects of the addition of extracts from sunflower and maize crop residue on the chemical composition, physicochemical properties, oxidative stability, microbiological status, and sensory acceptability of refrigerated stored pork liver pâtés, and to compare the observed effects with those of synthetic antioxidant (BHT) and control (without the addition of synthetic phenolic antioxidants and plant extracts) pork liver pâtés.

2. Materials and Methods

2.1. Materials and Chemicals

The sunflower and maize stalk residues were collected in October 2021 from the territory of the Autonomous Province of Vojvodina (Serbia). Butylated hydroxytoluene (≥99%, FCC, FG) was purchased from Sigma-Aldrich (Steinheim, Germany). Methanol (HPLC), ethanol (HPLC), sodium carbonate anhydrous (pro analysis), and potassium acetate anhydrous (pro analysis) were acquired from VWR Chemicals (Lutterworth, Leicestershire, UK). Glacial acetic acid (pro analysis) was purchased from Zorka Pharma (Šabac, Serbia), and sodium acetate trihydrate (pro analysis) from Centrohem (Stara Pazova, Serbia). Quercetin hydrate (HPLC) and 2,2′-azino-bis(3-ethylbenzothiazoline)-6-sulphonic acid (ABTS) (HPLC) were obtained from TCI Europe N.V. (Zwijndrecht, Belgium). 2,2-diphenyl-1-picrylhydrazyl (DPPH, HPLC), Folin-Ciocalteu reagent (pro analysis), galic acid (HPLC), iron(III) choride hexahydrate (pro analysis), iron(III) chloride heptahydrate (pro analysis), and TBA reagent (pro analysis) were acquired from Merck (Darmstadt, Germany). Potassium peroxydisulphate (pro analysis) and acetonitrile (HPLC-MS) were purchased from (ThermoFisher Scientific, Branchburg, NJ, USA). Aluminium nitrate nonahydrate (pro analysis) and 2,4,6-tris(2-piridyl)-s-triazine (TPTZ) (HPLC) were obtained from Fluka Chemie AG (Buchs, Switzerland), while formic acid (LC-MS) was acquired from Honeywell, Fluka (Seelze, Germany). For the standard solutions' preparation, tricin (PhytoLab GmbH & Co. KG, Vestenbergsgreuth, Germany), ferulic acid (Extrasynthese, Genay Cedex, France), chlorogenic acid (Carl Roth GmbH & Co. KG, Karlsruhe, Germany), p-coumaric acid, and 3,5-di-caffeoylquinic acid (Sigma-Aldrich, St. Louis, MO, USA) were used. For fatty acid analyses, a hexane/isopropanol mixture (ASE 200, Dionex, Potsdam, Germany) and standard FAME mixture (Supelco 37 Component FAME Mix, Supelco, Bellefonte, PA, USA) were used. The media used for the cultivation of microorganisms: peptone water, plate count agar—PCA, Violet Red Bile Glucose agar—VRGB, and De Man, Rogosa, Sharp Agar-MRS were purchased from Merck (KGaA, Darmstadt, Germany).

2.2. Preparation of Crop Residue Extracts

The sunflower and maize stalk residue extracts were obtained by two-step extraction using hexane and 96% ethanol, as previously described by Glišić et al. [17]. Briefly, a stainless-steel extractor was loaded with 3 kg of ground plant material and extracted with a six-fold weight of hexane for 1 h at 40 °C, followed by filtration and concentration using a DLAB RE 200 Pro industrial rotary evaporator (60 °C, 60 rpm, 216–200 mbar, 150 min). The residues were then extracted with six-fold weight of ethanol/H_2O (96:4, v/v) at 45 °C. The ethanolic extracts were further filtrated and concentrated using an evaporator under the same working conditions.

2.3. Evaluation of Total Phenol Content

The total phenol content in the extracts was evaluated according to Ahmad et al. [18]. Aliquots of 20 μL of the extract solution were mixed with 100 μL of Folin–Ciocalteu reagent in 96-well microplates and incubated for 6 min at room temperature. Then, 75 μL of sodium carbonate (Na_2CO_3) solution was added. After 2 h of incubation in the dark, the absorbance at 740 nm was read. The results were calculated using the gallic acid calibration curve (y = 7.9754x + 0.0187; R^2 = 0.9938) and presented as mg gallic acid equivalents per gram of dry extract (mg GAE/g DE).

2.4. Evaluation of Total Flavonoid Content

The total flavonoid content was estimated as previously described by Chatatikun and Chiabchalard [19], with modifications suggested by Sembering et al. [20]. The ethanol extracts were mixed with 96% ethanol, 10% aluminum nitrate nonahydrate ($Al(NO_3)_3 \times 9H_2O$), and 1M potassium acetate (CH_3COOK) in 96-microtiter plates and incubated for 40 min at room temperature. The absorbance was measured at 415 nm. The results were read from the quercetin calibration curve (y = 3.3703x + 0.2047; R^2 = 0.9837) and are expressed as mg quercetin equivalents per gram of dry extract (mg QE/g DE).

2.5. LC-MS Analysis of Ethanol Extracts Obtained from Corn and Sunflower Crop Residues

The contents of phenolic components were determined using the LC-MS System 1260/6130 (Agilent Technologies, Waldbronn, Germany), which consisted of a quaternary pump, degasser, autosampler, column compartment, DAD, ESI ion source, and single quadrupole mass detector. Prior to analysis, the extracts were dissolved in ethanol to achieve a concentration of 10 mg/mL. The prepared solutions were filtered using membrane filters with a pore size of 0.45 μm. Separation was carried out on a reversed-phase Zorbax SB-Aq column (3 × 150 mm, particle diameter 3.5 μm, Agilent Technologies). The mobile phase consisted of 0.1% aqueous formic acid (A) and acetonitrile (B). The gradient elution protocol was as follows: 10–90% B (0–25 min), 90–90% B (25–27 min), and 90–10% B (27–30 min). The flow rate was set at 0.35 mL/min. UV spectra were measured in the range of 190–640 nm, while mass spectra were obtained in negative ion mode within the m/z range from 100 to 1000 using fragmentor voltages of 100 and 250 V. The spray chamber parameters included a gas temperature of 350 °C, drying gas flow of 10 L/min, nebulizer pressure of 40 psig, and capillary voltage of 3500 V. The extract constituents were identified by comparing their retention times, UV, and MS spectra with the corresponding parameters of the commercially available reference compounds obtained under the same chromatographic conditions. Quantification was performed using the external standard method, and the contents are expressed as mg of compound per 1 g of dry extract. Tricin was determined at 350 nm (y = 43243.30x + 51.60, R^2 = 1, concentration range 0.0050–0.2500 mg/mL, LoD = 0.0017 mg/mL, LoQ = 0.0050 mg/mL), while chlorogenic acid (y = 3406.40x + 1.29, R^2 = 0.9982, 0.0047–0.0280 mg/mL, LoD = 0.0015 mg/mL, LoQ = 0.0047 mg/mL), ferulic acid (y = 44074.88x + 6.94, R^2 = 1, 0.0014–0.0700 mg/mL, LoD = 0.0003 mg/mL, LoQ = 0.0008 mg/mL), and *p*-coumaric acid (y = 98637.35x + 111.44, R^2 = 0.9997, 0.0004–0.1200 mg/mL, LOD = 0.0023 mg/mL, LoQ = 0.0071 mg/mL) were quantified at 320 nm.

2.6. Evaluation of Antioxidant Activity

The free radical scavenging activity was evaluated using the protocol developed by Prieto [21] and Xiao et al. [22] for 1,1-diphenyl-2-picrylhydrazyl (DPPH) and 2,2′-azino-bis(3-ethylbenzothiazoline)-6-sulfonic acid (ABTS), respectively. For the DPPH assay, serial dilutions of the extracts (100 μL) were prepared in methanol in 96-microtiter plates. The obtained working solutions of the extracts were mixed with 0.2 mM DPPH solution and incubated in the dark at room temperature for 30 min. Subsequently, the absorbance at 517 nm was read in comparison to a blank containing methanol instead of the extract. The ABTS assay was also performed in 96-microtiter plates. The stock solution of ABTS free radicals was prepared 12–16 h before the experiment by mixing an equal volume of 7 mM ABTS and 2.45 mM potassium persulfate ($K_2S_2O_8$) in an aqueous solution. After incubation in the dark, 4.3 mL was diluted to 100 mL in distilled water and the absorbance at 734 nm was adjusted to 0.74 ± 0.03. Serial dilutions of the extract were prepared in distilled water and mixed with 0.2 mL of ABTS working solution. The radical scavenging capacity (RSC) for both assays was calculated using the following equation:

$$RSC\ (\%) = [(A_C - A_S)/A_C] \times 100\%$$

where A_C and A_S are the absorbance values of the blank and test solutions, respectively. The rates of the scavenged free radicals as a function of the extract concentration were used to generate a graph. The results were read from the obtained graph, expressed as IC_{50} values (mg/mL), representing the extract concentration that reduces free radicals by 50%, and compared with standard antioxidants (BHT and *L*-ascorbic acid).

The ferric reducing/antioxidant power assay (FRAP) was performed as described by Benzie and Devaki [23] with slight modifications. The aqueous working solutions were prepared from the extract and mixed with FRAP reagent (0.3 M Na-acetate buffer (pH 3.6): 10 mM TPTZ solution in 20 mM HCl: 20 mM $FeCl_3$ in ratio 10:1:1 (*v*/*v*/*v*)) in a final volume of 0.2 mL. After incubation at 37 °C for 10 min, the absorbances were measured at 595 nm. The results were determined from the ferrous sulphate heptahydrate calibration curve and are expressed as μmol Fe^{2+} equivalents per mg of dry extract (μmol Fe^{2+} equivalents/mg DE).

2.7. Production Process of Pork Liver Pâtés

The commercial fresh shoulder lean pork meat, livers, and backfat used in the production of the pork liver pâtés were obtained from the local slaughter plant in Pećinci (Srem district, Serbia) at 24 h postmortem, all originating from Yorkshire × Landrace crossbreed pigs. The pork liver pâtés were produced in the experimental meat-processing facility at the Department of Food Hygiene and Technology, Faculty of Veterinary medicine, University of Belgrade. All pork liver pâtés were elaborated based on the same recipe similar to the commercial product [24], and the ingredients in the basic formulation per 1 kg of meat emulsion are presented in Table 1. In this study, four experimental groups of pork liver pâtés were prepared: a control batch (CON) of pork liver pâtés without added synthetic phenolic antioxidants or crop residue extracts; a batch formulated with the synthetic antioxidant butylated hydroxytoluene (BHT), added up to the highest concentration allowed by Serbian regulation [25] for this type of meat product (0.02%); a batch formulated with 1% of sunflower stalk residue extract (SSRE); and a batch formulated with 1% of maize stalk residue extract (MSRE). The selection of these extract concentrations was based on a preliminary study conducted in our laboratory (unpublished data), primarily considering the exhibited crop extracts' antioxidant and antimicrobial activity and the sensory acceptability of the final product. For each formulation, 3.5 kg of raw material was used to produce the pork liver pâtés, the batches were made in triplicate, and these were elaborated independently in three production processes.

The pork backfat and livers were chopped into cubes approximately 1.5 cm^3 in size, while the meat was processed in a mincer equipped with a plate featuring a hole with a 6 mm diameter. The fat was pre-cooked at 65 °C for 15 min. Initially, the livers and meat were comminuted in a cutter (Misowy 5.31 Sirman C6 Cutter, Polska) for 3 min. Subsequently, cubes of fat cooled at room temperature were added to the cutter along with the remaining ingredients and an appropriate amount of crop extract or synthetic antioxidant according to the formulation. All components were mixed in the cutter until a homogeneous raw batter was achieved. The meat batter was then vacuum-packed (100 g per bag, 35 bags per experimental group) and cooked by immersion in a hot water bath until it reached the meat core temperature of 72 °C for 15 min (aprox. 30 min). After the pâté samples were allowed to cool at room temperature, they were stored refrigerated at 4 ± 1 °C in the darkness for 90 days from the day of the manufacture (day 0). Units from each formulation of each replicate of pork liver pâtés were randomly taken at days 0, 20, 40, 60, and 90 after production for the following analysis: proximate chemical composition, fatty acid profile, lipid oxidative stability, physicochemical properties, instrumental color measurement, and microbiological analysis. A sensory analysis was carried out on day 30 of storage.

Table 1. Liver pâté formulations (g/kg).

Ingredients	Batches			
	CON	BHT	SSRE	MSRE
Liver	330	330	330	330
Subcutaneous fat	300	300	300	300
Lean meat	207	207	207	207
Water	125	125	125	125
Soy protein isolate	15	15	15	15
Curing salt (0.6% nitrite in NaCl)	15	15	15	15
Sodium tripolyphosphate	5	5	5	5
Mix of spices	3	3	3	3
BHT	-	0.2	-	-
Sunflower stalk residues extract	-	-	10	-
Maize stalk residues extract	-	-	-	10

CON—control pâtés; BHT—pâtés with butylated hydroxytoluene added; SSRE—pâtés with sunflower stalk ethanol extract added; MSRE—pâtés with maize stalk ethanol extract added; the concentrations of BHT and crop residue extracts were calculated on the basis of total ingredients.

2.8. Determination of Pork Liver Pâtés' Chemical Composition

The proximate chemical composition of the pork liver pâtés, including the contentsof fat [26], moisture [27], protein [28], ash [29], nitrite [30], and NaCl [31], was evaluated in triplicate according to official methods. All results are expressed as g/100 g of pork liver pâtés. The content of total phosphorus (natural and added) was estimated based on the official method [32] using photometric measurements conducted with a Halo DB-20/DB-20S (Dynamica, Livingston, UK). The energy value (Kcal/100 g pork liver pâté) was determined according to the Atwater coefficients corresponding to proteins (4.02 kcal/g), lipids (9.00 kcal/g), and carbohydrates (3.87 kcal/g).

2.9. Determination of Pork Liver Pâtés' Physicochemical Properties

The pH of the pork liver pâtés was directly measured at room temperature (25 °C) using a pH meter (Testo 205, Testo AG, Lenzkirch, Germany) equipped with a digital identification system, temperature compensation sensor, and a proper glass electrode for penetration. The pH of the pork liver pâtés, three samples from each replicate of four groups, was measured three times by the same individual using the same pH meter consistently throughout the experiment. Before each series of measurements, the pH meter was calibrated at room temperature (25 °C) with pH 4.00 and 7.00 standard buffer solutions. Between each pH measurement, the pH meter electrode was rinsed immediately and thoroughly with distilled water and dried carefully with a clean paper towel.

The water activity (a_w) of the pork liver pâtés was measured at room temperature (25 °C) using an a_w-meter (FAst/1, GBX Scientific Instruments, Romans, France), following the guidelines in the instruction manual.

2.10. Determination of Pork Liver Pâtés' Color

The instrumental color of the pork liver pâtés was automatically measured at room temperature (25 °C) after approximately 30 min of blooming time using a portable colorimeter (NR110, 3NH Technology Co., Ltd., Shenzhen, China) with the following settings: 8 mm aperture, 2° viewing angle, and D65 illuminant. Before measurement, the colorimeter was calibrated according to the manufacturer's instructions. During the color measurement, the pork liver pâtés were placed on a white background, while the spectrally pure glass was put between the samples and the colorimeter. The color was measured on three pâté samples from each replicate of each formulation at three different randomly selected points. The results are expressed as lightness (L*), redness (a*), yellowness (b*), hue (h°), and chroma (C*) values. The numerical total color difference (ΔE) between the samples (S) was calculated with regard to the control group of pork liver pâtés (CON) on day 0 and day 90 based on Formula (1) and within the same pork liver pâté group between day 0 and day 90 (ΔE*) according to Formula (2) [33]:

$$\Delta E = \sqrt{(a_s - a_c)^2 + (b_s - b_c)^2 + (L_s - L_c)^2} \quad (1)$$

$$\Delta E^* = \sqrt{\left(a^*_{90} - a^*_0\right)^2 + \left(b^*_{90} - b^*_0\right)^2 + \left(L^*_{90} - L^*_0\right)^2} \quad (2)$$

2.11. Determination of Pork Liver Pâtés' Fatty Acid Profile

The fatty acid profile of pork liver pâtés was determined following the methods previously described by Spirić et al. [34] and Glisic et al. [35]. Total lipids extraction from the pork liver pâtés was performed by the accelerated solvent extraction method with a hexane/isopropanol mixture. For fatty acid methyl esters (FAMEs) preparation, trimethylsulfonium hydroxide was used [36]. The separation and quantification of the FAMEs were performed using a gas chromatograph (Shimadzu 2010, Kyoto, Japan) equipped with a flame ionization detector (FID), split/split less injector, fused silica cyanopropyl HP-88 capillary column (100 m × 0.25 mm × 0.20 mm, J&W Scientific, Folsom, CA, USA), and workstation (Shimadzu GC Solution ver. 2.3). The characteristics and conditions of the separation of the FAMEs using gas chromatography were as follows: (i) the injector and detector temperatures were set at 250 °C and 280 °C, respectively; (ii) nitrogen was used as the carrier gas at a 1.33 mL/min flow rate and injector split ratio of 1:50; and (iii) the initial oven temperature of 125 °C increased at 10 °C/min to 175 °C and was held for 10 minutes, increased at 5 °C/min to 210 °C, held for 5 min, and then increased at 2 °C/min to a final temperature of 230 °C. Individual FAMEs were identified by comparing their retention times with a known standard FAME mixture. The determination of the fatty acid profiles was performed in triplicate at the beginning of refrigerated storage. Quantification was performed using heneicosanoic acid methyl ester (C21:0) as an internal standard. Data regarding fatty acid profiles are expressed as mg per 100 g total fatty acids. The proportions of polyunsaturated (PUFA), monounsaturated (MUFA), and saturated (SFA) fatty acid contents and the ratios of PUFA/SFA and n-6/n-3 were also calculated. The health lipid indices, atherogenic index (AI), thrombogenic index (TI), and the index of hypocholesterolemic/hypercholesterolemic fatty acids (HH), were computed as follows [37]:

AI = (C12:0 + 4 × C14:0 + C16:0)/(n-3 PUFA + n-6 PUFA + MUFA);

TI = (C14:0 + C16:0 + C18:0)/[0.5 × MUFA + 0.5 × n-6 PUFA + 3 × n-3 PUFA + (n-3 PUFA/n-6 PUFA)];

HH = (18:1n-9 + 18:2n-6 + 18:3n-3 + 20:4n-6 + 20:5n-3 + 22:6n-3)/(14:0 + 16:0)

2.12. Determination of Pork Liver Pâtés' Lipid Oxidative Stability

The oxidative stability of the pork liver pâtés was determined in triplicate during the storage time based on the Thiobarbituric Acid Reactive Substances (TBARs) test using the combined method according to Tarladgis et al. [38] and Holland [39]. Briefly, 20 g of the pork liver pâtés was homogenized in distilled water and treated with HCl so a final pH of 1.5 was reached. The acidified homogenates were distilled using a distiller with a 50 mL bulb to collect the distillate. Afterwards, 5 mL of collected distillate and 5 mL of TBA reagent (0.3% 2-thiobarbituric acid in 90% acetic acid) were vortexed. After vortexing, the samples were heated in a constant-temperature boiling water bath at 95 °C for 35 min, and, thereafter, were cooled for 10 min at room temperature (25 °C). The reaction mixture absorbance was read at 532 nm using a Cecil CE 2021 spectrophotometer (Cecil Instruments Ltd., Cambridge, UK), and the calculated value is expressed as mg malonaldehyde per kilogram (mg MDA/kg) of pork liver pâté. The determination of the peroxide value (POV) was performed based on the ISO standard [40], as outlined in Glisic et al. [40], and the results are expressed as millimoles of active oxygen per kilogram of pork liver pâté (mmol O_2/kg).

2.13. Microbiological Analysis of Pork Liver Pâtés

Microbiological analyses were performed in triplicate to evaluate the sanitary conditions of the pork liver pâtés from day 0 to day 90 of storage. On each examination day, 10 g from every pork liver pâté formulation was aseptically excised from the interior, transferred into sterile stomacher bags, homogenized with 90 mL of 0.1% of sterile peptone water, and blended in stomacher (Colworth 400, Seward, London, UK) for two minutes. Afterwards, appropriate serial decimal dilutions of each sample were prepared in 0.1% peptone water and inoculated onto the corresponding agar plates. The total viable count (TVC) and psychrotrophic bacteria count were determined on PCA by pour plates aerobic incubation at 30 °C for 72 h and at 7 °C for 10 days, respectively. *Enterobacteriaceae* microorganisms were isolated and counted on a VRBG agar by pour plates aerobic incubation at 37 °C for 24 h. The isolation and determination of lactic acid bacteria (LAB) were carried out on a MRS agar following incubation in anaerobic conditions at 30 °C for 72 h. The results are expressed as logarithms of colony-forming units per gram of pork liver pâté (log CFU/g).

2.14. Sensory Analysis of Pork Liver Pâtés

A sensory analysis of four pork liver pâté formulations was performed after 30 days of storage in three replicates by a trained panel with 7 experienced sensorists (4 females and 3 males, 35–55 years of age) selected from staff members of the Department of Food Hygiene and Technology (Faculty of Veterinary medicine, University of Belgrade). All panelists were regular consumers of pork liver pâté. Before testing, the panelists underwent a training session to define the attributes and descriptors to be evaluated in the pâtés. The test took place in the tasting room of the Department of Food Hygiene and Technology (Faculty of Veterinary Medicine, University of Belgrade), equipped with individual cabins. Rectangular pieces of about 1.5 cm × 2 cm were cut from the center of slices, coded with randomly selected 3-digit numbers, and served at room temperature in a randomized order to each panelist. Bread and tap water were also provided to the panelists to cleanse their palates between samples.

According to a quantitative descriptive analysis (QDA), the intensity of eight sensorial attributes was rated using a lineal structured scale from 0 (minimum intensity) to 7 (maximum intensity), with the ability to give semi-scores. The descriptors of the attributes were as follows: color (from light to dark); odor (from imperceptible to intense); cohesiveness (from low to high); homogeneity (from smooth to coarse); aroma (from imperceptible to intense); flavor, including saltiness, fattiness, and rancid flavor (from imperceptible to intense); adhesiveness (from low to high); and juiciness (from dry to juicy). At the end of the test, the panelists were asked to give a score from 0 to 7 for the products' overall acceptability.

2.15. Statistical Analysis

The chemical composition and fatty acid profile were determined after the pâtés' elaboration (day 0), and their sensory characteristics were evaluated after one month of storage, with three repetitions for each experimental replicate (4 treatments × 1 time period × 3 batches × 3 experimental replications). Throughout the storage period (0, 20, 40, 60, and 90 days), color, pH, a_w, microbiology, POV, and TBARs were determined in three repetitions for each experimental replicate (4 treatments × 5 time periods × 3 batches × 3 experimental replications).

Statistical analyses were conducted with the SPSS software package version 20.0 (SPSS Inc., Chicago, IL, USA). Values from the crop residue extracts' antioxidant capacity, proximate composition, fatty acid profile, and intensity of sensory attributes were evaluated using a General Linear Mixed Model (GLMM), with treatments as a fixed effect and the replicates as a random effect. For the pH, a_w, microbiology, color, and lipid oxidation parameters' data, the treatments and storage time were considered as fixed effects and the replicates as a random effect. No significant differences ($p > 0.05$) were observed in the variations among the replicates for any of the examined features. Except for the total phenol and flavonoid contents in the crop extracts, for which the independent samples *t*-test was used, the intergroup comparisons were appraised by one-way analysis of variance (ANOVA) followed by Tukey's multiple comparison tests. Statistical significance was set at $p < 0.05$ or $p < 0.01$.

3. Results and Discussion

3.1. TPC, TFC, Antioxidant Activity, and Phenolic Compounds of Sunflower and Maize Stalk Residue Extracts

The total phenol and flavonoid contents in the maize stalk residue extract were higher than those in the sunflower stalk ethanolic extract (Table 2). However, this difference was significant only in the total phenolic compounds values (20.44 vs. 15.83 mg GAE/g dry basis; $p < 0.05$).

Table 2. The total phenol and flavonoid contents in the ethanol extracts from sunflower and maize waste material (mean ± SD).

	SSRE	MSRE
TPC (mg GAE/g) [1]	15.83 ± 0.30 [a]	20.44 ± 0.23 [b]
TFC (mg QE/g) [2]	8.98 ± 0.34	9.37 ± 0.83

TPC—total phenol content; TFC—total flavonoid content; SSRE—sunflower stalk ethanol extract; and MSRE—maize stalk ethanol extract. The results were obtained from three different experiments performed in triplicates; [a,b] Different superscripts within the same row indicate a significant difference ($p < 0.05$). [1] mg of gallic acid equivalents per g of dry extract. [2] mg of quercetin equivalents per g of dry extract.

In previous studies, moderate to high TPCs [41], with values ranging from approximately 29 to about 390 mg GAE/g, have been observed in different plant extracts (grape seed, green tea leaves, chestnut leaves, beer residue, peanut skin, annatto seed flour, and persimmon flour) used as natural antioxidants in pork pâtés [9,42–45]. Contrastingly, Martín-Sánchez et al. [46] and Lorenzo et al. [47] formulated pork pâtés with date palm and seaweed extracts, which showed slightly lower but still moderate levels of total phenolic contents (14.5 and 12.8 mg GAE/g), consistent with the results of the present study.

The literary data on the contents of phenolic compounds in different parts of the sunflower and corn plants are limited [48–51]. Furthermore, to the best of our knowledge, there is no available information regarding the total phenol and flavonoid contents in sunflower crop residue, while the study by Vazquez-Olivo et al. [49] is one of the few where data on the phenolic compounds content in corn stover can be found.

Gai et al. [50] showed that the total phenolic content in the methanol extracts of sunflower aerial parts in the late stage of flowering was 21.7 mg GAE/g, a result comparable to the findings of the current study. Kamal [48] found a notably higher phenol and flavonoid content in sunflower methanol extracts, with the lowest average two-year values in stems (228.35 mg GAE/g and 55.33 mg QE/g, respectively), followed by roots, while the highest amount was reported in the leaves extract.

In the present study, the levels of *p*-coumaric acid, ferulic acid, and tricin in the ethanol extract of maize crop residues were determined using LC-DAD-ESI-MS and found to be 1.64 mg/g, 1.21 mg/g, and 1.99 mg/g, respectively. Additionally, chlorogenic acid was detected in the ethanol extract of sunflower field waste at a concentration of 2.40 mg/g. Previous studies have revealed that chlorogenic acid is the most prevalent phenolic acid in extracts of sunflower kernels, shells, sunflower seed flour, and seed cake [52–54]. However, in the late flowering sunflower plant extract, the predominant compound was 3,5-di-*O*-caffeoylquinic acid (12.82 mg/g), followed by chlorogenic acid (7.04 mg/g) [50]. Considering that the efficiency of extraction is influenced by the type of solvents and the composition of the raw material, previous studies have demonstrated that an ethanol solution is suitable for extracting phenolic compounds from the sunflower matrix [54,55]. Alexandrino et al. [54] further indicated that ethanol extracts of sunflower flour also contain other soluble compounds, such as sugars and proteins. Corroborating the literature data presented, in the present study, chlorogenic acid was identified in the investigated sunflower extract, however, an approximately three times lower content in our sample was observed than that previously reported [50]. Furthermore, a compound with the molecular ion at m/z 515 was detected and a retention time corresponding to that of the reference standard of 3,5-di-*O*-caffeoylquinic acid. Nevertheless, we were unable to confirm its identity with certainty, due to low-quality UV spectra resulting from its very low concentration.

Previously, Vazquez-Olivo et al. [49], by evaluating the distribution of the phenolic compounds in different parts of maize, showed that TPC in maize stem extract was 9.34 mg GAE/g DW, indicating the significant effect of plant organ on TPC, as well as on the bound and free phenolic amounts. In accordance with these results, Adekunle et al. [51] also found a lower TPC in the ethanol extract of corn stalk (just below 2 mg GAE/g) compared to the results of our study. On the contrary, in Vijayalaxmi et al.'s [10] study, higher total polyphenols and total flavonoids contents were found (426.5 mg GAE/g and 87.2 mg QE/g, respectively) in a corn husk extract obtained using 50% ethanol. Further, by examining different agro-industrial byproducts, Abbasi-Parizad et al. [56] determined a TPC of 12 mg GAE/g DW in the aqueous ethanolic extract of red corn cob, which is also lower compared to the result obtained for our maize stalk residues. These variations could be explained by the fact that different organs are exposed to environmental stress differently, among which corn husks are the most exposed, followed by the cob, stalk, or grain [49]. In Abbasi-Parizad et al.'s [56] study, phenolic acids constituted 28.9% of the total phenols, whereas flavonoids accounted for the remaining 71.1%. The most prevalent phenolic acids were chlorogenic acid and ferulic acid, while catechin and epicatechin were the predominant ones among flavonoids. Galeana-López et al. [57] discovered that the most abundant phenolic compound in corn husk ethanolic extract was ferulic acid (12.93 mg/g), followed by *p*-coumaric acid (5.74 mg/g). In our sample, the content of *p*-coumaric acid was higher than that of ferulic acid, and both acids were present in lower amounts compared to the mentioned study. Additionally, we showed that the flavon tricin occurred in an appreciable quantity. In contrast to the findings reported by Abbasi-Parizad et al. [56], we did not detect chlorogenic acid, catechin, or epicatechin in the examined maize extract.

As can be seen in Table 3, the determined in vitro antioxidant activity was related to the concentration of the phenolic compounds of the crop residue ethanolic extracts. Namely, the radical scavenging capacity evaluated through DPPH and ABTS assays was more than 2-fold greater for maize stalk residue extract (IC_{50} were 0.41 and 1.72 mg/mL, in DPPH and ABTS tests, $p < 0.05$, respectively) than that for sunflower stalk residue extract. Further, the ferric ions reducing power of MSRE was also significantly higher compared to that of SSRE (0.25 and 0.14 μmol Fe^{2+}/g, $p < 0.05$, respectively). In comparison with standard compounds, the synthetic antioxidants BHT and *L*-ascorbic acid, significantly higher concentrations of MSRE and SSRE were required to achieve a 50% decrease in the initial DPPH and ABTS levels ($p < 0.05$). Further, the FRAP values for both extracts were significantly lower ($p < 0.05$) than those of the positive control. In contrast to the present results, the 50% methanolic extract of corn husk showed a higher reducing power and better DPPH scavenging activity (IC_{50} 0.024 mg/mL) compared to the positive control ascorbic acid (0.038 mg/mL) [10].

Gai et al. [50] demonstrated that the antioxidant scavenging activity of sunflower extract changes during different growth stages, with the late flowering plant extract showing results of 0.18 mg/mL and 0.34 μmol Fe^{2+}/mg for DPPH and FRAP, respectively. Despite lower antioxidant activity, our results are consistent with a positive correlation between the TPC and antioxidant activity of the sunflower extract presented in this study. However, Abbasi-Parizad et al. [56] found that the total antioxidant activity did not align with the trend of TPC across various agro-industrial byproduct extracts, including red corn cobs. Vazquez-Olivo et al. [49] reported variations in the antioxidant capacity among different maize organs. For example, the highest antioxidant capacity for DPPH was found in the free phenolic extracts from the stalk, while the highest values for the bound phenolic extract were identified in the cob and husk.

Table 3. Antioxidant activity of the ethanol extracts from sunflower and maize waste material (mean ± SD).

Tests	Extract/Positive Control			
	SSRE	MSRE	BHT	*L*-AA
DPPH IC_{50} (mg/mL)	1.02 ± 0.01 [a]	0.41 ± 0.02 [b]	0.038 ± 0.001 [c]	0.0057 ± 0.0003 [d]
ABTS IC_{50} (mg/mL)	4.53 ± 1.20 [a]	1.72 ± 0.10 [b]	0.04 ± 0.002 [c]	0.0316 ± 0.0003 [d]
FRAP (μmol Fe^{2+}/g)	0.14 ± 0.01 [a]	0.25 ± 0.01 [b]	0.41 ± 0.03 [c]	1.618 ± 0.004 [d]

SSRE—sunflower stalk ethanol extract; MSRE—maize stalk ethanol extract; BHT—butylated hydroxytoluene; and *L*-AA—*L*-ascorbic acid. The results were obtained from three different experiments performed in triplicates; [a,b,c,d] Different superscripts within the same row indicate a significant difference ($p < 0.05$).

The compounds quantified in the examined extracts are known to display antioxidant effects in experimental settings [58–60]. Therefore, it can be concluded that the total antioxidant activity arises, at least in part, from their presence.

Finally, the differences in the antioxidant activity of the sunflower and corn extracts observed in the mentioned studies were due to their different phytochemical profiles, which depends on numerous factors, including the environmental, agronomic, and meteorological conditions and the extraction procedure (solvent polarity, temperature, and extraction time). Additionally, these differences also result from the methods used for the evaluation of antioxidant activity, since they are based on different molecular mechanisms such as electron transfer and hydrogen donation [10,49,57].

3.2. *Proximate Composition of Pork Liver Pâtés*

The proximate compositions and energy values of the control pork liver pâtés and pâtés made with BHT, sunflower, and maize residue extracts are given in Table 4. No significant differences ($p > 0.05$) were found among the different formulations regarding moisture, protein, fat, ash, NaCl, nitrite, and P level (expressed as P_2O_5), as well as energy value. Given the same raw material composition across all batches, these results were anticipated. Namely, all the formulations were in the accordance with the Serbian legislation, which defines a minimum 9% protein content in liver pâtés and sausages [61]. The present results corroborate previous studies that have evaluated similar liver pâté formulations, where the addition of plant extracts and essential oils as natural antioxidants did not affect the chemical composition of the meat products [6,44,62].

Table 4. Proximate composition and energy value of four pork liver pâté formulations (mean ± SD).

Item	Batches				*p* Value
	CON	BHT	SSRE	MSRE	
Moisture (%)	54.77 ± 0.31	54.67 ± 0.24	54.50 ± 0.71	54.44 ± 0.69	0.6875
Protein (%)	14.39 ± 0.04	14.50 ± 0.17	14.40 ± 0.14	14.42 ± 0.12	0.3777
Fat (%)	27.95 ± 0.31	27.97 ± 0.21	28.32 ± 0.53	28.29 ± 0.71	0.3941
Ash (%)	2.86 ± 0.02	2.81 ± 0.01	2.86 ± 0.01	2.83 ± 0.07	0.0970
NaCl (g/100 g)	1.75 ± 0.13	1.61 ± 0.02	1.59 ± 0.08	1.78 ± 0.19	0.0564
$NaNO_2$ (mg/kg)	26.11 ± 2.92	27.64 ± 1.94	29.35 ± 1.48	27.40 ± 1.95	0.1057
P_2O_5 (g/kg)	6.70 ± 0.16	6.64 ± 0.11	6.69 ± 0.03	6.76 ± 0.05	0.2727
Energy value (Kcal/100 g)	309.10 ± 2.75	309.70 ± 1.94	312.50 ± 5.16	312.30 ± 6.26	0.4455

CON—control pâtés; BHT—pâtés with butylated hydroxytoluene added; SSRE—pâtés with sunflower stalk ethanol extract added; MSRE—pâtés with maize stalk ethanol extract added; n = 3 for each group of pork liver pâtés in triplicate.

3.3. Fatty Acid Composition of Pork Liver Pâtés

The fatty acid (FA) composition and key nutritional indices of both the control and pork liver pâtés prepared with crop residue extracts at the beginning of refrigerated storage are presented in Table 5. The main FA fraction in all the pâté formulations was composed of MUFAs (ranging from 44,880–45,086 mg/100 g), followed by SFAs and PUFAs. The primary fatty acids determined were C18:1 > C16:0 > C18:0 > C18:2, with a significantly higher content of linoleic acid in the SSRE and MSRE pâtés, consequently leading to an increased n-6 content in these pâtés ($p < 0.01$). These results are consistent with those reported by other authors [42,45,63], where pork liver pâtés were primarily composed of MUFAs, with oleic acid identified as the predominant fatty acid. Further, corroborating the present findings, the incorporation of natural extracts from seaweed or various agro-industrial residues (beer residue, chestnut leaves, and peanut skin) did not affect or lead to minimal deviations in the FA content of the pork liver pâtés during processing [44,64]. However, unlike our results, Munekata et al. [62] did not find differences in the n-6 and total FA contents between the negative and positive controls (BHT) of ground sheep patties, as well as patties with an added 1% peanut skin extract, both at the beginning and after a 20-day storage period.

Table 5. Fatty acid composition (mg per 100 g total fatty acids) of the pâté formulations (mean ± SD).

Fatty Acid	Batches				p Value
	CON	BHT	SSRE	MSRE	
C14:0	1107 ± 19.04	1113 ± 31.41	1110 ± 14.14	1095 ± 18.71	0.5042
C15:0	35.00 ± 10.49	32.67 ± 3.27	35.50 ± 5.01	30.00 ± 2.76	0.4166
C16:0	27,397 ± 138.70	27,349 ± 80.52	27,380 ± 33.47	27,265 ± 97.31	0.1120
C16:1	1380 ± 48.58	1384 ± 32.00	1370 ± 23.66	1365 ± 24.29	0.7457
C17:0	266.7 ± 15.06	260.0 ± 38.99	265.0 ± 22.58	265.0 ± 18.71	0.9718
C18:0	15,093 ± 158.80	14,982 ± 33.12	15,085 ± 69.50	15,092 ± 74.94	0.1583
C18:1cis-9	43,542 ± 237.90	43,627 ± 159.00	43,510 ± 137.00	43,598 ± 158.20	0.6631
C18:2n-6	10,105 ± 122.60 [a]	10,192 ± 68.82 [ab]	10,275 ± 86.89 [b]	10,292 ± 59.47 [b]	0.0053
C18:3n-3	368.3 ± 19.41	380.0 ± 14.14	375.0 ± 28.11	375.0 ± 30.82	0.8692
C20:0	213.3 ± 20.66	208.3 ± 14.38	205.0 ± 18.71	225.0 ± 27.39	0.3878
C20:2	366.7 ± 13.66	365.0 ± 17.61	345.0 ± 18.71	365.0 ± 24.29	0.1843
C20:3n-3	35.83 ± 4.36	39.17 ± 2.64	39.67 ± 2.87	40.00 ± 2.45	0.1197
SFA	44,112 ± 282.00	43,945 ± 83.34	44,081 ± 60.51	43,972 ± 129.60	0.2478
MUFA	44,972 ± 198.10	45,086 ± 130.90	44,880 ± 131.30	44,963 ± 166.80	0.2009
PUFA	10,876 ± 126.20 [a]	10,976 ± 60.62 [ab]	11,035 ± 100.50 [b]	11,072 ± 73.23 [b]	0.0096
n-6	10,105 ± 122.60 [a]	10,192 ± 68.82 [ab]	10,275 ± 86.89 [b]	10,292 ± 59.47 [b]	0.0053
n-3	404.2 ± 17.75	419.2 ± 15.84	414.7 ± 28.18	415.0 ± 32.19	0.7481
n-6/n-3	25.05 ± 1.27	24.35 ± 1.00	24.87 ± 1.54	24.93 ± 1.97	0.8525
PUFA/SFA	0.2450 ± 0.005 [a]	0.2498 ± 0.001 [ab]	0.2503 ± 0.002 [b]	0.2518 ± 0.001 [b]	0.0064
AI	0.5737 ± 0.005	0.5710 ± 0.003	0.5726 ± 0.001	0.5684 ± 0.004	0.1291
TI	1.516 ± 0.016	1.503 ± 0.005	1.512 ± 0.005	1.505 ± 0.006	0.0841
HH	1.897 ± 0.019	1.907 ± 0.008	1.901 ± 0.004	1.913 ± 0.011	0.1115

CON—control pâtés; BHT—pâtés with butylated hydroxytoluene added; SSRE—pâtés with sunflower stalk ethanol extract added; MSRE—pâtés with maize stalk ethanol extract added; AI—Atherogenic index; TI—Thrombogenic index; and HH—Hypocholesterolemic/hypercholesterolemic; [a,b] Different superscripts within the same row indicate a significant difference ($p < 0.05$); n = 3 for each group of pork liver pâtés in triplicate.

The addition of sunflower and maize crop residue extracts significantly increased the PUFA content and PUFA/SFA ratio of the pork liver pâtés ($p < 0.01$); however, this ratio was lower than the optimal recommended values (0.5–0.7) in the Mediterranean diet [65]. In contrast to these results, the addition of green tea extract slightly decreased the PUFA/SFA ratio in liver pâtés [42]. The n-6/n-3 ratios in our pâté formulations (around 25) exceeded those reported in other pork liver pâtés (from 7 to 18.69) [42,63,66,67] and were several times higher than the nutritionally recommended limit of 4 [68].

By calculating the health lipid indices, it can be noticed that, despite a significantly higher n-6 content in the pâtés with added crop residue extracts, this increase was not sufficient to reflect a difference in the AI. The absence of differences in the n-3 content between different pâtés batches resulted in an expected TI value of around 1.5 in all formulations. Meat products with a lower AI and TI and higher HH values are characterized by a healthier lipid composition, with a potentially favorable effect on the cardiovascular system [69]. Following the recommended levels for AI and TI, which should be below 1 [69], our pâtés, featuring a fat content of around 28%, typical for commercial meat products incorporating pork liver [63], met this criterion exclusively for AI, while HH was approximately 1.9.

3.4. Water Activity (a_w) Values of Pork Liver Pâtés

The water activity was influenced by the inclusion of different extracts and storage times, with the interaction between both factors also proving to be significant (Table 6, $p < 0.01$). During the first 20 days of storage, the a_w values did not differ between pâté formulations, except for a significantly lower a_w value in the pâtés with sunflower extract added on day 0 (0.941). The crop residue extracts and BHT led to an increase in the a_w values on days 40 and 90 ($p < 0.01$), while, on day 60, only the pâtés containing maize crop residue and BHT demonstrated higher a_w values compared to the control ($p < 0.01$). The water activity of each pâté formulation showed a tendency to increase from the beginning of the experiment, starting at approximately 0.945, until day 40 of storage. However, after two months of storage, a continuous decline in a_w values was observed, reaching approximately 0.907 by the day 90. Comparable results were reported by Martín-Sánchez et al. [46], who reported that the addition of date palm paste, either alone or in combination with annatto extract, exhibited an increasing trend in a_w values during the initial storage period across all formulations of spreadable pâtés, with a noticeable decline observed after 3 weeks of storage. Furthermore, Amaral et al. [70] found a similar behavior in reducing the a_w values of lamb pâtés packaged in a polyamide casing system during a 90-day storage period. Contrastingly, other authors reported minimal or no alterations in a_w values in liver pâtés with natural ingredients added [8,45].

Table 6. The water activity (a_w) of pork liver pâtés with added synthetic antioxidant and crop residue extracts (mean ± SD).

a_w	Batches				Group
	CON	BHT	SSRE	MSRE	p Value
0 day	0.946 ± 0.001^{aA}	0.946 ± 0.001^{aA}	0.941 ± 0.001^{bA}	0.947 ± 0.0001^{aA}	<0.0001
20 day	0.956 ± 0.003^{B}	0.959 ± 0.001^{B}	0.957 ± 0.005^{B}	0.958 ± 0.005^{B}	0.4577
40 day	0.935 ± 0.001^{aC}	0.950 ± 0.010^{bA}	0.951 ± 0.007^{bcB}	0.962 ± 0.005^{cB}	<0.0001
60 day	0.921 ± 0.001^{aD}	0.926 ± 0.001^{bC}	0.923 ± 0.001^{aC}	0.926 ± 0.001^{bC}	<0.0001
90 day	0.902 ± 0.003^{aE}	0.906 ± 0.001^{bD}	0.910 ± 0.001^{cD}	0.910 ± 0.001^{cD}	<0.0001
Time p Value	<0.0001	<0.0001	<0.0001	<0.0001	
Group × Time p Value	<0.0001	<0.0001	<0.0001	<0.0001	

CON—control pâtés; BHT—pâtés with butylated hydroxytoluene added; SSRE—pâtés with sunflower stalk ethanol extract added; and MSRE—pâtés with maize stalk ethanol extract added; [a,b,c] Different lower-case superscript letters within the row indicate a significant difference between groups on the same day of storage ($p < 0.05$); [A,B,C,D,E] Different upper-case superscript letters within the column indicate differences within the same group on different days of storage ($p < 0.05$); n = 3 for each group of pork liver pâtés in triplicate.

3.5. pH of Pork Liver Pâtés

Table 7 displays the evolution of the pH values in the pork liver pâtés stored under refrigeration at 4 °C. The pH was significantly affected by the formulation, storage time, and interaction of these factors ($p < 0.01$). The inclusion of crop extracts resulted in a significant decrease in pH compared to both the control and BHT pâtés ($p < 0.01$) over the 90-day storage period. Despite these statistically significant differences, variations in this parameter were small, ranging from 6.33 to 6.52, which is considered to be usual for this kind of product [71]. This aligns with the findings of Martín-Sánchez et al. [46] and Lorenzo et al. [47], who reported the addition of grape seed extract, green tea extract, and date palm coproducts paste, rich in organic acids, to decrease the pH value in raw minced porcine patties and spreadable pork liver pâtés. Similar patterns in pH evolution within each batch during the 24-week storage period were observed by Pateiro et al. [42] in pig liver pâtés with 0.1% of tea, chestnut, and grape seed extracts. Additionally, Pateiro et al. [9] showed a dose-dependent pH reduction in pig liver pâtés with increasing concentrations of grape and tea extracts. However, significant modifications in the pH of pork liver pâtés elaborated with different natural extracts from seaweed, beer residue, and chestnut leaves were not found, indicating that predominant active compounds in these extracts, mainly phenolic acids, did not affect the pH of the pâté formulations [44,64].

Table 7. Physicochemical properties of pork liver pâtés during refrigerated storage (mean ± SD).

Days of Storage	Batches	Parameters L*	a*	b*	C*	h°	pH
0							
	CON	56.05 ± 1.71 ^{aA}	12.43 ± 0.66 ^{a}	12.15 ± 0.75 ^{a}	17.39 ± 0.77 ^{aA}	44.38 ± 2.27 ^{aAB}	6.48 ± 0.02 ^{aAB}
	BHT	53.68 ± 4.45 ^{aA}	13.24 ± 1.38 ^{bABC}	12.56 ± 0.91 ^{aA}	18.26 ± 1.45 ^{bAC}	44.05 ± 1.16 ^{aA}	6.46 ± 0.02 ^{bA}
	SSRE	50.92 ± 2.86 ^{bA}	12.89 ± 0.57 ^{abA}	11.89 ± 0.39 ^{aA}	17.46 ± 0.53 ^{abA}	42.45 ± 1.46 ^{aA}	6.42 ± 0.01 ^{cA}
	MSRE	55.87 ± 2.70 ^{aAB}	11.22 ± 0.61 ^{cA}	14.08 ± 1.22 ^{bA}	18.04 ± 0.83 ^{abA}	51.36 ± 3.71 ^{bA}	6.42 ± 0.01 ^{cA}
20							
	CON	59.00 ± 1.71 ^{aB}	12.90 ± 0.34 ^{ac}	12.31 ± 0.44 ^{a}	17.96 ± 0.69 ^{aAB}	43.66 ± 0.76 ^{aA}	6.50 ± 0.01 ^{aB}
	BHT	58.09 ± 1.43 ^{aB}	12.80 ± 0.56 ^{aAB}	11.23 ± 0.33 ^{bB}	17.04 ± 0.41 ^{bBC}	41.30 ± 1.73 ^{bB}	6.53 ± 0.01 ^{bB}
	SSRE	52.17 ± 3.48 ^{bA}	13.22 ± 0.36 ^{cA}	11.95 ± 0.52 ^{aAC}	17.73 ± 0.74 ^{aA}	42.10 ± 0.93 ^{abA}	6.45 ± 0.01 ^{cBD}
	MSRE	57.18 ± 2.02 ^{aAB}	10.86 ± 0.49 ^{bA}	14.70 ± 1.05 ^{cA}	18.30 ± 0.84 ^{aA}	53.46 ± 3.01 ^{cA}	6.49 ± 0.01 ^{dB}
40							
	CON	54.03 ± 3.90 ^{aA}	12.54 ± 0.70 ^{a}	12.66 ± 0.67 ^{a}	17.73 ± 0.74 ^{aAB}	45.27 ± 1.77 ^{aAB}	6.43 ± 0.01 ^{aC}
	BHT	56.36 ± 3.52 ^{abAB}	12.56 ± 0.34 ^{aA}	11.26 ± 0.37 ^{bB}	16.87 ± 0.41 ^{bB}	41.88 ± 1.47 ^{bB}	6.47 ± 0.01 ^{bA}
	SSRE	57.57 ± 0.84 ^{bB}	14.26 ± 0.49 ^{bB}	12.84 ± 1.00 ^{aB}	19.19 ± 1.09 ^{cB}	41.93 ± 1.25 ^{bA}	6.33 ± 0.06 ^{cC}
	MSRE	57.91 ± 3.33 ^{bA}	11.24 ± 0.47 ^{cA}	14.13 ± 0.57 ^{cA}	18.06 ± 0.57 ^{aA}	51.48 ± 1.37 ^{cA}	6.39 ± 0.01 ^{dC}
60							
	CON	59.48 ± 3.46 ^{aB}	12.70 ± 0.52 ^{a}	12.34 ± 0.39 ^{a}	17.71 ± 0.48 ^{aAB}	44.16 ± 1.58 ^{aAB}	6.49 ± 0.04 ^{acB}
	BHT	57.57 ± 2.06 ^{acB}	13.37 ± 0.51 ^{bBC}	11.62 ± 0.46 ^{bB}	17.69 ± 0.55 ^{aC}	40.99 ± 1.46 ^{bB}	6.51 ± 0.05 ^{aC}
	SSRE	54.45 ± 1.82 ^{bC}	14.01 ± 0.58 ^{cB}	12.49 ± 0.44 ^{aBC}	18.72 ± 0.75 ^{bB}	41.49 ± 1.15 ^{bA}	6.46 ± 0.01 ^{bD}
	MSRE	55.54 ± 1.82 ^{bcB}	12.58 ± 0.50 ^{aB}	14.29 ± 0.52 ^{cA}	19.03 ± 0.69 ^{bB}	48.67 ± 0.79 ^{cB}	6.47 ± 0.01 ^{bcD}
90							
	CON	59.36 ± 3.57 ^{aB}	12.53 ± 0.55 ^{a}	12.71 ± 0.76 ^{a}	18.08 ± 0.90 ^{aB}	45.92 ± 3.19 ^{aB}	6.49 ± 0.01 ^{aB}
	BHT	55.06 ± 2.64 ^{bA}	13.68 ± 0.59 ^{bC}	13.77 ± 0.74 ^{bC}	19.41 ± 0.82 ^{bD}	45.18 ± 1.38 ^{abA}	6.52 ± 0.01 ^{bBC}
	SSRE	51.51 ± 1.46 ^{cA}	14.79 ± 0.39 ^{cC}	14.34 ± 0.37 ^{bD}	20.60 ± 0.37 ^{cC}	44.11 ± 1.08 ^{bB}	6.43 ± 0.01 ^{cAB}
	MSRE	55.56 ± 1.70 ^{bB}	12.13 ± 0.41 ^{aB}	15.87 ± 1.01 ^{cB}	19.98 ± 0.94 ^{bcC}	52.55 ± 1.55 ^{cA}	6.44 ± 0.01 ^{dE}
Group	p Value	<0.0001	<0.0001	<0.0001	<0.0001	<0.0001	<0.0001
Time	p Value	<0.0001	<0.0001	<0.0001	<0.0001	<0.0001	<0.0001
Group × Time	p Value	<0.0001	<0.0001	<0.0001	<0.0001	<0.0001	<0.0001

	ΔE 0	ΔE 90	ΔE*
CON	-	-	3.35
BHT	2.54	4.57	1.89
SSRE	5.16	8.33	3.15
MSRE	2.29	4.96	2.03

CON—control pâtés; BHT—pâtés with butylated hydroxytoluene added; SSRE—pâtés with sunflower stalk ethanol extract added; MSRE—pâtés with maize stalk ethanol extract added; ΔE 0 and 90-color differences with respect to control pâtés on days 0 and 90; and ΔE*—color differences within the same group on days 0 and 90; ^{a,b,c,d} Different lower-case superscript letters within the column indicate a significant difference between groups on the same day of storage ($p < 0.05$); ^{A,B,C,D,E} Different upper-case superscript letters within the column indicate differences within the same group on different days of storage ($p < 0.05$); n = 3 for each group of pork liver pâtés in triplicate on every day of storage.

3.6. Color of Pork Liver Pâtés

The color characteristics of the pâtés were affected by incorporating crop residue extracts (Table 7). Initially, pâtés with sunflower crop extract were significantly darker (L* = 50.92), whereas maize crop extract reduced the a* value and increased the b* and hue values of the pâtés ($p < 0.01$). The pâtés with added BHT were significantly redder (a* = 13.24), with a higher chroma value (C* = 18.26) compared to the control (C* = 17.39) ($p < 0.01$). These differences between formulations persisted after 20 days of storage ($p < 0.01$). However, after two and three months of storage, both formulations with added extracts were significantly darker compared to the control pâtés ($p < 0.01$). At the end of storage, the addition of sunflower crop extract significantly increased the a*, b*, and chroma values and decreased the hue value of the pâtés, while the maize extract significantly increased the proportion of yellow (b*) and decreased the proportion of red color (a*) in the pâtés ($p < 0.01$). Following these findings, Pateiro et al. [9] also showed that the color of tea and grape seed extracts incorporated into the pig liver pâtés led to the different color modifications.

Color modification due to the addition of BHT and crop residue extracts could be regarded as observable, since the total color difference (ΔE) compared to the control pâtés on days 0 and 90 was greater than 2 [72], with the most pronounced changes observed in the SSRE pâtés (5.16 and 8.33, respectively). During storage, color changes in the pâtés were also observed, with the highest values of the total color difference (ΔE*) from 0 to 90 days being found in the control (3.35) pâtés and pâtés with sunflower crop extract added (3.15).

While no changes in a* and b* values were observed in the control pâtés during storage, these pâtés were significantly lighter after two months of storage ($p < 0.01$). On the other hand, the evolution of L* values during storage in formulations with added BHT or crop extracts was somewhat inconsistent. The a* and chroma values significantly increased after 2 and 3 months of storage compared to day 0 in formulations with added crop extracts ($p < 0.01$), while the b* value significantly increased on day 90 of the study (SSRE and MSRE, 14.34 and 15.87, respectively), with a variable trend in this parameter during storage. In contrast to these results, Pateiro et al. [9] demonstrated that, with increasing amounts of natural antioxidants (grape and tea extracts), both the a* and b* values of the pâtés were lower.

Initial differences in the lightness (L*) and yellowness (b*) of the SSRE and MSRE pâtés in comparison with the control pâtés could be attributed to the incorporation of 1% of brownish-yellow components of the sunflower and maize crop extracts and the operational process. The mentioned differences were even more pronounced after three months of storage, and the changes were particularly evident in the pâtés with sunflower crop extract, which also exhibited the lowest oxidative stability. Specifically, previous studies have reported that the oxidation and oxygenation of myoglobin during the preparation and storage of cooked meat products lead to color changes, primarily, an increase in the b* value and the development of a brownish-grey color [5,73]. However, the highest total color change value after the storage of the control pâtés suggests that, in addition to oxidative processes, other factors such as chemical composition, structural modifications of the meat emulsion, and heat treatment could also be responsible for changes in pâté color parameters [73,74]. These results are consistent with previous studies that have investigated the impact of storage and the addition of plant antioxidants on the color and oxidation parameters in various pâté formulations [5,43,70,74].

3.7. TBARs and Peroxide Value (POV)

Significant effects of treatment, storage, and their interaction on lipid oxidation parameters were observed. Therefore, the results, categorized by groups of pâtés and different examination days, are presented in Table 8.

A significantly higher peroxide value was observed in the SSRE and MSRE pâtés compared to the control and BHT pâtés ($p < 0.01$), continuously increasing in all formulations from day 0 to day 60 of storage. On the last day of storage, a decrease in POV was recorded, where the difference between the control and pâtés with maize crop residue extract (1.275 and 1.195 mmol/kg, respectively, $p > 0.05$) was not determined, while the SSRE pâtés had the highest POV (2.515 mmol/kg). Corroborating our results, Pateiro et al. [42] showed that BHT most effectively prevented the formation of primary lipid oxidation products that occur during the elaboration and thermal treatment of the pâtés. In contrast, after 24 weeks of storage, the same authors observed more than a 2.5 times lower peroxide value in pâtés enriched with chestnut and grape seed extracts than that in the control group. Likewise, Van Cuong and Chin [43] found that an increasing concentration of annatto seed powder (from 0.1 to 0.5%) resulted in a reduced peroxide value in pork patties throughout storage. Limited research has been conducted on the analysis of peroxide value in liver pâtés, and its validity as an indicator of lipid oxidation is questionable due to the low stability of peroxides, which readily break down into non-peroxide compounds [43]. According to the Codex Alimentarius, the peroxide value of animal fats intended for consumption should be below 10 mEq of active oxygen/kg [75].

Table 8. Evolution of TBARS and peroxide values of pork liver pâté formulations during refrigerated storage time (mean ± SD).

Days of Storage	Batches	Parameters	
		POV (mmol/kg)	TBARs (MDA mg/kg)
0			
	CON	0.312 ± 0.053 ^{aA}	0.088 ± 0.012 ^{aA}
	BHT	0.018 ± 0.005 ^{bA}	0.015 ± 0.004 ^{bA}
	SSRE	1.442 ± 0.085 ^{cA}	0.088 ± 0.010 ^{aA}
	MSRE	1.665 ± 0.143 ^{dA}	0.045 ± 0.010 ^{cA}
20			
	CON	0.421 ± 0.058 ^{aA}	0.090 ± 0.007 ^{aA}
	BHT	0.183 ± 0.031 ^{bB}	0.088 ± 0.007 ^{aB}
	SSRE	2.355 ± 0.119 ^{cB}	0.195 ± 0.007 ^{bB}
	MSRE	2.345 ± 0.051 ^{cB}	0.145 ± 0.009 ^{cB}
40			
	CON	0.515 ± 0.093 ^{aA}	0.135 ± 0.011 ^{aB}
	BHT	0.474 ± 0.123 ^{aC}	0.125 ± 0.004 ^{bC}
	SSRE	2.660 ± 0.597 ^{bB}	0.225 ± 0.011 ^{cC}
	MSRE	2.500 ± 0.071 ^{bC}	0.195 ± 0.012 ^{dC}
60			
	CON	2.045 ± 0.469 ^{aB}	0.145 ± 0.004 ^{aB}
	BHT	1.175 ± 0.059 ^{bD}	0.130 ± 0.008 ^{aC}
	SSRE	4.335 ± 0.607 ^{cC}	0.275 ± 0.011 ^{bD}
	MSRE	2.645 ± 0.174 ^{dD}	0.205 ± 0.038 ^{cC}
90			
	CON	1.275 ± 0.152 ^{aC}	0.340 ± 0.044 ^{acC}
	BHT	1.142 ± 0.088 ^{bD}	0.195 ± 0.027 ^{bD}
	SSRE	2.515 ± 0.120 ^{cB}	0.450 ± 0.021 ^{cE}
	MSRE	1.195 ± 0.048 ^{aE}	0.345 ± 0.033 ^{aD}
Group	*p* Value	<0.0001	<0.0001
Time	*p* Value	<0.0001	<0.0001
Group × Time	*p* Value	<0.0001	<0.0001

CON—control pâtés; BHT—pâtés with butylated hydroxytoluene added; SSRE—pâtés with sunflower stalk ethanol extract added; and MSRE—pâtés with maize stalk ethanol extract added; ^{a,b,c,d} Different lower-case superscript letters within the column indicate a significant difference between groups on the same day of storage ($p < 0.05$); ^{A,B,C,D,E} Different upper-case superscript letters within the column indicate differences within the same group on different days of storage ($p < 0.05$); n = 3 for each group of pork liver pâtés in triplicate on every day of storage.

During refrigerated storage, an increasing trend in TBARs values was evident across all pâté formulations. In the initial two months, the pâtés elaborated with crop residue extracts exhibited significantly higher TBARs values, with the SSRE group had the highest TBARs (ranging from 0.195 to 0.275 mg MDA/kg). After 90 days of storage, the BHT pâtés showed the lowest TBARs value (0.195 mg MDA/kg), and no significant differences were observed between the control and MSRE pâtés ($p > 0.05$). Meanwhile, the SSRE pâtés maintained the highest value at 0.450 mg MDA/kg.

Based on the in vitro tests of the present study and previously published findings on the same extracts [17], it appears that the ethanolic sunflower and maize residue extracts could exhibit antioxidant and antimicrobial activity. Namely, it was expected that the addition of sunflower and maize stalk residue extracts, considering that these extracts contain moderate TPC levels [41], would reduce lipid oxidation in the pork liver pâtés. However, throughout the storage period, the lowest TBARs value was observed in the pâtés with added BHT, followed by the control pâtés. In other words, the results suggest the opposite—the ethanol extracts used not only failed to provide a sufficient protective effect, but also induced a pro-oxidative effect in the pork liver pâtés. Since there are no available literature data regarding the incorporation of sunflower and maize residue extracts into food matrices, it is challenging to define all factors that could contribute to such an effect.

Many types of plant extracts used as natural antioxidants in pâtés have demonstrated substantial antioxidant activity in both, in vitro tests and in preventing lipid oxidation in these diverse meat products during cold storage [9,42,43,47,64,76]. However, Munekata et al. [44] noted no impact of agro-industrial by-product extracts on TBARs' generation. Moreover, several studies have reported increased lipid oxidation in pork liver pâtés containing extracts rich in phenolic compounds. Specifically, the inclusion of date palm co-product paste in quantities of 2.5% and 7.5% heightened the lipid oxidation in spreadable pork liver pâtés [46]. Chestnut aqueous extract compromised the lipid stability of modified atmosphere-packaged pork patties [47]. Lucas-González et al. [3] observed significantly higher TBARs values in pork liver pâtés enriched with persimmon flour. The pro-oxidative effect of mushroom flours in liver pâtés was also determined by Cerón-Guevara et al. [77]. Furthermore, it has been shown that higher concentrations of rosemary essential oil (600 ppm) lead to increased oxidative deterioration in frankfurters made with meat from intensively reared white pigs [78].

The obtained results can be partially attributed to the likelihood that the antioxidant–pro-oxidant outcomes of crop residue extracts are influenced by the chemical characteristics of the ingredients in the complex emulsified biphasic system of the pork pâté and the overall manufacturing process [46,79]. Indeed, the interaction between extracts' compounds, mainly phenolics, and other active substances present in the pâté matrix, such as sugars, nitric oxide, and Fe^{3+}, could promote lipid peroxidation [3,80]. Furthermore, besides the fact that sunflower and maize stalk residue extracts had a low miscibility in the pâtés' emulsion, the use of a cutter during ingredient comminution and mixture modifies the lipid state and iron availability in the pâtés, thereby also influencing the efficiency of the phenolic compounds in the extracts [44,46]. Additionally, the thermal treatment applied during pâté production may lead to the degradation of phenolic compounds in crop extracts, reducing their concentration and antioxidant potential [76]. Concurrently, heat releases pro-oxidative agents such as iron from myoglobin, which catalyzes lipid oxidation and leads to the formation of MDA from polyunsaturated fatty acids [77].

Our study's results affirm earlier assertions that 0.009% nitrites are effective in preserving low TBARs values [79], as evidenced by all the pâté formulations, including the control, exhibiting TBARs values below the critical threshold of 1 mg MDA/kg, indicating the chemical deterioration of the product [43]. Furthermore, in the present study, as all formulations contained added nitrites, data related to the independent effect of extracts in meat systems are not available. Therefore, the antioxidative effect must be considered through the interaction of nitrites and crop residue extracts. One of the primary mechanisms through which nitrites exert an antioxidative effect in meat products is by binding NO to myoglobin, forming nitrosylmyoglobin, thus preventing the release and oxidation of Fe^{2+} [81]. In a study of the interaction of nitrites with green tea extract in pepperoni, Lin et al. [79] proposed that, given NO's potential roles as a reducing, oxidizing, and nitrosating agent, there is a likelihood of its interaction with reductants, such as our crop residue ethanolic extracts. Moreover, various in vitro studies have documented the capability of plant extracts to scavenge or inhibit the production of peroxynitrite and NO [82,83]. Although there is still insufficient evidence regarding how the combination of plant extracts and nitrites may either reduce or potentially enhance lipid oxidation, in agreement with our results, similar effects have been observed in other meat products, such as Milano salami and emulsion-type pork sausage [84,85]. The interaction between crop residue extract and nitrite could also contribute to alterations in the fatty acid composition of pâtés, particularly the significantly higher levels of n-6 and PUFAs in the SSRE and MSRE pâtés, potentially accelerating the lipid oxidation in these formulations during storage [79].

Finally, aligned with our microbiological analysis results showing significantly higher microbial counts in the control and BHT pâté formulations compared to those enriched with the sunflower and maize crop residue extracts, one might infer that the degradation of lipid oxidation products occurred due to the presence of microbial catalysts, resulting in a reduced TBARs value in the pâtés without extracts added [86].

3.8. Microbiological Analyses

The microbial count was influenced by both the formulation of the pâtés and the storage time, with a significant interaction observed between these factors ($p < 0.01$) (Table 9). During the storage period, analyses revealed that *Enterobacteriaceae* were not detected in any of the pâté samples, confirming the results from previous studies [46,74], where it was suggested that the absence of these bacteria results from effective thermal treatment and good storage conditions, preventing post-processing microbial contamination.

Table 9. Microbial counts (total viable count (TVC), lactic acid bacteria (LAB), and psychrotrophic bacteria) in pork liver pâtés during chilling storage (log CFU/g) (mean ± SD).

Days of Storage	Batches	Microbial Counts TVC	LAB	Psychrotrophic
0				
	CON	1.80 ± 0.19^{aA}	1.53 ± 0.13^{aA}	1.51 ± 0.14^{aA}
	BHT	1.75 ± 0.21^{aA}	1.60 ± 0.11^{aA}	1.40 ± 0.16^{abA}
	SSRE	1.47 ± 0.17^{bA}	1.41 ± 0.13^{bA}	1.30 ± 0.11^{bA}
	MSRE	1.55 ± 0.15^{bA}	1.35 ± 0.11^{bA}	1.36 ± 0.13^{bA}
20				
	CON	2.37 ± 0.15^{aB}	2.34 ± 0.13^{aB}	2.20 ± 0.14^{B}
	BHT	2.44 ± 0.14^{aB}	2.22 ± 0.13^{acB}	2.24 ± 0.12^{B}
	SSRE	2.19 ± 0.09^{bB}	2.01 ± 0.27^{bB}	2.15 ± 0.10^{B}
	MSRE	2.21 ± 0.08^{bB}	2.10 ± 0.28^{bcB}	2.14 ± 0.14^{B}
40				
	CON	3.37 ± 0.10^{aC}	2.96 ± 0.44^{aC}	3.18 ± 0.08^{aC}
	BHT	3.44 ± 0.14^{aC}	2.74 ± 0.32^{abC}	3.10 ± 0.11^{abC}
	SSRE	3.12 ± 0.13^{bC}	2.56 ± 0.33^{bC}	3.03 ± 0.09^{bC}
	MSRE	3.20 ± 0.12^{bC}	2.70 ± 0.37^{abC}	3.03 ± 0.10^{bC}
60				
	CON	4.52 ± 0.14^{aD}	4.22 ± 0.26^{aD}	4.10 ± 0.19^{aD}
	BHT	4.41 ± 0.09^{bD}	3.95 ± 0.24^{bD}	4.00 ± 0.11^{aD}
	SSRE	3.98 ± 0.14^{cD}	3.75 ± 0.33^{bD}	3.80 ± 0.11^{bD}
	MSRE	3.81 ± 0.06^{dD}	3.70 ± 0.32^{bD}	3.83 ± 0.13^{bD}
90				
	CON	5.05 ± 0.16^{aE}	4.71 ± 0.19^{aE}	4.60 ± 0.17^{aE}
	BHT	5.11 ± 0.15^{aE}	4.63 ± 0.22^{aE}	4.49 ± 0.14^{aE}
	SSRE	4.69 ± 0.16^{bE}	4.14 ± 0.17^{bE}	4.00 ± 0.14^{bE}
	MSRE	4.73 ± 0.11^{bE}	4.16 ± 0.24^{bE}	4.07 ± 0.14^{bE}
Group	*p* Value	<0.0001	<0.0001	<0.0001
Time	*p* Value	<0.0001	<0.0001	<0.0001
Group × Time	*p* Value	<0.0001	<0.0001	<0.0001

CON—control pâtés; BHT—pâtés with butylated hydroxytoluene added; SSRE—pâtés with sunflower stalk ethanol extract added; and MSRE—pâtés with maize stalk ethanol extract added; ^{a,b,c,d} Different lower-case superscript letters within the column indicate a significant difference between groups on the same day of storage ($p < 0.05$); ^{A,B,C,D,E} Different upper-case superscript letters within the column indicate differences within the same group on different days of storage ($p < 0.05$); n = 3 for each group of pork liver pâtés in triplicate on every day of storage.

The initial TVC, LAB, and psychrotrophic bacteria counts were < 2 log CFU/g in all formulations, with a significant decrease in the pâtés elaborated with crop residue extracts compared to the control and BHT pâtés ($p < 0.01$). Consistent with findings reported by Van Cuong and Chin [43], the microbial count increased with an increasing storage time. The number of aerobic bacteria and LAB remained similar throughout storage, reaching values above 5 log CFU/g only in the control and BHT pâtés for TVC, while the LAB count was below 5 log CFU/g in all pâté formulations. At the end of the storage period, a slightly lower psychrotrophic bacteria count was found, ranging from 4 to 4.6 log CFU/g for the SSRE and control pâtés, respectively.

Similar findings were observed in the study by Munekata et al. [62], where peanut skin extract did not impact the initial TVC in pâtés. However, over the storage period, formulations with BHT and peanut extracts exhibited lower TVC and LAB counts compared to the control. Additionally, high concentrations of date paste, either alone or combined with annatto extract, have been demonstrated to effectively inhibit microbial growth in pork liver pâtés [46]. Consistent with our results, patties with added BHT, green tea, and grape seed extracts displayed reduced TVC, LAB, and psychotropic aerobic bacterial counts relative to the control [47].

In a prior study, Glišić et al. [17] demonstrated the antimicrobial potential of sunflower and maize stalk residue ethanolic extracts, attributing it to the presence of phenolic bioactive compounds. The minimal inhibitory concentration (MIC) for these extracts was found to be 320 µg/mL for *Salmonella* Typhimurium, *Salmonella* Enteritidis, *Staphylococcus aureus*, *Escherichia coli*, *Listeria monocytogenes*, and *Yersinia enterocolitica*. Similarly, Alexandrino et al. [54] reported the antibacterial activity of the ethanolic extract of sunflower flour against certain Gram-positive and Gram-negative bacteria, primarily ascribing this effect to the presence of chlorogenic acid in the extract. Specifically, chlorogenic acid has the ability to bind to bacterial membranes, causing damage to the membrane integrity, potentially leading to bacterial death [87]. Conversely, some studies have found that various plant extracts, such as seaweed and annatto seed, did not impact the TVC and LAB in pork liver pâtés during storage [43,64]. In contrast, Cerón-Guevara et al. [77] observed that the inclusion of mushroom flours in liver pâtés led to an elevated total microbial count (>5 log CFU/g) due to the contamination of spore-forming bacteria by mushroom flours that survived thermal treatment.

The phenolics identified in the examined extract have scientifically confirmed activity against a wide range of microorganisms, such as Gram-positive and Gram-negative bacteria [54,58–60,87]. Hence, it can be expected that they contributed to the observed activity.

It can be inferred that the microbial growth in all our pâté formulations did not surpass the critical value for this type of product (6 log CFU/g), suggesting the absence of product spoilage after three months of storage [88]. This outcome is likely attributed to the synergistic effects of heat treatment, the incorporation of additives such as sodium chloride and nitrite, and the adherence to hygienic processing conditions and refrigerated storage [70]. Moreover, the microbiological quality of the pork liver pâtés was enhanced by the addition of the sunflower and maize stalk residue extracts.

3.9. Sensory Analyses

In previous studies addressing the incorporation of plant extracts into matrices of diverse types of liver pâté, the primary focus was on monitoring physicochemical characteristics and oxidative stability. The evaluation of product appearance predominantly relied on instrumental color measurement [5,9,42–44,46,47,64]. Given that the success of introducing new ingredients into meat and meat products is substantially dependent on consumer acceptability, gauged by the deviation in main sensory attributes compared to conventional products [89], there are limited studies that have conducted sensory evaluations of pâtés with added plant extracts as natural antioxidants [3,8,62,77,90].

The findings from the sensory analyses of four different pâté formulations are synthesized in Figure 1. Trained panelists noted significant impact on color, odor, aroma, and flavor due to the addition of crop residue extracts ($p < 0.05$). Specifically, regarding color, the SSRE pâtés were assessed as being significantly darker compared to both, the control and BHT pâtés ($p < 0.05$). This observation aligns with objective color measurements, validating that the inclusion of sunflower stalk residue extract, due to the extract's coloration, increased the redness (a*) and reduced the lightness (L*) of the pâtés. Comparable color alterations were documented by other researchers [3,43] when incorporating annatto seed and persimmon powders, underscoring color intensity as the most affected sensory attribute. Given the paramount importance of color in meat products, such deviations can significantly impact the overall acceptability of pâtés among consumers [91].

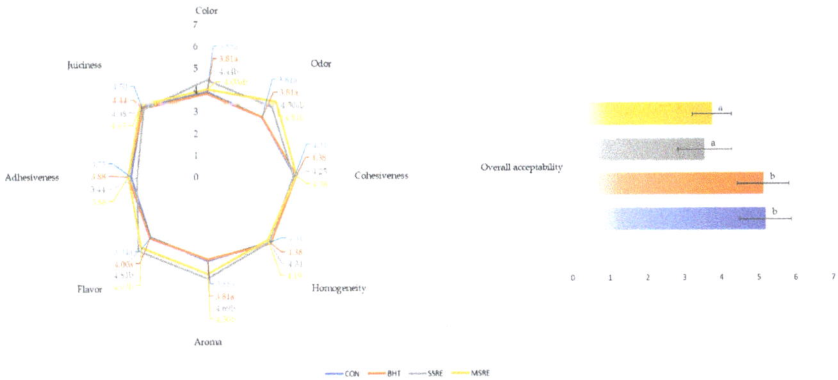

Figure 1. Intensity of the sensory attributes of pork liver pâtés and scores for overall acceptability. CON—control pâtés; BHT—pâtés with butylated hydroxytoluene added; SSRE—pâtés with sunflower stalk ethanol extract added; and MSRE—pâtés with maize stalk ethanol extract added. [a,b] Different lower-case superscript letters for each attribute indicate a significant difference between groups ($p < 0.05$); n = 3 for each group of pork liver pâtés in triplicate.

It was also shown that the formulations with the addition of crop residue extracts had a significantly more intense flavor and aroma ($p < 0.01$), while heightened odor intensity was only observed in the MSRE pâtés ($p < 0.01$). These sensory characteristics scores could be related to the lipid oxidation results and the presence of fat-oxidized metabolites. Namely, the higher POV and TBARs values in the SSRE and MSRE formulations could be responsible for the poor flavor ratings of these pâtés [43,71]. Other authors have also linked lipid oxidation to the presence of off-odor and off-flavor in patties, however, unlike our results, in their studies, natural plant extracts were effective in reducing rancid odor and flavor development [62,90].

The inclusion of crop ethanol extracts had no adverse impact on other sensory attributes, such as the cohesiveness, homogeneity, adhesiveness, and juiciness of the pâtés ($p > 0.05$), indicating the successful incorporation of the residue extract paste into the pâtés' emulsion. Nevertheless, despite these aspects, the observed alterations in color, aroma, and flavor resulted in lower overall acceptance scores for the pâtés with crop residue extracts compared to the control and BHT formulations. Notably, the SSRE pâtés received the lowest ratings, approaching the acceptability threshold at 3.56. Earlier studies have indicated that the incorporation of plant extracts, while impacting color, did not yield significant differences in other sensory attributes. In fact, overall preferences, in some instances, were rated even more favorably than control pâté formulations [3,8]. The challenges arising from the introduction of new ingredients with potential negative effects on the sensory quality of pork liver pâtés can be effectively addressed by employing additives and seasonings. These components play a crucial role in controlling the color and modulating the intense flavor and aroma of pâtés, ultimately enhancing consumer acceptability [43,77].

4. Conclusions

The in vitro tests demonstrated that the ethanolic extracts from sunflower and maize crop residues exhibited antioxidant capacity to some extent; nevertheless, this was significantly lower than synthetic antioxidants. When incorporated into pork liver pâtés, these extracts exhibited a noticeable antimicrobial effect; however, they proved ineffective in preventing lipid oxidation. Specifically, the crop residue extracts displayed a pro-oxidant effect when compared to the use of only nitrites and a nitrites/BHT combination in the pork liver pâtés during storage. The addition of crop residue extracts influenced the instrumental color and intensified the flavor and aroma of the pork liver pâtés, resulting in lower acceptability scores for the product. The work is still in progress to completely elucidate the phenolic profiles of sunflower and maize stalk residue extracts and whether their composition could be influenced by the modification of processing or extraction to increase the amount of antioxidant compounds. To validate the potential of these agricultural residues as a source of natural preservatives for meat product applications, forthcoming studies should explore the mechanisms of interaction between the extract compounds and the components of pork liver pâtés. Additionally, investigations into the structural and functional changes that occur during processing and conservation treatments are essential for understanding the observed pro-oxidative effect and its detrimental impact on sensory characteristics.

Author Contributions: Conceptualization, methodology, formal analysis, project administration, writing and editing—original draft preparation, M.G. and M.B.C., these authors contributed equally to this work and share the first authorship; Formal analysis, writing—original draft preparation, N.Č.; Formal analysis, data curation, M.S.; Methodology, formal analysis, writing—review and editing S.S. and I.V.; Conceptualization, supervision, project administration, funding acquisition, writing—review and editing, Z.M. All authors have read and agreed to the published version of the manuscript.

Funding: The research was supported by the Science Fund of the Republic of Serbia, No. 7752847, "Value-Added Products from Maize, Wheat and Sunflower Waste as Raw Materials for Pharmaceutical and Food Industry–PhAgroWaste" and by the Ministry of Education, Science and Technological Development of the Republic of Serbia (Contract number: 451-03-66/2024-03/200143).

Institutional Review Board Statement: The ethics self-assessment (regarding Humans and Data Protection) evaluated by the Ethics Committee of the Faculty of Veterinary Medicine, University of Belgrade was conducted.

Informed Consent Statement: Informed consent was obtained from all subjects involved in the study.

Data Availability Statement: The original contributions presented in the study are included in the article, further inquiries can be directed to the corresponding author.

Acknowledgments: The authors would like to thank Milan Ž. Baltić for his expertise and valuable support throughout all aspects of our study.

Conflicts of Interest: The authors declare no conflicts of interest.

References

1. Estévez, M.; Ventanas, J.; Cava, R.; Puolanne, E. Characterisation of a traditional Finnish liver sausage and different types of Spanish liver pâtés: A comparative study. *Meat Sci.* **2005**, *71*, 657–669. [CrossRef] [PubMed]
2. Terrasa, A.M.; Staffolo, M.D.; Tomás, M.C. Nutritional improvement and physicochemical evaluation of liver pâté formulations. *LWT* **2016**, *66*, 678–684. [CrossRef]
3. Lucas-González, R.; Pellegrini, M.; Viuda-Martos, M.; Pérez-Álvarez, J.Á.; Fernández-López, J. Persimmon (*Diospyros kaki* Thunb.) coproducts as a new ingredient in pork liver pâté: Influence on quality properties. *Int. J. Food Sci. Technol.* **2019**, *54*, 1232–1239. [CrossRef]
4. Estévez, M.; Cava, R. Lipid and protein oxidation, release of iron from heme molecule and colour deterioration during refrigerated storage of liver pâté. *Meat Sci.* **2004**, *68*, 551–558. [CrossRef] [PubMed]
5. Estévez, M.; Ventanas, S.; Cava, R. Effect of natural and synthetic antioxidants on protein oxidation and colour and texture changes in refrigerated stored porcine liver pâté. *Meat Sci.* **2006**, *74*, 396–403. [CrossRef] [PubMed]
6. Estévez, M.; Ramírez, R.; Ventanas, S.; Cava, R. Sage and rosemary essential oils versus BHT for the inhibition of lipid oxidative reactions in liver pâté. *LWT-Food Sci. Technol.* **2007**, *40*, 58–65. [CrossRef]
7. World Health Organization. *The Global Prevalence of Anaemia in 2011*; World Health Organization: Geneva, Switzerland, 2015. Available online: https://iris.who.int/bitstream/handle/10665/177094/9789241564960_eng.pdf?sequence=1 (accessed on 12 January 2024).
8. Martín-Sánchez, A.M.; Ciro-Gómez, G.; Sayas, E.; Vilella-Esplá, J.; Ben-Abda, J.; Pérez-Álvarez, J.Á. Date palm by-products as a new ingredient for the meat industry: Application to pork liver pâté. *Meat Sci.* **2013**, *93*, 880–887. [CrossRef]

9. Pateiro, M.; Lorenzo, J.M.; Vázquez, J.A.; Franco, D. Oxidation stability of pig liver pâté with increasing levels of natural antioxidants (grape and tea). *Antioxidants* **2015**, *4*, 102–123. [CrossRef]
10. Vijayalaxmi, S.; Jayalakshmi, S.K.; Sreeramulu, K. Polyphenols from different agricultural residues: Extraction, identification and their antioxidant properties. *J. Food Sci. Technol.* **2015**, *52*, 2761–2769. [CrossRef]
11. European Commission, Directorate-General for Research and Innovation. *A Sustainable Bioeconomy for Europe—Strengthening the Connection between Economy, Society and the Environment—Updated Bioeconomy Strategy*; Publications Office of the European Union: Luxembourg, 2018. Available online: https://data.europa.eu/doi/10.2777/792130 (accessed on 12 January 2024).
12. United States Department of Agriculture (USDA); The Foreign Agricultural Service (FAS). World Agricultural Production. 2023. Available online: https://ipad.fas.usda.gov/cropexplorer/cropview/commodityView.aspx?cropid=2224000 (accessed on 28 January 2024).
13. Santana-Méridas, O.; González-Coloma, A.; Sánchez-Vioque, R. Agricultural residues as a source of bioactive natural products. *Phytochem. Rev.* **2012**, *11*, 447–466. [CrossRef]
14. Sadh, P.K.; Duhan, S.; Duhan, J.S. Agro-industrial wastes and their utilization using solid state fermentation: A review. *Bioresour. Bioprocess.* **2018**, *5*, 1. [CrossRef]
15. Faustino, M.; Veiga, M.; Sousa, P.; Costa, E.M.; Silva, S.; Pintado, M. Agro-food byproducts as a new source of natural food additives. *Molecules* **2019**, *24*, 1056. [CrossRef] [PubMed]
16. Leyva-López, N.; Lizárraga-Velázquez, C.E.; Hernández, C.; Sánchez-Gutiérrez, E.Y. Exploitation of agro-industrial waste as potential source of bioactive compounds for aquaculture. *Foods* **2020**, *9*, 843. [CrossRef] [PubMed]
17. Glišić, M.; Bošković Cabrol, M.; Čobanović, N.; Baltić, M.Ž.; Vranešević, J.; Samardžić, S.; Maksimović, Z. Antimicrobial activity of ethanolic extracts from wheat, sunflower and maize crop residues. *Arch. Vet. Sci.* **2023**, *16*, 53–67. [CrossRef]
18. Ahmad, I.; Yanuar, A.; Mulia, K.; Mun'im, A. Application of ionic liquid as a green solvent for polyphenolics content extraction of *Peperomia pellucida* (L) Kunth Herb. *J. Young Pharm.* **2017**, *9*, 486–490. [CrossRef]
19. Chatatikun, M.; Chiabchalard, A. Phytochemical screening and free radical scavenging activities of orange baby carrot and carrot (*Daucus carota* Linn.) root crude extracts. *J. Chem. Pharm. Res.* **2013**, *5*, 97–102.
20. Sembring Novia, E.; Elya, B.; Sauriasari, R. Phytochemical screening, total flavonoid and total phenolic content and antioxidant activity of different parts of *Caesalpinia bonduc* (L.) Roxb. *Pharmacogn. J.* **2018**, *10*, 123–127. [CrossRef]
21. Prieto, J.M. Procedure: Preparation of DPPH Radical, and Antioxidant Scavenging Assay. Dr. Prieto's DPPH Microplate Protocol. 2012, pp. 7–9. Available online: https://www.researchgate.net/file.PostFileLoader.html?id=503cd1c9e39d5ead1100004 3&assetKey=AS:271744332435456@1441800305338 (accessed on 21 February 2024).
22. Xiao, F.; Tao, X.; Baiyi, L.; Ruihai, L. Guidelines for antioxidant assays for food components. *Food Front.* **2020**, *1*, 60–69. [CrossRef]
23. Benzie, I.F.; Devaki, M. The ferric reducing/antioxidant power (FRAP) assay for non-enzymatic antioxidant capacity: Concepts, procedures, limitations and applications. In *Measurement of Antioxidant Activity & Capacity: Recent Trends and Applications*; Apak, R., Capanoglu, E., Shahidi, F., Eds.; Wiley: Hoboken, NJ, USA, 2018; pp. 77–106. [CrossRef]
24. Delgado-Pando, G.; Cofrades, S.; Rodríguez-Salas, L.; Jiménez-Colmenero, F. A healthier oil combination and konjac gel as functional ingredients in low-fat pork liver pâté. *Meat Sci.* **2011**, *88*, 241–248. [CrossRef]
25. Serbian Regulation 53/2018-22. Regulation on the Food Additives. *Official Gazette of the Republic of Serbia*. 2018. Available online: https://www.pravno-informacioni-sistem.rs/SlGlasnikPortal/eli/rep/sgrs/ministarstva/pravilnik/2018/53/1/reg (accessed on 12 January 2024).
26. *ISO 1444:1996*; Meat and Meat Products—Determination of Free Fat Content. International Organization for Standardization: Geneva, Switzerland, 1996.
27. *ISO 1442:1997*; Meat and Meat Products—Determination of Moisture Content (Reference Method). International Organization for Standardization: Geneva, Switzerland, 1997.
28. *ISO 937:1978*; Meat and Meat Products—Determination of Nitrogen Content (Reference Method). International Organization for Standardization: Geneva, Switzerland, 1978.
29. *ISO 936:1998*; Meat and Meat Products—Determination of Total Ash. International Organization for Standardization: Geneva, Switzerland, 1998.
30. *ISO 2918:1975*; Meat and Meat Products—Determination of Nitrite Content (Reference Method). International Organization for Standardization: Geneva, Switzerland, 1975.
31. *ISO 1841-1:1996*; Meat and Meat Products—Determination of Chloride Content—Part 1: Volhard Method. International Organization for Standardization: Geneva, Switzerland, 1996.
32. *ISO 23776:2021*; Meat and Meat Products—Determination of Total Phosphorous Content. International Organization for Standardization: Geneva, Switzerland, 2021.
33. Yudd, D.B.; Wyszecki, G. *Color in Business, Science and Industry*, 3rd ed.; Wiley: New York, NY, USA, 1975.
34. Spiric, A.; Trbovic, D.; Vranic, D.; Djinovic, J.; Petronijevic, R.; Matekalo-Sverak, V. Statistical evaluation of fatty acid profile and cholesterol content in fish (common carp) lipids obtained by different sample preparation procedures. *Anal. Chim. Acta* **2010**, *672*, 66–71. [CrossRef]
35. Glisic, M.; Baltic, M.; Glisic, M.; Trbovic, D.; Jokanovic, M.; Parunovic, N.; Suvajdzic, B.; Boskovic, M.; Vasilev, D. Inulin-based emulsion-filled gel as a fat replacer in prebiotic-and PUFA-enriched dry fermented sausages. *Int. J. Food Sci. Technol.* **2019**, *54*, 787–797. [CrossRef]

36. *ISO 5509:2000*; Animal and Vegetable Fats and Oils—Preparation of Methyl Esters of Fatty Acids. International Organization for Standardization: Geneva, Switzerland, 2000.
37. Fuchs, R.H.B.; Ribeiro, R.P.; Matsushita, M.; Tanamati, A.A.C.; Bona, E.; de Souza, A.H.P. Enhancement of the nutritional status of Nile tilapia (*Oreochromis niloticus*) croquettes by adding flaxseed flour. *LWT* **2013**, *54*, 440–446. [CrossRef]
38. Tarladgis, B.G.; Pearson, A.M.; Dugan, L.R. Chemistry of the 2-thibarbituric acid test for determination oxidative rancidity in foods. II Formation of the TBA malonaldehyde complex without acid-heat treatment. *J. Sci. Food Agric.* **1964**, *15*, 602–607. [CrossRef]
39. Holland, C.D. Determination of malonaldehyde as an index of rancidity in nut meats. *J. AOAC* **1971**, *54*, 1024–1026. [CrossRef]
40. *ISO 3960:2017*; Animal and Vegetable Fats and Oils—Determination of Peroxide Value—Iodometric (Visual) Endpoint Determination. International Organization for Standardization: Geneva, Switzerland, 2017.
41. Tibbetts, S.M.; Milley, J.E.; Lall, S.P. Nutritional quality of some wild and cultivated seaweeds: Nutrient composition, total phenolic content and in vitro digestibility. *J. Appl. Phycol.* **2016**, *28*, 3575–3585. [CrossRef]
42. Pateiro, M.; Lorenzo, J.M.; Amado, I.R.; Franco, D. Effect of addition of green tea, chestnut and grape extract on the shelf-life of pig liver pâté. *Food Chem.* **2014**, *147*, 386–394. [CrossRef] [PubMed]
43. Van Cuong, T.; Chin, K.B. Effects of annatto (*Bixa orellana* L.) seeds powder on physicochemical properties, antioxidant and antimicrobial activities of pork patties during refrigerated storage. *Korean J. Food Sci. Anim. Resour.* **2016**, *36*, 476–486. [CrossRef] [PubMed]
44. Munekata, P.E.S.; Domínguez, R.; Campagnol, P.C.B.; Franco, D.; Trindade, M.A.; Lorenzo, J.M. Effect of natural antioxidants on physicochemical properties and lipid stability of pork liver pâté manufactured with healthy oils during refrigerated storage. *J. Food Sci. Technol.* **2017**, *54*, 4324–4334. [CrossRef]
45. Lucas-González, R.; Pérez-Álvarez, J.Á.; Viuda-Martos, M.; Fernández-López, J. Pork liver pâté enriched with persimmon coproducts: Effect of in vitro gastrointestinal digestion on its fatty acid and polyphenol profile stability. *Nutrients* **2021**, *13*, 1332. [CrossRef]
46. Martín-Sánchez, A.M.; Ciro-Gómez, G.L.; Zapata-Montoya, J.E.; Vilella-Esplá, J.; Pérez-Álvarez, J.A.; Sayas-Barberá, E. Effect of date palm coproducts and annatto extract on lipid oxidation and microbial quality in a pork liver pâté. *J. Food Sci.* **2014**, *79*, M2301–M2307. [CrossRef]
47. Lorenzo, J.M.; Sineiro, J.; Amado, I.R.; Franco, D. Influence of natural extracts on the shelf life of modified atmosphere-packaged pork patties. *Meat Sci.* **2014**, *96*, 526–534. [CrossRef]
48. Kamal, J. Quantification of alkaloids, phenols and flavonoids in sunflower (*Helianthus annuus* L.). *Afr. J. Biotechnol.* **2011**, *10*, 3149–3151.
49. Vazquez-Olivo, G.; López-Martínez, L.X.; Contreras-Angulo, L.; Heredia, J.B. Antioxidant capacity of lignin and phenolic compounds from corn stover. *Waste Biomass Valor.* **2019**, *10*, 95–102. [CrossRef]
50. Gai, F.; Karamać, M.; Janiak, M.A.; Amarowicz, R.; Peiretti, P.G. Sunflower (*Helianthus annuus* L.) plants at various growth stages subjected to extraction-Comparison of the antioxidant activity and phenolic profile. *Antioxidants* **2020**, *9*, 535. [CrossRef] [PubMed]
51. Adekunle, A.E.; Rabeya, T.; Jehadin, F.; Asad, M.A.; Ayodele, O.O.; Islam, M.S. Compositional and microstructural changes in compressed hot water pretreated corn stalk. *Curr. Res. Green. Sustain. Chem.* **2021**, *4*, 100057. [CrossRef]
52. Weisz, G.M.; Kammerer, D.R.; Carle, R. Identification and quantification of phenolic compounds from sunflower (*Helianthus annuus* L.) kernels and shells by HPLC-DAD/ESI-MSn. *Food Chem.* **2009**, *115*, 758–765. [CrossRef]
53. Skałecki, I.; de Espíndola Sobczyk, A.; Marczak, L.D.F.; Sarkis, J. Optimization of ultrasound assisted extraction of phenolic compounds from sunflower seed cake using response surface methodology. *Waste Biomass Valor.* **2019**, *10*, 33–44. [CrossRef]
54. Alexandrino, T.D.; da Silva, M.G.; Ferrari, R.A.; Ruiz, A.L.T.G.; Duarte, R.M.T.; Simabuco, F.M.; Bezzera, R.M.N.; Pacheco, M.T.B. Evaluation of some in vitro bioactivities of sunflower phenolic compounds. *Curr. Res. Food Sci.* **2021**, *4*, 662–669. [CrossRef]
55. Karamać, M.; Kosińska, A.; Estrella, I.; Hernández, T.; Duenas, M. Antioxidant activity of phenolic compounds identified in sunflower seeds. *Eur. Food Res. Technol.* **2012**, *235*, 221–230. [CrossRef]
56. Abbasi-Parizad, P.; De Nisi, P.; Scaglia, B.; Scarafoni, A.; Pilu, S.; Adani, F. Recovery of phenolic compounds from agro-industrial by-products: Evaluating antiradical activities and immunomodulatory properties. *Food Bioprod. Process.* **2021**, *127*, 338–348. [CrossRef]
57. Galeana-López, J.A.; Lizárraga-Velázquez, C.E.; Hernández, C.; Leyva-López, N.; Heredia, J.B. Corn husk phenolics modulate hepatic antioxidant response in Nile Tilapia (*Oreochromis niloticus*) exposed to hypoxia. *Molecules* **2021**, *26*, 6161. [CrossRef]
58. Boz, H. *p*-Coumaric acid in cereals: Presence, antioxidant and antimicrobial effects. *Int. J. Food Sci. Technol.* **2015**, *50*, 2323–2328. [CrossRef]
59. Naveed, M.; Hejazi, V.; Abbas, M.; Kamboh, A.A.; Khan, G.J.; Shumzaid, M.; Ahmad, F.; Babazadeh, D.; FangFang, X.; Modarresi-Ghazani, F.; et al. Chlorogenic acid (CGA): A pharmacological review and call for further research. *Biomed. Pharmacother.* **2018**, *97*, 67–74. [CrossRef]
60. Halpani, C.G.; Mishra, S. Design, synthesis, characterization of ferulic acid and p-coumaric acid amide derivatives as an antibacterial/antioxidant agent. *Pharm. Sci. Adv.* **2024**, *2*, 100023. [CrossRef]

61. Serbian Regulations 94/2015, 104/2015, 19/2017. Regulation on the Quality of Ground Meat, Meat Preparations and Meat Products. Official Gazette of the Republic of Serbia. 2017. Available online: https://www.pravno-informacioni-sistem.rs/SlGlasnikPortal/eli/rep/sgrs/ministarstva/pravilnik/2015/94/2 (accessed on 12 January 2024).
62. Munekata, P.E.S.; Fernandes, R.D.P.P.; de Melo, M.P.; Trindade, M.A.; Lorenzo, J.M. Influence of peanut skin extract on shelf-life of sheep patties. *Asian Pac. J. Trop. Biomed.* **2016**, *6*, 586–596. [CrossRef]
63. Echarte, M.; Conchillo, A.; Ansorena, D.; Astiasarán, I. Evaluation of the nutritional aspects and cholesterol oxidation products of pork liver and fish patés. *Food Chem.* **2004**, *86*, 47–53. [CrossRef]
64. Agregán, R.; Franco, D.; Carballo, J.; Tomasevic, I.; Barba, F.J.; Gómez, B.; Muchenje, V.; Lorenzo, J.M. Shelf life study of healthy pork liver pâté with added seaweed extracts from *Ascophyllum nodosum*, *Fucus vesiculosus* and *Bifurcaria bifurcata*. *Food Res. Int.* **2018**, *112*, 400–411. [CrossRef]
65. Ulbricht, T.L.V.; Southgate, D.A.T. Coronary heart disease: Seven dietary factors. *Lancet* **1991**, *338*, 985–992. [CrossRef]
66. Estévez, M.; Morcuende, D.; Ramírez, R.; Ventanas, J.; Cava, R. Extensively reared Iberian pigs versus intensively reared white pigs for the manufacture of liver pâté. *Meat Sci.* **2004**, *67*, 453–461. [CrossRef]
67. D'arrigo, M.; Hoz, L.; Cambero, I.; Lopez-Bote, C.J.; Pin, C.; Ordóñez, J.A. Production of n-3 fatty acid enriched pork liver pâté. *LWT-Food Sci. Technol.* **2004**, *37*, 585–591. [CrossRef]
68. Food and Agriculture Organization. Fat and fatty acid requirements for adults. In *Fats and Fatty Acids in Human Nutrition—Report of an Expert Consultation*; FAO (Food and Agriculture Organization of the United Nations): Rome, Italy, 2010; pp. 55–62.
69. Vlaicu, P.A.; Panaite, T.D.; Turcu, R.P. Enriching laying hens eggs by feeding diets with different fatty acid composition and antioxidants. *Sci. Rep.* **2021**, *11*, 20707. [CrossRef]
70. Amaral, D.S.; Silva, F.A.P.; Bezerra, T.K.A.; Arcanjo, N.M.O.; Guerra, I.C.D.; Dalmás, P.S.; Madruga, M.S. Effect of storage time and packaging on the quality of lamb pâté prepared with 'variety meat'. *Food Packag. Shelf Life* **2015**, *3*, 39–46. [CrossRef]
71. Lazárková, Z.; Kratochvílová, A.; Salek, R.N.; Polášek, Z.; Šiška, L.; Pětová, M.; Buňka, F. Influence of heat treatment on the chemical, physical, microbiological and sensorial properties of pork liver pâté as affected by fat content. *Foods* **2023**, *12*, 2423. [CrossRef]
72. Francis, F.J.; Clydesdale, F.M. *Food Colorimetry: Theory and Applications*; AVI Publishing Co., Inc.: Westport, CT, USA, 1975.
73. Fernández-López, J.; Pérez-Alvarez, J.A.; Aranda-Catalá, V. Effect of mincing degree on colour properties in pork meat. *Color. Res. Appl.* **2000**, *25*, 376–380. [CrossRef]
74. Fernández-López, J.; Sayas-Barberá, E.; Sendra, E.; Pérez-Alvarez, J.A. Quality characteristics of ostrich liver pâté. *J. Food Sci.* **2004**, *69*, snq85–snq91. [CrossRef]
75. Food and Agriculture Organization of the United Nations; World Health Organization. *Codex-Standard 211-1999. Codex Standard for Named Animal Fats. Section 3. Codex Standard for Fats and Oils from Animal Sources. The Codex Alimentarius. Fats, Oils and Related Products*, 2nd ed.; Food and Agriculture Organization of the United Nations; World Health Organization: Rome, Italy, 2001; Volume 8, pp. 1–7.
76. Munekata, P.E.S.; Calomeni, A.D.V.; Rodrigues, C.E.D.C.; Fávaro-Trindade, C.S.; Alencar, S.M.D.; Trindade, M.A. Peanut skin extract reduces lipid oxidation in cooked chicken patties. *Poult. Sci.* **2015**, *94*, 442–446. [CrossRef]
77. Cerón-Guevara, M.I.; Santos, E.M.; Lorenzo, J.M.; Pateiro, M.; Bermúdez-Piedra, R.; Rodríguez, J.A.; Rosas, J.C.; Rangel-Vargas, E. Partial replacement of fat and salt in liver pâté by addition of Agaricus bisporus and Pleurotus ostreatus flour. *Int. J. Food Sci. Technol.* **2021**, *56*, 6171–6181. [CrossRef]
78. Estévez, M.; Cava, R. Effectiveness of rosemary essential oil as an inhibitor of lipid and protein oxidation: Contradictory effects in different types of frankfurters. *Meat Sci.* **2006**, *72*, 348–355. [CrossRef]
79. Lin, Y.; Huang, M.; Zhou, G.; Zou, Y.; Xu, X. Prooxidant effects of the combination of green tea extract and sodium nitrite for accelerating lipolysis and lipid oxidation in pepperoni during storage. *J. Food Sci.* **2011**, *76*, C694–C700. [CrossRef]
80. Yen, G.C.; Chen, H.Y.; Peng, H.H. Antioxidant and pro-oxidant effects of various tea extracts. *J. Agric. Food Chem.* **1997**, *45*, 30–34. [CrossRef]
81. Pegg, R.B.; Shahidi, F. *Nitrite Curing of Meat, the N-Nitrosamine Problem and Nitrite Alternatives*; Food & Nutrition Press, Inc.: Trumbull, CT, USA, 2000; pp. 23–104.
82. Radi, R.; Denicola, A.; Alvarez, B.; Ferrer-Sueta, G.; Rubbo, H. The biological chemistry of peroxynitrite. In *Nitric Oxide, Biology and Pathobiology*; Ignarro, L., Ed.; Academic Press: San Diego, CA, USA, 2000; pp. 57–82.
83. Tsai, P.J.; Tsai, T.H.; Yu, C.H.; Ho, S.C. Comparison of NO-scavenging and NO-suppressing activities of different herbal teas with those of green tea. *Food Chem.* **2007**, *103*, 181–187. [CrossRef]
84. Ghiretti, G.P.; Zanardi, E.; Novelli, E.; Campanini, G.; Dazzi, G.; Madarena, G.; Chizzolini, R. Comparative evaluation of some antioxidants in salame Milano and mortadella production. *Meat Sci.* **1997**, *47*, 167–176. [CrossRef] [PubMed]
85. Seo, K.Y.; Yun, I.R.; Kim, G.D.; Jung, E.Y.; Joo, S.T.; Yang, H.S. Effect of puerariae radix extracts on pH, color, 2-thiobarbituric acid reactive substances (TBARS) and reduced nitrite content of emulsion-type pork sausage during storage. In Proceedings of the 56th International Congress of Meat Science and Technology (ICoMST), Jeju, Republic of Korea, 15–20 August 2010.
86. Wójciak, K.M.; Dolatowski, Z.J. Oxidative stability of fermented meat products. *Acta Sci. Pol. Technol. Aliment.* **2012**, *11*, 99–109.
87. Lou, Z.; Wang, H.; Zhu, S.; Ma, C.; Wang, Z. Antibacterial activity and mechanism of action of chlorogenic acid. *J. Food Sci.* **2011**, *76*, M398–M403. [CrossRef]

88. Centre for Food Safety. *Microbiological Guidelines for Food (For Ready-to-Eat Food in General and Specific Food Items)*; Food and Environmental Hygiene Department: Hong Kong, China, 2014.
89. Sun-Waterhouse, D.; Wadhwa, S.S. Industry-relevant approaches for minimising the bitterness of bioactive compounds in functional foods: A review. *Food Bioprocess Technol.* **2013**, *6*, 607–627. [CrossRef]
90. Colindres, P.; Susan Brewer, M. Oxidative stability of cooked, frozen, reheated beef patties: Effect of antioxidants. *J. Sci. Food Agric.* **2011**, *91*, 963–968. [CrossRef] [PubMed]
91. Boles, J.A.; Pegg, R. *Meat Color*; Montana State University and Saskatchewan Food Product Innovation Program; University of Saskatchewan: Saskatoon, SK, Canada, 2010.

Disclaimer/Publisher's Note: The statements, opinions and data contained in all publications are solely those of the individual author(s) and contributor(s) and not of MDPI and/or the editor(s). MDPI and/or the editor(s) disclaim responsibility for any injury to people or property resulting from any ideas, methods, instructions or products referred to in the content.

foods

MDPI

Article

Exploring the Impact of Infusion Parameters and In Vitro Digestion on the Phenolic Profile and Antioxidant Capacity of Guayusa (*Ilex guayusa* Loes.) Tea Using Liquid Chromatography, Diode Array Detection, and Electrospray Ionization Tandem Mass Spectrometry

Hasim Kelebek [1], Hatice Kubra Sasmaz [1], Ozge Aksay [1], Serkan Selli [2], Ozan Kahraman [3,*] and Christine Fields [3]

[1] Department of Food Engineering, Faculty of Engineering, Adana Alparslan Turkes Science and Technology University, 01250 Adana, Turkey; hkelebek@atu.edu.tr (H.K.); haticemedine95@hotmail.com (H.K.S.); ozgegurler2@gmail.com (O.A.)

[2] Department of Food Engineering, Faculty of Engineering, University of Cukurova, 01330 Adana, Turkey; sselli@cu.edu.tr

[3] Applied Food Sciences Inc., 675-B Town Creek Road, Kerrville, TX 78028, USA; cfields@appliedfoods.com

* Correspondence: ozan@appliedfoods.com

Abstract: Guayusa tea is derived from the leaves of the *Ilex guayusa* Loes. plant, which is native to the Amazon rainforest. Beyond its pleasant sensory properties, Guayusa tea is rich in antioxidants, phenolics, and minerals. In this study, the effects of infusion time, temperature, and solvent conditions on the color, antioxidant capacity, total phenolic content, phenolic profile, and antimicrobial activity of Guayusa (*Ilex guayusa* Loes.) tea were investigated. Guayusa tea samples were prepared using two different solvents, ethanol and water, with 4, 6, and 8-h infusions at 60 and 70 °C. Liquid chromatography, diode array detection, and electrospray ionization tandem mass spectrometry (LC-DAD-ESI-MS/MS) were used to determine a comprehensive profile of phenolic compounds and to detect differences due to infusion conditions. Moreover, after the Guayusa tea infusion with the highest bioactive properties was determined, the effects of in vitro gastrointestinal digestion on the total phenolic content, antioxidant capacity, and phenolic compounds of the Guayusa tea infusion were measured. Phenolic profile analysis identified 29 compounds, among which chlorogenic acid and its derivatives were predominant. The increase in infusion time was correlated with an elevation in total phenolic content. Significant differences were observed between water and ethanol infusions of Guayusa in terms of phenolics and antioxidants. The total amount of phenolic compounds in the samples prepared with both solvents was found to increase after oral intake, depending on the digestion stage; meanwhile, the amounts of flavonoid compounds and di-O-caffeoylquinic acid derivatives decreased during digestion.

Keywords: extractions; antioxidant capacity; LC–MS/MS; Guayusa tea; in vitro digestion; antimicrobial activity

1. Introduction

Tea, a beverage with a rich cultural history, has garnered attention for its diverse flavors and potential health benefits attributed to bioactive compounds. Guayusa (*Ilex guayusa* Loes.), a traditional Amazonian tea, has recently gained interest for its unique flavor profile and purported health-promoting properties. Guayusa is an evergreen South American tree that grows between southern Colombia and northern Peru, especially in the upper Amazon basin of Ecuador. It has been used since ancient times for medical benefits such as pain relief and preventing adverse central nervous system effects, especially by Amazonian indigenous tribes [1–3]. Commercially, this interest is expected to drive the global herbal tea market to reach a Compound Annual Growth Rate (CAGR) of 4.5%, as estimated for 2018–2024 [4]. Noteworthy for its secure use in traditional foods, Guayusa tea has been designated Generally Recognized as Safe (GRAS) by the United States Food and Drug Administration (GRN 869, 870, 883) [5].

Scientific findings have shown that Guayusa leaf extracts have antioxidant, anti-glycemic, antifungal, antibacterial, and anti-inflammatory effects [2,6]. Studies have reported that caffeine, theobromine, and theophylline concentrations are present in leaf alcoholic extracts, and caffeine and chlorogenic acids were found to be the major compounds [7,8]. The phenolic and carotenoid contents of Guayusa teas were analyzed, and 14 phenolic and 5 carotenoid compounds were identified. It was found that chlorogenic acid and its derivatives from phenolic acids and quercetin-3-hexose, as well as derivatives from flavonoids, were dominant. The researchers also reported that Guayusa leaves were rich in α-carotene, luteolin, and violaxanthin + neoxanthin [2].

A critical factor influencing the bioactive composition of tea is the infusion process, where various parameters such as temperature, time, and solvent can significantly impact the extraction of phytochemicals. The properties of most phenolic substances vary from polar to non-polar in nature; many extraction factors, especially the solvent system, are very challenging for extraction efficiency [3,9]. Therefore, understanding the phenolic profile of Guayusa tea and its alterations under different infusion conditions (time, temperature, type of solvent, etc.) is important for optimizing its health-promoting attributes. Phenolic compounds are also affected by the digestion process and can be released under the action of enzymes, and studies have reported that some flavonoid compounds are released during in vitro digestion. Phenolic compounds have been found to be mainly released in the digestive stages of the stomach and small intestine. Digestion processes are an important factor affecting the biological activity of phenolic compounds in the body [10]. Moreover, the often-neglected aspect of how these compounds transform the digestive process holds pivotal significance, as it dictates their bioavailability and consequential physiological impacts. However, no comprehensive study has investigated the effects of infusion conditions on the phenolic compositions of Guayusa teas.

This study aims to investigate both water and alcohol-based extractions to provide a comprehensive characterization of Guayusa tea leaves. While water-based extracts are common, we aimed to identify phenolic compounds that are insoluble in water but soluble in ethyl alcohol. This study aims to delve into the effects of the intricate interplay between infusion parameters and in vitro digestion on the phenolic profile and antioxidant capacity (2,2-Diphenyl-1-picrylhydrazyl (DPPH) assay, and 2,2′-azino-bis (3-ethylbenzothiazoline-6-sulfonic acid) (ABTS) assay) of Guayusa tea. In the scope of this study, the Guayusa tea infusion with the highest bioactive properties was identified, and the effects of infusion conditions on total phenolic content, antioxidant capacity, and phenolic compounds during in vitro gastrointestinal digestion were evaluated.

2. Materials and Methods

2.1. Standards and Chemicals

The HPLC standards for chlorogenic acid, caffeine, neochlorogenic acid, quercetin, kaempferol, 3-o-feruloylquinic acid, and kaempferol 3-glucoside were purchased from Sigma Co (St. Louis, MO, USA). Acetonitrile, formic acid, 2,2′-azino-bis-(3-ethyl-benzothiazoline-6-sulfonic acid) diammonium salt (ABTS), Trolox ((±)-6-Hydroxy-2,5,7,8-tetramethylchromane-2-carboxylic acid), gallic acid, sodium carbonate, potassium persulfate, and 2,2-diphenyl-1-picryl hydrazyl (DPPH) were obtained from Merck (Gernsheim, Germany). α-amylase (EC 3.2.1.1), pepsin (EC 3.4.23.1), pancreatin (EC 232.468.9) and bile salts, potassium chloride, sodium bicarbonate, potassium dihydrogen phosphate, sodium chloride, magnesium chloride, ammonium carbonate, and calcium chloride were purchased from Sigma Co. (St. Louis, MO, USA). The standards were prepared daily, and all chemicals used were analytical grade or higher.

2.2. Sample Preparations

Guayusa tea infusions were prepared using a water bath (Labotec, Johannesburg, South Africa) with different solvents at varying temperatures and durations. This study involved weighing 3 g of the sample and adding 30 mL of water, which was then infused at 60 °C for 4, 6, and 8 h. The same infusion time intervals were also repeated at 70 °C. Additionally, samples were infused with ethanol (70%) using the same time intervals and temperatures. After each infusion period, the tea was rapidly cooled to room temperature, centrifuged at 6500 rpm for 15 min at 4 °C, and filtered through filter paper (Whatman #1). Each infusion was then further filtered through a 0.20 µm PTFE syringe filter (26 mm, 0.20 µm, Phenomenex, Bologna, Italy) and used for analyses [11,12].

2.3. Color Analysis

The color of tea infusions was measured using a Konica Minolta CM-5 (Konica Minolta Optics Inc. Osaka, Japan) colorimeter. The results were presented using five color parameters (L*, a*, b*, C, and h). The L* indicates brightness (whiteness/lightness/darkness), a* represents redness (positive values) or greenness (negative values), b* indicates yellowness (positive values) or blueness (negative values), C denotes the chroma value, and h° signifies the hue angle value [13].

2.4. Analysis of Phenolic Compounds by LC-DAD-ESI-MS/MS

Phenolic compound analysis was conducted following the procedure outlined by Kelebek et al. [14] utilizing LC-DAD-ESI-MS/MS in the negative ionization mode. Tea infusion samples were passed through a 0.45 µm membrane filter and directly injected into a HPLC system (Agilent 1100, Agilent Technologies, Palo Alto, CA, USA) connected to a Windows NT 4.0-based ChemStation software system. The LC–MS system employed in this study included a binary pump, degasser, and autosampler. A Phenomenex reversed-phase C-18 column (4.6 mm × 250 mm, 5 µm) (Torrance, CA, USA) was employed, along with a diode array detector (DAD). Two distinct solvents were used: water/formic acid (99:1; *v*/*v*) as Solvent A, and acetonitrile/solvent A (60:40; *v*/*v*) as Solvent B. Phenolic compounds were eluted under the following conditions: 0.5 mL/min flow rate at 25 °C; isocratic conditions from 0 to 5 min with 0% B; gradient conditions for all the following steps: from 0% to 5% B in 20 min, from 5% to 15% B in 18 min, from 15% to 25% B in 14 min, from 25 to 50% B in 31 min, and from 50 to 100% B in 3 min; and washing and reconditioning of the column. Curves were generated using commercial standards, with extracts ranging from approximately 1 to 200 mg/L concentrations and R^2 values exceeding 0.99. When a reference compound was unavailable, analogous elements were calibrated alongside the molecular weight correction factor. Limits of quantification (LOQ) and detection (LOD) were calculated based on signal-to-noise ratios of 10 and 3, respectively [14].

2.5. Antioxidant Capacity and Total Phenolic Content Analysis

The antioxidant capacity was assessed using two distinct methods: 2,2-diphenyl-1-picrylhydrazyl (DPPH) and 2,2′-azino-bis (3-ethylbenzothiazoline-6-sulphonic acid (ABTS) assays. The samples' DPPH and ABTS-scavenging activities were evaluated according to the procedures outlined in the work completed by Kelebek et al. [15]. In both antioxidant capacity analyses (DPPH and ABTS), a Trolox standard solution was used at various concentrations to obtain the standard curve (3.125–200 µmol). The results were expressed as mM TE (Trolox equivalent)/L.

The total phenolic content (TPC) of tea infusion samples was analyzed via the Folin–Ciocalteu method [16]. A total of 200 µL of extract/standard solution and 1.5 mL of Folin–Ciocalteu reagent (1:10 ratio) were added to the spectrophotometer cuvette. Five minutes later, 1.5 mL of 6% sodium carbonate solution was added, and the mixture was kept in the dark at room temperature for 90 min. The absorbance values were then measured at 765 nm using an UV-Vis spectrophotometer (BMG Labtech, Spectrostar Nano, Ortenberg, Germany). Gallic acid was used as a standard. The total phenolic content of the sample was calculated using a calibration curve prepared with the standard, and the results were expressed as mg GAE (gallic acid equivalent)/L.

2.6. Antimicrobial Activity Test

Staphylococcus aureus ATCC 29213, *Bacillus subtilis* ATCC 11774, *Klebsiella pneumoniae* ATCC 13883, and *Escherichia coli* ATCC 25922 were used as test organisms. These microorganisms were spread-plated on tryptic soy agar and incubated at 35 °C for 12–18 h. After incubation, the turbidity of the bacterial suspension was adjusted to 108 CFU/mL (0.5 McFarland unit) with a sterile saline solution. The antimicrobial activities of extracts on the selected test organisms were determined with the agar well diffusion technique [17]. Muller–Hinton agar plates were prepared by spreading 100 µL of the inoculum over the entire agar surface. Then, a 6 to 8 mm diameter well was punched aseptically with a sterile cork borer, and 30 to 50 µL of extract was added to the wells. Agar plates were incubated under suitable conditions (35 °C for 18–24 h), depending on the test microorganism. Each experiment was carried out in duplicate. After incubation, inhibition zones were measured three times, and the average was calculated.

2.7. In Vitro Digestion

The in vitro digestion model of Infogest, which includes the sequential simulation of oral, gastric, and intestinal digestion, was performed according to the method described in the study of Brodkorb et al. [18]. The total phenolic content, antioxidant capacity (DPPH and ABTS), and total phenolic substance (LC-DAD-ESI-MS/MS) samples collected from the oral, gastric, and intestinal phases were determined with three replicates.

2.8. Statistical Analysis

The results obtained were analyzed with a confidence level of 95% ($p < 0.05$) using the SPSS software program (version 24, SPSS Inc., Chicago, IL, USA), after evaluations for normality and variance homogeneity. Duncan's multiple comparison test was used to determine whether or not there were significant differences. In addition, a heatmap was used in the XLSTAT programme (trial version, 2023).

3. Results

3.1. Color Characteristics of Infusions

Color is one of the most powerful aspects influencing the consumer acceptability of herbal teas [19,20]. The impact of varying time, temperature, and solvent conditions on the color of Guayusa tea infusions was determined. The results of the color analysis are provided in Table 1. The color parameters (L^*, a^*, b^*, C, and h°) exhibited significant variations based on infusion times and temperature ($p < 0.05$). It was determined that the L^* value gradually decreased with increasing infusion time, and the color became darker. The L^* value was low in infusions prepared with water, resulting in a darker color than the ethanolic extracts (Table 1). As the infusion time increased, there was a greater transfer of substances into the water, resulting in a decrease in the L^* value, which in turn could have contributed to the darkening of the color. It was also found that phenolic compounds were higher in water-phase extracts with a lower L^* value, i.e., a darker color. Therefore, the increasing amount of water-soluble phenolic compounds caused the extract to be darker in color. No study has been conducted in the existing literature to explore the infusion of Guayusa teas employing diverse solvents, altering infusion durations, and varying temperature conditions. Consequently, a comparison with the literature data was not attainable. Previous studies have determined that the L^* value decreases as the infusion time increases in black tea infusions, leading to a darker color [13,21]. Another study reported that the L^* value decreased with increasing infusion time and temperature in black tea samples, and the color became darker with longer infusion times [12]. Moreover, Uzlasir et al. [11] found that the L^* value decreased with infusion time while exploring the phenolic compositions, antioxidant properties, and colour characteristics of elderberry flowers' methanol, ethanol, and aqueous extracts.

Table 1. The effect of infusion time and temperature on the colour of tea infusions.

	L*	a*	b*	C	h°
Gw-60 °C-4 h	23.43 ± 0.01 [a]	36.17 ± 0.01 [a]	40.28 ± 0.02 [a]	54.13 ± 0.02 [a]	48.08 ± 0.01 [a]
Gw-60 °C-6 h	23.36 ± 0.01 [a]	36.22 ± 0.01 [a]	40.17 ± 0.05 [a]	54.09 ± 0.04 [a]	47.97 ± 0.04 [b]
Gw-60 °C-8 h	23.19 ± 0.10 [b]	36.09 ± 0.04 [a]	39.87 ± 0.14 [b]	53.78 ± 0.13 [b]	47.85 ± 0.07 [c]
Gw-70 °C-4 h	22.89 ± 0.10 [c]	35.94 ± 0.11 [b]	39.32 ± 0.15 [c]	53.27 ± 0.18 [c]	47.58 ± 0.04 [d]
Gw-70 °C-6 h	22.52 ± 0.15 [d]	35.80 ± 0.13 [b]	38.69 ± 0.23 [d]	52.71 ± 0.25 [d]	47.22 ± 0.07 [e]
Gw-70 °C-8 h	22.09 ± 0.02 [e]	35.86 ± 0.02 [b]	37.98 ± 0.06 [e]	52.23 ± 0.05 [e]	46.65 ± 0.04 [f]
Get-60 °C-4 h	34.57 ± 0.01 [a]	41.07 ± 0.01 [c]	59.44 ± 0.04 [a]	72.25 ± 0.03 [a]	55.36 ± 0.02 [a]
Get-60 °C-6 h	33.26 ± 0.04 [b]	41.53 ± 0.04 [b]	57.29 ± 0.09 [b]	70.75 ± 0.09 [b]	54.06 ± 0.03 [b]
Get-60 °C-8 h	32.62 ± 0.03 [c]	41.52 ± 0.04 [b]	56.20 ± 0.13 [c]	69.87 ± 0.13 [c]	53.54 ± 0.04 [c]
Get-70 °C-4 h	32.25 ± 0.04 [d]	41.47 ± 0.03 [b]	55.55 ± 0.06 [d]	69.32 ± 0.07 [d]	53.26 ± 0.01 [e]
Get-70 °C-6 h	31.97 ± 0.15 [e]	40.89 ± 0.13 [d]	55.06 ± 0.29 [e]	68.59 ± 0.31 [e]	53.40 ± 0.06 [d]
Get-70 °C-8 h	31.14 ± 0.01 [f]	41.69 ± 0.02 [a]	53.67 ± 0.01 [f]	67.97 ± 0.02 [f]	52.16 ± 0.02 [f]

[a–f] Different letters in the columns represent statistically significant differences ($p < 0.05$). Gw: Guayusa water extraction; Get: Guayusa ethanol extraction.

3.2. Antioxidant Capacity and Total Phenolic Content Analysis

Herbal infusions have been taken as beverages for generations due to their high antioxidant capacity. These infusions have a lot of promise as important natural sources of antioxidants, with the ability to reduce diseases caused by oxidative stress [22].

Assays utilizing the ABTS radical cation are widely employed in assessing antioxidant capacity, alongside the commonly used DPPH assay. The ABTS method involves the spectrophotometric measurement of changes in the ABTS cation radical concentration resulting from its reaction with antioxidants. Meanwhile, the DPPH assay offers a rapid, straightforward, and cost-effective means of evaluating antioxidant activity. It relies on the reduction of DPPH, a stable free radical compound [23]. The antioxidant capacity of the infusions was evaluated using the DPPH and ABTS methods, and the results are presented in Table 2.

It was found that the antioxidant capacity of Guayusa tea infusion samples increased directly with increasing infusion time and infusion temperature ($p < 0.05$). Among the Guayusa tea infusion samples, the infusion prepared with water exhibited the highest DPPH and ABTS antioxidant capacities. The Guayusa tea infusion sample prepared with water at 70 °C for 8 h (Gw-70 °C-8 h) exhibited the highest antioxidant capacity, with DPPH and ABTS values of 86.12 mM TE/L and 88.19 mM TE/L, respectively. Compared to ethyl alcohol infusions, water infusions showed better antioxidant potential across all time and temperature ranges. In both solvents, increasing the temperature and duration increased the antioxidant potential.

In a study by Pardau et al. [24], antioxidant capacity values for Guayusa, determined by the ORAC method, ranged between 798 and 2019 μmol TE/g. The researchers concluded that Guayusa is a good source of phenolic compounds with antioxidant properties. In another study, Garca-Ruiz et al. [2] assessed the antioxidant capacity of Guayusa green leaves using DPPH and ORAC techniques. The antioxidant capacity of Guayusa green leaves was 32.98 mmol Trolox 100 g/DW according to the DPPH assay and 154.03 mmol Trolox 100 g/DW according to the ORAC test. These findings suggest that Guayusa tea leaves have a high antioxidant capacity. On the other hand, by evaluating the antioxidant capacities of black tea derived from the Camellia sinensis plant under various brewing conditions, the DPPH antioxidant capacity was found to range from 1505.20 to 2454.17 mmol TE/L, while the ABTS analysis yielded values between 1965.14 and 3214.96 mmol TE/L [12]. Comparing our findings from Guayusa tea infusions to those of black tea showed that black tea exhibited a higher antioxidant capacity.

The total phenolic contents (TPCs) of Guayusa tea infusion samples are given in Table 2. Significant differences depended on temperature, infusion time, and solvents ($p < 0.05$).

The highest TPC was observed in the Gw-70 °C-8 h (19,467.58 mg GAE/L) sample. The results showed that the TPCs of Guayusa tea infusion samples increased depending on infusion time and temperature. High extraction temperatures increased the permeability of cell walls to solvents and components, thereby increasing extraction efficiency. The solubility of tea components increased with infusion time and temperature [13]. In a study investigating infusion-dependent changes in the phenolic, antioxidant, and color properties of St. John's wort (*Hypericum perforatum* L.) teas, findings indicated a substantial increase in both phenolic compound levels and antioxidant activity under extended infusion periods across three distinct tea extractions [25]. The elevation in the solubility of tea components could have contributed to the augmentation of both antioxidant capacity and TPC.

Table 2. The effect of infusion time and temperature on the antioxidant capacity of infusions.

	DPPH (mM TE/L)	ABTS (mM TE/L)	TPC (mg GAE/L)
Gw-60 °C-4 h	70.40 ± 0.91 [e]	80.65 ± 3.16 [b]	16,152.51 ± 248.88 [e]
Gw-60 °C-6 h	71.49 ± 1.28 [d]	77.38 ± 2.13 [e]	16,692.24 ± 208.04 [d]
Gw-60 °C-8 h	73.09 ± 1.01 [c]	78.45 ± 0.80 [c]	18,064.84 ± 33.03 [b]
Gw-70 °C-4 h	69.53 ± 0.51 [f]	74.26 ± 1.29 [f]	16,116.89 ± 63.98 [f]
Gw-70 °C-6 h	74.41 ± 1.28 [b]	77.73 ± 0.85 [d]	17,310.50 ± 63.70 [c]
Gw-70 °C-8 h	86.12 ± 1.24 [a]	88.19 ± 1.46 [a]	19,467.58 ± 38.49 [a]
Get-60 °C-4 h	54.83 ± 3.00 [f]	71.51 ± 2.86 [f]	14,757.08 ± 37.40 [f]
Get-60 °C-6 h	65.30 ± 1.57 [e]	73.74 ± 2.70 [e]	15,604.11 ± 63.93 [e]
Get-60 °C-8 h	76.96 ± 1.17 [c]	74.62 ± 1.65 [d]	16,706.85 ± 42.07 [c]
Get-70 °C-4 h	72.34 ± 2.91 [d]	75.55 ± 0.29 [c]	16,340.64 ± 113.30 [d]
Get-70 °C-6 h	78.08 ± 2.09 [b]	77.24 ± 0.06 [b]	17,985.39 ± 63.87 [b]
Get-70 °C-8 h	79.21 ± 1.26 [a]	83.09 ± 2.86 [a]	18,852.05 ± 27.89 [a]

[a–f] Different letters in the columns represent statistically significant differences ($p < 0.05$). Gw: Guayusa water infusions; Get: Guayusa ethanol infusions.

Correlation analysis was performed to determine the relationships between antioxidant values (DPPH, ABTS), TPC, and phenolic compounds in samples after different infusion and extraction conditions (Figures 1 and 2). Figure 1 shows a high and positive correlation was found between Gw (r = 0.92 for TPC and DPPH; r = 0.78 for TPC and ABTS) and Get (r = 0.89 for TPC and DPPH; r = 0.93 for TPC and ABTS) infusions, as supported by the research findings (Figures 1 and 2).

Correlation analysis to determine the relationships between antioxidant activity, TPC, and phenolic compounds in Guayusa tea samples is not available in the literature. Rodriguez Vaquero et al. [26] found high correlations between the TPC and DPPH capacities of tea infusions. Furthermore, in a study evaluating the relationship between TPC and the antioxidant capacity of boiled brew and tea plant infusions via the Pearson correlation matrix, TPC was reported to be positively and significantly correlated with DPPH capacity for all infusion times. It was reported that the strongest positive correlation was observed between TPC and DPPH capacity in mint and linden leaf infusions [26,27].

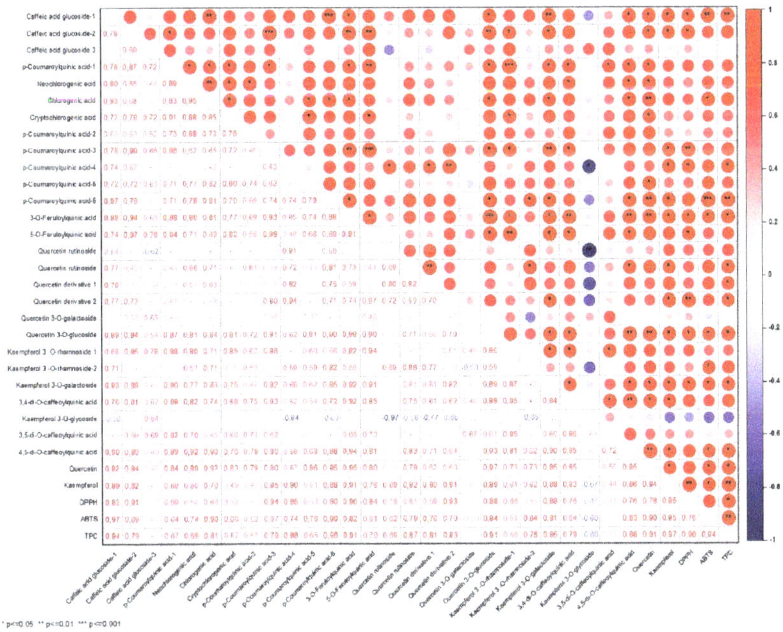

Figure 1. The correlation matrix of the antioxidant activity and phenolic profile of the Guayusa ethanol-water (Get) infusions.

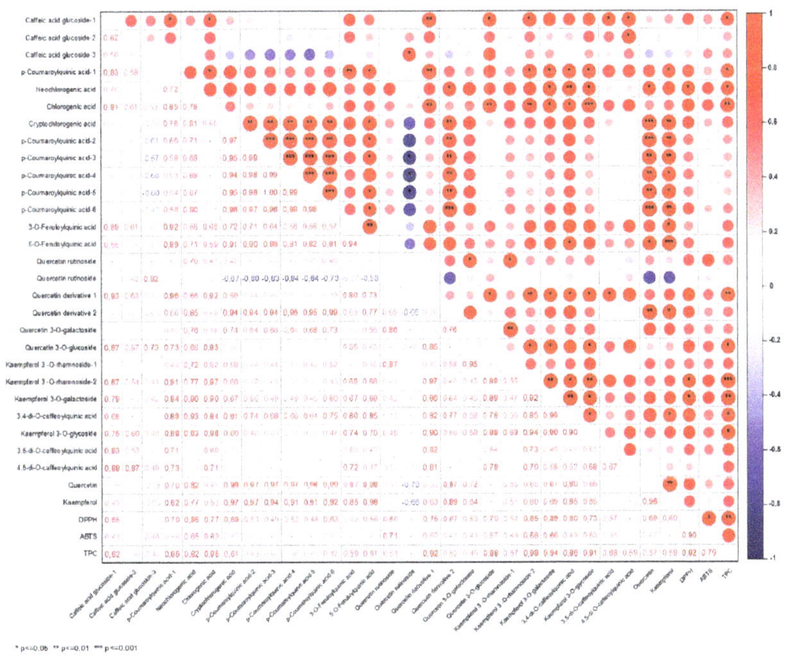

Figure 2. The correlation matrix of the antioxidant activity and phenolic profile of the Guayusa water (Gw) infusions.

3.3. Phenolic Profile of Guayusa Infusions

Data representing the retention time, λmax in the UV region, molecular ion, main fragment ions in MS^2, and tentative compound identification obtained by HPLC-DAD–ESI-MS/MS analyses are presented in Table 3. LC–ESI-MS/MS multiple reaction monitoring (MRM) chromatograms of some of the identified phenolic compounds in Guayusa tea infusions are given in Figures 3–6. A total of 29 phenolic compounds were identified and quantified. Similar phenolic profiles were observed in infusions at both temperatures and times, but the amounts of phenolic compounds showed significant increases with increasing time and temperature ($p < 0.05$). An increase in infusion time also led to an increase in total phenolic content. Another important piece of data that draws attention in this study shows that the amount of phenolic compounds is significantly higher in infusions made with water than in infusions made with ethyl alcohol ($p < 0.05$). When phenolic profiles were evaluated, chlorogenic acid and its derivatives (CGAs) were dominant in all infusions. CGAs are esters of hydroxycinnamic acids (HCAs) such as caffeic acid (CFA), ferulic acid (FA), p-coumaric acid (p-CoA), and sinapic acid (SA) to quinic acid (QA or 1L-1(OH),3,4/5-tetrahydroxycyclohexane carboxylic acid). These complex compounds exhibit a wide range of physicochemical properties due to positional esterification on the quinic acid moiety, forming regio-derivatives [28].

Caffeic acid derivatives: Three caffeic acid derivatives were identified at $m/z = 341$ (Peaks 1–3). These molecules, identified as caffeic acid hexosides, produced the same fragmentation ions corresponding to hexose moiety loss (162 Da) but with slightly varied abundances. The same $[M\text{-}H]^-$ at $m/z = 341$ is also formed by caffeoyl hexoses, in which caffeic acid is coupled to the sugar moiety by an ester bond rather than an ether bond; however, fragments indicative of sugar moiety fragmentation are seen. The fourth chemical with a $m/z = 341$ fragment ion was recognized as a caffeic acid O-glucoside derivative. Caffeic acid glucoside-1 was identified as the predominant compound within this group. Its concentration ranged from 115.30 to 127.78 mg/L in water-based infusions and 117.10 to 140.14 mg/L in ethyl-alcohol-based infusions.

Caffeoylquinic acid derivatives: Guayusa infusions include three positional derivatives of caffeoylquinic acids. Peaks 6–8 in the ESI-MS/MS in negative ion mode produced the same $[M\text{-}H]^-$ ion at m/z 353 as the chemical formula $C_{16}H_{18}O_9$ predicted. In MS/MS, the molecular ions $[M\text{-}H]^-$ produced four peaks at m/z 191, 179, 173, and 135. Peaks 6, 7, and 8 had the structures of neochlorogenic acid, cryptochlorogenic acid, and chlorogenic acid, respectively. Caffeic acid and quinic acid have been esterified to generate chlorogenic acids. In ESI MS in negative ion mode, the diagnostic fragmentation patterns of chlorogenic acid derivatives included the cleavage of intact caffeoyl and quinic acid fragments. Chlorogenic acid was determined to be the dominant compound in both infusion conditions and varied between 6428.28 and 6557.41 mg/L in water-based infusions and 4484.53 and 5145.29 mg/L in ethyl-alcohol-based infusions. Neochlorogenic acid was the other dominant compound after chlorogenic acid. These two compounds constitute a significant part of the total amount of phenolic compounds. García-Ruiz et al. [2] reported that chlorogenic and neochlorogenic acids were also dominant compounds in Guayusa teas.

One study reported that the chlorogenic acid content of a cup of coffee (200 mL) varied between 20 and 675 mg of chlorogenic acid in the coffee content, but these values varied according to the coffee type and brewing method [29]. Moreover, in another study to determine the chlorogenic acid content of green coffee infusions, it was reported that the amount of chlorogenic acid varied between 628 and 1040 mg/L in C. arabica infusions and between 682 and 1210 mg/L in C. canephora infusions [30]. The data we obtained on the chlorogenic acid amounts of Guayusa tea in water-based and ethanol infusions were higher than in coffee samples, compared to previous studies.

Coumaroylquinic acid derivatives: The hierarchical scheme keys for the $LC\text{-}MS^n$ identification of CGAs were used to identify the metabolite with a molecular ion $[M\text{-}H]^-$ at m/z 337, identified as 5-coumaroylquinic acid, because it formed a fragment ion at m/z 191 [QA-H]- indicating loss of a coumaroyl moiety (Peak 4, Peaks 8–12) (Table 3).

Six compounds were identified in the structure of the coumaroylquinic acid isomer, with p-Coumaroylquinic acid-2 identified as the dominant one. These compounds increased with rising time and temperature, reaching higher concentrations in ethyl-alcohol-based infusions than in water infusions (Tables 4 and 5).

Feruloylquinic acid derivatives: The same approach was used to identify two feruloylquinic acid (FQA) derivatives (Peaks 13 and 14), which were recognized by their precursor ion $[M-H]^-$ at m/z 367 and based on the fragmentation patterns and Rt given in Table 3. Despite their differing strengths, the two FQA regio-derivatives were discovered in Guayusa infusions. The base peaks at m/z 193 [FA-H]- and m/z 173 [QA-H-H2O]- were employed as diagnostic peaks for 3-FQA and 5-FQA, respectively, as specified in the hierarchical scheme keys for the LC-MS^n identification of CGAs. FQA yields m/z 134 $[FA-H-CO_2-CH_3]^-$ as well. As a result, Molecules (13) and (14) were labeled as 3-FQA acid and 5-FQA, respectively. The amounts of the 3-FQA compound ranged between 70.69 and 119.38 mg/L, and that of the 5-FQA compound ranged between 72.51 and 130.08 mg/L in water-based infusions. In ethyl-alcohol-based infusions of these compounds, the amounts were found between 85.88 and 129.19 mg/L and between 103.91 and 165.17 mg/L, respectively. Ethyl alcohol infusions provided higher solubility.

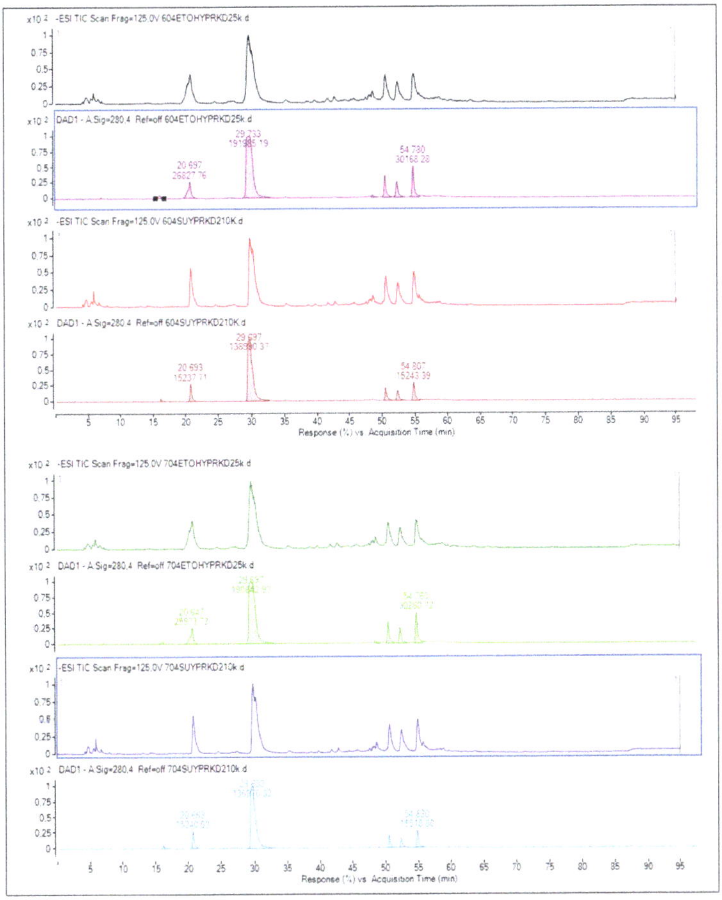

Figure 3. Chromatogram of phenolic compounds identified by LC-DAD and LC–ESI-MS/MS in Get infusions.

Dicaffeoylquinic acids: Three di-CQA compounds were detected in the prepared tea infusions. These molecules exhibit spectrum features with UV maxima at 242.6 and 327.0 nm and retention times (Rt) of 50.53, 52.37, and 54.85 min. The ESI-MS/MS spectra revealed fragment ions $[M-H]^-$ at m/z 515, $[M-C_9H_6O_3]$ at m/z 353, and $[M-H-2C_9H_6O_3]$ at m/z 191 (Table 3). The identification of these compounds, which exhibited identical spectral data, was confirmed using standard substances. 3,4-di-O-caffeoylquinic acid, 3,5-di-O-caffeoylquinic acid, and 4,5-di-O-caffeoylquinic acid were found to be the dominant compounds in the study, and it was determined that the infusions prepared with ethyl alcohol were richer in terms of these compounds.

Flavonoids: Twelve flavonols were detected in tea infusions, with quercetin derivatives found as the predominant flavonols in all samples. Seven peaks were identified as quercetin derivatives based on their UV spectra and MS fragmentation, ultimately producing the quercetin aglycone at m/z 301 in negative mode (Table 3). There are some peaks with the same $[M-H]^-$, as shown in Table 3. Peaks 19 and 20 both exhibited $[M-H]^-$ at m/z 463, resulting in a fragment at m/z 301 (hexose moiety loss). The MS^2 spectra of m/z 301 yielded quercetin-like ions at m/z 179 and 151 in both cases. The two peaks were confirmed by comparing their absorption spectra and retention times to those of legitimate standards. Peak 19 was tentatively identified as quercetin 3-O-galactoside, and Peak 20 was tentatively recognized as quercetin 3-O-glucoside. Quercetin 3-O-glucoside was found to be the predominant compound in the group of quercetin-derived compounds, followed by quercetin 3-O-galactoside. The amounts of these two compounds and other quercetin derivative compounds were higher in ethyl-alcohol-based infusions than in water-based infusions (Tables 4 and 5).

Based on their UV spectra and MS fragmentation, five peaks were recognized as kaempferol derivatives, leading to the kaempferol aglycone at m/z 285 in the negative mode. The $[M-H]^-$ of a kaempferol-hexose conjugate is represented by the ion at m/z 447, and the resultant MS^2 fragment at m/z 285 ($[M-H]^-$ 162) is a kaempferol. Peak 25 was identified as kaempferol-3-O-glucoside and had $[M-H]^-$ at m/z 447 with a fragment at m/z 285 (loss of 162 amu, hexose moiety). Peaks 23 and 25 were also confirmed when their absorption spectra and retention durations were compared to the standards. Peak 25 had the same MS spectra as Peak 23, as indicated in Table 3. Galactosides elute before matching glucosides, and a kaempferol galactoside was discovered in this manner [31].

Table 3. Retention time, mass spectral characteristics, and identity of phenolic compounds present in tea infusion.

Compound	RT (min)	Formula	Mass $[M-H]^-$	Fragment Ions	λ max (nm)	Proposed Compound
1	17.09	$C_{15}H_{17}O_9$	341	179, 161	324, 296sh	Caffeic acid glucoside-1
2	24.59	$C_{15}H_{17}O_9$	341	179, 161	324, 296sh	Caffeic acid glucoside-2
3	27.9	$C_{15}H_{17}O_9$	341	179, 161	324, 290sh	Caffeic acid glucoside-3
4	27.07	$C_{16}H_{17}O_8$	337	191, 163	310	*p*-Coumaroylquinic acid-1
5	20.8	$C_{16}H_{17}O_9$	353	191, 179, 135	316, 290sh	Neochlorogenic acid
6	29.67	$C_{16}H_{17}O_9$	353	707, 191, 179	322, 296sh	Chlorogenic acid
7	35.52	$C_{16}H_{17}O_9$	353	707, 191, 179	324, 296sh	Cryptochlorogenic acid
8	38.57	$C_{16}H_{17}O_8$	337	191, 163	310	*p*-Coumaroylquinic acid-2
9	39.71	$C_{16}H_{17}O_8$	337	191, 163	311	*p*-Coumaroylquinic acid-3
10	45.41	$C_{16}H_{17}O_8$	337	191, 163	311	*p*-Coumaroylquinic acid-4
11	57.09	$C_{16}H_{17}O_8$	337	191, 163	311	*p*-Coumaroylquinic acid-5
12	58.4	$C_{16}H_{17}O_8$	337	191, 163	311	*p*-Coumaroylquinic acid-6
13	41.67	$C_{17}H_{19}O_9$	367	191, 134	324, 296sh	3-*O*-Feruloylquinic acid
14	42.88	$C_{17}H_{19}O_9$	367	191, 134	324, 296sh	5-*O*-Feruloylquinic acid
15	45.37	$C_{27}H_{29}O_{16}$	609	301, 300, 271	253, 343	Quercetin rutinoside
16	45.66	$C_{27}H_{29}O_{16}$	609	301, 300, 271	254, 349	Quercetin rutinoside
17	47.05	-	609	301, 300, 271	254, 349	Quercetin derivative 1
18	58.72	-	609	301, 300, 271	254, 349	Quercetin derivative 2

Table 3. *Cont.*

Compound	RT (min)	Formula	Mass [M-H]$^-$	Fragment Ions	λ max (nm)	Proposed Compound
19	48.23	$C_{21}H_{19}O_{12}$	463	301, 300, 271	254, 349	Quercetin 3-O-galactoside
20	48.57	$C_{21}H_{19}O_{12}$	463	301, 300, 271	254, 351	Quercetin 3-O-glucoside
21	47.54	$C_{27}H_{29}O_{15}$	593	285	264, 340	Kaempferol 3-O-rhamnoside-1
22	47.96	$C_{27}H_{29}O_{15}$	593	285, 255, 227	264, 346	Kaempferol 3-O-rhamnoside-2
23	50.96	$C_{21}H_{19}O_{11}$	447	342, 285, 255	264, 342	Kaempferol 3-O-galactoside
24	50.53	$C_{25}H_{23}O_{12}$	515	353, 191, 179	325, 296sh	3,4-di-O-caffeoylquinic acid
25	52.49	$C_{21}H_{19}O_{11}$	447	285, 255	264, 334	Kaempferol 3-O-glycoside
26	52.37	$C_{25}H_{23}O_{12}$	515	353, 191, 179	326, 296sh	3,5-di-O-caffeoylquinic acid
27	54.85	$C_{25}H_{23}O_{12}$	515	353, 191, 179	326, 296sh	4,5-di-O-caffeoylquinic acid
28	63.66	$C_{15}H_9O_7$	301	273, 257, 179, 151	254, 349	Quercetin
29	75.93	$C_{15}H_9O_6$	285	285, 229, 185, 151	264, 346	Kaempferol

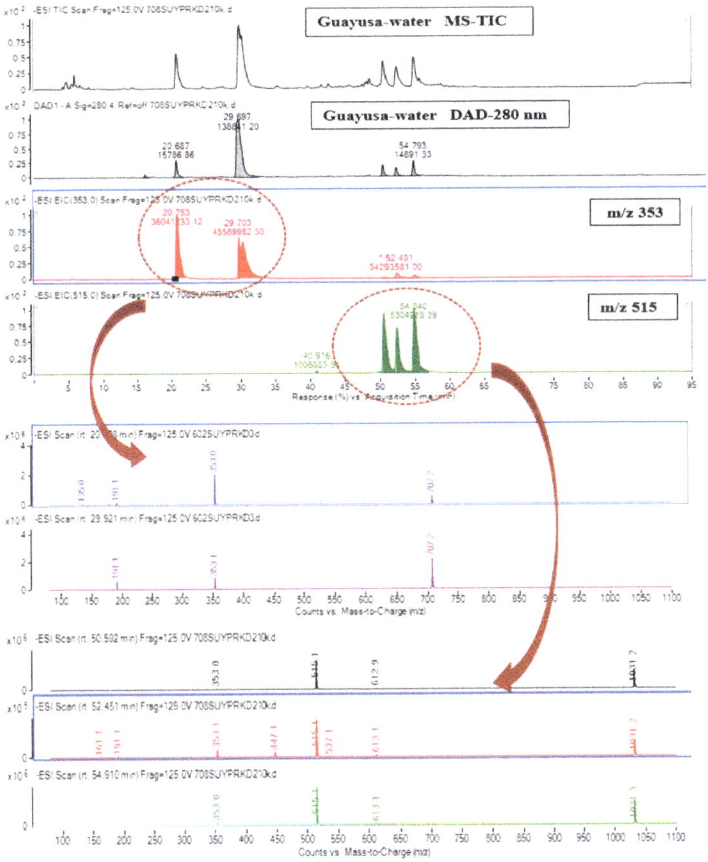

Figure 4. ESI-MS spectrum of chlorogenic acids and di-O-caffeoylquinic acids; LC-MS/MS chromatogram in negative MRM mode with mass transitions.

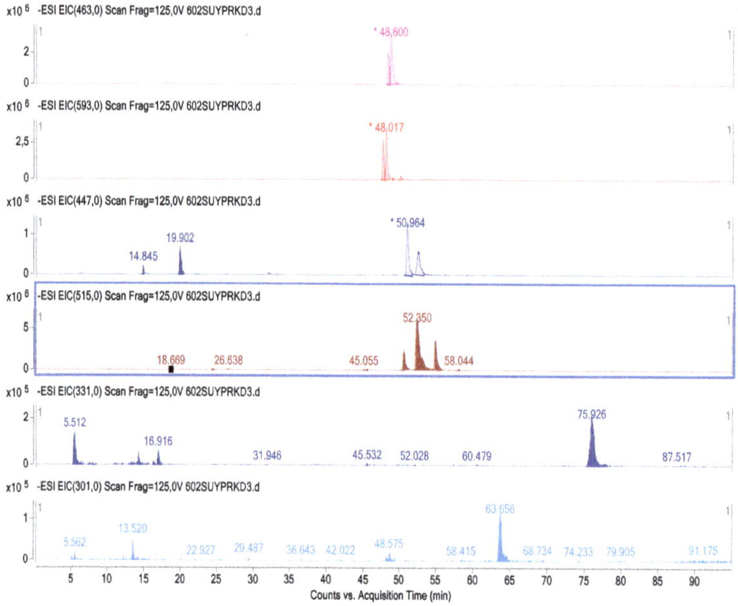

Figure 5. Chromatogram of phenolic compounds identified by LC-DAD and LC–ESI-MS/MS in Gw-60 °C tea infusion.

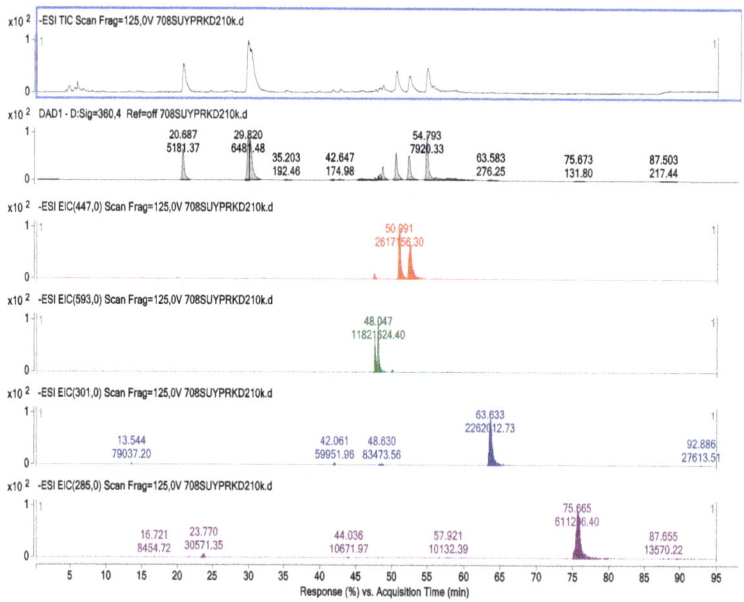

Figure 6. Chromatogram of phenolic compounds identified by LC-DAD and LC–ESI-MS/MS in Gw-70 °C tea infusion.

Table 4. Concentration of phenolic compounds (mg/L) in Guayusa water (Gw) infusions.

Proposed Compound	Gw-60 °C-4 h	Gw-60 °C-6 h	Gw-60 °C-8 h	Gw-70 °C-4 h	Gw-70 °C-6 h	Gw-70 °C-8 h
Caffeic acid glucoside-1	115.38 ± 0.56 d	124.59 ± 0.61 b	126.76 ± 1.17 a	118.99 ± 0.58 c	120.47 ± 0.59 c	127.78 ± 0.62 a
Caffeic acid glucoside-2	64.72 ± 0.32 c	72.33 ± 0.35 b	94.91 ± 0.88 a	72.02 ± 0.15 b	71.38 ± 0.18 b	72.41 ± 0.30 b
Caffeic acid glucoside-3	69.85 ± 0.34 d	75.36 ± 0.32 b	78.19 ± 0.52 a	42.36 ± 0.20f	65.46 ± 0.22 e	71.07 ± 0.21 c
p-Coumaroylquinic acid-1	46.98 ± 0.23 f	60.25 ± 0.29 e	75.50 ± 0.70 b	65.53 ± 0.12 d	67.67 ± 0.18 c	80.60 ± 0.19 a
Neochlorogenic acid	3462.19 ± 10.90 b	3452.12 ± 13.85 b	3503.15 ± 12.31 b	3466.74 ± 12.92 b	3563.44 ± 10.40 a	3587.26 ± 14.21 a
Chlorogenic acid	6428.28 ± 31.38 b	6447.19 ± 31.47 ab	6547.63 ± 60.38 a	6422.63 ± 31.35 ab	6494.11 ± 31.70 ab	6557.41 ± 32.01 a
Cryptochlorogenic acid	225.42 ± 1.10 f	238.84 ± 1.17 e	280.10 ± 2.58 d	342.01 ± 1.67 c	348.82 ± 1.70 b	395.38 ± 1.93 a
p-Coumaroylquinic acid-2	40.77 ± 0.20 f	51.23 ± 0.25 e	55.33 ± 0.51 c	83.55 ± 0.41 c	86.89 ± 0.42 b	88.49 ± 0.43 a
p-Coumaroylquinic acid-3	56.86 ± 0.28 e	62.63 ± 0.31 d	65.14 ± 0.60 c	96.99 ± 0.47 b	98.64 ± 0.48 a b	99.39 ± 0.49 a
p-Coumaroylquinic acid-4	26.61 ± 0.13 f	34.27 ± 0.17 d	29.31 ± 0.27 e	83.46 ± 0.41 c	86.80 ± 0.42 b	92.83 ± 0.45 a
p-Coumaroylquinic acid-5	12.24 ± 0.06 e	17.28 ± 0.08 d	25.24 ± 0.23 c	148.75 ± 0.73 b	148.08 ± 0.42 b	154.93 ± 0.76 a
p-Coumaroylquinic acid-6	17.22 ± 0.08 f	20.66 ± 0.10 e	29.86 ± 0.28 d	140.98 ± 0.69 c	186.79 ± 0.91 b	204.35 ± 1.00 a
3-O-Feruloylquinic acid	70.69 ± 0.35 e	100.38 ± 0.49 d	123.33 ± 1.14 a	114.56 ± 0.56 c	118.28 ± 0.58 b	119.38 ± 0.58 b
5-O-Feruloylquinic acid	72.51 ± 0.23 e	94.77 ± 0.46 d	114.60 ± 1.06 c	121.61 ± 0.59 b	123.54 ± 0.60 b	130.08 ± 0.64 a
Quercetin rutinoside	40.20 ± 0.20 b	36.27 ± 0.18 e	38.74 ± 0.36 c	37.71 ± 0.18 b	39.77 ± 0.19 b	40.82 ± 0.20 a
Quercetin rutinoside	61.92 ± 0.30 b	58.99 ± 0.29 c	63.04 ± 0.58 a	53.15 ± 0.26 e	56.49 ± 0.28 d	57.15 ± 0.28 d
Quercetin derivative 1	15.14 ± 0.07 e	17.51 ± 0.09 c	20.04 ± 0.18 b	17.06 ± 0.08 d	17.62 ± 0.09 c	20.83 ± 0.10 a
Quercetin derivative 2	38.28 ± 0.19 d	35.98 ± 0.18 e	38.37 ± 0.35 d	54.08 ± 0.26 c	65.62 ± 0.32 b	68.77 ± 0.34 a
Quercetin 3-O-galactoside	161.02 ± 0.79 b	146.93 ± 0.72 c	161.30 ± 1.49 b	161.95 ± 0.79 b	166.65 ± 0.81 a	168.48 ± 0.82 a
Quercetin 3-O-glucoside	401.32 ± 1.96 c	422.93 ± 2.06 b	440.25 ± 4.06 a	392.52 ± 1.92 c	426.33 ± 2.08 b	441.89 ± 2.16 a
Kaempferol 3-O-rhamnoside-1	65.85 ± 0.32 d	54.57 ± 0.27 e	68.61 ± 0.63 b	63.99 ± 0.31 d	68.33 ± 0.33 b	70.33 ± 0.34 a
Kaempferol 3-O-rhamnoside-2	73.46 ± 0.36 e	76.78 ± 0.37 d	84.12 ± 0.78 b	75.12 ± 0.37 d	79.07 ± 0.39 c	87.77 ± 0.43 a
Kaempferol 3-O-galactoside	172.75 ± 0.84 f	212.74 ± 1.04 d	236.22 ± 2.18 c	181.57 ± 0.89 e	248.58 ± 1.21 b	287.40 ± 1.40 a
3,4-di-O-caffeoylquinic acid	2540.74 ± 12.40 d	2568.04 ± 12.54 cd	2596.23 ± 23.94 abc	2567.76 ± 12.54 cd	2618.47 ± 12.78 ab	2631.11 ± 12.84 a
Kaempferol 3-O-glycoside	95.00 ± 0.46 e	100.63 ± 0.49 d	140.20 ± 1.29 b	100.08 ± 0.49 d	128.28 ± 0.63 c	142.37 ± 0.70 a
3,5-di-O-caffeoylquinic acid	2298.18 ± 7.22 c	2351.86 ± 10.48 b	2382.19 ± 13.97 ab	2344.89 ± 10.11 b	2283.03 ± 11.02 c	2403.95 ± 8.74 a
4,5-di-O-caffeoylquinic acid	3450.31 ± 16.84 d	3626.28 ± 17.70 bc	3718.92 ± 34.30 a	3533.06 ± 17.25 c	3565.06 ± 17.40 c	3609.36 ± 17.62 bc
Quercetin	60.83 ± 0.30 f	63.31 ± 0.31 e	67.01 ± 0.62 d	79.05 ± 0.39 c	82.75 ± 0.40 b	88.88 ± 0.43 a
Kaempferol	41.40 ± 0.20 f	47.82 ± 0.23 e	54.21 ± 0.50 d	60.60 ± 0.30 c	63.03 ± 0.31 b	65.43 ± 0.32 a
Total	20.226.12 ± 98.74 d	20.672.56 ± 100.92 c	21.258.52 ± 196.05 b	21.042.79 ± 102.73 b	21.489.44 ± 104.91 b	21.965.91 ± 107.24 a

a–f Different letters in the rows represent statistically significant differences ($p < 0.05$). Gw: Guayusa water infusions.

Table 5. The concentration of phenolic compounds (mg/L) in Guayusa ethanol-water (Get) infusions.

Proposed Compound	Get-60 °C-4 h	Get-60 °C-6 h	Get-60 °C-8 h	Get-70 °C-4 h	Get-70 °C-6 h	Get-70 °C-8 h
Caffeic acid glucoside-1	117.10 ± 0.57 d	119.47 ± 0.58 d	127.28 ± 0.62 b	125.17 ± 0.61 c	127.58 ± 0.62 b	140.14 ± 1.24 a
Caffeic acid glucoside-2	47.39 ± 0.23 e	77.19 ± 0.38 c	98.22 ± 0.24 a	73.04 ± 0.36 d	84.02 ± 0.41 b	97.59 ± 0.87 a
Caffeic acid glucoside-3	63.25 ± 0.31 e	86.99 ± 0.42 b	98.15 ± 0.48 a	64.30 ± 0.31 e	69.83 ± 0.34 d	79.12 ± 0.70 c
p-Coumaroylquinic acid-1	73.28 ± 0.36 d	81.42 ± 0.40 c	108.17 ± 0.53 a	74.93 ± 0.37 d	84.20 ± 0.41 b	109.50 ± 0.97 a
Neochlorogenic acid	3067.05 ± 11.97 c	3060.35 ± 14.94 c	3176.38 ± 12.51 b	2979.04 ± 14.54 d	3091.19 ± 12.09 c	3319.35 ± 23.45 a
Chlorogenic acid	4484.53 ± 21.89 d	4518.18 ± 22.06 d	4688.97 ± 22.89 b	4496.08 ± 21.95 d	4606.45 ± 22.49 c	5145.29 ± 25.66 a
Cryptochlorogenic acid	239.05 ± 1.17 d	261.03 ± 1.27 c	270.03 ± 1.32 b	240.83 ± 1.18 d	244.00 ± 1.19 d	284.67 ± 2.53 a

Table 5. *Cont.*

Proposed Compound	Get-60 °C-4 h	Get-60 °C-6 h	Get-60 °C-8 h	Get-70 °C-4 h	Get-70 °C-6 h	Get-70 °C-8 h
p-Coumaroylquinic acid-2	105.82 ± 0.52 [d]	110.94 ± 0.24 [c]	113.39 ± 0.55 [b]	87.01 ± 0.32 [e]	111.89 ± 0.55 [b c]	124.50 ± 1.10 [a]
p-Coumaroylquinic acid-3	104.69 ± 0.51 [d]	111.01 ± 0.54 [c]	118.43 ± 0.58 [a]	111.83 ± 0.35 [c]	114.38 ± 0.56 [b]	117.24 ± 1.04 [a]
p-Coumaroylquinic acid-4	52.82 ± 0.26 [f]	56.50 ± 0.28 [e]	83.48 ± 0.41 [d]	106.27 ± 0.51 [c]	120.35 ± 0.59 [a]	110.14 ± 0.98 [b]
p-Coumaroylquinic acid-5	85.42 ± 0.42 [f]	114.05 ± 0.56 [b]	103.37 ± 0.50 [c]	91.63 ± 0.25 [e]	99.34 ± 0.48 [d]	129.30 ± 1.15 [a]
p-Coumaroylquinic acid-6	79.73 ± 0.39 [f]	94.42 ± 0.46 [e]	102.57 ± 0.28 [c]	99.97 ± 0.49 [d]	114.81 ± 0.56 [b]	144.74 ± 1.28 [a]
3-*O*-Feruloylquinic acid	85.88 ± 0.42 [f]	101.29 ± 0.49 [d]	119.94 ± 0.59 [b]	97.55 ± 0.48 [e]	117.50 ± 0.57 [c]	129.19 ± 1.15 [a]
5-*O*-Feruloylquinic acid	103.91 ± 0.51 [f]	133.89 ± 0.65 [d]	174.87 ± 0.85 [a]	129.25 ± 0.63 [e]	141.03 ± 0.69 [c]	165.17 ± 1.47 [b]
Quercetin rutinoside	29.61 ± 0.14 [c]	28.74 ± 0.14 [d]	28.67 ± 0.14 [d]	38.07 ± 0.19 [b]	41.46 ± 0.20 [a]	37.31 ± 0.33 [b]
Quercetin rutinoside	65.67 ± 0.32 [b]	60.73 ± 0.30 [d]	66.54 ± 0.32 [b]	63.12 ± 0.31 [c]	79.58 ± 0.39 [a]	80.36 ± 0.71 [a]
Quercetin derivative 1	24.12 ± 0.12 [d]	19.36 ± 0.09 [e]	23.74 ± 0.12 [d]	25.05 ± 0.12 [c]	29.12 ± 0.14 [b]	29.94 ± 0.27 [a]
Quercetin derivative 2	60.26 ± 0.29 [d]	63.96 ± 0.31 [c]	83.22 ± 0.41 [b]	85.41 ± 0.42 [a]	86.44 ± 0.42 [a]	85.99 ± 0.76 [a]
Quercetin 3-O-galactoside	164.21 ± 0.80 [c]	163.69 ± 0.80 [c]	184.80 ± 0.90 [a]	165.21 ± 0.81 [c]	172.85 ± 0.84 [b]	159.35 ± 1.41 [d]
Quercetin 3-O-glucoside	379.63 ± 1.85 [e]	420.77 ± 2.05 [c]	451.28 ± 2.20 [b]	405.73 ± 1.98 [d]	448.43 ± 2.19 [b]	479.54 ± 4.26 [a]
Kaempferol 3-O-rhamnoside-1	71.39 ± 0.35 [e]	76.20 ± 0.37 [c]	97.61 ± 0.48 [a]	73.66 ± 0.36 [d]	77.29 ± 0.38 [c]	92.13 ± 0.82 [b]
Kaempferol 3-O-rhamnoside-2	85.75 ± 0.42 [d]	87.13 ± 0.43 [c]	74.58 ± 0.36 [e]	85.51 ± 0.42 [d]	105.91 ± 0.52 [b]	117.92 ± 1.05 [a]
Kaempferol 3-O-galactoside	250.18 ± 1.22 [e]	260.85 ± 1.27 [d]	313.65 ± 1.53 [b]	284.10 ± 1.39 [c]	288.41 ± 1.41 [c]	328.08 ± 2.91 [a]
3,4-di-O-caffeoylquinic acid	2463.61 ± 12.03 [c]	2465.04 ± 12.03 [c]	2663.39 ± 13.00 [a]	2439.60 ± 11.91 [c]	2622.91 ± 12.80 [b]	2665.55 ± 23.65 [a]
Kaempferol 3-O-glycoside	190.16 ± 0.93 [c]	212.53 ± 1.04 [b]	217.93 ± 1.06 [a]	111.33 ± 0.54 [f]	115.75 ± 0.57 [e]	128.72 ± 1.14 [d]
3,5-di-O-caffeoylquinic acid	2162.81 ± 10.56 [c]	2138.01 ± 10.44 [c]	2286.31 ± 11.16 [a]	2085.66 ± 10.18 [d]	2188.69 ± 10.68 [b]	2209.15 ± 19.60 [b]
4,5-di-O-caffeoylquinic acid	3524.99 ± 17.21 [c]	3531.31 ± 17.24 [c]	3803.41 ± 18.57 [b]	3516.01 ± 17.16 [c]	3772.46 ± 18.42 [b]	3977.99 ± 35.30 [a]
Quercetin	70.65 ± 0.34 [f]	80.02 ± 0.39 [d]	85.83 ± 0.42 [c]	74.14 ± 0.36 [e]	87.44 ± 0.43 [b]	101.24 ± 0.90 [a]
Kaempferol	41.09 ± 0.20 [f]	44.03 ± 0.21 [e]	52.90 ± 0.26 [c]	49.71 ± 0.24 [d]	57.10 ± 0.28 [b]	58.15 ± 0.52 [a]
Total	18.294.08 ± 89.31 [d]	18.579.09 ± 90.70 [d]	19.817.10 ± 96.75 [b]	18.279.20 ± 89.24 [d]	19.300.41 ± 94.22 [c]	20.647.34 ± 183.21 [a]

[a–f] Different letters in the rows represent statistically significant differences ($p < 0.05$). Get: Guayusa ethanol infusions.

Five compounds in the structures of kaempferol derivatives were determined. Kaempferol 3-O-galactoside was the dominant compound in water (172.75–287.40 mg/L) and ethyl alcohol (250.18–328.08 mg/L)-based infusions. As with other flavonoid compounds, the amount of this compound was higher in ethyl alcohol infusions, and significant increases were determined depending on increasing time and temperature ($p < 0.05$).

A heatmap was used to visualize the distribution of phenolic compounds in samples after different infusion and extraction conditions. Figures 7 and 8 show heatmaps generated from the data in Tables 4 and 5, which give an overview of the magnitude of the numeric differences observed in all phenolic compounds in Guayusa samples compared to infusion conditions for each compound. The average concentration of each phenolic was marked by a different color on the heatmap, changing between blue and red. Darker red tones indicate major abundance, while darker blue tones indicate minor quantities. As can be seen in Figures 7 and 8, Guayusa teas were divided into two clusters based on the infusion temperatures of 60 °C and 70 °C, which were grouped into separate categories. Samples clustered in the same category show high similarity and correlation.

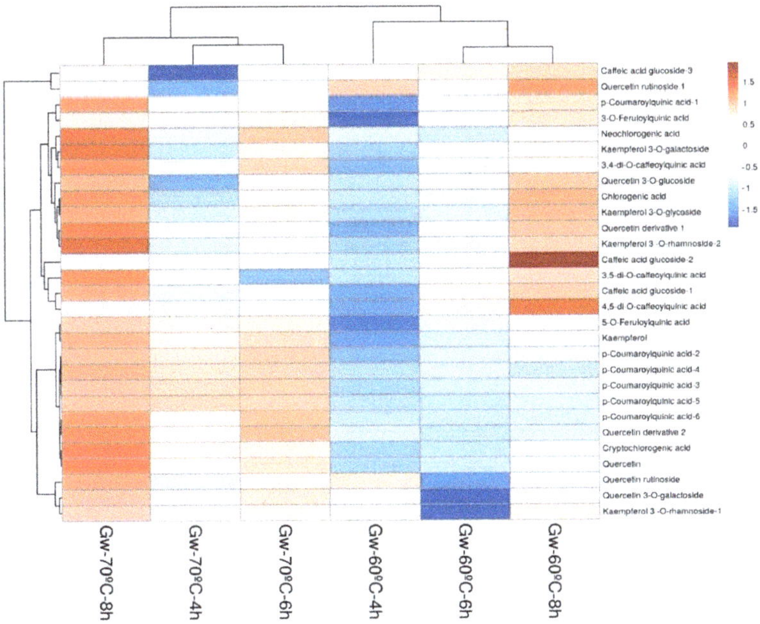

Figure 7. Heatmap of phenolic compounds in Guayusa water (Gw) infusions. Rows are centered; unit variance scaling is applied to rows. Both rows (29 rows; phenolics) and columns (6 columns; infusions) are clustered using correlation distance and average linkage.

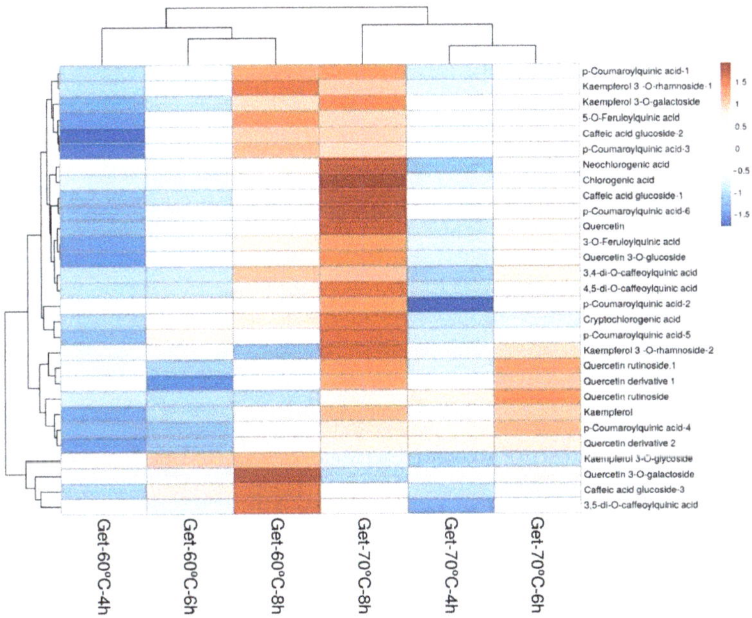

Figure 8. Heatmap of phenolic compounds in Guayusa ethanol-water (Get) infusions. Rows are centered; unit variance scaling is applied to rows. Both rows (29 rows; phenolics) and columns (6 columns; infusions) are clustered using correlation distance and average linkage.

As a result of the comprehensive evaluations in this study, which explored different solvents, infusion times, and temperatures, the highest antioxidant capacity and phenolic compound potential were identified in water- and ethyl-alcohol-based infusions at a temperature of 70 °C for 8 h. In the second stage of the study, changes in antioxidant potential and phenolic compounds were investigated during the antimicrobial effect and in vitro digestion stages of these samples.

3.4. Antimicrobial Activity

The antimicrobial effects of Guayusa tea infusion against two Gram-positive (*S. aureus* ATCC 29213 and *B. subtilis* ATCC 11774) and two Gram-negative (*K. pneumoniae* ATCC 13883 and *E. coli* ATCC 25922) pathogenic microorganisms were investigated (Table 6). Greater zone diameters indicate a higher antimicrobial effect. As can be seen in Table 6, water-based infusions with an elevated phenolic and antioxidant potential demonstrated a stronger antimicrobial effect compared to ethyl-alcohol-based infusions. It was determined that teas generally exhibited a high antimicrobial effect on various microorganisms, except for *E. coli* ATCC 25922, where no significant effect was observed ($p > 0.05$). Gram-negative bacteria, such as *E. coli*, are more resilient to polyphenols than Gram-positive bacteria due to their distinct cell wall compositions. One study found no antimicrobial activity against *E. coli*, possibly due to a pair of membranes surrounding each bacterial cell and a unique outer membrane that provides some form of resistance to these bacteria [32]. A study investigating the antibacterial activity of green tea, black tea, and different oolong teas reported antibacterial activity against three pathogenic bacteria, namely, *S. aureus* ATCC 29213, *E. faecalis* ATCC 29212, and *P. aeruginosa* ATCC 27853, as well as antifungal activity against *C. albicans* ATCC 10231. Previous studies have shown that green and black tea have antibacterial effects against pathogens, including *E. faecalis*, *S. aureus*, *C. albicans*, and *P. aeruginosa*. Furthermore, green tea has been shown to have antimicrobial effects against both Gram-positive and Gram-negative bacteria (e.g., *E. coli*, *Salmonella* spp., *S. aureus*, and *Enterococcus* spp.) [33–35].

Table 6. Antimicrobial effects of Guayusa infusions.

	Inhibition Zone Diameter (mm)	
Test Microorganisms	**Gw-70 °C-8 h**	**Get-70 °C-8 h**
S. aureus ATCC 29213	21.27 ± 0.42 [a]	17.40 ± 0.26 [b]
B. subtilis ATCC 11774	17.75 ± 0.46 [a]	14.51 ± 0.40 [b]
K. pneumoniae ATCC 13883	20.98 ± 0.51 [a]	14.19 ± 0.62 [b]
E. coli ATCC 25922	-	-

[a,b] Different letters in the rows represent statistically significant differences ($p < 0.05$).

3.5. Impact of In Vitro Digestion on Bioactive Compound Profiles

The Guayusa tea infusion with the highest bioactive properties was determined (Gw-70 °C-8 h and Get-70 °C-8 h) and the effects of in vitro gastrointestinal digestion on total phenolic content, antioxidant capacity, and phenolic compounds were investigated.

Effect of in vitro digestion on antioxidant capacity and total phenolic content: The antioxidant capacity (DPPH and ABTS) and total phenolic content of the samples at the in vitro gastrointestinal digestion stages are given in Table 7. After simulated in vitro digestion, an increase in DPPH, ABTS, and TPC results was observed. In the infusion sample of Gw-70 °C-8 h, the DPPH antioxidant capacity was 37.82, 45.72, and 158.42 mM Trolox/L, whereas the ABTS antioxidant capacity was 51.58, 61.84, and 66.81 mM Trolox/L in the oral, gastric, and intestinal phases, respectively, as presented in Table 7. Similarly, in the infusion sample of Get-70 °C-8 h, the DPPH antioxidant capacity was 43.34, 46.64, and 61.36 mM Trolox/L, while the ABTS antioxidant capacity was 61.04, 67.08, and 70.01 mM Trolox/L in the oral, gastric, and intestinal phases, respectively, as shown in Table 7.

Bioactive compounds are obtained using different types of solvents that affect the biological properties of the extracts [36]. Ethanol and water are two different solvents used to extract bioactive compounds. These two solvents may have affected in vitro digestion, and their bioavailability may have differed.

Studies have found that the antioxidant capacity varies throughout the process of digestion. This is due to the fact that a significant degree of the radical scavenging capacity of phenolic compounds relies on the pH of the environment. As a result, the antioxidant capacity can fluctuate while undergoing digestion. Additionally, phenolic compounds can experience structural changes while undergoing gastrointestinal transit due to the ionization of hydroxyl groups. This may lead to an increase in antioxidant capacity at higher pH values [37]. Enzymes (e.g., α-amylase, pepsin, and pancreatin), temperature (37 °C), and changes in pH during in vitro digestion (e.g., 2.5 or 7) can affect the release of antioxidant compounds in different ways. These effects depend on the food matrix and its interaction with other compounds such as proteins, carbohydrates, lipids, fiber, or minerals. Enzymes used at various stages of in vitro digestion have been shown to influence the behavior of various molecules and their degradation/formation. These interactions and the degradation/formation of molecules in in vitro digestion play an important role in changing the bioavailability of antioxidants [38,39].

Table 7. Effect of in vitro digestion model on antioxidant activity (DPPH and ABTS) and total phenolic compounds (TPC).

	ABTS (mM/L)	DPPH (mM/L)	TPC (mg/L)
Gw-70 °C-8 h	34.14 ± 1.96 ^{d}	36.63 ± 1.74 ^{d}	9468.42 ± 103.53 ^{cd}
Oral	37.82 ± 1.84 ^{cd}	51.58 ± 0.08 ^{c}	9293.86 ± 115.91 ^{d}
Gastric	45.72 ± 1.11 ^{b}	61.84 ± 0.88 ^{b}	12.352.63 ± 131.89 ^{b}
Intestinal	58.42 ± 1.90 ^{a}	66.81 ± 1.26 ^{a}	14.856.14 ± 59.85 ^{a}
Get-70 °C-8 h	37.69 ± 2.26 ^{e}	45.08 ± 0.83 ^{e}	10.340.35 ± 83.73 ^{e}
Oral	43.34 ± 0.86 ^{cd}	61.04 ± 1.02 ^{c}	11.882.46 ± 98.62 ^{d}
Gastric	46.64 ± 2.85 ^{bc}	67.08 ± 1.27 ^{b}	14.114.04 ± 368.44 ^{cd}
Intestinal	61.36 ± 2.86 ^{a}	70.01 ± 1.94 ^{a}	17.112.28 ± 220.88 ^{a}

$^{a–e}$ Different letters in the columns represent statistically significant differences ($p < 0.05$).

Effect of in vitro digestion on phenolic compounds: The bioaccessibility of phenolic compounds in Guayusa tea infusion was determined by Infogest static in vitro gastrointestinal food digestion simulation. The phenolic compound profiles in oral, gastric, and intestinal samples was determined using HPLC-DAD-ESI-MS/MS and significant differences were observed between oral, gastric, and intestinal samples after simulated in vitro digestion ($p < 0.05$) (Table 8). During the gastrointestinal digestion of Guayusa water and ethanol infusion samples, a total of 27 phenolic compounds were identified, including caffeic acid glucoside, neochlorogenic acid, 3-O-feruloylquinic acid, 5-O-feruloylquinic acid, quercetin rutinoside, kaempferol 3-O-galactoside, 3,4-di-O-caffeoylquinic acid, kaempferol 3-O-glycoside, 3,5-di-O-caffeoylquinic acid, and 4,5-di-O-caffeoylquinic acid (Table 8).

Table 8. Effect of in vitro digestion model on phenolic compounds (mg/L).

Proposed Compound	Ethanol Infusions							Water Infusions						
	Initial	Oral	%	Gastric	%	Intestinal	%	Initial	Oral	%	Gastric	%	Intestinal	%
Caffeic acid glucoside-1	140.14 ± 1.24 ^{c}	107.58 ± 0.59 ^{d}	−23.2	120.10 ± 1.00 ^{e}	−14.3	144.08 ± 2.78 ^{a}	2.8	127.78 ± 0.62 ^{d}	119.06 ± 1.33 ^{e}	−6.8	127.25 ± 4.17 ^{f}	−0.4	147.97 ± 1.18 ^{a}	15.8
Caffeic acid glucoside-2	97.59 ± 0.87 ^{e}	85.60 ± 0.56 ^{f}	−12.3	104.87 ± 2.16 ^{d}	7.5	124.70 ± 0.80 ^{a}	27.8	72.41 ± 0.30 ^{g}	65.06 ± 0.77 ^{h}	−10.2	109.72 ± 2.71 ^{c}	51.5	122.04 ± 1.64 ^{b}	68.5
Caffeic acid glucoside-3	79.12 ± 0.70 ^{e}	77.25 ± 1.07 ^{f}	−2.4	86.51 ± 1.12 ^{d}	9.3	112.57 ± 1.45 ^{c}	42.3	71.07 ± 0.21 ^{h}	73.68 ± 1.01 ^{g}	3.7	116.08 ± 2.15 ^{b}	63.3	135.46 ± 0.48 ^{a}	90.6
p-Coumaroylquinic acid-1	109.50 ± 0.97 ^{d}	62.07 ± 1.31 ^{h}	−43.3	106.48 ± 0.88 ^{e}	−2.8	122.45 ± 2.20 ^{c}	11.8	80.60 ± 0.19 ^{f}	69.16 ± 1.19 ^{g}	−14.2	129.78 ± 2.57 ^{b}	61.0	152.33 ± 0.39 ^{a}	89.0
Neochlorogenic acid	3319.35 ± 23.45 ^{f}	3137.15 ± 2.61 ^{h}	−5.5	3556.73 ± 7.88 ^{e}	7.2	4143.54 ± 3.42 ^{b}	24.8	3587.26 ± 14.21 ^{d}	3315.00 ± 1.42 ^{g}	−7.6	3592.75 ± 3.18 ^{c}	0.2	4327.97 ± 1.46 ^{a}	20.6
Chlorogenic acid	5145.29 ± 25.66 ^{g}	4909.03 ± 2.78 ^{h}	−4.6	5270.98 ± 2.49 ^{f}	2.4	6791.18 ± 2.17 ^{c}	32.0	6557.41 ± 32.01 ^{d}	6413.88 ± 72.29 ^{e}	−2.2	6822.25 ± 2.11 ^{b}	4.0	7882.65 ± 8.02 ^{a}	20.2
Cryptochlorogenic acid	284.67 ± 2.53 ^{f}	205.23 ± 4.71 ^{h}	−27.9	247.99 ± 1.15 ^{g}	−12.9	327.07 ± 2.03 ^{e}	14.9	395.38 ± 1.93 ^{b}	388.08 ± 2.72 ^{d}	−1.8	394.93 ± 1.51 ^{a}	−0.1	424.28 ± 0.96 ^{a}	7.3
p-Coumaroylquinic acid-2	124.50 ± 1.10 ^{c}	102.99 ± 0.65 ^{d}	−17.3	136.82 ± 1.10 ^{b}	9.9	165.76 ± 3.03 ^{a}	33.1	88.49 ± 0.43 ^{g}	88.87 ± 1.60 ^{g}	0.4	94.22 ± 1.16 ^{f}	6.5	115.88 ± 1.32 ^{d}	31.0
p-Coumaroylquinic acid-3	117.24 ± 1.04 ^{c}	100.34 ± 2.35 ^{e}	−14.4	119.79 ± 2.80 ^{b}	2.2	186.48 ± 2.29 ^{a}	59.1	99.39 ± 0.49 ^{f}	89.82 ± 1.67 ^{g}	−9.6	106.21 ± 1.01 ^{d}	6.9	119.95 ± 1.18 ^{b}	20.7
p-Coumaroylquinic acid-4	110.14 ± 0.98 ^{c}	97.12 ± 1.25 ^{f}	−11.8	115.93 ± 1.25 ^{b}	5.3	142.98 ± 3.60 ^{a}	29.8	92.83 ± 0.45 ^{g}	91.55 ± 0.93 ^{h}	−1.4	100.12 ± 0.26 ^{e}	7.9	108.29 ± 0.72 ^{d}	16.7
p-Coumaroylquinic acid-5	129.30 ± 1.15 ^{f}	82.07 ± 1.31 ^{h}	−36.5	115.86 ± 0.20 ^{g}	−10.4	134.28 ± 1.08 ^{e}	3.9	154.93 ± 0.76 ^{d}	168.57 ± 2.02 ^{c}	8.8	181.13 ± 4.26 ^{b}	16.9	185.09 ± 0.06 ^{a}	19.5
p-Coumaroylquinic acid-6	144.74 ± 1.28 ^{f}	89.29 ± 1.00 ^{h}	−38.3	136.80 ± 3.57 ^{g}	−5.5	165.60 ± 1.75 ^{e}	14.4	204.35 ± 1.00 ^{c}	197.06 ± 1.33 ^{d}	−3.6	244.12 ± 1.98 ^{b}	19.5	257.78 ± 0.87 ^{a}	26.1
3-O-Feruloylquinic acid	129.19 ± 1.15 ^{c}	127.03 ± 1.37 ^{cd}	−1.7	145.67 ± 1.03 ^{b}	12.8	163.48 ± 1.41 ^{a}	26.5	119.38 ± 0.58 ^{e}	118.32 ± 0.96 ^{d}	−0.9	121.56 ± 1.19 ^{d}	1.8	147.34 ± 0.94 ^{b}	23.4
5-O-Feruloylquinic acid	165.17 ± 1.47 ^{b}	126.59 ± 1.99 ^{g}	−23.4	154.27 ± 2.87 ^{c}	−6.6	184.70 ± 1.38 ^{a}	11.8	130.08 ± 0.64 ^{f}	126.24 ± 1.07 ^{g}	−3.0	133.02 ± 0.90 ^{f}	2.3	151.02 ± 0.25 ^{d}	16.1
Quercetin rutinoside	37.31 ± 0.33 ^{c}	34.21 ± 0.41 ^{d}	−8.3	32.78 ± 1.16 ^{e}	−1.1	25.43 ± 0.79 ^{e}	−31.8	40.82 ± 0.20 ^{a}	39.49 ± 0.72 ^{b}	−3.3	36.02 ± 0.02 ^{c}	−11.8	28.71 ± 0.51 ^{f}	−29.7
Quercetin rutinoside	80.36 ± 0.71 ^{a}	79.92 ± 0.12 ^{a}	−0.5	75.96 ± 0.77 ^{b}	−5.5	74.18 ± 1.66 ^{b}	−7.7	57.15 ± 0.28 ^{c}	55.27 ± 0.33 ^{d}	−3.3	51.30 ± 0.65 ^{e}	−10.2	45.74 ± 1.49 ^{f}	−20.0
Quercetin derivative 1	29.94 ± 0.27 ^{a}	28.80 ± 0.28 ^{a}	−3.8	25.07 ± 0.75 ^{b}	−16.3	22.70 ± 0.88 ^{c}	−24.2	20.83 ± 0.10 ^{d}	19.26 ± 0.37 ^{e}	−7.5	18.67 ± 0.75 ^{f}	−10.4	18.83 ± 0.30 ^{f}	−9.6
Quercetin derivative 2	85.99 ± 0.76 ^{a}	85.24 ± 1.08 ^{a}	−0.9	82.02 ± 1.10 ^{b}	−4.6	80.95 ± 0.42 ^{c}	−5.9	68.77 ± 0.34 ^{d}	66.28 ± 0.40 ^{e}	−3.6	62.85 ± 0.21 ^{f}	−8.6	58.73 ± 0.61 ^{g}	−14.6
Quercetin 3-O-galactoside	159.35 ± 1.41 ^{b}	134.48 ± 0.74 ^{d}	−15.6	123.54 ± 0.65 ^{e}	−22.5	125.15 ± 1.61 ^{e}	−21.5	168.48 ± 0.82 ^{a}	159.26 ± 1.05 ^{b}	−5.5	146.74 ± 0.66 ^{c}	−12.9	133.36 ± 0.58 ^{d}	−20.8
Quercetin 3-O-glucoside	479.54 ± 4.26 ^{a}	464.11 ± 1.26 ^{b}	−3.2	374.90 ± 1.55 ^{f}	−21.8	366.96 ± 1.21 ^{g}	−23.5	441.89 ± 2.16 ^{c}	437.51 ± 0.86 ^{d}	−1.0	393.60 ± 0.70 ^{e}	−10.9	366.84 ± 0.64 ^{g}	−17.0

Table 8. *Cont.*

Proposed Compound	Ethanol Infusions							Water Infusions						
	Initial	Oral	%	Gastric	%	Intestinal	%	Initial	Oral	%	Gastric	%	Intestinal	%
Kaempferol 3-O-rhamnoside-1	92.13 ± 0.82 [a]	87.80 ± 0.28 [b]	−4.7	65.79 ± 0.87 [d]	−28.6	60.72 ± 1.07 [e]	−34.1	70.33 ± 0.34 [c]	64.62 ± 0.54 [d]	−8.1	59.53 ± 0.66 [e]	−15.4	55.66 ± 0.71 [f]	−20.9
Kaempferol 3-O-rhamnoside-2	117.92 ± 1.05 [a]	106.36 ± 0.91 [b]	−9.8	90.16 ± 1.47 [d]	−23.5	95.20 ± 1.03 [c]	−19.3	87.77 ± 0.43 [e]	82.32 ± 0.45 [f]	−6.2	77.75 ± 0.92 [g]	−11.4	63.30 ± 0.08 [h]	−27.9
Kaempferol 3-O-galactoside	328.08 ± 2.91 [a]	273.94 ± 0.09 [d]	−16.5	237.67 ± 1.03 [g]	−27.6	226.90 ± 1.42 [h]	−30.8	287.40 ± 1.40 [b]	280.20 ± 1.12 [c]	−2.5	262.94 ± 1.50 [e]	−8.5	253.73 ± 0.52 [f]	−11.7
3,4-di-O-caffeoylquinic acid	2665.55 ± 23.65 [a]	2209.55 ± 3.47 [d]	−17.1	2141.66 ± 1.00 [e]	−19.7	2048.97 ± 2.72 [g]	−23.1	2631.11 ± 12.84 [a]	2478.75 ± 1.76 [b]	−5.8	2390.81 ± 1.68 [c]	−9.1	2093.59 ± 0.58 [f]	−20.4
Kaempferol 3-O-glycoside	128.72 ± 1.14 [c]	118.29 ± 1.00 [d]	−8.1	108.97 ± 1.18 [f]	−15.3	100.97 ± 0.71 [g]	−21.6	142.37 ± 0.70 [a]	132.04 ± 0.59 [b]	−7.3	128.40 ± 1.69 [c]	−9.8	110.87 ± 0.67 [e]	−22.1
3,5-di-O-caffeoylquinic acid	2209.15 ± 19.60 [b]	2067.49 ± 2.14 [d]	−6.4	1793.84 ± 0.62 [f]	−18.8	1613.11 ± 2.10 [g]	−27.0	2403.95 ± 8.74 [a]	2152.15 ± 1.20 [c]	−10.5	1901.43 ± 0.66 [e]	−20.9	1740.44 ± 0.48 [f]	−27.6
4,5-di-O-caffeoylquinic acid	3977.99 ± 35.30 [a]	3770.45 ± 2.19 [b]	−5.2	3645.80 ± 1.99 [c]	−8.4	3238.45 ± 3.72 [g]	−18.6	3609.36 ± 17.62 [d]	3486.66 ± 15.08 [e]	−3.4	3439.14 ± 1.85 [f]	−4.7	3221.54 ± 0.47 [h]	−10.7
Total	20.487.9 ± 22.9 [f]	18.769.97 ± 37.5 [h]	−8.4	19.216.95 ± 37.8 [g]	−6.2	20.988.60 ± 47.1 [d]	2.4	21.811.60 ± 22.45 [b]	20.778.16 ± 81.55 [e]	−4.7	21.242.31 ± 35.2 [c]	−2.6	22.469.37 ± 26.9 [a]	3.0

[a–h] Different letters in the columns represent statistically significant differences ($p < 0.05$). % values show the change in digestion stages according to the initial concentration.

When the oral, gastric, and intestinal digestion phases were compared in both infusion samples, it was found that there was an increase in the amount of total phenolic compounds. However, when a comparison was made according to pre-digestion, it was found that there was an 8.4% decrease in the oral phase, a 6.2% decrease in the gastric phase, and a 2.4% increase in the intestinal phase of ethyl-alcohol infusions. Similar changes were found in infusions prepared with water. There was a 4.7% decrease in the oral phase, a 2.6% decrease in the gastric phase, and a 3% increase in the intestinal phase. The highest bioavailability in water and ethanol infusions was observed in the intestinal phase. After simulated gastric digestion, many phenolic compounds' (quercetin rutinoside, quercetin derivative, quercetin 3-O-galactoside, quercetin 3-O-glucoside, kaempferol 3-O-rhamnoside, kaempferol 3-O-galactoside, 3,4-di-O-caffeoylquinic acid, kaempferol 3-O-glycoside, 3,5-di-O-caffeoylquinic acid, and 4,5-di-O-caffeoylquinic acid) amounts were found to decrease. Neochlorogenic, chlorogenic, cryptochlorogenic, and p-coumaroylquinic acids increased after the gastric phase. The increase in caffeoylquinic acid derivatives can be attributed to the decrease in di-caffeoylquinic acid compounds resulting from fragmentation and the subsequent formation of these compounds [40]. It has been reported that the stability of di-O-caffeoylquinic acids is low, and depending on pH and temperature conditions, diCQAs can isomerise with each other and transform into mono-CQAs, caffeic acid, and compounds with the formula $C_{15}H_{14}O_6$ [41]. Changes in pH levels and the presence of bile salts during simulated digestion may significantly impact the reduction of the quantity of phenolic compounds and increase the monomeric CQAs [42]. In previous studies, it was observed that most of the polyphenols in herbal infusions decreased markedly during the gastric stage and increased after the intestinal stage. It has been suggested that this may be due to the pH levels and other environmental conditions of the gastric and intestinal phases being unsuitable, resulting in damage to certain polyphenols [43,44]. The study conducted by Ozkan et al. [37] investigated the stability and bioaccessibility of phenolics during digestion by using a static in vitro digestion protocol on plant infusions from different regions of Turkey. The results showed that there was a significant increase in the content of several polyphenols, including gallic acid, protocatechuic acid, epicatechin, chlorogenic acid, rutin, and syringic acid, after in vitro intestinal digestion. However, the study also found that most of the polyphenols in herbal infusions decreased significantly during the gastric phase and increased after the intestinal phase. Some polyphenols were even undetectable after in vitro digestion.

4. Conclusions

This study determined the effects of infusion time, temperature, and solvent type on the color, antioxidant capacity, total phenolic content, and phenolic profile, as well as the impact of in vitro gastrointestinal digestion on the phenolic content, antioxidant capacity, and phenolic compounds of Guayusa tea. In general, when Guayusa tea infusions were evaluated, it was observed that the L* value decreased as the infusion time increased, and the color became darker. Antioxidant capacity and total phenol content analysis showed that Guayusa tea infusions had significant antioxidant properties. DPPH and ABTS results showed that the antioxidant capacity of the infusions increased with increasing infusion time and temperature. The phenolic profile of the infusions was determined and found to be dominated by chlorogenic acid and its derivatives. In Guayusa water infusions, the sample containing the highest phenolic content was found to be Gw-70 °C-8 h, and in ethanol infusions, it was Get-70 °C-8 h. Antimicrobial activity was determined against *Staphylococcus aureus* ATCC 29213, *Bacillus subtilis* ATCC 11774, *Klebsiella pneumoniae* ATCC 13883, and *Escherichia coli* ATCC 25922 test organisms. Antimicrobial activity analysis showed that Guayusa tea infusions generally have high activity against Gram-positive bacteria.

Depending on the digestion stages, the total amount of phenolic compounds in the samples prepared with both solvents increased after oral intake. When the profiles of phenolic compounds were analyzed, neochlorogenic acid, chlorogenic acid, cryptochlorogenic acid, 3-O-feruloylquinic acid, 5-O-feruloylquinic acid, and p-coumaroylquinic acids derivatives increased significantly ($p < 0.05$). The amounts of flavonoid compounds and di-O-caffeoylquinic acid derivatives decreased in the digestion stages.

In conclusion, this study revealed that Guayusa tea has antimicrobial effects, high antioxidant properties, and phenolic compounds. It also aligns with the broader objective of advancing tea science by integrating traditional knowledge with modern analytical techniques. The outcomes are poised to contribute valuable insights to the burgeoning field of functional foods and beverages, with implications for the tea industry and public health. In the future, optimizing infusion parameters and exploring novel extraction techniques could enhance the extraction of bioactive compounds from Guayusa tea. Further characterization of phenolic compounds, elucidation of their mechanisms of action, and assessments of bioavailability are crucial for understanding their health benefits. Additionally, research should focus on sustainable cultivation practices, formulation development, and global market opportunities, while fostering multidisciplinary collaboration and integrating traditional knowledge with modern science to promote cultural appreciation and environmental protection.

Author Contributions: H.K.: conceptualization, data curation, formal analysis, funding acquisition, writing—original draft, writing—review & editing. S.S.: conceptualization, data curation, formal analysis, funding acquisition, writing—original draft, writing—review & editing. H.K.S.: conceptualization, data curation, formal analysis, funding acquisition, writing—original draft, writing—review & editing. O.A.: data curation, formal analysis. O.K.: conceptualization, data curation, formal analysis, funding acquisition, writing—original draft, writing—review & editing. C.F.: formal analysis, funding acquisition—review & editing. All authors have read and agreed to the published version of the manuscript.

Funding: This research received no external funding.

Institutional Review Board Statement: Not applicable.

Informed Consent Statement: Not applicable.

Data Availability Statement: The original contributions presented in the study are included in the article, further inquiries can be directed to the corresponding author.

Conflicts of Interest: O.K. and C.F. are employees of Applied Food Sciences Inc. Applied Food Sciences provided the funding for this study. The company's authors had no influence on the correctness and fairness of the research results. The remaining authors declare that the research was conducted in the absence of any commercial relationships.

References

1. Kapp, R.W., Jr.; Mendes, O.; Roy, S.; McQuate, R.S.; Kraska, R. General and genetic toxicology of guayusa concentrate (*Ilex guayusa*). *Int. J. Toxicol.* **2016**, *35*, 222–242. [CrossRef]
2. García-Ruiz, A.; Baenas, N.; Benítez-González, A.M.; Stinco, C.M.; Meléndez-Martínez, A.J.; Moreno, D.A.; Ruales, J. Guayusa (*Ilex guayusa* L.) new tea: Phenolic and carotenoid composition and antioxidant capacity. *J. Sci. Food Agric.* **2017**, *97*, 3929–3936. [CrossRef] [PubMed]
3. Gan, R.Y.; Zhang, D.; Wang, M.; Corke, H. Health Benefits of Bioactive Compounds from the Genus Ilex, a Source of Traditional Caffeinated Beverages. *Nutrients* **2018**, *10*, 682. [CrossRef] [PubMed]
4. Infinium Global Research. Available online: https://www.infiniumglobalresearch.com/reports/global-herbal-tea-market (accessed on 22 August 2023).
5. GRAS. Notification for Ilex guayusa Leaf Extract (869), Ilex guayusa Leaf Hydroethanolic Extract (870, 883). 2019. Available online: https://www.fda.gov/food/gras-notice-inventory/recently-published-gras-notices-and-fda-letters (accessed on 22 August 2023).
6. Sequeda-Castañeda, L.; Modesti Costa, G.; Celis, C.; Gamboa, F.; Gutiérrez, S.; Luengas, P. *Ilex guayusa* loes (Aquifoliaceae): Amazon and Andean native plant. *Pharmacol. Online* **2016**, *3*, 193–202.
7. Negrin, A.; Long, C.; Motley, T.J.; Kennelly, E.J. LC-MS metabolomics and chemotaxonomy of caffeine-containing holly (Ilex) species and related taxa in the Aquifoliaceae. *J. Agric. Food Chem.* **2019**, *67*, 5687–5699. [CrossRef] [PubMed]
8. Oellig, C.; Schunck, J.; Schwack, W. Determination of caffeine, theobromine and theophylline in Mate beer and Mate soft drinks by high-performance thin-layer chromatography. *J. Chromatogr. A* **2018**, *1533*, 208–212. [CrossRef] [PubMed]

9. Arteaga-Crespo, Y.; Radice, M.; Bravo-Sanchez, L.R.; García-Quintana, Y.; Scalvenzi, L. Optimisation of ultrasound-assisted extraction of phenolic antioxidants from *Ilex guayusa* Loes. leaves using response surface methodology. *Heliyon* **2020**, *6*, e03043. [CrossRef] [PubMed]
10. Stanisavljević, N.; Samardžić, J.; Janković, T.; Šavikin, K.; Mojsin, M.; Topalović, V.; Stevanović, M. Antioxidant and antiproliferative activity of chokeberry juice phenolics during in vitro simulated digestion in the presence of food matrix. *Food Chem.* **2015**, *175*, 516–522. [CrossRef]
11. Uzlasir, T.; Kadiroglu, P.; Selli, S.; Kelebek, H. LC-DAD-ESI-MS/MS characterization of elderberry flower (*Sambucus nigra*) phenolic compounds in ethanol, methanol, and aqueous extracts. *J. Food Process. Preserv.* **2021**, *45*, e14478. [CrossRef]
12. Kelebek, H. LC-DAD–ESI-MS/MS characterization of phenolic constituents in Turkish black tea: Effect of infusion time and temperature. *Food Chem.* **2016**, *204*, 227–238. [CrossRef]
13. Liang, Y.; Lu, J.; Zhang, L.; Wu, S.; Wu, Y. Estimation of black tea quality by analysis of chemical composition and colour difference of tea infusions. *Food Chem.* **2003**, *80*, 283–290. [CrossRef]
14. Kelebek, H.; Sevindik, O.; Uzlasir, T.; Selli, S. LC-DAD/ESI MS/MS characterization of fresh and cooked Capia and Aleppo red peppers (*Capsicum annuum* L.) phenolic profiles. *Eur. Food Res. Technol.* **2020**, *246*, 1971–1980. [CrossRef]
15. Kelebek, H.; Selli, S.; Kadiroğlu, P.; Kola, O.; Kesen, S.; Uçar, B.; Çetiner, B. Bioactive compounds and antioxidant potential in tomato pastes as affected by hot and cold break process. *Food Chem.* **2017**, *220*, 31–41. [CrossRef]
16. Velioglu, Y.; Mazza, G.; Gao, L.; Oomah, B. Antioxidant activity and total phenolics in selected fruits, vegetables, and grain products. *J. Agric. Food Chem.* **1998**, *46*, 4113–4117. [CrossRef]
17. Kadiroğlu, P. FTIR spectroscopy for prediction of quality parameters and antimicrobial activity of commercial vinegars with chemometrics. *J. Sci. Food Agric.* **2018**, *98*, 4121–4127. [CrossRef]
18. Brodkorb, A.; Egger, L.; Alminger, M.; Alvito, P.; Assunção, R.; Ballance, S.; Bohn, T.; Bourlieu-Lacanal, C.; Boutrou, R.; Carrière, F.; et al. INFOGEST static in vitro simulation of gastrointestinal food digestion. *Nat. Protoc.* **2019**, *14*, 991–1014. [CrossRef]
19. Andrés-Bello, A.; Barreto-Palacios, V.; García-Segovia, P.; Mir-Bel, J.; Martínez-Monzó, J. Effect of pH on color and texture of food products. *Food Eng. Rev.* **2013**, *5*, 158–170. [CrossRef]
20. Jin, L.; Li, X.-B.; Tian, D.-Q.; Fang, X.-P.; Yu, Y.-M.; Zhu, H.-Q.; Ge, Y.-Y.; Ma, G.-Y.; Wang, W.-Y.; Xiao, W.-F.; et al. Antioxidant properties and color parameters of herbal teas in China. *Ind. Crops Prod.* **2016**, *87*, 198–209. [CrossRef]
21. Polat, A.; Kalcıoğlu, Z.; Müezzinoğlu, N. Effect of infusion time on black tea quality, mineral content and sensory properties prepared using traditional Turkish infusion method. *Int. J. Gastron. Food Sci.* **2022**, *29*, 100559. [CrossRef]
22. Kamara, B.I.; Brandt, E.V.; Ferreira, D.; Joubert, E. Polyphenols from Honeybush tea (*Cyclopia intermedia*). *J. Agric. Food Chem.* **2003**, *51*, 3874–3879. [CrossRef]
23. Sridhar, K.; Charles, A.L. In vitro antioxidant activity of Kyoho grape extracts in DPPH and ABTS assays: Estimation methods for EC50 using advanced statistical programs. *Food Chem.* **2019**, *275*, 41–49. [CrossRef]
24. Pardau, M.D.; Pereira, A.S.P.; Apostolides, Z.; Serem, J.C.; Bester, M.J. Antioxidant and anti-inflammatory properties of Ilex guayusa tea preparations: A comparison to Camellia sinensis teas. *Food Funct.* **2017**, *8*, 4601–4610. [CrossRef]
25. Kelebek, H.; Sevindik, O.; Selli, S. LC-DAD-ESI-MS/MS-based phenolic profiling of St John's Wort Teas and their antioxidant activity: Eliciting infusion induced changes. *J. Liq. Chromatogr. Relat. Technol.* **2019**, *42*, 15–19. [CrossRef]
26. Rodríguez Vaquero, M.J.; Tomassini Serravalle, L.R.; Manca de Nadra, M.C.; Strasser de Saad, A.M. Antioxidant capacity and antibacterial activity of phenolic compounds from argentinean herbs infusions. *Food Control* **2010**, *21*, 779–785. [CrossRef]
27. Dobrinas, S.; Soceanu, A.; Popescu, V.; Carazeanu Popovici, I.; Jitariu, D. Relationship between Total Phenolic Content, Antioxidant Capacity, Fe and Cu Content from Tea Plant Samples at Different Brewing Times. *Processes* **2021**, *9*, 1311. [CrossRef]
28. Clifford, M.N.; Jaganath, I.B.; Ludwig, I.A.; Crozier, A. Chlorogenic acids and the acyl-quinic acids: Discovery, biosynthesis, bioavailability and bioactivity11Electronic supplementary information (ESI). *Nat. Prod. Rep.* **2017**, *34*, 1391–1421. [CrossRef]
29. Rojas-González, A.; Figueroa-Hernández, C.Y.; González-Rios, O.; Suárez-Quiroz, M.L.; González-Amaro, R.M.; Hernández-Estrada, Z.J.; Rayas-Duarte, P. Coffee Chlorogenic Acids Incorporation for Bioactivity Enhancement of Foods: A Review. *Molecules* **2022**, *27*, 3400. [CrossRef]
30. Macheiner, L.; Schmidt, A.; Schreiner, M.; Mayer, H.K. Green coffee infusion as a source of caffeine and chlorogenic acid. *J. Food Compos. Anal.* **2019**, *84*, 103307. [CrossRef]
31. Del Rio, D.; Stewart, A.J.; Mullen, W.; Burns, J.; Lean, M.E.; Brighenti, F.; Crozier, A. HPLC-MSn analysis of phenolic compounds and purine alkaloids in green and black tea. *J. Agric. Food Chem.* **2004**, *52*, 2807–2815. [CrossRef]
32. Zaidan, M.; Noor Rain, A.; Badrul, A.; Adlin, A.; Norazah, A.; Zakiah, I. In vitro screening of five local medicinal plants for antibacterial activity using disc diffusion method. *Trop Biomed.* **2005**, *22*, 165–170.
33. Archana, S.; Abraham, J. Comparative analysis of antimicrobial activity of leaf extracts from fresh green tea, commercial green tea and black tea on pathogens. *J. Appl. Pharm. Sci.* **2011**, *01*, 149–152.
34. Tea, G. Anti-infective properties of Epigallocatechin-3-gallate (EGCG), a component of. *Br. J. Pharmacol.* **2013**, *168*, 1059–1073. [CrossRef]
35. Jigisha, A.; Nishant, R.; Navin, K.; Pankaj, G. Green tea: A magical herb with miraculous outcomes. *Int. Res. J. Pharm* **2012**, *3*, 139–148.
36. Li, X.; Zhu, F.; Zeng, Z. Effects of different extraction methods on antioxidant properties of blueberry anthocyanins. *Open Chem.* **2021**, *19*, 138–148. [CrossRef]

37. Ozkan, G.; Sakarya, F.B.; Tas, D.; Yurt, B.; Ercisli, S.; Capanoglu, E. Effect of In Vitro Digestion on the Phenolic Content of Herbs Collected from Eastern Anatolia. *ACS Omega* **2023**, *8*, 12730–12738. [CrossRef]
38. Reboredo-Rodríguez, P.; González-Barreiro, C.; Martínez-Carballo, E.; Cambeiro-Pérez, N.; Rial-Otero, R.; Figueiredo-González, M.; Cancho-Grande, B. Applicability of an in vitro digestion model to assess the bioaccessibility of phenolic compounds from olive-related products. *Molecules* **2021**, *26*, 6667. [CrossRef]
39. Cilla, A.; Bosch, L.; Barberá, R.; Alegría, A. Effect of processing on the bioaccessibility of bioactive compounds–A review focusing on carotenoids, minerals, ascorbic acid, tocopherols and polyphenols. *J. Food Compos. Anal.* **2018**, *68*, 3–15. [CrossRef]
40. Cañas, S.; Rebollo-Hernanz, M.; Braojos, C.; Benítez, V.; Ferreras-Charro, R.; Dueñas, M.; Aguilera, Y.; Martín-Cabrejas, M.A. Understanding the Gastrointestinal Behavior of the Coffee Pulp Phenolic Compounds under Simulated Conditions. *Antioxidants* **2022**, *11*, 1818. [CrossRef]
41. Li, Y.-J.; Zhang, C.-F.; Ding, G.; Huang, W.-Z.; Wang, Z.-Z.; Bi, Y.-A.; Xiao, W. Investigating the thermal stability of six caffeoylquinic acids employing rapid-resolution liquid chromatography with quadrupole time-of-flight tandem mass spectrometry. *Eur. Food Res. Technol.* **2015**, *240*, 1225–1234. [CrossRef]
42. Governa, P.; Manetti, F.; Miraldi, E.; Biagi, M. Effects of in vitro simulated digestion on the antioxidant activity of different *Camellia sinensis* (L.) Kuntze leaves extracts. *Eur. Food Res. Technol.* **2022**, *248*, 119–128. [CrossRef]
43. Ozkan, G.; Franco, P.; De Marco, I.; Capanoglu, E.; Esatbeyoglu, T. Investigating the effects of supercritical antisolvent process and food models on antioxidant capacity, bioaccessibility and transepithelial transport of quercetin and rutin. *Food Funct.* **2022**, *13*, 4469–4477. [CrossRef] [PubMed]
44. Gonçalves, J.; Ramos, R.; Luís, Â.; Rocha, S.; Rosado, T.; Gallardo, E.; Duarte, A.P. Assessment of the Bioaccessibility and Bioavailability of the Phenolic Compounds of *Prunus avium* L. by in Vitro Digestion and Cell Model. *ACS Omega* **2019**, *4*, 7605–7613. [CrossRef]

Disclaimer/Publisher's Note: The statements, opinions and data contained in all publications are solely those of the individual author(s) and contributor(s) and not of MDPI and/or the editor(s). MDPI and/or the editor(s) disclaim responsibility for any injury to people or property resulting from any ideas, methods, instructions or products referred to in the content.

foods

Article

Efficient and Selective Extraction of Rhamnogalacturonan-I-Enriched Pectic Polysaccharides from Tartary Buckwheat Leaves Using Deep-Eutectic-Solvent-Based Techniques

Ding-Tao Wu [1,*], Jing Lei [1], Jie Li [1], Mei-Mei Qu Mo [2], Wen-Bing Li [2], Yu-Jing Huang [1], Yi-Chen Hu [1], Ai-Li Wang [1] and Liang Zou [1]

[1] Key Laboratory of Coarse Cereal Processing (Ministry of Agriculture and Rural Affairs), Sichuan Engineering & Technology Research Center of Coarse Cereal Industralization, School of Food and Biological Engineering, Chengdu University, Chengdu 610106, China

[2] Tibetan Plateau Ethnic Medicinal Resources Protection and Utilization Key Laboratory of National Ethnic Affairs Commission of the People's Republic of China, Sichuan Provincial Qiang-Yi Medicinal Resources Protection and Utilization Technology Engineering Laboratory, Southwest Minzu University, Chengdu 610225, China

* Correspondence: wudingtao@cdu.edu.cn; Tel./Fax: +86-028-84616653

Abstract: Tartary buckwheat green leaves are considered to be among the most important by-products in the buckwheat industry. Although Tartary buckwheat green leaves are abundant in pectic polysaccharides, their potential applications in the food industry are quite scarce. Therefore, to promote their potential applications as functional or fortified food ingredients, both deep-eutectic-solvent-assisted extraction (DESE) and high-pressure-assisted deep eutectic solvent extraction (HPDEE) were used to efficiently and selectively extract pectic polysaccharides from Tartary buckwheat green leaves (TBP). The results revealed that both the DESE and HPDEE techniques not only improved the extraction efficiency of TBP but also regulated its structural properties and beneficial effects. The primary chemical structures of TBP extracted using different methods were stable overall, mainly consisting of homogalacturonan and rhamnogalacturonan-I (RG-I) pectic regions. However, both the DESE and HPDEE methods could selectively extract RG-I-enriched TBP, and the proportion of the RG-I pectic region in TBP obviously improved. Additionally, both the DESE and HPDEE methods could improve the antioxidant and anti-glycosylation effects of TBP by increasing its proportion of free uronic acids and content of bound polyphenolics and reducing its molecular weight. Moreover, both the DESE and HPDEE methods could partially intensify the immunostimulatory effect of TBP by increasing its proportion of the RG-I pectic region. These findings suggest that DES-based extraction techniques, especially the HPDEE method, can be promising techniques for the efficient and selective extraction of RG-I-enriched TBP.

Keywords: Tartary buckwheat; deep eutectic solvent; high-pressure processing; rhamnogalacturonan-I-enriched pectic polysaccharide; biological activity

1. Introduction

Pectic polysaccharides are abundant inside plant cell walls, which consist of various structurally distinct regions or domains, mainly homogalacturonan (HG), rhamnogalacturonan-I (RG-I), and rhamnogalacturonan-II (RG-II) [1–3]. The HG region usually consists of a linear backbone of →4)-α-D-GalA*p*-(1→ residues that are partially methyl-esterified and O-acetylated, accounting for about 65% of pectic polysaccharides. In addition, the RG-I region is usually composed of a backbone of →4)-α-D-GalA*p*-(1→2,4)-α-L-Rha*p*-(1→ disaccharide residues with neutral side chains, accounting for about 20–35% of pectic polysaccharides [4]. Generally, commercial pectic polysaccharide mainly consists of an

HG region with a small amount of RG-I domain, which is widely utilized as a thickener and gel in the food, medicine, and cosmetics industries [5,6]. However, recent experimental results have demonstrated that RG-I-enriched pectic polysaccharides exhibit several unique functions superior to HG-enriched pectic polysaccharides, such as anti-cancer, immunomodulatory, and prebiotic properties [2,6]. Therefore, the preparation of RG-I-enriched pectic polysaccharides from edible plants and their favorable applications in the functional food industry are becoming increasingly attractive.

Tartary buckwheat (*Fagopyrum tataricum* Gaertn) is a gluten-free pseudocereal of the genus *Fagopyrum*, which is widely distributed in China and Russia [7,8]. Its seeds constitute a highly nutritional cereal, which has a plethora of functional components (e.g., dietary polysaccharides, polyphenols, and proteins) [7,8]. As a consequence, its seeds have various health-promoting effects owing to their abundant functional ingredients [8]. In China, Tartary buckwheat seeds are widely consumed as raw materials of functional tea beverages [8]. In fact, like Tartary buckwheat seeds, Tartary buckwheat green leaves, some of the most important by-products in the buckwheat industry, also have abundant functional components, and these leaves are also utilized as raw materials in functional tea beverages. Interestingly, compared to the seeds, the green leaves are much richer in pectic polysaccharides, primarily consisting of HG and RG-I domains. Moreover, their main constituent monosaccharides are galacturonic acid (GalA), galactose (Gal), arabinose (Ara), and rhamnose (Rha), and their molecular masses are in the range of 5.803×10^5–6.975×10^5 Da [9]. Furthermore, pectic polysaccharides from Tartary buckwheat green leaves also exhibit health-promoting benefits, e.g., in vitro antioxidant, antiglycation, anticancer, prebiotic, and anti-hyperlipidemia effects [9]. Nevertheless, studies on pectic polysaccharides from Tartary buckwheat green leaves and their potential applications as functional or fortified food ingredients are still scant. So, to promote their potential application, it is necessary to develop green and efficient methods for preparing pectic polysaccharides from Tartary buckwheat green leaves.

Generally, the conventional hot water extraction (CHWE) method is considered the common approach for industrialized pectic polysaccharide preparation; nevertheless, it always requires a long extraction time to attain a good extraction yield [4]. Although alkaline and acidic buffers are usually added into the water to reduce processing time and regulate the physicochemical properties of pectic polysaccharides, these buffers are always environmentally unfriendly [3,4]. Deep eutectic solvents (DESs) are attracting increasing interest in relation to their application as an alternative extraction solvent for efficiently preparing pectic polysaccharides from edible plants [10–13]. A DES is a new green extraction solvent that is composed of a hydrogen bond acceptor and a hydrogen bond donor, providing additional interactions with plant cell walls and subsequently allowing more pectic polysaccharides to be extracted compared to conventional extraction solvents [10,13–15]. Furthermore, DESs also offer several advantages, e.g., inexpensiveness, biodegradability, and eco-friendliness [11,16], and pectic polysaccharides extracted by DESs usually provide superior health benefits [13,15,17,18]. Furthermore, apart from the extraction solvent used, several innovative methods, e.g., ultrasound extraction, microwave extraction, and high-pressure processing extraction, are also applied to improve the extraction rates and health benefits of pectic polysaccharides [4]. In particular, as an important non-thermal, eco-friendly, and green processing technique, high-pressure processing extraction can enhance the extraction rate for targeting RG-I-enriched pectic polysaccharides from fruits and their by-products [19–21]. Therefore, we speculated that the combined utilization of a DES and high pressure could be a promising approach for the efficient and selective extraction of RG-I-enriched pectic polysaccharides from edible plants.

This study aimed to optimize the extraction conditions of both deep-eutectic-solvent-assisted extraction (DESE) and high-pressure-assisted deep eutectic solvent extraction (HPDEE) for the efficient preparation of pectic polysaccharides from Tartary buckwheat green leaves (TBP) and to further reveal the corresponding influences on the structural properties and biological effects of TBP. The findings can provide innovative approaches for the selective extraction of rhamnogalacturonan-I-enriched TBP and are beneficial to enhancing their potential applications as functional or fortified food ingredients.

2. Materials and Methods

2.1. Materials and Chemicals

Tartary buckwheat green leaves (*F. tataricum* cv. Chengku No. 2) were collected from buckwheat plants after two months of seeding in the buckwheat breeding and cultivation base of Chengdu University (GPS coordinate 104°56′59″ N, 30°32′90″ E). Tartary buckwheat green leaves were vacuum freeze-dried at −80 °C for 48 h and then ground into a powder and sieved through a 50-mesh screen.

2,2′-azino-bis-(3-ethylbenzothiazoline-6-sulfonic acid) (ABTS), 2,2-diphenyl-1-picrylhydrazyl (DPPH), and 1-pheny-3-methyl-5-pyrazolone (PMP) were acquired from Aladdin-E (Shanghai, China). Monosaccharide standards, namely, rhamnose (Rha), mannose (Man), glucuronic acid (GlcA), galacturonic acid (GalA), glucose (Glc), galactose (Gal), xylose (Xyl), and arabinose (Ara), were acquired from Sigma-Aldrich (St. Louis, MO, USA). ELISA kits for the determination of tumor necrosis factor-alpha (TNF-α) and interleukin-6 (IL-6) levels and characteristics were acquired from Elabscience (Wuhan, China).

2.2. Extraction and Isolation of Pectic Polysaccharides from Tartary Buckwheat Green Leaves

2.2.1. Pre-Treatment of Raw Materials

In order to remove ethanol-soluble substances in the Tartary buckwheat green leaves before the extraction of pectic polysaccharides, the powder of dried Tartary buckwheat green leaves was mixed with ethanol at a concentration of 80% (*v*/*v*) in a ratio of 1:10 (g/mL). Subsequently, the mixtures were extracted using an ultrasonic cleaning machine (SB-800DTD, Ningbo Scientz Biotechnology Co., Ltd., Ningbo, China) operated at 480 W for 30 min to eliminate the majority of ethanol-soluble constituents. Afterward, the extracted residues were dried in a drying oven at 45 °C (Shanghai Boxun Industry and Commerce Co., Ltd., Shanghai, China) and subsequently utilized for the extraction of pectic polysaccharides in the following steps.

2.2.2. Conventional Hot-Water Extraction (CHWE)

CHWE was performed based on a previously documented method, with minor modifications [9]. Accordingly, the extracted residues of Tartary buckwheat green leaves (5.0 g) were treated with deionized water (1:30, g/mL) at 95 °C for 180 min. After centrifugation (4000× *g*, 15 min, 4 °C), the supernatant was obtained and concentrated, and the starch was removed via enzymatic degradation with both α-amylase (5 U/mL and 6 h) and glucoamylase (5 U/mL and 8 h). After the removal of starch, the supernatant (about 100 mL) was applied for the precipitation of pectic polysaccharides by adding four volumes (about 400 mL) of 95% (*v*/*v*) ethanol. Furthermore, the precipitations were dissolved in ultrapure water. To completely eliminate small molecules (e.g., monosaccharides, oligosaccharides, and hydrolysates derived from starch), ultrafiltration (molecular weight cutoff, 3500 Da) was carried out, and the supernatant was ultra-filtered three times. Finally, pectic polysaccharides from Tartary buckwheat green leaves (TBP) extracted via CHWE were pre-frozen at −80 °C ± 2 °C for 8 h, dried using vacuum freeze-drying at −80 °C ± 2 °C for 48 h, and named TBP-W.

2.2.3. Deep-Eutectic-Solvent-Assisted Extraction (DESE)

A deep eutectic solvent (DES) storage solution was synthesized by mixing choline chloride and ethylene glycol (molar ratio, 1:3) based on a previously documented method [16]. Furthermore, the DES storage solution was diluted by adding certain volumes of deionized water to prepare a DES extraction solvent. The extracted residues of Tartary buckwheat green leaves (about 5.0 g) were mixed with different volumes of DES extraction solution and then extracted at different temperatures for specific extraction times. Afterward, the next steps in the preparation of pectic polysaccharides from Tartary buckwheat green leaves (TBP) were conducted, constituting the same steps as discussed in Section 2.2.2, and TBP extracted via DESE was named TBP-D. For the optimization of DESE, single-factor experimental design (SFD) was applied to assess the impact of the water content in the DES solvent (10–70%), extraction time (90–210 min), liquid-to-solid ratio (20–60 mL/g), and extraction temperature (60–100 °C) on the extraction yield of TBP-D. Moreover, a Box–Behnken design (BBD) was also employed to assess the impacts of independent factors on the yields of TBP-D. The experimental runs and corresponding results are shown in Table 1. Finally, the experimental results were investigated and modeled using a second-order polynomial equation.

Table 1. The Box–Behnken designs with independent variables and observed yields for deep-eutectic-solvent-assisted extraction (DESE) and high-pressure-assisted deep eutectic solvent extraction (HPDEE).

Experiments	Levels of Extraction Parameters (DESE) [a]				Yields (mg/g)	Levels of Extraction Parameters (HPDEE) [b]				Yields (mg/g)
	X_{11}	X_{12}	X_{13}	X_{14}		X_{21}	X_{22}	X_{23}	X_{24}	
1	0 (180)	1 (70)	−1 (30)	0 (90)	48.6	1 (13)	0 (55)	0 (40)	−1 (90)	28.3
2	−1 (150)	0 (55)	0 (40)	−1 (80)	35.9	0 (10)	1 (70)	−1 (30)	0 (110)	17.0
3	1 (210)	0 (55)	0 (40)	1 (100)	66.1	0 (10)	0 (55)	0 (40)	0 (110)	38.8
4	0 (180)	−1 (40)	0 (40)	1 (100)	60.4	−1 (7)	0 (55)	1 (50)	0 (110)	28.8
5	1 (210)	−1 (40)	0 (40)	0 (90)	58.9	0 (10)	0 (55)	−1 (30)	1 (130)	23.9
6	1 (210)	1 (70)	0 (40)	0 (90)	53.9	1 (13)	0 (55)	1 (50)	0 (110)	30.1
7	0 (180)	0 (55)	0 (40)	0 (90)	74.5	0 (10)	0 (55)	0 (40)	0 (110)	39.4
8	0 (180)	0 (55)	1 (50)	−1 (80)	30.7	1 (13)	1 (70)	0 (40)	0 (110)	22.3
9	0 (180)	0 (55)	1 (50)	1 (100)	57.1	−1 (7)	−1 (40)	0 (40)	0 (110)	31.6
10	0 (180)	1 (70)	0 (40)	1 (100)	49.8	−1 (7)	0 (55)	0 (40)	1 (130)	29.8
11	0 (180)	0 (55)	0 (40)	0 (90)	73.8	0 (10)	−1 (40)	−1 (30)	0 (110)	26.3
12	−1 (150)	0 (55)	0 (40)	1 (100)	61.9	0 (10)	0 (55)	0 (40)	0 (110)	39.5
13	0 (180)	−1 (40)	0 (40)	−1 (80)	41.1	1 (13)	0 (55)	0 (40)	1 (130)	30.4
14	0 (180)	1 (70)	1 (50)	0 (90)	36.8	0 (10)	0 (55)	−1 (30)	−1 (90)	22.9
15	1 (210)	0 (55)	1 (50)	0 (90)	57.6	0 (10)	−1 (40)	1 (50)	0 (110)	30.9
16	−1 (150)	−1 (40)	0 (40)	0 (90)	56.8	1 (13)	0 (55)	−1 (30)	0 (110)	27.1
17	−1 (150)	0 (55)	−1 (30)	0 (90)	63.2	0 (10)	0 (55)	1 (50)	1 (130)	29.5
18	0 (180)	0 (55)	0 (40)	0 (90)	72.9	0 (10)	0 (55)	0 (40)	0 (110)	38.5
19	0 (180)	0 (55)	0 (40)	0 (90)	75.8	0 (10)	−1 (40)	0 (40)	1 (130)	32.7
20	1 (210)	0 (55)	−1 (30)	0 (90)	61.1	0 (10)	1 (70)	1 (50)	0 (110)	18.9
21	0 (180)	0 (55)	−1 (30)	1 (100)	64.5	−1 (7)	0 (55)	0 (40)	−1 (90)	24.6
22	0 (180)	0 (55)	0 (40)	0 (90)	72.3	1 (13)	−1 (40)	0 (40)	0 (110)	32.1
23	0 (180)	−1 (40)	−1 (30)	0 (90)	55.5	−1 (7)	1 (70)	0 (40)	0 (110)	19.9
24	0 (180)	−1 (40)	1 (50)	0 (90)	53.6	−1 (7)	0 (55)	−1 (30)	0 (110)	23.7
25	0 (180)	1 (70)	0 (40)	−1 (80)	24.3	0 (10)	1 (70)	0 (40)	1 (130)	19.3
26	0 (180)	0 (55)	−1 (30)	−1 (80)	39.7	0 (10)	0 (55)	0 (40)	0 (110)	39.2
27	−1 (150)	0 (55)	1 (50)	0 (90)	50.9	0 (10)	0 (55)	1 (50)	−1 (90)	25.1
28	−1 (150)	1 (70)	0 (40)	0 (90)	40.9	0 (10)	−1 (40)	0 (40)	−1 (90)	27.5
29	1 (210)	0 (55)	0 (40)	−1 (80)	43.3	0 (10)	1 (70)	0 (40)	−1 (90)	18.7

[a] DESE: X_{11}, extraction time (min); X_{12}, water content in DES solvent (%, *v*/*v*); X_{13}, liquid/solid ratio (mL/g); X_{14}, extraction temperature (°C). [b] HPDEE: X_{21}, extraction time (min); X_{22}, water content in DES solvent (%, *v*/*v*); X_{23}, liquid/solid ratio (mL/g); X_{24}, extraction pressure (MPa).

2.2.4. High-Pressure-Assisted Deep Eutectic Solvent Extraction (HPDEE)

The HPDEE was carried out using an experimental-level ATS high-pressure homogenizer (AH-NANO, ATS Engineering Limited (Suzhou), Suzhou, China) at room temperature. The DES extraction solvent was prepared in accordance with Section 2.2.3. The extracted residues of Tartary buckwheat green leaves (about 5.0 g) were mixed with different volumes of DES extraction solution and then extracted under different pressures for certain extraction time. Afterward, the next steps for the preparation of pectic polysaccharides from Tartary buckwheat green leaves (TBP), the same steps as those discussed in Section 2.2.2, were carried out, and the TBP extracted by HPDEE was named TBP-PD. For the optimization of HPDEE, the SFD was also used to assess the impacts of extraction time (7–19 min), water content in DES solvent (10–70%), liquid-to-solid ratio (20–60 mL/g), and extraction pressure (50–130 MPa) on the extraction yield of TBP-PD. Similarly, a BBD design was also employed to evaluate the impacts of extraction factors on the yields of TBP-PD. The experimental runs and corresponding results are shown in Table 1. The experimental results were investigated and modeled by using a second-order polynomial equation.

2.3. Physicochemical and Structural Characterization of Pectic Polysaccharides from Tartary Buckwheat Green Leaves Extracted Using Different Techniques

The quantification of total polysaccharides, acidic sugars, proteins, and bound polyphenolics in TBP-W, TBP-D, and TBP-PD was conducted using colorimetric methods performed as described in previous studies [9,22]. The molecular masses (M_w) and molecular mass distributions (M_w/M_n) of TBP-W, TBP-D, and TBP-PD were measured via size exclusion chromatography combined with multiangle laser light scattering detection and refractive index detection (Wyatt Technology Co., Santa Barbara, CA, USA), as previously documented [22]. The primary chemical structures of TBP-W, TBP-D, and TBP-PD were elucidated using Fourier transform infrared spectroscopy (FT-IR) analysis and nuclear magnetic resonance spectroscopy (NMR) analysis, as previously reported [9,22]. Briefly, the monosaccharide units of TBP-W, TBP-D, and TBP-PD were measured via PMP-derivatization followed by liquid chromatography analysis (LC, L-20A, Shimadzu, Japan). The chemical groups and esterified degrees of TBP-W, TBP-D, and TBP-PD were analyzed using FT-IR analysis (Spectrum Two, PerkinElmer, Waltham, MA, USA). The glycosidic linkages of TBP-W, TBP-D, and TBP-PD were analyzed utilizing a Bruker Ascend 600 MHz spectrometer (Bruker, Rheinstetten, Germany).

2.4. Evaluation of Beneficial Effects of Pectic Polysaccharides from Tartary Buckwheat Green Leaves Extracted Using Different Techniques

The free-radical-scavenging abilities (ABTS and DPPH) and ferric-reducing antioxidant powers (FRAP) of TBP-W, TBP-D, and TBP-PD were determined to assess their antioxidant effects [9,22]. The scavenging abilities of TBP extracted via different techniques were assessed at various concentrations, and the IC_{50} values (mg/mL) of TBP-W, TBP-D, and TBP-PD were calculated separately. In addition, the inhibitory effects of TBP-W, TBP-D, and TBP-PD against the generation of advanced glycation end-products (AGEs) in an in vitro bovine serum albumin–glucose model were determined to assess their anti-glycosylation effects [9]. Furthermore, the immunostimulatory properties of TBP-W, TBP-D, and TBP-PD were assessed by using an in vitro cell model [22]. Effects of TBP-W, TBP-D, and TBP-PD on the release of nitric oxide (NO), IL-6, and TNF-α from RAW 264.7 macrophages were assessed.

2.5. Statistical Analysis

All data were analyzed using Design Expert software (v8.0.6.1, Stat-Ease Inc., Minneapolis, MN, USA) and IBM SPSS Statistics software (v22.0, SPSS, Inc., Chicago, IL, USA). One-way analysis of variance and Student's *t*-test were carried out to assess statistical significance ($p < 0.05$).

3. Results and Discussion

3.1. Extraction Optimization of Deep-Eutectic-Solvent-Assisted Extraction (DESE) and High-Pressure-Assisted Deep Eutectic Solvent Extraction (HPDEE)

3.1.1. Extraction Optimization for the DESE Method

Figure 1 illustrates the impacts of various extraction factors, namely, time, temperature, the content of water in the deep eutectic solvent (DES), and the liquid-to-solid ratio, on the extraction yields of TBP-D. Specifically, the yields of TBP-D exhibited an upward trend as the extraction time increased from 90 to 180 min, but the yields declined past this point as time went on (Figure 1A). This result might be due to the thermal degradation of pectic polysaccharides during the long heating process [4]. In addition, temperature was also an important factor influencing the yield of pectic polysaccharides [5]. The extraction yield reached a maximum value when the temperature was 90 °C (Figure 1B). It is important to note that a higher temperature might also result in the thermal degradation of pectic polysaccharides. Furthermore, the water content in a DES significantly affects the physical properties of a DES solution, resulting in various extraction yields of pectic polysaccharides [13]. As presented in Figure 1C, the extraction yields of TBP-D exhibited a positive correlation with the content of water ranging from 10% to 55%, while the yields of TBP-D declined with the increased water content. Generally, the low water content in a DES can make it difficult to penetrate into plant cell walls due to its high viscosity, while high water content can block the interactions between pectic polysaccharides and the DES [13,23]. Moreover, as represented in Figure 1D, the yield of TBP-D reached a maximum value when the liquid-to-solid ratio was 40 mL/g. Finally, the optimal time, temperature, water content in the DES, and liquid-to-solid ratio were determined to be 180 min, 90 °C, 55%, and 40 mL/g, respectively.

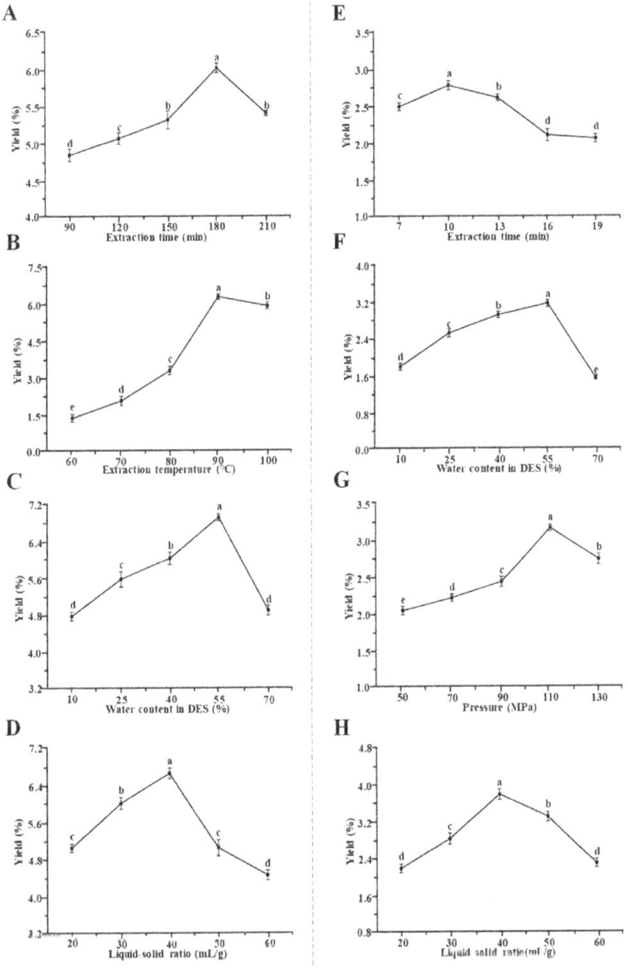

Figure 1. Impacts of various extraction factors of deep-eutectic-solvent-assisted extraction (**A–D**) and high-pressure-assisted deep eutectic solvent extraction (**E–H**) methods on the yields of pectic polysaccharides from Tartary buckwheat green leaves. (**A–D**) indicate the effects of extraction time, extraction temperature, water content in DES, and liquid-solid ratio on the yields of pectic polysaccharides, respectively. (**E–H**) indicate the effects of extraction time, water content in DES, extraction pressure, and liquid-solid ratio on the yields of pectic polysaccharides, respectively. The error bars are standard deviations. Significant differences ($p < 0.05$) among the yields obtained under different extraction conditions are indicated by data bearing different letters (a–e).

Table 1 displays the BBD matrix and outcomes, and the predictive mathematical model is a second-order polynomial equation:

$$\begin{aligned} Y_1 = 7.39 + 0.2608X_{11} - 0.06000X_{12} - 0.3825X_{13} + 1.21X_{14} + 0.2725X_{11}X_{12} \\ +0.2200X_{11}X_{13} - 0.0800X_{11}X_{14} - 0.2475X_{12}X_{13} \\ +0.1550X_{12}X_{14} + 0.0400X_{13}X_{14} - 0.6142X_{11}^2 - 1.49X_{12}^2 \\ -1.00X_{13}^2 - 1.56X_{14}^2 \end{aligned} \quad (1)$$

where Y_1 is the yield of TBP-D, while X_{11}, X_{12}, X_{13}, and X_{14} are the time (min), water content in the DES (%), the liquid-to-solid ratio (mL/g), and temperature (°C), respectively.

According to the outcomes presented in Table 2, the predictive model for the DESE method was significant and could adequately predict the yields of TBP-D [23,24]. In addition, the predictive model had good repeatability, good reliability, and sufficient precision according to the values of the coefficient of variation, adequate precision, R^2, and adjusted R^2 (Table 2) [25,26]. Furthermore, two-dimensional (2D) contour plots of the predictive model are shown in Figure 2. As shown in Figure 2, the interaction effects between the extraction time and water content in the DES, extraction time and the liquid-to-solid ratio, and the water content in the DES and liquid-to-solid ratio were significant, constituting the same effects as those shown in Table 2. By analyzing the equations, the following predictive conditions were obtained: a time of 189.71 min, a water content in the DES of 51.73% (*v*/*v*), a liquid-to-solid ratio of 36.68 mL/g, and a temperature of 92.61 °C. In fact, the validation experiments were conducted using a time of 190.0 min, a water content in the DES of 52.0% (*v*/*v*), a liquid-to-solid ratio of 37.0 mL/g, and a temperature of 93.0 °C, and the observed yield of TBP-D was 7.59% ± 0.09% (n = 3), demonstrating a satisfactory level of concordance with the predictive yield (7.61%). Furthermore, the yield of TBP-D extracted via DESE was extremely higher than that of TBP-W (3.21% ± 0.06%) extracted via CHWE (Table 3), which might be attributed to the fact that the DES could provide additional interactions with the plant cell walls and, subsequently, that more pectic polysaccharides could be extracted [10,13]. These findings indicate that the DESE method can be used as an effective approach for the preparation of TBP.

Table 2. Analysis of the variance for the fitted second-order polynomial models for deep-eutectic-solvent-assisted extraction (DESE) and high-pressure-assisted deep eutectic solvent extraction (HPDEE).

	DESE					HPDEE				
	Sum of Squares	*df*	Mean Square	*F* Value	*p* Value	Sum of Squares	*df*	Mean Square	*F* Value	*p* Value
Model	52.4800	14	3.7500	142.9300	<0.0001	12.3700	14	0.8839	534.4000	<0.0001
X_{11} (X_{21})	0.8164	1	0.8164	31.1300	<0.0001	0.1180	1	0.1180	71.3500	<0.0001
X_{12} (X_{22})	4.3200	1	4.3200	164.7200	<0.0001	3.5200	1	3.5200	2128.7700	<0.0001
X_{13} (X_{23})	1.7600	1	1.7600	66.9400	<0.0001	0.4181	1	0.4181	252.8100	<0.0001
X_{14} (X_{24})	17.4700	1	17.4700	666.2000	<0.0001	0.2852	1	0.2852	172.4400	<0.0001
$X_{11}X_{12}$ ($X_{21}X_{22}$)	0.2970	1	0.2970	11.3300	0.0046	0.0090	1	0.0090	5.4600	0.0349
$X_{11}X_{13}$ ($X_{21}X_{23}$)	0.1936	1	0.1936	7.3800	0.0167	0.0110	1	0.0110	6.6700	0.0217
$X_{11}X_{14}$ ($X_{21}X_{24}$)	0.0256	1	0.0256	0.9761	0.3399	0.0240	1	0.0240	14.5300	0.0019
$X_{12}X_{13}$ ($X_{22}X_{23}$)	0.2450	1	0.2450	9.3400	0.0085	0.0182	1	0.0182	11.0200	0.0051
$X_{12}X_{14}$ ($X_{22}X_{24}$)	0.0961	1	0.0961	3.6600	0.0763	0.0529	1	0.0529	31.9800	<0.0001
$X_{13}X_{14}$ ($X_{23}X_{24}$)	0.0064	1	0.0064	0.2440	0.6290	0.0289	1	0.0289	17.4700	0.0009

Table 2. *Cont.*

	DESE					HPDEE				
	Sum of Squares	*df*	Mean Square	*F* Value	*p* Value	Sum of Squares	*df*	Mean Square	*F* Value	*p* Value
X_{11}^2 (X_{21}^2)	2.4500	1	2.4500	93.3100	<0.0001	1.2200	1	1.2200	740.1300	<0.0001
X_{12}^2 (X_{22}^2)	14.3600	1	14.3600	547.6000	<0.0001	4.4500	1	4.4500	2689.8500	<0.0001
X_{13}^2 (X_{23}^2)	6.5400	1	6.5400	249.4300	<0.0001	3.5600	1	3.5600	2151.4900	<0.0001
X_{14}^2 (X_{24}^2)	15.8000	1	15.8000	602.2700	<0.0001	2.6100	1	2.6100	1578.4900	<0.0001
Residual	0.3672	14	0.0262			0.0232	14	0.0017		
Lack of fit	0.2919	10	0.0292	1.5500	0.3573	0.0161	10	0.0016	0.9082	0.5922
Pure error	0.0753	4	0.0188			0.0071	4	0.0018		
Correlation	52.8500	28				12.4000	28			

DESE: $R^2 = 0.9931$, $R^2_{adj} = 0.9861$, coefficient of variation (CV) = 2.97%, and adeq. precision = 43.0161; X_{11}, extraction time (min); X_{12}, water content in DES (%, v/v); X_{13}, liquid-solid ratio (mL/g); X_{14}, extraction temperature (°C). HPDEE: $R^2 = 0.9981$, $R^2_{adj} = 0.9963$, coefficient of variation (CV) = 1.44%, and adeq. precision = 76.2316; X_{21}, extraction time (min); X_{22}, water content in DES (%, v/v); X_{23}, liquid-solid ratio (mL/g); X_{24}, extraction pressure (MPa).

Table 3. Chemical compositions, molecular weights (M_w), and constituent monosaccharides of pectic polysaccharides extracted from Tartary buckwheat green leaves using different methods.

	TBP-W	TBP-D	TBP-PD
Yields and chemical compositions			
Extraction yields (mg/g)	32.13 ± 0.55^{c}	75.93 ± 0.91^{a}	39.37 ± 1.03^{b}
Total polysaccharides (mg/100 mg)	88.75 ± 0.95^{b}	92.63 ± 3.01^{ab}	94.48 ± 1.97^{a}
Total uronic acids (mg/100 mg)	47.67 ± 0.74^{a}	42.29 ± 0.41^{b}	38.68 ± 1.41^{c}
Total proteins (mg/100 mg)	1.95 ± 0.11^{a}	1.49 ± 0.08^{b}	1.88 ± 0.07^{a}
TPC (mg GAE/g)	4.36 ± 0.21^{c}	23.63 ± 0.69^{a}	13.68 ± 0.25^{b}
Degree of esterification (%)	42.13 ± 0.11^{a}	24.24 ± 0.67^{b}	21.65 ± 0.16^{c}
Molecular weight and its distribution			
$M_w \times 10^5$ (Da, error)	1.29 ± 0.01^{a}	0.67 ± 0.01^{c}	0.88 ± 0.01^{b}
M_w/M_n	2.29	2.09	1.82
Monosaccharides and molar ratios			
Galacturonic acid (GalA)	3.35	2.34	1.71
Galactose (Gal)	1.83	1.61	1.56
Arabinose (Ara)	1.00	0.98	0.97
Rhamnose (Rha)	1.00	1.00	1.00
Xylose (Xyl)	0.14	0.16	0.17
Glucose (Glc)	0.22	0.19	0.24
Glucuronic acid (GlcA)	0.32	0.25	0.22
Mannose (Man)	0.32	0.31	0.31
MR1, Rha/GalA	0.30	0.43	0.58
MR2, (Gal + Ara)/Rha	2.83	2.59	2.53

TBP-W, TBP-D, and TBP-PD indicate pectic polysaccharides extracted from Tartary buckwheat green leaves using conventional hot water extraction, deep-eutectic-solvent-assisted extraction, and high-pressure-assisted deep eutectic solvent extraction, respectively. TPC indicates total polyphenolic content. mg GAE/g, mg of gallic acid equivalent per gram of polysaccharides. Superscripts (a–c) differ significantly ($p < 0.05$) among TBP-W, TBP-D, and TBP-PD.

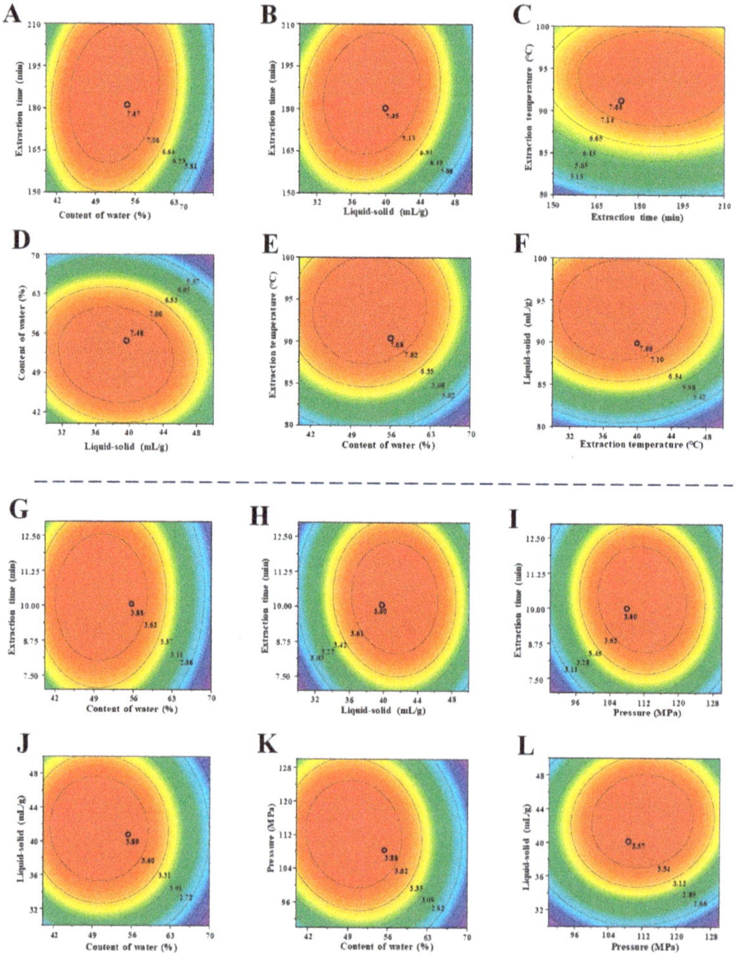

Figure 2. Two-dimensional contour plots of deep-eutectic-solvent-assisted extraction (**A–F**) and high-pressure-assisted deep eutectic solvent extraction (**G–L**). (**A–F**) indicate interactions among extraction time, water content in DES, extraction temperature, and liquid/solid ratio, respectively. (**G–L**) indicate interactions among extraction time, water content in DES, extraction pressure, and liquid–solid ratio, respectively.

3.1.2. Extraction Optimization for the HPDEE Method

Figure 1 also illustrates the impacts of various extraction factors, namely, time, water content in the DES, liquid-to-solid ratio, and extraction pressure, on the yields of TBP-PD. The results revealed that the yields of TBP-PD extracted using HPDEE under different extraction times, water proportions in the DES, and liquid-to-solid ratios exhibited similar change trends to those of TBP-D extracted via DESE. In addition, the effect of high pressure on the yields were determined, as presented in Figure 1G. The yields of TBP-PD enhanced when the pressure increased from 50 MPa to 110 MPa, while obviously declining with a further increase in pressure. This result could be due to the physical degradation of pectic polysaccharides under excessively high pressure [27]. Finally, the optimal time, water content in the DES, pressure, and solid/liquid ratio were determined to be 10 min, 55%, 110 MPa, and 40 mL/g, respectively.

In addition, based on the experimental outcomes (Table 1), the predictive mathematical model was also a second-order polynomial equation, as shown below:

$$\begin{aligned} Y_2 = 3.91 + 0.0992X_{21} - 0.5417X_{22} + 0.1867X_{23} + 0.1542X_{24} + 0.0475X_{21}X_{22} - 0.0525X_{21}X_{23} \\ -0.0775X_{21}X_{24} - 0.0675X_{22}X_{23} - 0.1150X_{22}X_{24} + 0.0850X_{23}X_{24} - 0.04344X_{21}^2 \\ -0.8282X_{22}^2 - 0.7407X_{23}^2 - 0.6344X_{24}^2 \end{aligned} \quad (2)$$

where Y_2 is the yield of TBP-PD, and X_{21}, X_{22}, X_{23}, and X_{24} are the time (min), water content in the DES solvent (%), liquid-to-solid ratio (mL/g), and pressure (MPa), respectively.

Furthermore, the predictive model for the HPDEE method was also highly statistically significant ($p < 0.0001$) according to the outcomes presented in Table 2. Furthermore, the predictive equation for the HPDEE method also had good repeatability, good reliability, and sufficient precision, which allowed it to adequately predict the yields of TBP-PD. Furthermore, according to the combination of the ANOVA analysis results (Table 2) and the 2D contour plots (Figure 2) of the predictive model, the interactions among different extraction parameters ($X_{21}X_{22}$, $X_{21}X_{23}$, $X_{22}X_{23}$, $X_{21}X_{24}$, $X_{22}X_{24}$, and $X_{23}X_{24}$) were significant ($p < 0.05$). All the data showed that the time, water content in the DES, liquid-to-solid ratio, and pressure were decisive factors for the TBP-PD extraction. By analyzing the experimental outcomes, the predictive conditions were determined, namely, a time of 9.46 min, a water content in the DES of 49.20% (v/v), a liquid-to-solid ratio of 40.99 mL/g, and a pressure of 117.05 MPa. Indeed, under the optimal conditions (a time of 10.0 min, a water content in the DES of 50.0% (v/v), a liquid-to-solid ratio of 41.0 mL/g, and a pressure of 110.0 MPa), the observed yield of TBP-PD was 3.94% ± 0.10% (n = 3), exhibiting a favorable level of concordance with the predicted yield of 3.98%. In addition, the yield of TBP-PD was also higher than that of TBP-W (3.21% ± 0.06%) in this study (Table 3). Meanwhile, the extraction time for the HPDEE method was only 10.0 min, which was significantly shorter than that of the CHWE (180 min), indicating that the optimized HPDEE method exhibited superior efficiency in comparison to the CHWE method. Furthermore, although the extraction yield of TBP-PD was lower than that of TBP-D, the time required for the HPDEE method (10 min) was also much shorter than that of the DESE method (190 min).

3.2. *Physicochemical and Structural Properties of Pectic Polysaccharides from Tartary Buckwheat Green Leaves Extracted Using Different Methods*

3.2.1. Comparison of Chemical Components of TBP-W, TBP-D, and TBP-PD

Generally, a favorable extraction technology can not only improve the yield but also regulate the physicochemical properties and improve the functional properties of natural polysaccharides. Therefore, to reveal the influences of various extraction methods on the physicochemical and structural properties of TBP, the chemical components of TBP-W, TBP-D, and TBP-PD were detected and analyzed. The findings showed that various extraction methods not only notably affected the extraction yield of TBP but also obviously influenced its chemical components (Table 3). More specifically, it was revealed that the levels of polysaccharides in TBP-W, TBP-D, and TBP-PD ranged from 88.75 mg/100 mg to 94.48 mg/100 mg, and both TBP-D and TBP-PD had higher total quantities of polysaccharides than TBP-W. Furthermore, the total uronic acids in TBP-W, TBP-D, and TBP-PD ranged from 38.68 mg/100 mg to 47.67 mg/100 mg, and the lowest content was found in TBP-PD. A previous study revealed that TBP is mainly composed of HG and RG-I domains [9]. Therefore, compared to TBP-W, the lower uronic acid levels observed in TBP-PD might be attributed to the increased proportion of the RG-I pectic domain, suggesting that the HPDEE method might selectively extract RG-I-enriched pectic polysaccharides. In fact, previous studies have demonstrated that the high-pressure-assisted extraction method can accelerate the release of RG-I-enriched pectic polysaccharides from edible plants [19,20]. TBP-PD was identified as constituting RG-I-enriched pectic polysaccharides in later studies (Section 3.2.2). In addition, minor proteins were found in TBP-W, TBP-D, and TBP-PD, with the levels ranging from 1.49 mg/100 mg to 1.95 mg/100 mg, indicating that the influence of proteins on the beneficial effect of TBP could be neglected. Moreover, although polyphenolics were removed through the combined use of ultrasound-assisted ethanol extraction, ethanol precipitation, and membrane fractionation, few bound polyphenolics were still found in TBP-W, TBP-D, and TBP-PD, with the levels ranging from 4.36 mg GAE/g to 23.63 mg GAE/g. Compared to TBP-W, the higher number of bound polyphenolics observed in TBP-D might be due to the favorable solubility of polyphenolics in the DES [28]. In fact, the bound polyphenolics had positive effects on the biological functions of pectic polysaccharides [29,30].

3.2.2. Comparison of Molecular Masses and Monosaccharide Compositions of TBP-W, TBP-D, and TBP-PD

To further understand the impacts of various extraction techniques on the structural properties of TBP, the molecular weights, monosaccharide units, chemical groups, and glycosidic bonds in TBP-W, TBP-D, and TBP-PD were detected and compared. The size exclusion chromatograms of TBP-W, TBP-D, and TBP-PD are shown in detail in Figure 3A. It can be seen that TBP-W, TBP-D, and TBP-PD exhibited distinct chromatograms. Compared to TBP-W, the chromatogram of TBP-D shows an obvious shift to the right, suggesting that the molecular weight of TBP-D extracted via DESE was lower than that of TBP-W. Notably, only TBP-PD possessed a symmetrical polysaccharide fraction, suggesting that the HPDEE method could selectively extract distinctive polysaccharide fraction with a reduced content of uronic acids from Tartary buckwheat green leaves (Table 3). In fact, TBP-PD was identified as consisting of RG-I-enriched pectic polysaccharides in the later studies. In addition, the molecular weights of TBP-W, TBP-D, and TBP-PD were determined to be 1.29×10^5 Da, 0.67×10^5 Da, and 0.88×10^5 Da, respectively. These findings show that the molecular mass of TBP-W was notably higher than the molecular masses of TBP-D and TBP-PD, similar to previous reports showing that polysaccharides extracted using DES-based methods had lower molecular weights than that of CHWE [13,15,21]. Furthermore, the polydispersity (M_w/M_n) of TBP-PD was lower than that of TBP-W and TBP-D, suggesting that TBP-PD exhibited a relatively uniform molecular weight distribution.

A monosaccharide composition analysis was carried out to enable the structural characterization of TBP prepared using various methods. As displayed in Figure 3B, the monosaccharide units were identical for TBP-W, TBP-D, and TBP-PD, mainly consisting of GalA, Gal, Ara, and Rha, which confirmed that all the TBP types were pectic polysaccharides rich in homogalacturonan (HG) and rhamnogalacturonan-I (RG-I) domains. Nevertheless, the monosaccharide ratio was distinct for TBP extracted using different techniques. As presented in Table 3, the molar ratios of GalA, Gal, Ara, Rha, Xyl, Glc, GlcA, and Man in TBP-W, TBP-D, and TBP-PD were determined to be 3.35:1.83:1.00:1.00:0.14:0.22:0.32:0.32, 2.34:1.61:0.98:1.00:0.16:0.19:0.25:0.31, and 1.71:1.56:0.97:1.00:0.17:0.24:0.22:0.31, respectively. It is well known that the main monosaccharide unit in the HG domain is GalA, and the main monosaccharide units in the RG-I domain are GalA, Rha, Gal, and Ara [4,19]. Furthermore, the molar ratio of Rha/GalA (MR1) can reflect the proportion of RG-I and HG domains in pectic polysaccharides [3,4]. Notably, compared to TBP-W (GalA, 3.35; Rha/GalA, 0.30), the molar ratio of GalA in both TBP-D and TBP-PD decreased to 2.34 and 1.71, respectively, and their molar ratios of Rha/GalA increased to 0.43 and 0.58, respectively (Table 3). These findings showed that both the DESE and HPDEE methods could successfully applied to obtain RG-I-enriched TBP. The improvement in the RG-I domain in TBP-D could be due to the degradation of the HG backbone due to a β-elimination reaction [31]. Furthermore, the HG domain in pectic polysaccharides can also be destroyed under high-pressure treatment, resulting in an increase in the RG-I pectic domain [19–21]. Furthermore, the molar ratio of (Ara + Gal)/Rha (MR2) can reflect the extent of the branching degree of the RG-I pectic domain, and a high value indicates that the RG-I domain has many and/or long side chains [19]. Compared to TBP-W (2.83), the molar ratios of (Ara + Gal)/Rha in both TBP-D and TBP-PD slightly decreased to 2.59 and 2.53, indicating that both the DESE and HPDEE methods cause slight degradation of neutral side chains. Collectively, these findings suggest that the HPDEE technique can be employed as a promising approach for the targeted extraction of RG-I-enriched TBP.

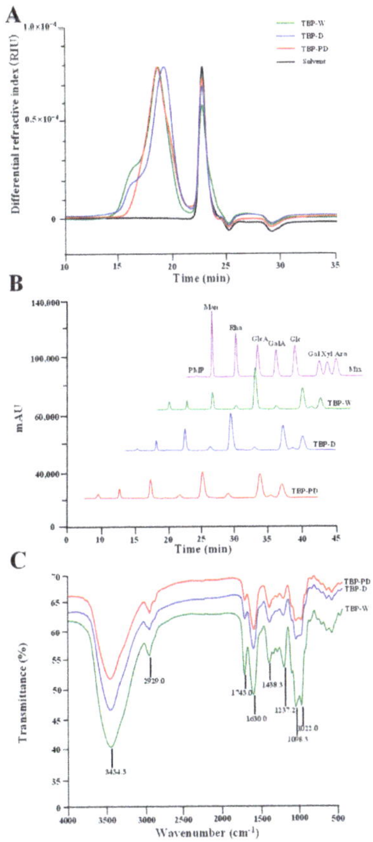

Figure 3. Size exclusion chromatography profiles (**A**), monosaccharide compositions (**B**), and Fourier transform infrared spectra (**C**) of TBP-W, TBP-D, and TBP-PD. TBP-W, TBP-D, and TBP-PD indicate pectic polysaccharides extracted from Tartary buckwheat green leaves via conventional hot water extraction, deep-eutectic-solvent-assisted extraction, and high-pressure-assisted deep eutectic solvent extraction, respectively. Man, mannose. Rha, rhamnose. GlcA, glucuronic acid. GalA, galacturonic acid. Glc, glucose. Gal, galactose. Xyl, xylose. Ara, arabinose.

3.2.3. Comparison of Molecular Weights and Monosaccharide Units of TBP-W, TBP-D, and TBP-PD

Both the FT-IR and NMR spectra of TBP-W, TBP-D, and TBP-PD were investigated and compared to reveal the influences of various methods on their primary chemical structures. The FT-IR spectra of TBP-W, TBP-D, and TBP-PD are presented in Figure 3C. The results revealed that similarly typical adsorption bands (e.g., 3434.3, 2929.0, 1743.0, 1630.0, 1438.3, 1237.2, 1098.3, and 1022.0 cm^{-1}) that corresponded to complex pectic polysaccharides were found in TBP-W, TBP-D, and TBP-PD, suggesting that their chemical groups were not affected by different extraction methods. The absorption bands around 1800–1600 cm^{-1} are due to the vibrations of carbonyl double bonds (C=O) [15,19], which can be utilized for the estimation of the degree of esterification of complex pectic polysaccharides. Generally, the adsorption bands at 1743.0 cm^{-1} and 1630.0 cm^{-1} indicate the vibrations of COO-R groups and COO- groups, respectively [19,20]. The esterification degrees of TBP-W, TBP-D, and TBP-PD were estimated to be 42.13%, 24.24%, and 21.65%, respectively, according to the ratio of adsorption peaks at 1743.0 cm^{-1} and 1630.0 cm^{-1}. These findings reveal that the DES-based extraction techniques could reduce the DE values of pectic polysaccharides, probably due to the promoted RG-I proportion. In addition, the differences found in the FT-IR spectra observed in the 1200–1000 cm^{-1} region indicate differences in the content of GalA [20]. Compared to TBP-D and TBP-PD, the content of GalA was higher in TBP-W, resulting in intense adsorption peaks at 1098.3 cm^{-1} and 1022.0 cm^{-1}.

To better understand the effect of various extraction methods on TBP's chemical structures, 1D NMR analysis was performed. The ^{1}H and ^{13}C NMR spectra of TBP prepared using various methods are presented in Figure 4. The results show that their 1D NMR spectra were extremely similar, indicating that the different extraction methods did not significantly affect their primary chemical structures. Obviously, the typical 1D NMR signals of HG and RG-I domains were observed in all samples. For instance, the ^{1}H signals ranging from 4.97 ppm to 5.44 ppm indicate α-D-GalA*p*, α-L-Rha*p*, and α-L-Ara*f* residues; the ^{1}H signals ranging from 4.46 ppm to 4.63 ppm indicate β-D-Gal*p* residues; and the ^{13}C signals ranging from 100.41 ppm to 109.24 ppm indicate α-D-GalA*p*, β-D-Gal*p*, and α-L-Ara*f* residues. The typical signals of GalA-OCH_3 were observed at 3.81 ppm and 52.86 ppm, further confirming that TBP-W, TBP-D, and TBP-PD were esterified pectic polysaccharides [19,20,32]. Furthermore, several weak signals corresponding to O-acetyl groups were found in the range of 2.00–2.17 ppm, suggesting that these pectic polysaccharides were partially O-acetylated in positions 2 and 3 [1,32]. The typical signals of 1,4-α-D-GalAMe*p* were found at 4.97 ppm (H-1), 100.41 ppm (C-1), and 170.82 ppm (C-6), while the signals of 1,4-α-D-GalA*p* residue were found at 5.01 ppm (H-1), 99.57 ppm (C-1), and 173.34 ppm (C-6) [22,32]. The typical signals of 1,2-α-L-Rha*p* and 1,2,4-α-L-Rha*p* were detected at 5.31 ppm and 5.24 ppm (H-1) as well as 1.24 ppm and 1.30 ppm (H-6), respectively [32]. These findings reveal that both HG and RG-I regions existed in all samples. In addition, the typical signals of T-α-L-Ara*f*, 1,3-α-L-Ara*f*, and 1,5-α-L-Ara*f* were found at 5.15 (H-1)/109.24 ppm (C-1), 5.44 (H-1) ppm, and 5.09 (H-1)/107.36 (C-1) ppm, respectively [22,32]. The typical signals of 1,4-β-D-Gal*p*, 1,3-β-D-Gal*p*, and 1,3,6-β-D-Gal*p* were observed at 4.63 (H-1)/104.27 (C-1) ppm, 4.52 (H-1)/103.26 (C-1) ppm, and 4.46 (H-1)/103.26 (C-1) ppm, respectively [32–35]. These findings indicate that arabinan, type I arabinogalactan (AG-I), and type II arabinogalactan (AG-II) may exist as the branched chains of the RG-I pectic region. In conclusion, all the results show that HG and RG-I regions were observed to be the main pectic polysaccharides in TBP-W, TBP-D, and TBP-PD, and the proportion of RG-I was higher in TBP-PD than in TBP-W and TBP-D based on their monosaccharide ratios, chemical groups, and glycosidic linkages. Nevertheless, due to the limitation of 1D NMR analysis, further 2D NMR analysis is required to reveal more detailed chemical structures of TBP in future studies.

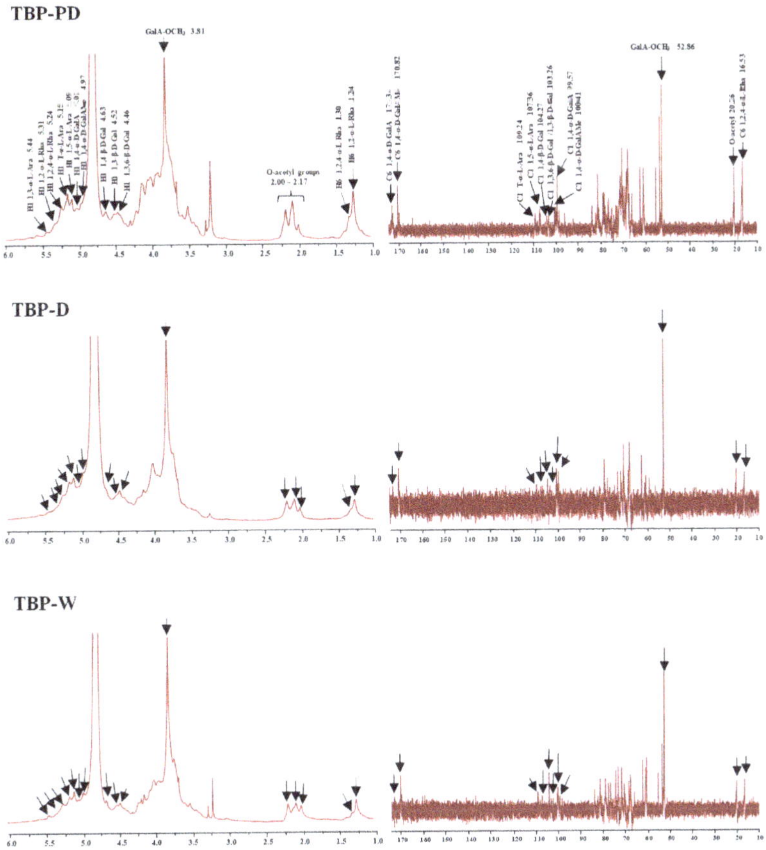

Figure 4. ^{1}H and ^{13}C nuclear magnetic resonance spectra of TBP-W, TBP-D, and TBP-PD. The sample codes are the same as those in Figure 3.

3.3. Comparison of Antioxidant Activities of TBP-W, TBP-D, and TBP-PD In Vitro

Previous studies have shown that dietary polysaccharides derived from Tartary buckwheat seeds, sprouts, and green leaves can scavenge various free radicals, e.g., ABTS, DPPH, and nitric oxide (NO) radicals [9,36,37]. Therefore, the present research examines the antioxidant capacities of TBP-W, TBP-D, and TBP-PD. Figure 5A–C present the antioxidant capacities of TBP prepared using various extraction methods. The results show that all the samples had notable antioxidant properties when compared to the positive controls. Specifically, the IC_{50} values of the scavenging ability of TBP-W, TBP-D, and TBP-PD against ABTS radicals were 4.626 ± 0.087 mg/mL, 0.776 ± 0.015 mg/mL, and 1.353 ± 0.052 mg/mL, respectively. The IC_{50} values of the scavenging ability of TBP-W, TBP-D, and TBP-PD against DPPH radicals were 6.126 ± 0.209 mg/mL, 0.966 ± 0.021 mg/mL, and 1.581 ± 0.021 mg/mL, respectively. In addition, the FRAP values of TBP-D and TBP-PD at the concentrations of 0.5, 1.0, and 1.5 mg/mL were higher than those of TBP-W. Based on the above results, it is clear that the antioxidant activity of TBP-D and TBP-PD was significantly stronger than that of TBP-W, and the highest antioxidant activities among the three samples were observed in TBP-D. These findings indicate that the DES-based extraction approaches can extract pectic polysaccharides with superior antioxidant capacities [17,18]. In general, the antioxidant activities of pectic polysaccharides are frequently linked to their phenolic groups (e.g., bound phenolic acids and bound flavonoids), negative charges (e.g., free carboxyl groups and sulfate groups), and molecular weights [30]. Consequently, the highest antioxidant activities observed in TBP-D among the three samples could be partially associated with its higher free uronic acid and bound polyphenolics levels and lower molecular mass (Table 3).

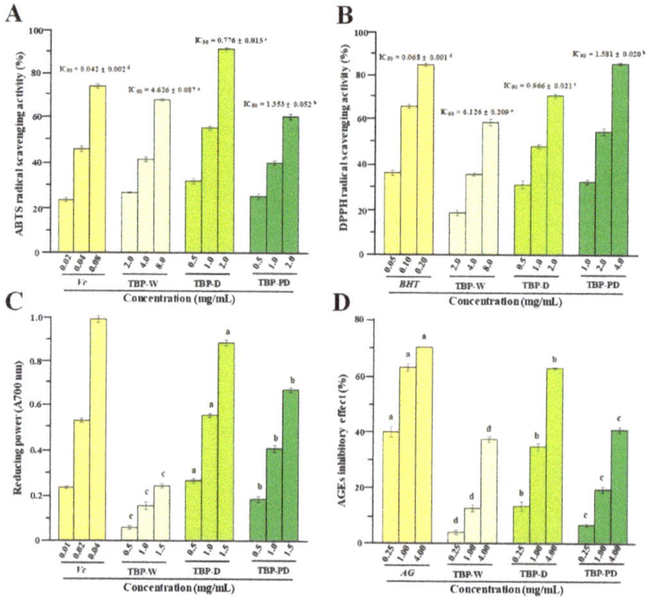

Figure 5. In vitro antioxidant (**A–C**) and anti-glycosylation (**D**) effects of TBP-W, TBP-D, and TBP-PD. The sample codes are the same as those in Figure 3. Vc, vitamin C. BHT, butylated hydroxytoluene. AG, Aminoguanidine. The error bars are standard deviations. Significant differences ($p < 0.05$) among different samples at the same concentration are shown by data bearing different letters (a–c). Significant differences ($p < 0.05$) among different samples and positive controls at the same concentration are shown by data bearing different letters (a–d).

3.4. Comparison of Anti-Glycosylation Activities of TBP-W, TBP-D, and TBP-PD In Vitro

Generally, advanced glycation end-products (AGEs) can be generated by complex aminocarbonyl reactions, and they can cause oxidative stress and the onset of a number of diseases [38]. Pectic polysaccharides can suppress the generation of AGEs during the Maillard reaction owing to their remarkable antioxidant properties [38]. Recent studies have unveiled that complex pectic polysaccharides from Tartary buckwheat sprouts and green leaves have potential anti-glycosylation effects [9,36]. Consequently, the in vitro anti-glycosylation activities of TBP prepared via various extraction methods were studied. Figure 5D presents the inhibitory effects of TBP-W, TBP-D, and TBP-PD against the generation of AGEs in the bovine serum albumin–glucose model, which revealed that all the samples exhibited potential anti-glycosylation effects. Indeed, the inhibitory effects of TBP-D against the generation of AGEs were greater than those of TBP-W and TBP-PD. More specifically, the IC_{50} values of TBP-W, TBP-D, and TBP-PD against the formation of AGEs were measured to be 8.086 ± 0.047 mg/mL, 2.355 ± 0.048 mg/mL, and 7.238 ± 0.135 mg/mL, respectively. Generally, pectic polysaccharides can eliminate free radicals during the generation of AGEs owing to their remarkable antioxidant effects [38]. Furthermore, it has been demonstrated that there is a positive correlation between the anti-glycosylation effect of pectic polysaccharides and their antioxidant activity [9,36]. Thus, the higher anti-glycosylation effect of TBP-D might be partially associated with its higher antioxidant capacity.

3.5. Comparison of Immunostimulatory Effects of TBP-W, TBP-D, and TBP-PD

Generally, the immune system comprises body defenses against foreign or probably dangerous invaders. Pectic polysaccharides can maintain host health by activating the immune system, directly or indirectly interacting with it, thus activating several cellular events [39]. As shown in Figure 6A–D, TBP-W, TBP-D, and TBP-PD had no cytotoxic effects on RAW 264.7 cells, all of which exhibited notable enhancements in the production of NO, IL-6, and TNF-α in a dose-dependent manner. In detail, at a concentration of 200 μg/mL, the levels of NO, IL-6, and TNF-α released from RAW 264.7 macrophages induced by TBP-W, TBP-D, and TBP-PD were in the ranges of 30.35 ± 1.01–37.06 ± 1.29 μM, 12,464.83 ± 337.054–23,265 ± 1289.17 pg/mL, and 4269.55 ± 183.18–9173.18 ± 465.00 pg/mL, respectively. These findings indicated that the immunostimulatory effects of TBP-D and TBP-PD were stronger than those of TBP-W, suggesting that the optimized DESE and HPDEE techniques could be potential extraction methods for extracting TBP with superior immunostimulatory effects.

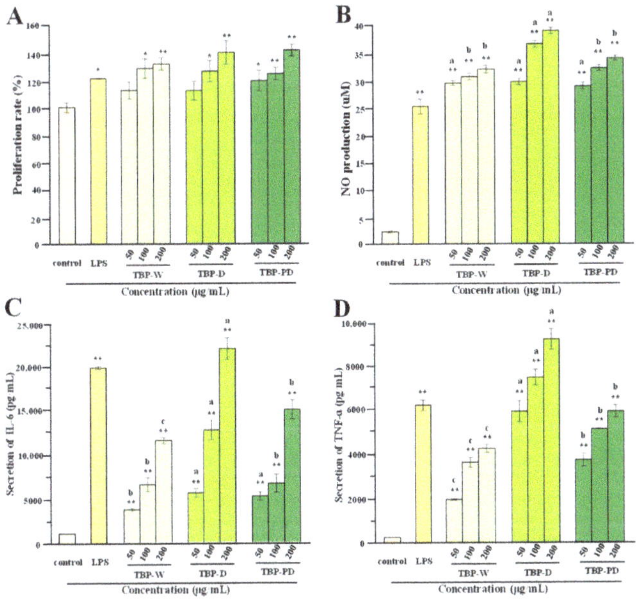

Figure 6. Immunostimulatory effects of TBP-W, TBP-D, and TBP-PD. The sample codes are the same as those in Figure 3. (**A–D**) indicate effects of TBP-W, TBP-D, and TBP-PD on the cell viability, NO production, secretion of IL-6, and secretion of TNF-α of RAW 264.7 macrophages, respectively. The error bars are standard deviations. Significant differences ($p < 0.05$) among different samples at the same concentration are shown by data bearing different letters (a–c). Significant differences between the blank control group and the sample group are shown by * $p < 0.05$ and ** $p < 0.01$.

Generally, the immunostimulatory effects of pectic polysaccharides are linked to their structural properties [39]. As a consequence, the differences found in the immunostimulatory effects of TBP-W, TBP-D, and TBP-PD are due to their complex chemical structures. Previous reports have demonstrated that the immunostimulatory effects of pectic polysaccharides are closely associated with the proportions of HG and RG-I regions [2,3]. A decrease in the HG domain in pectic polysaccharides, resulting in an increase in the RG-I domain, can improve their immunostimulatory effects [40,41]. Therefore, compared to TBP-W, the stronger immunostimulatory effects observed in TBP-D and TBP-PD might be partially attributed to the increased proportion of the RG-I domain, suggesting that both the DESE and HPDEE techniques could partially intensify the immunostimulatory effect of TBP by increasing RG-I content. Furthermore, the immunostimulatory effect of pectic polysaccharides is also associated with their molecular weight and degree of esterification. Therefore, a relatively low molecular mass and a suitable esterification degree might also contribute to the higher immunostimulatory effects of TBP-D and TBP-PD [42–44]. Nevertheless, due to the indigestibility of pectic polysaccharides in the upper gastrointestinal tract, further studies involving animal models should be carried out to confirm that whether TBP exerts immunostimulatory effects through the modulation of gut microbiota.

4. Conclusions

In this study, two DES-based extraction techniques (DESE and HPDEE) were developed for extracting TBP. Compared to the CHWE method, the DESE method could obviously enhance the extraction yield, while the HPDEE method could notably reduce the extraction time, resulting in an increase in extraction efficiency. Indeed, both the DESE and HPDEE methods could selectively extract RG-I-enriched TBP, and the proportion of RG-I domain in both TBP-D and TBP-DP obviously increased. Furthermore, both the DESE and HPDEE methods could promote the antioxidant and anti-glycosylation effects of TBP by increasing its proportion of free uronic acids and content of bound polyphenolics and reducing its molecular weight. Moreover, both the DESE and HPDEE methods could also intensify the immunostimulatory effect of TBP by increasing its proportion of RG-I domain. Collectively, these findings suggest that the developed DES-based extraction methods, especially the HPDEE method, are promising techniques for the selective extraction of RG-I-enriched TBP.

Author Contributions: Conceptualization, D.-T.W.; methodology, D.-T.W.; validation, J.L. (Jie Li) and M.-M.Q.M.; formal analysis, D.-T.W., J.L. (Jing Lei), J.L. (Jie Li), W.-B.L., Y.-C.H., A.-L.W. and L.Z.; investigation, J.L. (Jing Lei), J.L. (Jie Li), M.-M.Q.M. and Y.-J.H.; resources, Y.-C.H., A.-L.W. and L.Z.; data curation, D.-T.W. and J.L. (Jing Lei); writing—original draft preparation, D.-T.W.; writing—review and editing, W.-B.L.; supervision, D.-T.W.; project administration, D.-T.W.; funding acquisition, D.-T.W. All authors have read and agreed to the published version of the manuscript.

Funding: This work was supported by the Technology Innovation R&D Project of the Chengdu Science and Technology Bureau (No. 2022-YF05-00629-SN), the Scientific Research Fund Project of the Science and Technology Department of Sichuan Province (No. 2023YFN0011 and No. 24NSFSC0376), and the Scientific Research Foundation of the China Agriculture Research System (No. CARS-07-B-1).

Institutional Review Board Statement: Not applicable.

Informed Consent Statement: Not applicable.

Data Availability Statement: The original contributions presented in the study are included in the article, further inquiries can be directed to the corresponding author/s.

Conflicts of Interest: The authors declare no conflicts of interest.

References

1. Jin, M.-Y.; Li, M.-Y.; Huang, R.-M.; Wu, X.-Y.; Sun, Y.-M.; Xu, Z.-L. Structural features and anti-inflammatory properties of pectic polysaccharides: A review. *Trends Food Sci. Technol.* **2021**, *107*, 284–298. [CrossRef]
2. Wu, D.; Zheng, J.; Mao, G.; Hu, W.; Ye, X.; Linhardt, R.J.; Chen, S. Rethinking the impact of RG-I mainly from fruits and vegetables on dietary health. *Crit. Rev. Food Sci. Nutr.* **2020**, *60*, 2938–2960. [CrossRef]
3. Niu, H.; Dou, Z.; Hou, K.; Wang, W.; Chen, X.; Chen, X.; Chen, H.; Fu, X. A critical review of RG-I pectin: Sources, extraction methods, structure, and applications. *Crit. Rev. Food Sci. Nutr.* **2023**, *28*, 1–21. [CrossRef] [PubMed]
4. Mao, G.; Wu, D.; Wei, C.; Tao, W.; Ye, X.; Linhardt, R.J.; Orfila, C.; Chen, S. Reconsidering conventional and innovative methods for pectin extraction from fruit and vegetable waste: Targeting rhamnogalacturonan I. *Trends Food Sci. Technol.* **2019**, *94*, 65–78. [CrossRef]
5. Zhang, H.; Chen, J.; Li, J.; Yan, L.; Li, S.; Ye, X.; Liu, D.; Ding, T.; Linhardt, R.J.; Orfila, C.; et al. Extraction and characterization of RG-I enriched pectic polysaccharides from mandarin citrus peel. *Food Hydrocoll.* **2018**, *79*, 579–586. [CrossRef]
6. Hou, Z.; Hu, X.; Luan, L.; Yu, C.; Wang, X.; Chen, S.; Ye, X. Prebiotic potential of RG-I pectic polysaccharides from *Citrus subcompressa* by novel extraction methods. *Food Hydrocoll.* **2022**, *124*, e107213. [CrossRef]
7. Giménez-Bastida, J.A.; Zieliński, H. Buckwheat as a functional food and its effects on health. *J. Agric. Food Chem.* **2015**, *63*, 7896–7913. [CrossRef]

8. Zou, L.; Wu, D.T.; Ren, G.X.; Hu, Y.C.; Peng, L.X.; Zhao, J.L.; Garcia-Perez, P.; Carpena, M.; Prieto, M.A.; Cao, H.; et al. Bioactive compounds, health benefits, and industrial applications of Tartary buckwheat (*Fagopyrum tataricum*). *Crit. Rev. Food Sci. Nutr.* **2023**, *63*, 657–673. [CrossRef]
9. Song, Y.; Lei, J.; Li, J.; Wang, J.; Hu, J.-L.; Zheng, X.-Q.; Hu, Y.-C.; Zou, L.; Wu, D.-T. Structural properties and biological activities of soluble dietary fibers rich in pectic-polysaccharides from different buckwheat green leaves. *Int. J. Biol. Macromol.* **2023**, *253*, 126686. [CrossRef] [PubMed]
10. Shafie, M.H.; Yusof, R.; Gan, C.-Y. Deep eutectic solvents (DES) mediated extraction of pectin from *Averrhoa bilimbi*: Optimization and characterization studies. *Carbohydr. Polym.* **2019**, *216*, 303–311. [CrossRef] [PubMed]
11. Chen, M.; Lahaye, M. Natural deep eutectic solvents pretreatment as an aid for pectin extraction from apple pomace. *Food Hydrocoll.* **2021**, *115*, 106601. [CrossRef]
12. Zdanowicz, M.; Wilpiszewska, K.; Spychaj, T. Deep eutectic solvents for polysaccharides processing. A review. *Carbohydr. Polym.* **2018**, *200*, 361–380. [CrossRef] [PubMed]
13. Gu, J.Y.; Lin, L.Z.; Zhao, M.M. Demonstration of feasibility and effectiveness of deep eutectic solvent-water system extraction of RG-I type pectin from wolfberry based on target polysaccharide, solvent and their interactions. *Food Hydrocoll.* **2023**, *144*, 109027. [CrossRef]
14. Wu, D.T.; Fu, M.X.; Guo, H.; Hu, Y.C.; Zheng, X.Q.; Gan, R.Y.; Zou, L. Microwave-assisted deep eutectic solvent extraction, structural characteristics, and biological functions of polysaccharides from sweet tea (*Lithocarpus litseifolius*) leaves. *Antioxidants* **2022**, *11*, 1578. [CrossRef]
15. Zou, X.; Xiao, J.; Chi, J.; Zhang, M.; Zhang, R.; Jia, X.; Mei, D.; Dong, L.; Yi, Y.; Huang, F. Physicochemical properties and prebiotic activities of polysaccharides from *Zizyphus jujube* based on different extraction techniques. *Int. J. Biol. Macromol.* **2022**, *223*, 663–672. [CrossRef] [PubMed]
16. Wu, D.T.; Feng, K.L.; Huang, L.; Gan, R.Y.; Hu, Y.C.; Zou, L. Deep eutectic solvent-assisted extraction, partially structural characterization, and bioactivities of acidic polysaccharides from lotus leaves. *Foods* **2021**, *10*, 2330. [CrossRef] [PubMed]
17. Zhang, Y.; He, L.; Li, Q.; Cheng, J.; Wang, Y.; Zhao, J.; Yuan, S.; Chen, Y.; Shi, R. Optimization of ultrasonic-assisted deep eutectic solvent for the extraction of polysaccharides from *Indocalamus tessellatus* leaves and their biological studies. *Sustain. Chem. Pharm.* **2022**, *30*, 100855. [CrossRef]
18. Qu, H.; Wu, Y.; Luo, Z.; Dong, Q.; Yang, H.; Dai, C. An efficient approach for extraction of polysaccharide from abalone (*Haliotis Discus* Hannai Ino) viscera by natural deep eutectic solvent. *Int. J. Biol. Macromol.* **2023**, *244*, 125336. [CrossRef]
19. Hou, Z.Q.; Chen, S.G.; Ye, X.Q. High pressure processing accelarated the release of RG-I pectic polysaccharides from citrus peel. *Carbohydr. Polym.* **2021**, *263*, 118005. [CrossRef]
20. Ahmadi, S.; Yu, C.X.; Zaeim, D.; Wu, D.M.; Hu, X.X.; Ye, X.Q.; Chen, S.G. Increasing RG-I content and lipase inhibitory activity of pectic polysaccharides extracted from goji berry and raspberry by high-pressure processing. *Food Hydrocoll.* **2022**, *126*, 107477. [CrossRef]
21. Ke, Y.; Lin, L.Z.; Zhao, M.M. Lotus leaf polysaccharides prepared by alkaline water, deep eutectic solvent and high pressure homogenization-assisted dual enzyme extraction: A comparative study of structural features, prebiotic activities and functionalities. *Food Hydrocoll.* **2023**, *143*, 108870. [CrossRef]
22. Wu, D.T.; Li, J.; Wang, J.; Lei, J.; Gan, R.Y.; Qin, P.Y.; Hu, Y.C.; Wu, X.Y.; Zou, L. Comparison of soluble dietary fibers from various quinoa microgreens: Structural characteristics and bioactive properties. *Food Res. Int.* **2024**, *181*, 114108. [CrossRef]
23. Shang, X.-C.; Chu, D.; Zhang, J.-X.; Zheng, Y.-F.; Li, Y. Microwave-assisted extraction, partial purification and biological activity in vitro of polysaccharides from bladder-wrack (*Fucus vesiculosus*) by using deep eutectic solvents. *Sep. Purif. Technol.* **2021**, *259*, 118169. [CrossRef]
24. Luo, Y.; Peng, B.; Liu, Y.; Wu, Y.; Wu, Z. Ultrasound extraction of polysaccharides from guava leaves and their antioxidant and antiglycation activity. *Process Biochem.* **2018**, *73*, 228–234. [CrossRef]
25. Shang, H.; Wu, H.; Dong, X.; Shi, X.; Wang, X.; Tian, Y. Effects of different extraction methods on the properties and activities of polysaccharides from *Medicago sativa* L. and extraction condition optimization using response surface methodology. *Process Biochem.* **2019**, *82*, 179–188. [CrossRef]
26. Simayi, Z.; Aierken, W.; Rozi, P.; Ababaikeri, G.; Cao, B.; Zhou, C.L.; Askar, G.; Yang, X.J. Optimization of ultrasound-assisted extraction, structural, functional, and antioxidant properties of *Glycyrrhiza uralensis* seed protein. *Process Biochem.* **2023**, *124*, 1–12. [CrossRef]
27. Lei, J.; Li, W.; Fu, M.X.; Wang, A.Q.; Wu, D.T.; Guo, H.; Hu, Y.C.; Gan, R.Y.; Zou, L.; Liu, Y. Pressurized hot water extraction, structural properties, biological effects, and in vitro microbial fermentation characteristics of sweet tea polysaccharide. *Int. J. Biol. Macromol.* **2022**, *222*, 3215–3228. [CrossRef] [PubMed]
28. Ruesgas-Ramón, M.; Figueroa-Espinoza, M.C.; Durand, E. Application of deep eutectic solvents (DES) for phenolic compounds extraction: Overview, challenges, and opportunities. *J. Agric. Food Chem.* **2017**, *65*, 3591–3601. [CrossRef] [PubMed]
29. Huang, H.; Chen, J.; Hu, X.; Chen, Y.; Xie, J.; Ao, T.; Wang, H.; Xie, J.; Yu, Q. Elucidation of the interaction effect between dietary fiber and bound polyphenol components on the anti-hyperglycemic activity of tea residue dietary fiber. *Food Funct.* **2022**, *13*, 2710–2728. [CrossRef] [PubMed]
30. Fernandes, P.A.R.; Coimbra, M.A. The antioxidant activity of polysaccharides: A structure-function relationship overview. *Carbohydr. Polym.* **2023**, *314*, 120965. [CrossRef] [PubMed]

31. Yan, J.-K.; Wang, C.; Qiu, W.-Y.; Chen, T.-T.; Yang, Y.; Wang, W.-H.; Zhang, H.-N. Ultrasonic treatment at different pH values affects the macromolecular, structural, and rheological characteristics of citrus pectin. *Food Chem.* **2021**, *341*, 128216. [CrossRef]
32. Yao, H.Y.Y.; Wang, J.Q.; Yin, J.Y.; Nie, S.P.; Xie, M.Y. A review of NMR analysis in polysaccharide structure and conformation: Progress, challenge and perspective. *Food Res. Int.* **2021**, *143*, 110290. [CrossRef]
33. Redgwell, R.J.; Curti, D.; Wang, J.; Dobruchowska, J.M.; Gerwig, G.J.; Kamerling, J.P.; Bucheli, P. Cell wall polysaccharides of Chinese Wolfberry (*Lycium barbarum*): Part 2. Characterisation of arabinogalactan-proteins. *Carbohydr. Polym.* **2011**, *84*, 1075–1083. [CrossRef]
34. Liu, J.; Zhao, Y.; Wu, Q.; John, A.; Jiang, Y.; Yang, J.; Liu, H.; Yang, B. Structure characterisation of polysaccharides in vegetable "okra" and evaluation of hypoglycemic activity. *Food Chem.* **2018**, *242*, 211–216. [CrossRef]
35. Guo, W.; Rao, G.; Wen, X. Arabinogalactan in banana: Chemical characterization and pharmaceutical effects. *Int. J. Biol. Macromol.* **2021**, *167*, 1059–1065. [CrossRef]
36. Wu, D.T.; Wang, J.; Li, J.; Hu, J.L.; Yan, H.L.; Zhao, J.L.; Zou, L.; Hu, Y.C. Physicochemical properties and biological functions of soluble dietary fibers isolated from common and Tartary buckwheat sprouts. *LWT-Food Sci. Technol.* **2023**, *183*, 114944. [CrossRef]
37. Ji, X.L.; Han, L.; Liu, F.; Yin, S.; Peng, Q.; Wang, M. A mini-review of isolation, chemical properties and bioactivities of polysaccharides from buckwheat (*Fagopyrum* Mill). *Int. J. Biol. Macromol.* **2019**, *127*, 204–209. [CrossRef]
38. Jia, W.; Guo, A.; Zhang, R.; Shi, L. Mechanism of natural antioxidants regulating advanced glycosylation end products of Maillard reaction. *Food Chem.* **2023**, *404*, 134541. [CrossRef] [PubMed]
39. Ferreira, S.S.; Passos, C.P.; Madureira, P.; Vilanova, M.; Coimbra, M.A. Structure function relationships of immunostimulatory polysaccharides: A review. *Carbohydr. Polym.* **2015**, *132*, 378–396. [CrossRef] [PubMed]
40. Ho, G.T.T.; Ahmed, A.; Zou, Y.-F.; Aslaksen, T.; Wangensteen, H.; Barsett, H. Structure–activity relationship of immunomodulating pectins from elderberries. *Carbohydr. Polym.* **2015**, *125*, 314–322. [CrossRef] [PubMed]
41. Zou, Y.-F.; Fu, Y.-P.; Chen, X.-F.; Austarheim, I.; Inngjerdingen, K.T.; Huang, C.; Lei, F.-Y.; Song, X.; Li, L.; Ye, G.; et al. Polysaccharides with immunomodulating activity from roots of *Gentiana crassicaulis*. *Carbohydr. Polym.* **2017**, *172*, 306–314. [CrossRef]
42. Song, Y.-R.; Han, A.-R.; Lim, T.-G.; Lee, E.-J.; Hong, H.-D. Isolation, purification, and characterization of novel polysaccharides from lotus (*Nelumbo nucifera*) leaves and their immunostimulatory effects. *Int. J. Biol. Macromol.* **2019**, *128*, 546–555. [CrossRef] [PubMed]
43. Yuan, P.; Aipire, A.; Yang, Y.; Wei, X.; Fu, C.; Zhou, F.; Mahabati, M.; Li, J.; Li, J. Comparison of the structural characteristics and immunostimulatory activities of polysaccharides from wild and cultivated *Pleurotus feruleus*. *J. Funct. Food* **2020**, *72*, 104050. [CrossRef]
44. Li, W.; Li, J.; Wang, J.; He, Y.; Hu, Y.-C.; Wu, D.-T.; Zou, L. Effects of various degrees of esterification on antioxidant and immunostimulatory activities of okra pectic-polysaccharides. *Front. Nutr.* **2022**, *9*, 1025897. [CrossRef] [PubMed]

Disclaimer/Publisher's Note: The statements, opinions and data contained in all publications are solely those of the individual author(s) and contributor(s) and not of MDPI and/or the editor(s). MDPI and/or the editor(s) disclaim responsibility for any injury to people or property resulting from any ideas, methods, instructions or products referred to in the content.

foods

MDPI

Article

Effect of Fertilization on Phenolics of Rapeseeds and Their Antioxidant Potential

Ryszard Amarowicz [1,*], Bożena Cwalina-Ambroziak [2], Michał Adam Janiak [1], Marta Damszel [2], Arkadiusz Stępień [3], Katarzyna Sulewska [1], Magdalena Karamać [1] and Kamila Penkacik [1]

1 Institute of Animal Reproduction and Food Research, Polish Academy of Sciences, 10 Tuwima Street, 10-748 Olsztyn, Poland; m.janiak@pan.olsztyn.pl (M.A.J.); k.sulewska@pan.olsztyn.pl (K.S.); m.karamac@pan.olsztyn.pl (M.K.); k.penkacik@pan.olsztyn.pl (K.P.)

2 Department of Entomology, Phytopathology and Molecular Diagnostics, Faculty of Agriculture and Forestry, University of Warmia and Masury, 11-041 Olsztyn, Poland; bambr@uwm.edu.pl (B.C.-A.); marta.damszel@uwm.edu.pl (M.D.)

3 Department of Agroecosystems and Horticulture, Faculty of Agriculture and Forestry, University of Warmia and Mazury, 10-721 Olsztyn, Poland; arkadiusz.stepien@uwm.edu.pl

* Correspondence: r.amarowicz@pan.olsztyn.pl

Abstract: Three varieties of rapeseed (Castilla, California, and Nelson F1) were cultivated using medium–intensive (control), intensive, and economical (spare) technologies with different nitrogen and sulfur fertilization techniques. The antioxidant potential of rapeseeds was investigated using ABTS, FRAP, and DPPH assays. The content of total phenolic compounds was determined using the Folin–Ciocalteu phenol reagent. The profile of phenolic compounds was determined using high-performance liquid chromatography (HPLC). Diversifying fertilization in various ways influenced the content of phenolic compounds in extracts of rapeseed. In extracts from the Nelson F1 rapeseeds, intensive cultivation resulted in a lower content of phenolic compounds compared to the control group. Economic fertilization reduced the content of phenolic compounds in seeds from the California variety. HPLC chromatograms of the extracts were characterized by the presence of five (California and Castilla) and six (Nelson F1) main phenolic compounds. Two compounds were identified as sinapine and sinapic acid; others were classified as derivatives of sinapic acid. The effect of fertilization on the antioxidant activity of the seeds and their extracts varied depending on the plant variety and antioxidant assay. For the Castilla and California varieties, no differences were found in the results of the ABTS assay. The antiradical activity against $ABTS^{\bullet+}$ of extracts from the Nelson F1 intensive and spare cultivated seeds was higher than that of extracts from control seeds. The FRAP values of extracts/seeds from the Castilla variety cultivated using different methods did not differ significantly. The results of the DPPH assay were not affected by fertilization in the case of extracts from the California and Castilla varieties. However, the extracts from spare cultivated seeds of Nelson F1 exhibited stronger antiradical activity against $DPPH^{\bullet}$. These findings highlight the complex relationship between fertilization practices, phenolic compound accumulation, and antioxidant activity in rapeseed. Integrating varietal traits and cultivation practices is crucial for optimizing the nutritional benefits of rapeseed.

Keywords: rapeseed; fertilization; phenolic compounds; antioxidant activity; HPLC

1. Introduction

Phenolic compounds are widely distributed in the plant world. They differ in their chemical structure, which in turn affects their chemical properties and biological activity. They appear in all parts of plants, usually accumulated in cell vacuoles, paratomatal, epidermal, subepidermal leaf cells, shoots, and hairs [1], and take part in many plant life activities such as morphogenesis, sex determination, photosynthesis, respiration, regulation of gene expression, and regulation of the synthesis of growth hormones. One of the defense mechanisms of plants is the production of phenolics, which protect against herbivore attacks but also against pathogens and many abiotic stress factors. Many insects are sensitive to the effects of phenolic acids, flavonoids, tannins, and lignans [2,3]. Phenolics exist in various forms, such as soluble-free, soluble-bound, including esterified, etherified, or glycosylated, and insoluble-bound forms, based on their association with food matrices in plants [4].

The diverse biological properties of phenolics have resulted in a sharp increase in interest in them as bioactive food ingredients, which, as nutraceuticals, can contribute to the therapy of various disorders such as neurological diseases and cancer. Most phenolic compounds exhibit strong antioxidant properties. The antioxidant activity of phenolic compounds depends on their chemical structure and molecular weight and may occur through various mechanisms of action, including electron or hydrogen atom transfer, chelation of the pro-oxidant metal ions, and inhibition of enzymes that catalyze oxidation reactions [5].

Phenolic compounds with the ability to scavenge free radicals may participate in the non-enzymatic antioxidative protection system of the human body. Free radicals can be formed between others under the influence of xenobiotics, UV radiation, and ionizing radiation. Mutual interaction of free radicals with cellular biomolecules such as nucleic acids, proteins, lipids, and carbohydrates leads to various structural damages, for example in the DNA strand, which in turn results in mutations. The resulting changes in DNA may be a signal to initiate pathological cell proliferation and the cancer process. Free oxygen species and free radicals can also contribute to the development of many civilization diseases, for example, atherosclerosis, Parkinson's and Alzheimer's diseases, obesity, and chronic inflammation [6].

Rapeseed (*Brassica napus* L.) seeds are a rich source of oil, protein, and phenolic compounds [7–10]. Sinapine, the choline ester of sinapic acid, is the dominant phenolic compound of rapeseed. Sinapic acid in rapeseed also occurs as free phenolic acids or glycosides [11]. In rapeseed, small amounts of other phenolic acids, such as ferulic acid, caffeic acid, *p*-coumaric acid, vanillic acid, syringic acid, *p*-hydroxybenzoic acid, gentisic acid, and protocatechuic acid, have been reported [9]. The major phenolic compound of rapeseed oil is canolol [12], which is produced by the decarboxylation of sinapic acid during rapeseed thermal processing [13]. The presence of condensed tannins in rapeseed and canola hulls was reported by Naczk et al. [14] and Amarowicz et al. [15].

The antioxidant activity of phenolic compounds from rapeseed or canola was reported by several authors and determined through different methods showing free radical scavenging activity and reducing abilities of these compounds [15–20]. Extracts of phenolic compounds obtained from rapeseed were stronger than those of legumes and cereals [5].

The important accumulation of phenolic compounds in rapeseed results from climatic and soil conditions [21]. Utilizing appropriate production technologies to cultivate plants is an important element in modern, sustainable agriculture, and there are several goals to it, including optimizing mineral fertilization and protecting plants against pests. Nitrogen [22] and sulfur [23] fertilization is very important for many species, not only Brassicaceae plants; however, excess fertilization is harmful and impacts the cost of growing crops [24]. Together with the smaller yields of *Brassica rapa* L. chinensis, the ecological system showed a significantly higher level of phenolic compounds in plants compared to plants cultivated using the conventional system. According to Zhao et al. [25], this may result from the unavailability of nutrients.

The effect of fertilization on the plant phenolic compounds is not easily predictable, and the explanation of the different behaviors is not univocal or understandable on the basis of a biochemical pathway. The experiments of the effect of fertilization on the content of phenolic compounds in fruits (apple, black current, blueberry, grapes, marionberry, orange, peach, strawberry, yellow plum) and vegetables (cabbage, carrot, chicory, eggplant, lettuce, onion, pepper, potato, red beat, spinach, tomato) were finished with conclusions of "No differences between organic and conventional production", "Higher in biodynamic than in conventional production", and "Higher in organic than in conventional production". The carbon/nitrogen balance (CNB) pathways can be taken into account when studying the effect of nitrogen fertilization on the content and profile of plant phenolic compounds [26–28].

This research aimed to determine the effect of agro technology (different fertilization) on the antioxidant potential of rapeseeds and their phenolic profile. There are no publications on this important topic in the Scopus and Web of Science databases.

The aim of this research was related to the role of rapeseed as an oil plant and the possibility of using the material after oil extraction as a product to obtain natural oxidants.

2. Materials and Methods

2.1. Plant Material

Winter rape was cultivated in a 4-year monoculture in a close-field experiment in a random block system in triplicate. The experiment was performed in the Agricultural Production Plant in Bałcyny near Olsztyn in Poland (53°78′ N, 20°48′ E) on soil of a granulometric composition of loamy sand with a high content of phosphorus and an average amount of potassium and magnesium.

The following varieties of winter oilseed rape were included: two population varieties, California (tolerant of fungal diseases, with very good winter hardiness, moderately early in ripening) and Castilla (tolerant of fungal diseases, with very good winter hardiness, with even maturation of siliques), and the Nelson F1 hybrid variety (moderately resistant to fungal diseases, good winter hardiness, medium-early in ripening). Based on previous research [29,30], the following cultivation technologies were applied: medium–intensive (control), intensive, and economical (spare), differing in nitrogen and sulfur fertilization. The agrotechnical conditions of the experiment are shown in Table 1. Seeds were collected at the technical maturity stage.

Table 1. Schema of fertilization (kg/ha) during the field experiment.

Fertilization	Experimental Group		
	Control	Intensive	Spare
Phosphorus: Autumn	60	80	40
Potassium: Autumn	120	150	60
Nitrogen: Autumn	30	30	30
Nitrogen: Spring I	120	120	120
Nitrogen: Spring II	60	80	40
Sulphur: Spring	45	60	-

2.2. Chemicals

All solvents used were of analytical grade. Methanol, *n*-hexane, ferrous chloride, sinapic acid, sinapine, the Folin–Ciocalteu phenol reagent, 2,2-diphenyl-1-picrylhydrazyl (DPPH) radical, 2,2′-azinobis-(3-ethylbenzothiazoline-6-sulphonic acid) (ABTS), 2,4,6-tri(2-pyridyl)-*s*-triazine (TPTZ), and 6-hydroxy-2,5,7,8-tetramethyl-chroman -2-carboxylic acid (Trolox) were obtained from Sigma (St. Louis, MO, USA).

2.3. Extraction Procedure

Seeds were ground and defatted using a Soxhlet extractor and n-hexane as a solvent. Phenolic compounds present in defatted plant material were extracted with a methanol and water mixture (8:2; *v*/*v*) for 30 min at a temperature of 50 °C. The ratio of plant material to solvent was 1:10 (*w*/*v*) [31]. Extraction was carried out in a shaking water bath (Elpan 357, Wrocław, Poland). The extracts were filtrated, and the solid residue was extracted twice more. For the evaporation of methanol from the combined filtrates, a Büchi R-200 rotary evaporator (Büchi, Flawil, Switzerland) was used. The water residue was removed using a freeze dryer.

2.4. Determination of Total Phenolics

The total phenolic content in the rapeseed extracts was estimated using a colorimetric method with the Folin–Ciocalteu phenol reagent [32]. Briefly, a 0.25 mL aliquot of extract dissolved in methanol was added to a test tube containing 4 mL of distilled water. Then, 0.25 mL of the Folin–Ciocalteu phenol reagent and 0.25 mL of a saturated Na_2CO_3 solution were added. After 30 min, the absorbance was recorded at 725 nm using a Beckman DU 7500 diode array spectrophotometer (Beckman Coulter, Inc., Brea, CA, USA), and sinapic acid was used to prepare the standard curve. The results were expressed as mg of sinapic acid equivalents (SAE) per g of extract or g of seed fresh matter (FM).

2.5. Trolox Equivalent Antioxidant Capacity Determination

The ABTS assay described by Re et al. [33] was used to measure the Trolox equivalent antioxidant capacity (TEAC). The solution of $ABTS^{\bullet+}$ was prepared by mixing an ABTS solution in water with 2.45 mM sodium persulfate. After shaking for 12 h at room temperature, the obtained $ABTS^{\bullet+}$ stock solution was diluted with methanol to the desired concentration, giving an absorbance of 0.720 at 734 nm. For the colorimetric determination, the absorbance of the sample containing 2 mL of the $ABTS^{\bullet+}$ solution and 20 μL of rapeseed extract was read after 10 min at 734 nm. The calibration curve was plotted using the Trolox standard. The results were expressed as mmol of Trolox equivalents (TE) per g of extract or g of seed FM.

2.6. Ferric-Reducing Antioxidant Power Determination

A ferric-reducing antioxidant power (FRAP) assay was carried out according to the procedures of Benzie and Strain [34] and Amobonye et al. [35]. For the preparation of the working FRAP reagent, 300 mM acetate buffer (pH 3.6) was mixed with 10 mM TPTZ (in 40 mM HCl) and 20 mM $FeCl_3 \times 6\ H_2O$ in a ratio of 10:1:1 ($v/v/v$). For the colorimetric determination, the absorbance of the sample containing 2.25 mL of the FRAP reagent, 75 μL of methanolic solution of rapeseed extract, and 225 μL of deionized water was read at 593 nm after 30 min of incubation. The FRAP values were calculated using the calibration curve for $FeSO_4$. The results were expressed as mmol Fe^{2+} equivalents per g of extract or g of seed FM.

2.7. Determination of DPPH Radical Scavenging Activity

The antiradical activity of the extracts against $DPPH^{\bullet}$ was determined according to the Brand-Williams et al. [36] method, with modifications proposed by Sulewska et al. [37]. A 2 mL of deionized water was mixed with a 0.1 mL methanolic solution containing between 0.47 and 2.35 mg of rapeseed extract. Then, 0.25 m of 1 mM $DPPH^{\bullet}$ solution and 2 mL of deionized water were added. After 20 min, the absorbance of the solution was read at 517 nm. The curves of absorbance vs. extract concentration in the reaction mixture were plotted. The results were expressed as the half maximal scavenging concentration (SC_{50})—the concentration of antioxidants that scavenges half of the radicals.

2.8. HPLC Analysis

Rapeseed extract methanolic solution (5 mg/mL) was filtered (0.45 μm cellulose acetate filter; Millipore, Burlington, MA, USA), and phenolic compounds were analyzed using a Shimadzu HPLC system (Shimadzu Corp., Kyoto, Japan). This system consisted of two LC-10AD pumps, an SPD-M 10A photodiode array detector, and an SCL 10A system controller. Phenolic compounds were separated on a prepacked Luna C_{18} column (4 × 250 mm, 5 μm; Phenomenex, Torrance, CA, USA) for 50 min in a gradient system of 5–40% (v/v) acetonitrile in water adjusted to pH 2.5 with trifluoroacetic acid, at a flow rate of 1 mL/min [38]. The separation was monitored at 320 nm, and the injection volume was 20 μL. The content of sinapine and sinapic acid was calculated using the external standard method. The content of other phenolic compounds showing UV-DAD spectra typical for sinapic acid was expressed as equivalents of this acid.

2.9. Statistical Analysis

In this study, three samples of each rapeseed variety from each field variant were analyzed. Moreover, chemical determinations were triplicated. Results were reported as the mean and standard deviation. For evaluation of the significance of differences among means, analyses of variance (ANOVA) and the Duncan's post hoc test were performed at a level of $p < 0.05$ (GraphPad Prism version 6.04; GraphPad Software, San Diego, CA, USA). The principal component analysis (PCA) and Pearson correlation analysis were carried out using the Statistica 13.1 software (StatSoft Corp., Kraków, Poland).

3. Results

The content of total phenolic compounds in the rapeseed extracts ranged from 48.3 mg SAE/g extract (Castilla, intensive) to 60.9 mg SAE/g extract (California, control) (Table 2). The results expressed in relation to seeds ranged from 4.47 mg SAE/g seed FM (Castillo, spare) to 6.04 mg/g seed FM (Nelson F1, control). The results obtained for the three rapeseed varieties (control) varied as follows: California > Nelson F1 > Castilla for extracts, and Nelson F1 > California > Castilla for seed FM. Diversified fertilization in various ways influenced the total phenolic content of extracts/seeds from different cultivars. For extracts from the Nelson F1 rapeseeds, intensive fertilization resulted in a lower total phenolic content compared to the control group. Economic fertilization reduced the content of total phenolic compounds in extracts of the California variety. The total phenolic content in the California seeds was higher for the intensive group than for the spare group (Table 2).

Table 2. Effect of fertilization on the content of total phenolics and antioxidant potential determined as Trolox equivalent antioxidant capacity (TEAC) and ferric-reducing antioxidant potential (FRAP) of rapeseed fresh matter (FM) and extracts.

Assay	Unit	Cultivar	Experimental Group		
			Control	Intensive	Spare
Total Phenolics	mg SAE/g extract	California	60.9 ± 2.6 ab	59.5 ± 0.7 a	57.1 ± 0.8 b
		Castilla	48.9 ± 1.9 ab	48.3 ± 0.4 a	49.4 ± 0.5 b
		Nelson F1	53.9 ± 0.4 a	49.8 ± 1.2 b	53.2 ± 0.3 a
	mg SAE/g seed FM	California	5.21 ± 0.22 a	5.16 ± 0.06 a	4.60 ± 0.06 b
		Castilla	4.49 ± 0.17 a	4.52 ± 0.12 a	4.47 ± 0.05 a
		Nelson F1	6.04 ± 0.04 a	5.18± 0.12 b	5.26 ± 0.03 b
TEAC	mmol TE/g extract	California	0.468 ± 0.039 a	0.417 ± 0.035 a	0.469 ± 0.035 a
		Castilla	0.366 ± 0.033 a	0.360 ± 0.033 a	0.365 ± 0.032 a
		Nelson F1	0.411 ± 0.015 a	0.418 ± 0.020 a	0.432 ± 0.015 a
	mmol TE/g seed FM	California	0.040 ± 0.003 a	0.036 ± 0.003 a	0.038 ± 0.003 a
		Castilla	0.034 ± 0.003 a	0.034 ± 0.003 a	0.033 ± 0.003 a
		Nelson F1	0.033 ± 0.003 b	0.043 ± 0.002 a	0.043 ± 0.002 a
FRAP	mmol Fe^{2+}/g extract	California	1.61 ± 0.03 a	1.66 ± 0.05 a	1.43 ± 0.05 b
		Castilla	1.34 ± 0.0 a	1.30 ± 0.06 a	1.30 ± 0.08 a
		Nelson F1	1.35 ± 0.11 a	1.15 ± 0.05 b	1.31 ± 0.04 a
	mmol Fe^{2+}/g seed FM	California	0.138 ± 0.003 a	0.144 ± 0.004 a	0.115 ± 0.004 b
		Castilla	0.123 ± 0.007 a	0.122 ± 0.006 a	0.122 ± 0.007 a
		Nelson F1	0.151 ± 0.012 a	0.120 ± 0.005 b	0.129 ± 0.004 b

Means with different letters in the same row are significantly different ($p < 0.05$). SAE—sinapic acid equivalent; TE—Trolox equivalent.

The antioxidant activity determined using ABTS and FRAP assays also depended on the rapeseed variety (Table 2). The Castilla samples showed lower TEAC compared to the seeds and extracts of the other varieties. FRAP values were higher for the California extracts. When considering the effect of fertilization on TEAC, no differences were found between extracts for individual varieties and between seeds of different groups for the

California and Castilla varieties. In the case of the Nelson F1 variety, the TEAC of seeds from the intensive and spare groups (0.043 mmol TE/g seed FM) was higher than that of seeds from the control group (0.033 mmol TE/g seed FM). The obtained FRAP values for extracts of the Castilla variety did not differ from each other. In the California variant, the FRAP values for the spare group (1.43 Fe^{2+}/g extract; 0.115 mmol Fe^{2+}/g seed FM) were lower than the FRAP values for the control group (1.61 mmol Fe^{2+}/g extract; 0.138 mmol Fe^{2+}/g seed FM) and the intensive group (1.66 mmol Fe^{2+}/g extract; 0.144 mmol Fe^{2+}/g seed FM).

The results of the DPPH assay are depicted in Figures 1 and 2. For all extracts, dose-related antiradical activity against DPPH$^{\bullet}$ was observed. Comparing SC_{50} values, the effect of fertilization was not observed for extracts of the California and Castilla varieties. In turn, the extracts of the seeds from the Nelson F1 spare group exhibited stronger antiradical activity against DPPH$^{\bullet}$ than the extracts from the intensive and control groups.

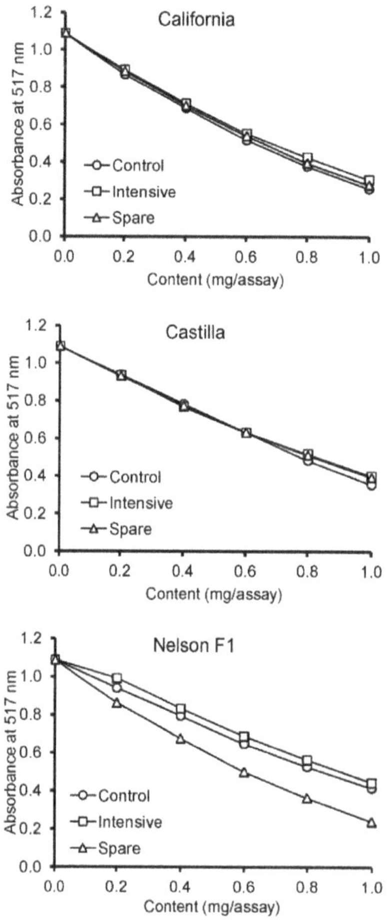

Figure 1. Antiradical activity of rapeseed extracts against DPPH radicals.

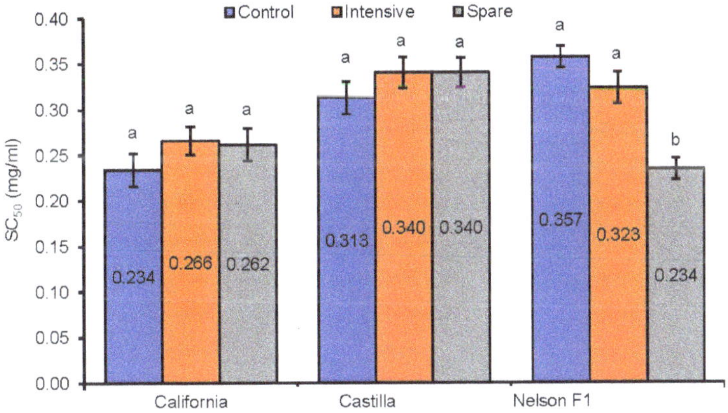

Figure 2. Antiradical activity of rapeseed extracts against DPPH radicals expressed as SC_{50}. Means with different letters are significantly different ($p < 0.05$).

The HPLC chromatograms of the extracts were characterized by the presence of five (California and Castilla) and six (Nelson F1) main phenolic compounds (Figure 3). Applying the original standards, compounds 1 and 4 were identified as sinapine and sinapic acid, respectively. Based on the DAD-UV spectra, it was tentatively found that compounds 2, 3, 5, and 6 were derivatives of sinapic acid.

The content of sinapine in the rapeseed extracts ranged from 64.5 mg/g extract (Nelson F1 intensive group) to 81.9 mg/g extract (California intensive group) (Table 3). The lowest content of sinapic acid was determined in the extracts from the Nelson F1 control group—3.8 mg/g extract, and the highest was in the extracts from the California control group—6.3 mg/g extract. The content of sinapine in seeds ranged from 6.2 mg/g seed FM (California spare group) to 7.4 mg/g seed FM (Nelson F1, control group). The lowest content of sinapic acid was found in the Nelson F1 intensively cultivated seeds—0.41 mg/g seed FM. The control and intensively cultivated seeds of the California and Castillo seeds in the control group were characterized by the highest content of sinapic acid—0.54 mg/g seed FM. Interestingly, compound 5 was quantified only in the seeds/extracts of the Nelson F1 cultivar. The content of this compound was relatively high in the range of 16.9–18.3 mg/g extract and 1.69–2.05 mg/g seed FM. Unlike sinapine and sinapic acid, for which no effect of fertilization was found, its content in control seeds was significantly higher than in intensively cultivated and spare seeds.

The collected data from the analyses of the total phenolic content and antioxidant assays were subjected to PCA to identify the relationships between these variables and rapeseed varieties cultivated in different ways. The PCA results computed for the values of variables expressed per extract are depicted in Figure 4 and for those expressed per seed FM in Figure 5. In the case of data relating to extracts, the first two components (PC1 and PC2) explained 92.45% of the total variance. The system was divided along PC1 into two groups. The first was formed by extracts from the California variety seeds cultivated in each method (A1–A3) and extracts from the Nelson F1 spare cultivated seeds (C3). Most of the variables (total phenolic content and results of ABTS and FRAP assays) described this group. The second group was formed by the rest of the samples, which were closely associated with the results of the DPPH assay. For data relating to seed FM, PC1 and PC2 significantly explained 96.66% of the total variance. Variables affected the A1 and A2, as well as seeds of the Nelson F1 variety from the control (C1) and spare (C3) cultivations. To determine direct correlations between variables, a Pearson correlation analysis was performed for data relating to the extracts. Total phenolic content significantly ($p < 0.05$) correlated with the results of the ABTS and FRAP assays (Table 4). A high correlation coefficient (0.672; $p < 0.05$) was also found for the correlation between TEAC and FRAP. These findings are in line with those shown by PCA (Figure 4B).

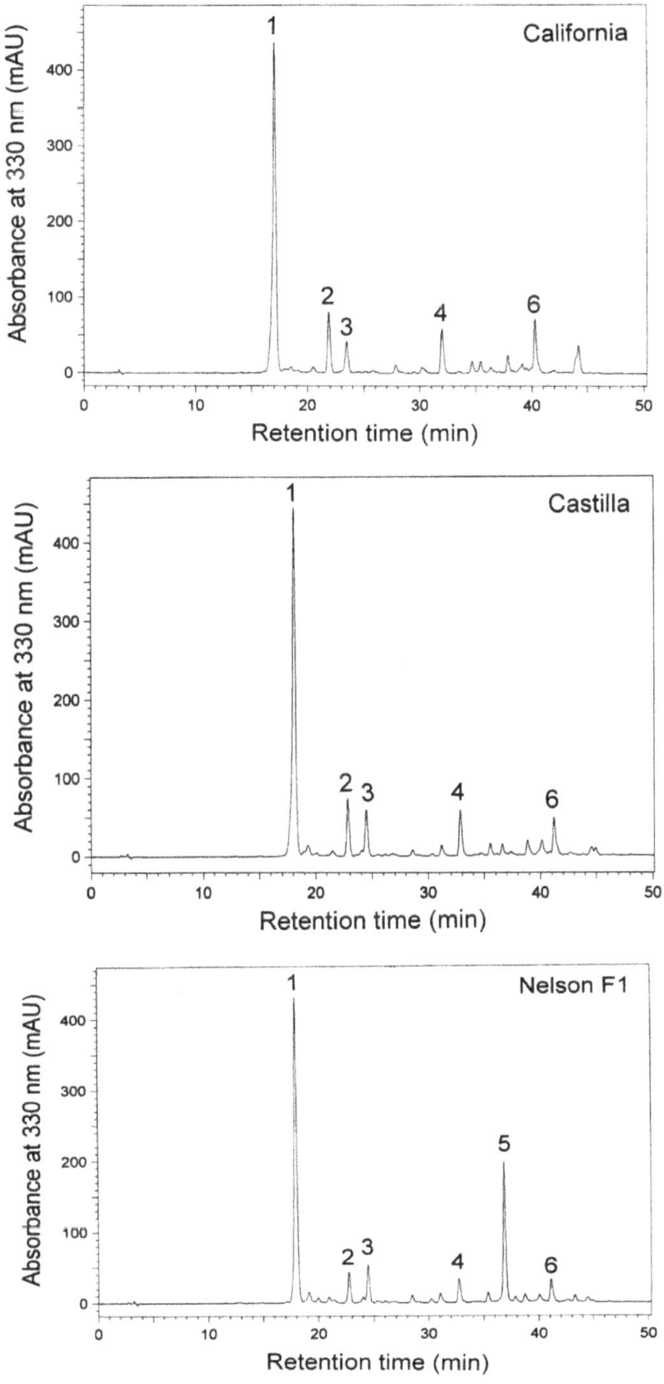

Figure 3. HPLC chromatograms of rapeseed extracts. The peak numbers correspond to the compounds listed in Table 3.

Table 3. Effect of fertilization on the content of individual phenolic compounds in rapeseed extracts and seeds.

Compound	Unit	Cultivar	Experimental Group		
			Control	Intensive	Spare
1 (Sinapine)	mg/g extract	California	80.1 ± 1.4 a	81.9 ± 5.1 a	76.2 ± 4.0 a
		Castilla	76.3 ± 3.4 a	73.0 ± 5.6 a	70.1 ± 1.4 a
		Nelson F1	65.8 ± 3.0 a	64.5 ± 3.3 a	70.8 ± 5.3 a
	mg/g seed FM	California	6.8 ± 0.1 a	7.0 ± 0.4 a	6.2 ± 0.3 a
		Castilla	7.0 ± 0.3 a	6.8 ± 0.5 a	6.3 ± 0.1 a
		Nelson F1	7.4 ± 0.3 a	6.7 ± 0.3 a	7.0 ± 0.5 a
2 (Sinapic acid derivative)	mg/g extract	California	8.4 ± 1.4 a	7.7 ± 0.6 a	7.6 ± 0.6 a
		Castilla	7.1 ± 0.9 a	7.1 ± 1.2 a	6.4 ± 0.2 a
		Nelson F1	4.0 ± 0.1 a	2.5 ± 0.2 b	3.1 ± 0.8 ab
	mg/g seed FM	California	0.72 ± 0.03 a	0.67 ± 0.05 a	0.61 ± 0.05 a
		Castilla	0.65 ± 0.08 a	0.66 ± 0.11 a	0.58 ± 0.02 a
		Nelson F1	0.45 ± 0.01 a	0.26 ± 0.02 b	0.31 ± 0.08 b
3 (Sinapic acid derivative)	mg/g extract	California	5.6 ± 0.3 a	5.1 ± 0.2 a	5.1 ± 0.3 a
		Castilla	5.9 ± 0.5 a	5.9 ± 1.0 a	5.4 ± 0.5 a
		Nelson F1	5.2 ± 0.1 a	4.9 ± 0.2 a	5.0 ± 0.6 a
	mg/g seed FM	California	0.48 ± 0.03 a	0.44 ± 0.02 a	0.41 ± 0.02 a
		Castilla	0.54 ± 0.06 a	0.49 ± 0.06 a	0.50 ± 0.01 a
		Nelson F1	0.58 ± 0.01 a	0.51 ± 0.02 b	0.50 ± 0.04 b
4 (Sinapic acid)	mg/g extract	California	6.3 ± 0.4 a	6.2 ± 0.2 a	6.1 ± 0.3 a
		Castilla	5.9 ± 0.7 a	5.2 ± 0.6 a	5.5 ± 0.1 a
		Nelson F1	3.8 ± 0.2 a	3.9 ± 0.5 a	4.4 ± 0.8 a
	mg/g seed FM	California	0.54 ± 0.04 a	0.54 ± 0.03 a	0.49 ± 0.02 a
		Castilla	0.54 ± 0.06 a	0.49 ± 0.06 a	0.50 ± 0.01 a
		Nelson F1	0.43 ± 0.02 a	0.41 ± 0.05 a	0.44 ± 0.08 a
5 (Sinapic acid derivative)	mg/g extract	California	-	-	-
		Castilla	-	-	-
		Nelson F1	18.3 ± 0.5 a	16.9 ± 1.0 a	17.1 ± 0.9 a
	mg/g seed FM	California	-	-	-
		Castilla	-	-	-
		Nelson F1	2.05 ± 0.06 a	1.76 ± 0.01 b	1.69 ± 0.09 b
6 (Sinapic acid derivative)	mg/g extract	California	8.5 ± 0.5 a	8.3 ± 0.2 a	7.6 ± 0.5 a
		Castilla	6.0 ± 0.7 a	5.7 ± 0.2 a	6.4 ± 1.5 a
		Nelson F1	3.8 ± 0.4 a	3.2 ± 0.4 a	3.5 ± 0.8 a
	mg/g seed FM	California	0.73 ± 0.04 a	0.72 ± 0.02 a	0.61 ± 0.04 a
		Castilla	0.55 ± 0.05 a	0.53 ± 0.05 a	0.58 ± 0.05 a
		Nelson F1	0.43 ± 0.04 a	0.33 ± 0.04 b	0.35 ± 0.08 b

Means with different letters in the same row are significantly different ($p < 0.05$). FM—fresh matter.

Table 4. Pearson correlations between total phenolic content, Trolox equivalent antioxidant capacity (TEAC), ferric-reducing antioxidant potential (FRAP), and DPPH radical scavenging activity of rapeseed of different varieties cultivated in different conditions.

Variables	Regression Equation	Correlation Coefficient
Total phenolics vs. TEAC	y = 0.0071x + 0.0327	0.814 *
Total phenolics vs. FRAP	y = 0.0293x + 0.5733	0.871 *
Total phenolics vs. DPPH	y = −0.0027x + 0.4418	−0.271
TEAC vs. FRAP	y = 1.740x + 0.667	0.672 *
TEAC vs. DPPH	y = −0.2182x + 0.3864	−0.190
FRAP vs. DPPH	y = −0.076x + 0.4016	−0.255

* significant at $p < 0.05$.

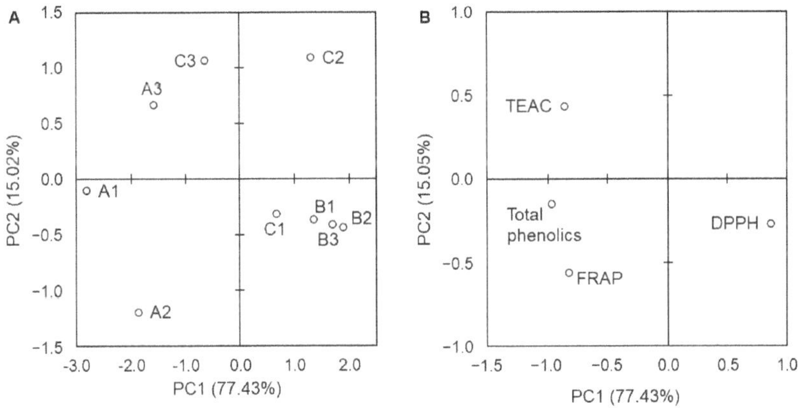

Figure 4. Plots of principal component analysis (PCA) of the objects (**A**) and variables (**B**) for data relating to extracts. A1—California control group; A2—California intensive group; A3—California spare group; B1—Castilla control group; B2—Castilla intensive group; B3—Castilla spare group; C1—Nelson F1 control group; C2—Nelson F1 intensive group; C3—Nelson F1 spare group; TEAC—Trolox equivalent antioxidant capacity; FRAP—ferric-reducing antioxidant potential; DPPH—DPPH radical scavenging activity.

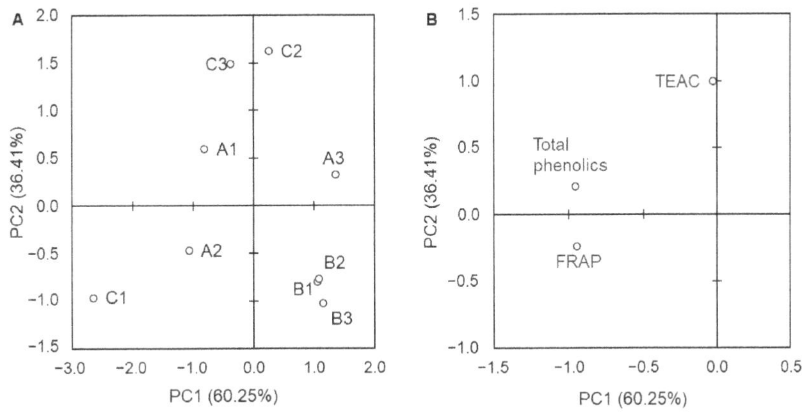

Figure 5. Plots of principal component analysis (PCA) of the objects (**A**) and variables (**B**) for data relating to seed FM. A1—California control group; A2—California intensive group; A3—California spare group; B1—Castilla control group; B2—Castilla intensive group; B3—Castilla spare group; C1—Nelson F1 control group; C2—Nelson F1 intensive group; C3—Nelson F1 spare group; TEAC—Trolox equivalent antioxidant capacity; FRAP—ferric-reducing antioxidant potential.

4. Discussion

The results of total phenolic and sinapine content in rapeseed extracts and seeds are comparable with those reported in the literature. Thiyam et al. [20] reported that the content of total phenolics ranged from 12 to 24 mg/g of rapeseed oil-free meal. According to Mert-Türk et al. [39], the sinapine content in five rapeseed cultivars ranged from 0.33 to 0.43 g/kg. Similar results were reported for sinapic acid content in rapeseed by Siger et al. [40], who found a significant variation in the content of this phenolic compound in rapeseed between the individual harvest years within the same varieties. The variety was a very important factor that influenced the content of phenolic compounds in broccoli [41,42].

Sinapine and sinapic acid are the well-known phenolic compounds of rapeseed. In our research, we also determined four other rapeseed phenolic compounds that were tentatively identified as sinapic acid derivatives. Identification was made based on the similarity of the shape of the UV-DAD spectra with a maxima at a wavelength of 328–330 nm of these compounds with that of sinapic acid. The presence of sinapic acid derivatives in rapeseed has been reported in the literature [43]. Among them, more polar compounds than sinapic acid, including sinaopyl glucose and sinaopyl malate, and less polar compounds, such as disinapic acid and its derivatives, were found. In our study, two sinapic acid derivatives (compounds **2** and **3**) were also more polar, and two others (compounds **5** and **6**) were eluted after sinapic acid. However, further research on determining the structure of these compounds using more advanced analytical techniques and their full identification is necessary. Ferulic acid and its derivatives, quercetin-sinapoyl di-hexosepentose, and kaempferol 3-sinapoylsophorotrioside-7-glucoside were another phenolic compound identified previously in rapeseeds [43]. Our study did not confirm the presence of these compounds in the analyzed varieties.

The effect of different fertilization systems used in our study on the content of total and individual phenolic compounds and the antioxidant potential of the rapeseeds and their extracts varied depending on the plant variety and antioxidant assay; however, clear trends of change were not found. The results of scientific research also do not indicate a clear effect of fertilization on the content of phenolic compounds in rapeseed and other plants belonging to the Brassicaceae family and their antioxidant potential. Increased nitrogen fertilization of rapeseed, as claimed by Butkutė et al. [22] and Mert-Türk et al. [39], increased the content of phenolic compounds in seeds. De Pascale et al. [44] demonstrated an increased accumulation of phenolic acids, including gallic acid, in *Brassica rapa* plants under the influence of increased sulfur fertilization. Fertilization of red cabbage with 150 kg N/ha (nitrogen in the form of ammonium nitrate) was most affected in terms of the content of phenolic compounds and the antioxidant capacity [45]. Sady et al. [46] found the highest content of phenolics in *Brassica oleracea* var. *capitata* alba L. fertilized with RSM (solution ammonium nitrate and urea 1:1). In contrast, according to Li et al. [47], a significant decrease in phenolic compounds in mustard leaves was caused by increasing the level of nitrogen fertilization. Broccoli showed a decrease in flavonoid contents with nitrogen fertilization [48]. The maximum concentration of phenolic compounds was found in broccoli [49] and Chinese cabbage [50] in nonfertilized N plants.

In the ecological system used for *Brassica rapa* L. chinensis plant cultivation, the level of phenolic compounds was higher compared to the conventional system [51]. Johnson et al. [52] reported that a liquid organic nitrogen source should be applied rather than traditional mineral fertilization due to the increased accumulation of *p*-coumaric acid in pac choi. In the organic growing cauliflower *Brassica oleracea* L. subsp. Botrytis, compared to a conventional one, 24, 21, 13, 48, and 44% higher contents of ascorbic acid, polyphenols, carotenoids, and volatile substances were found, as well as a higher antioxidant capacity [53]. Conversa et al. [54] also showed a higher level of bioactive compounds in organic *Brassica rapa* L. The influence of different nitrogen fertilizations at doses of 0, 100, 200, and 300 mg/dm^3, as well as the type of soil, did not affect the total phenolic content and FRAP results in the two varieties of kale, *Brassica oleracea* L. var. acephala [55]. However, red kale, in comparison with green-leafed kale, exhibited a positive effect of N fertilization. In addition, in other research, the production system did not show any effect on the total phenolic content and total flavonoid content in cauliflower (except for single genotypes) [56] or broccoli [57]. In experiments by Conversa et al. [54], the modifying effect of conventional and organic growing systems on the antioxidant properties of crude or processed *Brassica rapa* L. subsp. Sylvestris was not observed. According to Ibrahim and Jaafar [58], Zhang et al. [59], and Deng et al. [60], nitrogen deficiency or lower nitrogen content can trigger ethylene signaling and upregulate the transcription factor MYB12 to stimulate genes involved in the biosynthesis of phenylpropanoids and flavonoids, including chalcone synthase and L-phenylalanine ammonia-lyase. The results of a meta-analysis by Sun et al. [61] suggest a reduction in the internal C/N ratio by nitrogen application and a lowering of the biosynthesis of phenolic compounds, such as phenolic acids, flavonoids, including anthocyanins, and condensed tannins. Regulation by N at the branch point of 3-dehydroshikimate is proposed by Salminen and Maarit [28] to explain the variation in the abundance pattern of phenylpropanoids.

The principal component analysis (PCA) was employed in our study to further understand the relationship within the system. For the data relating to extracts, the importance of PC2 was relatively low (15.02%), and PC1 explained the majority of the variance (77.43%). This significant component discriminated the samples according to the variety rather than the type of fertilization, although in the case of the Nelson F1 variety, the division into seeds from different cultivations occurred along PC1. Considering the PCA results for the seed data, both PC1 (60.25%) and PC2 (36.41) significantly explained the total variance, and they clearly discriminated the Castilla variety from the remaining varieties. For the California and Nelson F1 seeds, samples were discriminated along PC1 and PC2, but no clear clusters were observed based on cultivation type, so we concluded that the effect of fertilization was individual for each variety.

The data from the PCA and the Pearson correlation analysis showed that the results of total phenolics, ABTS, and FRAP assays correlated with each other. The same relation between similar variables was observed for false flax extracts [38] and extracts from seeds of different *Vitis* species [62].

5. Conclusions

This study delved into the impact of varied fertilization techniques on the phenolic compound profile and antioxidant activity of different rapeseed cultivars. We observed significant differences in phenolic content, with the California variety showing the highest content under control conditions. Intensive fertilization reduced the content of phenolic compounds in the Nelson F1 seeds, while economic fertilization decreased the phenolic content in the California variety. Antioxidant assessments through ABTS and FRAP assays revealed cultivar-specific responses. The California variety exhibited higher antioxidant activity in intensively cultivated seeds, whereas the Nelson F1 variety showed stronger antiradical activity in spare-cultivated seeds. HPLC analysis identified key phenolic compounds, predominantly sinapine and sinapic acid. These findings highlight the complex relationship between fertilization practices, phenolic compound accumulation, and the antioxidant activity of rapeseed. Integrating varietal traits and cultivation practices is crucial for optimizing the nutritional benefits of rapeseed. Future research should explore the biochemical mechanisms influencing phenolic compound synthesis in response to specific fertilization regimens.

Author Contributions: Conceptualization, B.C.-A., R.A. and M.A.J.; methodology, B.C.-A., M.D., A.S. and M.A.J.; software, M.A.J. and K.S.; validation, M.A.J. and K.S.; formal analysis, K.S., M.A.J., K.P., M.K., A.S. and M.D.; investigation, M.D. and K.S.; data curation, M.D., K.S. and B.C.-A.; writing—original draft preparation, R.A. and B.C.-A.; writing—review and editing, R.A., B.C.-A. and M.K.; visualization, M.A.J.; supervision, R.A.; project administration, R.A. and B.C.-A. All authors have read and agreed to the published version of the manuscript.

Funding: This research received no external funding.

Institutional Review Board Statement: Not applicable.

Informed Consent Statement: Not applicable.

Data Availability Statement: Data is contained within the article.

Conflicts of Interest: The authors declare no conflicts of interest.

References

1. Zagoskina, N.V.; Zubova, M.Y.; Nechaeva, T.L.; Kazantseva, V.V.; Goncharuk, E.A.; Katanskaya, V.M.; Baranova, E.N.; Aksenova, M.A. Polyphenols in plants: Structure, biosynthesis, abiotic stress, regulation, and practical applications. *Int. J. Mol. Sci.* **2023**, *24*, 13874. [CrossRef]

2. Pratyusha, S. Phenolic compounds in the plant development and defense: An overview. In *Plant Stress Physiology—Perspectives in Agriculture*; Hasanuzzaman, M., Nahar, K., Eds.; IntechOpen: London, UK, 2022; pp. 1–17. [CrossRef]
3. Singh, S.; Kaur, I.; Kariyat, R. The multifunctional roles of polyphenols in plant-herbivore interactions. *Int. J. Mol. Sci.* **2021**, *22*, 1442. [CrossRef]
4. Shahidi, F.; Hossain, A. Importance of insoluble-bound phenolics to the antioxidant potential is dictated by source material. *Antioxidants* **2023**, *12*, 203. [CrossRef]
5. Amarowicz, R.; Pegg, R.B. Natural antioxidants of plant origin. *Adv. Food Nutr. Res.* **2019**, *90*, 1–81. [CrossRef]
6. Rasouli, H.; Farzaei, M.H.; Khodarahmi, R. Polyphenols and their benefits: A review. *Int. J. Food Prop.* **2017**, *20*, 1700–1741. [CrossRef]
7. Loganes, C.; Ballali, S.; Minto, C. Main properties of canola oil components: A descriptive review of current knowledge. *Open Agric. J.* **2016**, *10*, 69–74. [CrossRef]
8. Mupondwa, E.; Li, X.; Wanasundara, J.P.D. Technoeconomic prospects for commercialization of *Brassica* (*Cruciferous*) plant proteins. *J. Am. Oil Chem. Soc.* **2018**, *95*, 903–922. [CrossRef]
9. Chmielewska, A.; Kozłowska, M.; Rachwał, D.; Wnukowski, P.; Amarowicz, R.; Nebesny, E.; Rosicka-Kaczmarek, J. Canola/rapeseed protein–nutritional value, functionality and food application: A review. *Crit. Rev. Food Sci. Nutr.* **2020**, *61*, 3836–3856. [CrossRef]
10. Naczk, M.; Amarowicz, R.; Sullivan, A.; Shahidi, F. Current research developments on polyphenolics of rapeseed/canola: A review. *Food Chem.* **1998**, *62*, 489–502. [CrossRef]
11. Szydłowska-Czerniak, A. Rapeseed and its products—sources of bioactive compounds: A review of their characteristics and analysis. *Crit. Rev. Food Sci. Nutr.* **2013**, *53*, 307–330. [CrossRef]
12. Rękas, A.; Ścibisz, I.; Siger, A.; Wroniak, M. The effect of microwave pretreatment of seeds on the stability and degradation kinetics of phenolic compounds in rapeseed oil during long-term storage. *Food Chem.* **2017**, *222*, 43–52. [CrossRef]
13. Shrestha, K.; Stevens, C.V.; de Meulenaer, B. Isolation and identification of a potent radical scavenger (canolol) from roasted high erucic mustard seed oil from Nepal and its formation during roasting. *J. Agric. Food Chem.* **2012**, *60*, 7506–7512. [CrossRef]
14. Naczk, M.; Amarowicz, R.; Pink, D.; Shahidi, F. Insoluble condensed tannins of canola/rapeseed. *J. Agric. Food Chem.* **2000**, *48*, 1758–1762. [CrossRef]
15. Amarowicz, R.; Naczk, M.; Shahidi, F. Antioxidant activity of crude tannins of canola and rapeseed hulls. *J. Am. Oil Chem. Soc.* **2000**, *77*, 957–961. [CrossRef]
16. Szydłowska-Czerniak, A.; Tułodziecka, A. Antioxidant capacity of rapeseed extracts obtained by conventional and ultrasound-assisted extraction. *J. Am. Oil Chem. Soc.* **2014**, *91*, 2011–2019. [CrossRef]
17. Szydłowska-Czerniak, A.; Amarowicz, R.; Szłyk, E. Antioxidant capacity of rapeseed meal and rapeseed oils enriched with meal extract. *Eur. J. Lipid Sci. Technol.* **2010**, *112*, 750–760. [CrossRef]
18. Amarowicz, R.; Fornal, J.; Karamać, M. Effect of seed moisture on phenolic acids in rapeseed oil cake. *Grasas Y Aceites* **1995**, *46*, 354–356. [CrossRef]
19. Amarowicz, R.; Fornal, J.; Karamać, M.; Shahidi, F. Antioxidant activity of extracts of phenolic compounds from rapeseed oil cakes. *J. Food Lipids* **2001**, *8*, 65–74. [CrossRef]
20. Thiyam, U.; Stöckmann, H.; Zum Felde, T.; Schwarz, K. Antioxidative effect of the main sinapic acid derivatives from rapeseed and mustard oil by-products. *Eur. J. Lipid Sci. Technol.* **2006**, *108*, 239–248. [CrossRef]
21. Björkman, M.; Klingen, I.; Birch, A.N.E.; Bones, A.M.; Bruce, T.J.B.; Johansen, T.J.; Meadow, R.; Mølmann, J.; Seljåsen, R.; Smart, L.E.; et al. Phytochemicals of Brassicaceae in plant protection and human health—Influences of climate, environment and agronomic practice. *Phytochemistry* **2011**, *72*, 538–556. [CrossRef] [PubMed]
22. Butkutė, B.; Mašauskienė, A.; Šidlauskas, G.; Sliesaravičienė, L. The effect of agronomic factors and growth conditions on protein and fat content in the seed of spring oilseed rape and on the variation of fatty acids. *Agriculture* **2000**, *70*, 160–175.
23. Zhao, F.M.; McGrath, S.P.; Blake-Kalff, M.M.A.; Link, A.; Tucker, M. Crop responses to sulphur fertilization in Europe. Biogeochemistry of sulphur in agricultural systems. Part II. Fertilizers and fertilization. *Proc. Int. Fertl. Soc.* **2003**, *16*, 26–51.
24. Podleśna, A.; Cacak–Pietrzak, G.; Sowiński, M. Response of winter wheat to fertilizing S in the field experiment. *Fert. Fertil.* **2003**, *4*, 169–179.
25. Zhao, X.; Nechols, J.R.; Williams, K.; Wangc, W.; Carey, E.E. Comparison of phenolic acids in organically and conventionally grown pac choi (*Brassica rapa* L. *chinensis*). *J. Sci. Food. Agric.* **2009**, *89*, 940–946. [CrossRef]
26. Romani, A.; Ieri, F. Plant polyphenol content, soil fertilization and agricultural management: A review. *Eur. Food Res. Technol.* **2017**, *243*, 1107–1115. [CrossRef]
27. Nguyen, P.M.; Niemeyer, E.D. Effects of nitrogen fertilization on the phenolic composition and antioxidant properties of basil (*Ocinum basilicum* L.). *J. Agric. Food Chem.* **2008**, *56*, 8685–8691. [CrossRef]
28. Salminen, J.P.; Maarit, K. Chemical ecology of tannins and other phenolics: We need a change in approach. *Func. Ecol.* **2011**, *25*, 325–338. [CrossRef]
29. Cwalina-Ambroziak, B.; Stępień, A.; Kurowski, T.P.; Głosek-Sobieraj, M.; Wiktorski, A. The health status and yield of winter rapeseed (*Brassica napus* L.) grown in monoculture and in crop rotation under different agricultural production systems. *Arch. Agron. Soil Sci.* **2016**, *62*, 1722–1732. [CrossRef]

30. Stepień, A.; Wojtkowiak, K.; Pietrzak-Fiecko, R. Nutrient content, fat yield and fatty acid profile of winter rapeseed (*Brassica napus* L.) grown under different agricultural production systems. *Chilean J. Agric. Res.* **2017**, *77*, 266–272. [CrossRef]
31. Amarowicz, R.; Wanasundara, U.N.; Karamać, M.; Shahidi, F. Antioxidant activity of ethanolic extract of mustard seed. *Nahrung—Food* **1996**, *40*, 261–263. [CrossRef]
32. Naczk, M.; Shahidi, F. The effect of methanol-ammonia-water treatment on the content of phenolic acids of canola. *Food Chem.* **1989**, *31*, 159–164. [CrossRef]
33. Re, R.; Pellegrini, N.; Proteggente, A.; Pannala, A.; Yang, M.; Rice-Evans, C. Antioxidant activity applying an improved ABTS radical cation decolorization assay. *Free Rad. Biol. Med.* **1999**, *26*, 1231–1237. [CrossRef]
34. Benzie, I.F.; Strain, J.J. The ferric reducing ability of plasma (FRAP) as a measure of "antioxidant power": The FRAP assay. *Anal. Biochem.* **1999**, *239*, 70–76. [CrossRef]
35. Amobonye, A.; Bhagwat, P.; Ruzengwe, F.M.; Singh, S.; Pillai, S. Pear juice clarification using polygalacturonase from Beauveria bassiana: Effects on rheological, antioxidant and quality properties. *Pol. J. Food Nutr. Sci.* **2022**, *72*, 57–67. [CrossRef]
36. Brand-Williams, W.; Cuvelier, M.E.; Berset, C. Use of a free radical method to evaluate antioxidant activity. *LWT Food Sci. Technol.* **1995**, *28*, 25–30. [CrossRef]
37. Sulewska, K.; Rybarczyk-Płońska, A.; Karamać, M. Antioxidant capacity of lentil flour hydrolysates obtained with pancreatin. *Pol. J. Food Nutr. Sci.* **2022**, *72*, 381–391. [CrossRef]
38. Karamać, M.; Gai, F.; Peiretti, P.G. Effect of the growth stage of false flax (*Camelina sativa* L.) on the phenolic compound content and antioxidant potential of the aerial part of the plant. *Pol. J. Food Nutr. Sci.* **2020**, *70*, 189–198. [CrossRef]
39. Mert-Türk, F.; Kemal, Ü.L.; Egesel, Ö. Nitrogen and fungicide applications against *Erysiphe cruciferarum* affect quality components of oilseed rape. *Mycopathologia* **2008**, *165*, 27–35. [CrossRef] [PubMed]
40. Siger, A.; Nogala-Kałucka, M.M.; Lampart-Szczapa, E.E.; Hoffmann, A.A. Phenolic compound contents in new rape varieties. *Oilseed Crops* **2004**, *25*, 263–274.
41. Eberhardt, M.V.; Kobira, K.; Keck, A.S.; Juvik, J.A.; Jeffery, E.H. Correlation analyses of phytochemical composition, chemical, and cellular measures of antioxidant activity of broccoli (*Brassica oleracea* L. var. *italica*). *J. Agric. Food Chem.* **2005**, *53*, 7421–7431. [CrossRef]
42. Singh, A.; Upadhyay, K.; Prasad, A.; Bahadur, A.; Rai, M. Variability of carotenes, vitamin C, E and phenolics in brassica vegetables. *J. Food Comp. Anal.* **2007**, *20*, 106–112. [CrossRef]
43. Reungoat, V.; Allais, F.; Ducatel, H.; Ioannou, I. Extraction and purification processes of sinapic acid derivatives from rapeseed and mustard seed by-products. *Separ. Purific. Rev.* **2022**, *51*, 521–544. [CrossRef]
44. De Pascale, S.; Maggio, A.; Pernice, R.; Fogliano, V.; Barbieri, G. Sulphur fertilization may improve the nutritional value of *Brassica rapa* L. subsp. *sylvestris*. *Eur. J. Agron.* **2007**, *26*, 418–424. [CrossRef]
45. Biesiada, A.; Nawirska-Olszańska, A.; Kucharska, A.; Sokół-Łętowska, K.; Kędra. The effect of nitrogen fertilization on nutritive value and antioxidative activity of red cabbage. *Acta Sci Pol. Hortorum Cultus* **2010**, *9*, 13–21.
46. Sady, W.; Domagała-Świątek, I.; Rożek, S. Effect of nitrogen fertilization on sugars, ascorbic acid and phenolic compounds in white cabbage (*Brassica oleracea* var. *capitata alba* L.). *Acta Sci. Pol. Hortorum Cultus* **2010**, *9*, 41–51.
47. Li, J.; Zhu, Z.; Gerendas, J. Effects of nitrogen and sulfur on total phenolics and antioxidant activity in two genotypes of leaf mustard. *J. Plant Nutr.* **2008**, *31*, 1642–1655. [CrossRef]
48. Fortier, E.; Desjardins, Y.; Tremblay, N.; Bělec, C.; Côtč, M. Influence of irrigation and nitrogen fertilization on broccoli polyphenolics concentration. *Acta Hortic.* **2010**, *856*, 55–62. [CrossRef]
49. Jones, R.B.; Imsic, M.; Franz, P.; Tomkins, B.T. High nitrogen during growth reduced glucoraphanin and flavonol content in broccoli (*Brassica oleracea* var. *italica*) heads. *Aust. J. Exp. Agric.* **2007**, *47*, 1498–1505. [CrossRef]
50. Zhu, W.; Lin, X.; Jin, C.; Zhang, Y.; Fang, P. Effects of nitrogen application rates on antioxidant contents and antioxidative activities in Chinese cabbage (*Brassica chinensis* L.). *J. Zhejiang Univ. Agric. Life Sci.* **2009**, *35*, 299–306.
51. Zhao, X.; Carey, E.E.; Young, J.E.; Wang, W.; Iwamoto, T. Influences of organic fertilization, high tunnel environment, and postharvest storage on phenolic compounds in lettuce. *HortScience* **2007**, *42*, 71–76. [CrossRef]
52. Johnson, W.A.; Cloyd, R.A.; Nechols, J.R.; Williams, K.A.; Nelson, N.O.; Rotenberg, D.; Kennelly, M.M. Effect of nitrogen source on pac choi (*Brassica rapa* L.) chemistry and interactions with the diamondback moth (*Plutella xylostella* L.). *HortiScience* **2012**, *47*, 1457–1465. [CrossRef]
53. Lo Scalzo, R.; Picchi, V.; Migliori, C.A.; Campanelli, G.; Leteo, F.; Ferrari, V.; Di Cesare, L.F. Variations in the phytochemical contents and antioxidant capacity of organically and conventionally grown Italian cauliflower (*Brassica oleracea* L. subsp. botrytis): Results from a three-year field study. *J. Agric. Food Chem.* **2013**, *61*, 10335–10344. [CrossRef]
54. Conversa, G.; Bonasia, A.; Lazzizera, C.; Elia, A. Bio-physical, physiological, and nutritional aspects of ready-to-use cima di rapa (*Brassica rapa* L. subsp. sylvestris L. Janch. var. esculenta Hort.) as affected by conventional and organic growing systems and storage time. *Sci. Hortic.* **2016**, *213*, 76–86. [CrossRef]
55. Łata, B. Variability in enzymatic and non-enzymatic antioxidants in red and green-leafy kale in relation to soil type and N-level. *Sci. Hortic.* **2014**, *168*, 38–45. [CrossRef]
56. Picchi, V.; Migliori, C.; Lo Scalzo, R.; Campanelli, G.; Ferrari, V.; Di Cesare, L.F. Phytochemical content in organic and conventionally grown Italian cauliflower. *Food Chem.* **2012**, *130*, 501–509. [CrossRef]

57. Valverde, J.; Reilly, K.; Villacreces, S.; Gaffney, M.; Granta, J.; Brunton, N. Variation in bioactive content in broccoli (*Brassica oleracea* var. *italica*) grown under conventional and organic production systems. *J. Sci. Food Agric.* **2015**, *95*, 1163–11761. [CrossRef]
58. Ibrahim, M.H.; Jaafar, H.Z.E. Involvement of carbohydrate, protein and phenylanine ammonia lyase in up-regulation of secondary metabolites in *Labisia pumila* under various CO_2 and N_2 level. *Molecules* **2011**, *16*, 4172–4190. [CrossRef]
59. Zhang, Y.; Ma, X.-M.; Wang, X.-C.; Liu, J.-H.; Huang, B.-Y.; Guo, X.-Y.; Xiong, S.P.; La, G.-X. UPLC-QTOF analysis reveals metabolomic changes in the flag leaf of wheat (*Triticum aestivum* L.) under low-nitrogen stress. *Plant Physiol. Bioch.* **2017**, *111*, 30–38. [CrossRef]
60. Deng, B.; Li, Y.; Xu, D.; Ye, Q.; Liu, G. Nitrogen availability alters flavonoid accumulation in *Cyclocarya paliurus* via the effects on the internal carbon/nitrogen balance. *Sci. Rep.* **2019**, *9*, 2370. [CrossRef]
61. Sun, Y.; Guo, J.; Li, Y.; Luo, G.; Li, L.; Yuan, H.; Mur, L.A.J.; Guo, S. Negative effects of the simulated nitrogen deposition on plant phenolic metabolism: A meta-analysis. *Sci. Total Environ.* **2020**, *719*, 137442. [CrossRef] [PubMed]
62. Pozzo, L.; Grande, T.; Raffaelli, A.; Longo, V.; Weidner, S.; Amarowicz, R.; Karamać, M. Characterization of antioxidant and antimicrobial activity and phenolic compound profile of extracts from seeds of different *Vitis* species. *Molecules* **2023**, *28*, 4924. [CrossRef] [PubMed]

Disclaimer/Publisher's Note: The statements, opinions and data contained in all publications are solely those of the individual author(s) and contributor(s) and not of MDPI and/or the editor(s). MDPI and/or the editor(s) disclaim responsibility for any injury to people or property resulting from any ideas, methods, instructions or products referred to in the content.

foods MDPI

Article

In Vitro Antibacterial and Antioxidative Activity and Polyphenolic Profile of the Extracts of Chokeberry, Blackcurrant, and Rowan Berries and Their Pomaces

Kadrin Meremäe [1,*], Piret Raudsepp [1], Linda Rusalepp [1], Dea Anton [1], Uko Bleive [2] and Mati Roasto [1]

[1] Chair of Veterinary Biomedicine and Food Hygiene, Institute of Veterinary Medicine and Animal Sciences, Estonian University of Life Sciences, Fr. R. Kreutzwaldi 56/3, 51006 Tartu, Estonia; piret.raudsepp@emu.ee (P.R.); dea.anton@emu.ee (D.A.); mati.roasto@emu.ee (M.R.)
[2] Polli Horticultural Research Centre, Chair of Horticulture, Institute of Agricultural and Environmental Sciences, Estonian University of Life Sciences, Uus 2, 69108 Polli, Estonia; uko.bleive@emu.ee
* Correspondence: kadrin.meremae@emu.ee; Tel.: +372-731-3431

Abstract: The chemical composition of berries and berry pomaces is diverse, containing polyphenolic components that may have both antibacterial and antioxidative properties. In the present study, in vitro antibacterial effect of the extracts of chokeberry, blackcurrant, and rowan berries and berry pomaces against *L. monocytogenes*, *S. aureus*, *E. coli*, and *C. jejuni* was studied. In addition, the polyphenolic profile and antioxidant activity of these extracts were investigated. The polyphenolic profiles in the aqueous and 30% ethanolic extracts were determined chromatographically by HPLC-MS, and the total polyphenol content was estimated spectrophotometrically by HPLC-DAD-UV. The minimal inhibition concentrations (MICs) of the extracts against tested bacteria were determined by the broth microdilution method. The content of total polyphenols was highest and good antioxidative properties of the extracts were determined for chokeberry and blackcurrant berries and their pomaces. The highest proportions of total quercetin derivatives and anthocyanins were found in the extracts of chokeberry berry/pomace and blackcurrant berry/pomace, respectively. The sensitivity of tested microbes to the extracts of berries and berry pomaces was as follows: *S. aureus* > *L. monocytogenes* > *E. coli* and *C. jejuni*. In vitro antibacterial activity of tested extracts depended on the extraction solvent, mainly for the ethanolic extracts. Findings suggest that chokeberry and blackcurrant berries and their pomaces can be used as a good source of polyphenols with antioxidative properties, and they also have antibacterial activity against some foodborne pathogenic bacteria. It is important that the valuable compounds are extracted from juice press residues before their disposal.

Keywords: antibacterial effect; MIC; antioxidative activity; phenolic compounds; chokeberry; blackcurrant; rowan berries; berry pomaces

1. Introduction

In recent years, there has been an increasing trend in food technology to increase the use of various plant-origin materials, including vegetable, fruit, and berry pomaces, in the composition of food of animal origin [1–3]. The reasons for this are food safety and durability-improving properties [1] and the potential health benefits of plant-derived bioactive components to human health [4]. Also, it is in good agreement with the principles of the circular bioeconomy, because the use of berry pomaces reduces food waste and provides added value to the final products [5].

Among the berries, chokeberry [6], blackcurrant [7], and rowan berries [8] are rich sources of polyphenolic compounds, primarily anthocyanins, which act as good antioxidants for the human body [9]. In addition, polyphenolic compounds also have antimicrobial [10,11] and anti-inflammatory properties [12]. Bioactive compounds in berries include phenolic compounds such as phenolic acids, flavonoids, e.g., anthocyanins and flavonols, and tannins [13]. Berry pomaces, a by-product of juice production, are also of interest because they contain valuable phenolic compounds [14] and therefore have a great potential as a good dietary supplement in foods [15]. According to Heinonen [16] and Klavins et al. [14], the content of total polyphenols in berries is uneven, e.g., 10% in berry pulp, 28–35% in berry skin, and 60–70% in berry seeds, therefore the berry pomace, which contains berry seeds and skins, is a valuable source of polyphenols. According to our knowledge, the amount of berry pomaces is dependent on variety but is approximately 15–20% of the weight of fresh berries.

Several studies [17–19] have shown the antibacterial properties of berry pomaces against Gram-negative and Gram-positive bacteria. Furthermore, the good antibacterial properties of polyphenolic compounds and their use in food composition can contribute to the microbiological safety and quality of ready-to-eat foods of animal origin by inhibiting of spoilage microorganisms and essential foodborne pathogens [20]. Therefore, the possible antimicrobial properties of both berries and berry pomaces would be a compelling argument for the food industry for their wider use in food production. However, from the point of view of the development of health-promoting foods, it is also of interest for food producers to use plant materials incl. berries and their pomaces that have proven antibacterial properties in both in vitro and in vivo studies.

The main aim of this work is to determine in vitro antibacterial effect of aqueous and 30% ethanolic extracts of chokeberry, blackcurrant, and rowan berries and their pomaces against the growth of *L. monocytogenes*, *S. aureus*, *E. coli*, and *C. jejuni*. In addition, the polyphenolic profile and antioxidant activity of extracts of selected berries and their pomaces were investigated.

2. Materials and Methods

2.1. Plant Material and Preparation of Extracts

Chokeberry (*Aronia melanocarpa* (Michx.) Elliott), blackcurrant (*Ribes nigrum* L.), and rowan (*Sorbus aucuparia* L.) berries were harvested in Tartu County, Estonia. The berry pomaces were obtained as a by-product of berry juice production using the SmegSJF01CREU low-speed juicer (Smeg S.p.A, Guastalla, Italy) at the Polli Horticultural Research Centre, Estonian University of Life Sciences (Polli, Estonia). Plant materials were dried at 50 °C in a condensation dryer (CFD 1400 SS, Alpfrigo d.o.o., Logatec, Slovenia). The dry matter content of plant materials was in the range of 93.0–93.8%. The material was ground in a cutting mill (Retsch GM300) for 1 min at 1500 rpm and sieved using vibratory sieves (AS300 control, Retsch, Düsseldorf, Germany) with 1 mm Ø mesh elements for 4 min and the fraction obtained was ≤1 mm. Aqueous extracts (H_2O) and 30% aqueous ethanolic extracts (30% EtOH) were prepared. Plant powders (0.50 g) were weighed into tubes and 5 mL of ultrapure water or 30% EtOH was added, mixed, and shaken at 60 rpm for 60 min on a Multi RS-60 Multirotator (Biosan, Riga, Latvia). Then, the samples were centrifuged, the supernatants were removed, and the procedure was repeated to obtain 1:20 (*w*/*v*) clear extracts of the plant materials in water and in 30% EtOH. The samples were prepared in duplicate and stored at −40 °C in a freezer until further experiments.

2.2. Chromatographic Analyses

The extracts were chromatographically analyzed using a 1290 Infinity system (Agilent Technologies, Waldbronn, Germany), coupled to an Agilent 6450 Q-TOF mass spectrometer equipped with a Jetstream ESI source and to an Agilent 1290 Infinity Diode Array Detector.

The samples were subjected to a Zorbax 300SB-C18 column (2.1 × 150 mm; 5 μm; Agilent Technologies) kept at 40 °C. For the elution of the samples, a gradient of 5% water in acetonitrile (A) and 0.1% formic acid in water (B) was used as follows: 0–1.5 min 1.0% A, 21.50 min 25% A, 35.00–39.00 min 95% A, 39.10 min 1% A with the eluent flow rate set to 0.3 mL/min, regeneration time was 6 min with eluent flow rate set to 0.4 mL/min. The injection size was 2.0 μL. The mass spectrometer was working in negative ionization MS/MS mode with an MS acquisition rate of 1 spectrum/s and an MS/MS acquisition rate of 5 spectra/s. Fragmentor voltage was set to 135 V and the collision energy to 25 V. The diode array detector operated at a wavelength of 280 nm with 4 nm bandwidth. Data acquisition and initial data processing were carried out using the MassHunter software 10.0 (Agilent Technologies). Gallic acid was used to construct an external calibration graph for quantifying as gallic acid equivalent (GAE) using UV chromatograms at 280 nm. Compounds were identified by comparison of the m/z value, retention time, UV spectra, and MS/MS fragmentation patterns with standards or by comparing data from the literature or the METLIN database (Agilent Technologies).

2.3. Determination of Antioxidative Properties

Analyses of antioxidative (AO) properties of plant extracts (1:20 *w*/*v*) in ultrapure water or in 30% EtOH were performed using the DPPH free-radical-scavenging method on an Infinite 200 Pro M Plex Mono Cuvette plate reader instrument (Tecan Austria Gmbh, Grödig, Austria).

Measurements were performed at 515 nm, using 50 µL of sample and 150 µL of DPPH (1 mM) solution per well. Ultrapure water and 30% EtOH were used as respective blank samples.

For quantification of AO values in Trolox and gallic acid (GA) equivalents, respective calibration curves were used, measured in the same conditions as samples. For finding the linear range of the samples, 1:200, 1:400, and 1:800 (*w*/*v*) sample dilution were measured in duplicate. For calculations, aqueous extracts' 1:200 (*w*/*v*) dilution and 30% EtOH extracts' 1:400 (*w*/*v*) dilutions were used. Results are expressed as AO values in Trolox equivalents and GA equivalents in mM, showing standard deviations of the duplicate measurements.

2.4. Determination of Antimicrobial Activity

Both Gram-positive (*Listeria monocytogenes* ATCC 13929, *Staphylococcus aureus* ATCC 25923) and Gram-negative (*Escherichia coli* NCCB 100282, *Campylobacter jejuni* ATCC 33291) bacterial strains were used in an in vitro study. The isolates originated from the bacterial strain collection of the Food Hygiene and Safety Division of the Estonian University of Life Sciences. The bacteria were cultured on nonselective Brucella blood agar (Oxoid, Basingstoke, UK) plates and incubated at 37 °C for 24 h, except for *C. jejuni* (at 42 °C for 24 h). One colony (1 µL) of each strain was transferred to 4 mL of a 0.9% NaCl solution, and turbidity was adjusted to 0.5 McFarland standard. The bacterial inoculum was adjusted to a concentration of 10^6 colony-forming units per mL (cfu/mL).

The minimum inhibitory concentration (MIC) values of the extracts were determined against tested bacteria by the broth microdilution method in 96-well microplates, according to the EVS-EN ISO 20776-1:2020 [21] guidelines. Falcon® sterile microplate with a clear flat bottom and TC-treated plates were used. For the MIC test, 100 µL of the plant extract was added to the first well of the microplate. Other wells were inoculated with 50 µL of Mueller–Hinton Broth (MHB, Oxoid, Basingstoke, UK). Two-fold dilution series of 1:1 to 1:512 were made to obtain concentrations ranging from 0.001 to 0.82 mg GAE/mL in aqueous extracts and 0.001 to 1.49 mg GAE/mL in ethanolic extracts depending on the contents of total polyphenols in the tested extracts. Column 1 of the microplate contained the highest concentration of extracts, while column 10 contained the lowest. Then, 50 µL of bacterial suspension was dispensed into each well of a sterile 96-well plate. The final concentration of the bacterial inoculum in each well was 5×10^5 cfu/mL. Negative control wells (column 11) consisted of bacteria in MHB without extracts and positive control (column 12) contained gallic acid (1 mg GAE/mL). The plate was covered with a sterile lid and incubated for 18 h at 35 °C. The MIC value was defined as the lowest extract concentration able to completely inhibit visible growth of the target microorganism. Experiments were performed in triplicate and results were expressed as gallic acid equivalents per mg/mL (mg GAE/mL).

2.5. Statistical Analyses

Microsoft Excel 365 (Microsoft Corporation; Redmond, WA, USA) was used for data collection and statistical analyses including correlation analyses. The results were expressed as mean and standard deviation (SD). Correlations were considered strong if $r \geq 0.7$, moderate if $r > 0.3$, and weak if $r < 0.3$ [22,23]. Statistical analysis was also performed using IBM SPSS Statistics software, Version 29.0 (IBM Corp., Chicago, IL, USA). A one-way ANOVA and Tukey HSD test were utilized to evaluate the significant differences ($p < 0.05$) between the tested berry and berry pomace extracts.

3. Results and Discussion

3.1. Identification and Quantification of Phenolic Compounds

The total content of polyphenols in aqueous and ethanolic extracts of selected berries and berry pomaces is shown in Figure 1. The total content of polyphenols was higher in the ethanolic extracts than in the aqueous extracts and the difference between these extracts was statistically significant ($p < 0.05$). The content of total polyphenols was 1.49 ± 0.01 and 1.13 ± 0.02 mg GAE/mL in the ethanolic extracts of chokeberry berries (CB) and chokeberry pomace (CBP), respectively. Blackcurrant berries (BC) and pomace (BCP) and rowan berries (RB) and rowan berry pomace (RBP) contained 0.87 ± 0.01 and 0.54 ± 0.07 mg GAE/mL and 0.50 ± 0.03 and 0.29 ± 0.05 mg GAE/mL of polyphenols in the ethanolic extracts, respectively.

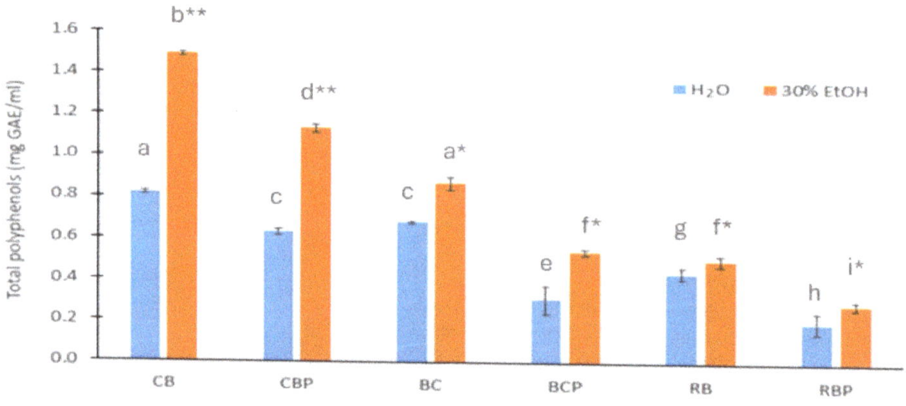

Figure 1. Total content of polyphenols (mg GAE/mL ± SD) in the aqueous (H_2O) and ethanolic extracts (30% EtOH) of chokeberry berries (CB), chokeberry pomace (CBP), blackcurrant berries (BC), blackcurrant pomace (BCP), rowan berries (RB), and rowan berry pomace (RBP) determined by HPLC-DAD-UV. Columns with the same letters (a–i) do not differ significantly ($p > 0.05$). * $p < 0.05$, ** $p < 0.01$ compared to the H_2O extract.

A total of 21 different polyphenolic compounds belonging to hydroxycinnamic acids, flavanols, flavonols, dihydrochalcones, and anthocyanins were tentatively identified from the extracts, of which 19 compounds were detected in BC and BCP extracts, and 16 compounds were found in both CB and RB and their pomace extracts (Table 1). Among hydroxycinnamic acids, protocatechuic acid was detected in all the studied extracts except for the ethanolic extract of RB. Caffeic acid was detected only in the extracts of BC and its pomace. Among flavanols, (epi)gallocatechins were found only in the extracts of BC and its pomace. A dihydrochalcone phloretin-di-C-hexoside was detected in all the tested extracts except for BC. Regarding anthocyanins, the most common compounds in the tested extracts were cyanidin hexosides. Delphinidin rutinoside and cyanidin rutinoside were detected only in the extracts of BC and BCP. Cyanidin pentosides were not found in the extracts of BC and its pomace or RB.

Table 1. Compounds tentatively identified by liquid chromatography–mass spectrometry from chokeberry, blackcurrant, and rowan berries and their pomace extracts.

Pseudomolecular ion Mass-to-Charge Ratio (*m*/*z*)	Compound	Detected in Aqueous Extracts	Detected in Ethanolic Extracts
191.0556	Quinic acid	+	+
Hydroxycinnamic acids:			
153.0193	Protocatechuic acid	+	CB, CBP, BC, BCP, RBP
337.0929	Coumaroylquinic acids	+	+
341.0878	Caffeic acid	BC, BCP	BC, BCP
353.0873	Chlorogenic acids	+	+
Flavanols:			
289.0718	Catechins	+	+
305.0700	(Epi)gallocatechins	BC, BCP	BC, BCP
577.1352	Procyanidin B type	+	+
Flavonols:			
301.0354	Quercetin	CB, CBP, BC, BCP, RB	+
433.0776	Quercetin pentosides	+	+
463.0882	Quercetin hexosides	+	+
595.1305	Quercetin pentosyl-hexosides	+	+
609.1461	Quercetin rhamnosyl hexoside	+	+
625.1410	Quercetin dihexosides	+	+
Dihydrochalcones:			
435.1297	Phloridzin	+	+
597.1825	Phloretin-di-C-hexoside	CB, CBP, BCP, RB, RBP	CB, CBP, BCP, RB, RBP
Anthocyanins:			
417.0827	Cyanidin pentosides	CB, CBP, RBP	CB, CBP, RBP
447.0928	Cyanidin hexoside 1	CB, CBP, RB, RBP	CB, CBP, RB, RBP
447.0928	Cyanidin hexoside 2	BC, BCP	BC, BCP
593.1506	Cyanidin rutinoside	BC, BCP	BC, BCP
609.1461	Delphinidin rutinoside	BC, BCP	BC, BCP

+, found in all tested extracts. Abbreviations: CB, chokeberry berries; CBP, chokeberry pomace; BC, blackcurrant berries, BCP, blackcurrant pomace; RB, rowan berries; RBP, rowan berry pomace.

Figure 2 shows the extracted ion chromatogram (EIC) peak areas of total anthocyanins and total quercetin derivatives in the aqueous and ethanolic extracts of the berries and their pomaces. The EIC peak areas of total anthocyanins were larger for ethanolic extracts compared to aqueous extracts ($p < 0.05$ except for RB). The EIC peak area of total anthocyanins was the largest in BC and CB samples followed by BCP and CBP. The EIC peak area of total quercetin derivatives was the largest in CB followed by CBP, RB, and RBP. Compared with tested berries, the differences in the EIC peak areas of total quercetin derivatives in aqueous and ethanol extracts were not significant ($p > 0.05$) for berry pomaces.

Polyphenolic profiles of selected berries and their pomaces obtained by extraction with water and 30% aq. ethanol are shown in Figures 3 and 4. Chlorogenic acids, cyanidin hexoside 1, quercetin hexosides, quercetin rhamnosyl hexoside, and quinic acid were found in all the extracts of both CB and RB and their pomaces (Figure 3).

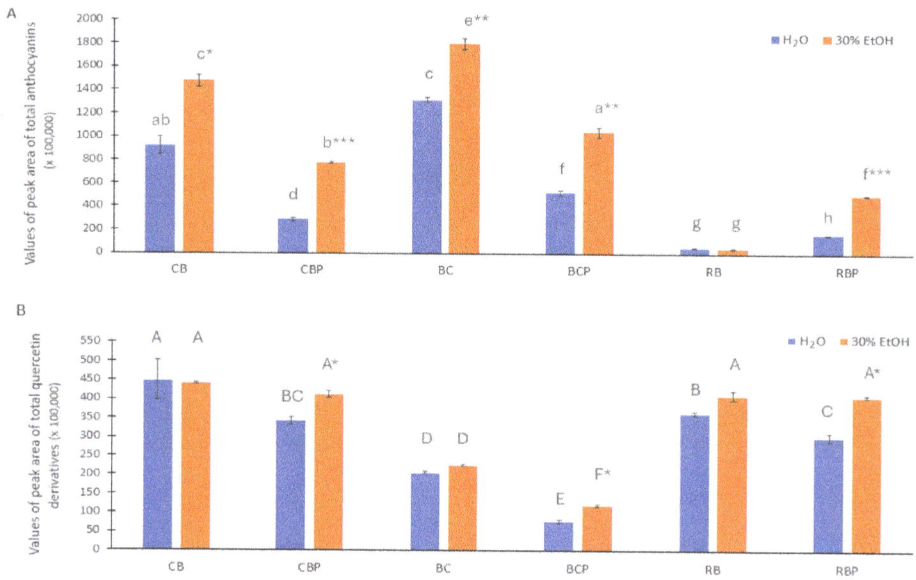

Figure 2. The extracted ion chromatogram peak areas (±SD) of total anthocyanins (**A**) and quercetin derivatives (**B**) obtained with water (H_2O) and 30% aq. ethanol (EtOH) extraction for the berries and berry pomaces. Abbreviations: CB, chokeberry berries; CBP, chokeberry pomace; BC, blackcurrant berries; BCP, blackcurrant pomace; RB, rowan berries; RBP, rowan berry pomace. Columns with the same letters (a–h in (**A**) and A–F in (**B**)) do not differ significantly ($p > 0.05$). * $p < 0.05$, ** $p < 0.01$, *** $p > 0.001$ compared to the H_2O extract.

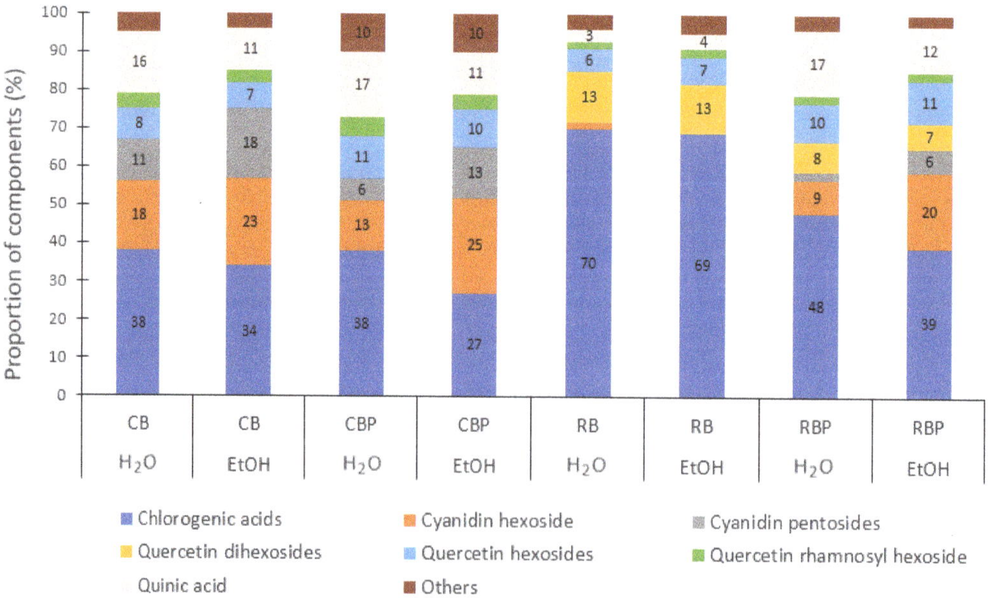

Figure 3. Polyphenolic profiles of chokeberry berries (CB), chokeberry pomace (CBP), rowan berries (RB), and rowan berry pomace (RBP) obtained with water (H_2O) and 30% aq. ethanol (EtOH) extraction.

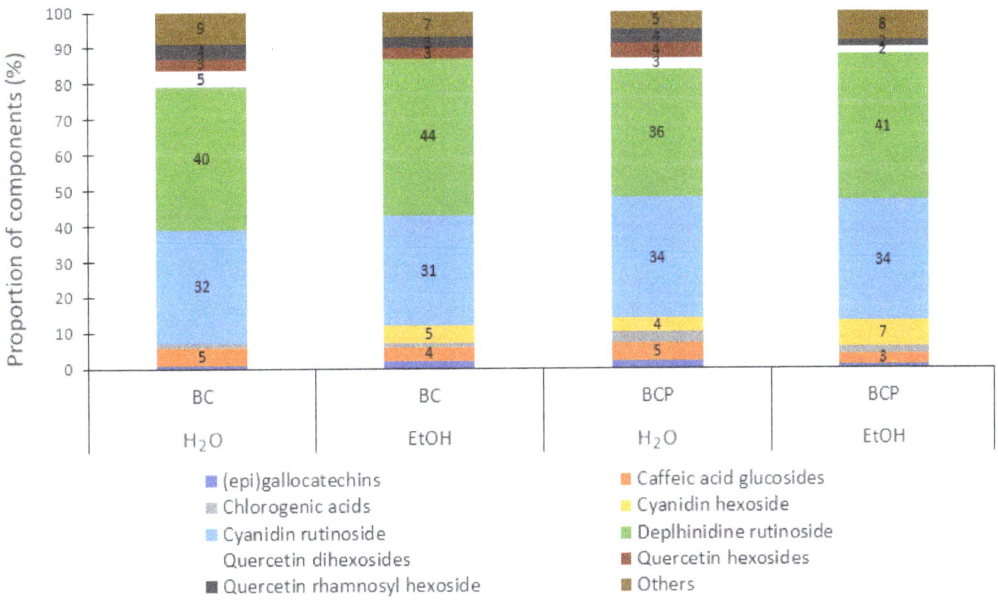

Figure 4. Polyphenolic profiles of blackcurrant berries (BC) and blackcurrant pomace (BCP) obtained with water (H_2O) and 30% aq. ethanol (EtOH) extraction.

In the aqueous and ethanolic extracts of CB, the proportion of identified polyphenolic compounds from highest to lowest were as follows: chlorogenic acids (34–38%) > cyanidin hexosides (18–23%) > cyanidin pentosides (11–18%) > quinic acid (11–16%) > quercetin hexosides (7–8%) > the rest of the compounds. A similar profile of polyphenols was also identified in CBP extracts, but the proportions of quercetin hexosides were higher (up to 3%) and the content of cyanidin pentosides was lower (up to 5%).

In both aqueous and ethanolic extracts of RB, the proportions of polyphenolic compounds from highest to lowest were similar: chlorogenic acids (69–70%) > quercetin dihexosides (13%) > quercetin hexosides (6–7%) > quinic acid (3–4%) > the rest of the compounds (Figure 3). A similar profile of polyphenols was also identified in RBP extracts, but the proportions of quinic acid (up to 14%), quercetin hexosides (up to 4%), and cyanidin hexoside 1 (up to 20%) were higher and content of chlorogenic acids was lower (up to 30%).

Polyphenolic profiles of BC and BCP obtained by extraction with water and 30% aq. ethanol are shown in Figure 4. A total of nine of the twenty-one detected compounds in the polyphenolic profiles of the aqueous and ethanolic extracts were higher in content than the other identified compounds according to the EIC peak areas (Figure 3). All nine compounds were identified in aqueous and ethanolic extracts except cyanidin hexoside 2 in berry aqueous extract, quercetin dihexoside in berry ethanolic extract, and quercetin hexosides in berry pomace ethanolic extracts. In the aqueous and ethanolic extracts of BC, the proportion of identified polyphenolic compounds from highest to lowest was as follows: delphinidin rutinoside (40–44%) > cyanidin rutinoside (31–32%) > quercetin rhamnosyl hexoside (3–4%) > quercetin hexosides (3%) > the rest of the compounds. A similar profile of polyphenols was also identified in BCP extracts, but the proportions of delphinidin rutinoside were lower (up to 4%) and that of cyanidin rutinoside was higher (up to 3%).

3.2. Antioxidative Properties

Antioxidative (AO) properties of the studied berries and their pomaces in 1:20 w/v extracts in water and 30% aq. ethanol are shown in Table 2. In the aqueous and 30% ethanolic extracts of berries and pomaces, the antioxidative properties from highest to lowest were as follows: CB > CBP > BC > BCP ≈ RB > RBP. The AO properties were statistically significantly ($p < 0.05$) higher in the ethanolic extracts than in the aqueous extracts. Rowan berry 30% ethanolic extract had equal AO properties to BCP 30% ethanolic extract ($p \geq 0.05$). Regarding all tested plant materials, 30% ethanolic extracts (1:20 *w*/*v*) showed better antioxidative properties compared with extracts made with water. Correlation analysis revealed that AO had a strong positive correlation with anthocyanin content in aqueous extracts ($r = 0.84$) and moderately positive correlation in 30% ethanolic ($r = 0.56$) extracts. The content of quercetin derivatives had a weak positive correlation with AO ($r = 0.1$–0.14) both in aqueous and 30% ethanolic extracts. The total content of polyphenols, however, correlated with AO strongly, showing strong positive correlations both in aqueous ($r = 0.92$) and in 30% ethanolic extracts ($r = 0.93$).

Table 2. Antioxidative (AO) properties (±SD) of 1:20 (*w*/*v*) aqueous (A) and 30% ethanolic (B) extracts of berries and their pomaces.

Extracts	AO in Trolox eq mM (A)	AO in GAE mM (A)	AO in Trolox eq mM (B)	AO in GAE eq mM (B)
CB	9.08 ± 0.10^{a}	0.004 ± 0.0000^{A}	$21.4 \pm 0.16^{b,***}$	$0.009 \pm 0.0001^{B,**}$
CBP	5.55 ± 0.17^{c}	0.003 ± 0.0001^{C}	$17.5 \pm 0.48^{d,***}$	$0.007 \pm 0.0002^{D,**}$
BC	8.73 ± 0.19^{a}	0.004 ± 0.0001^{A}	$16.0 \pm 0.11^{d,**}$	$0.007 \pm 0.0000^{E,**}$
BCP	4.57 ± 0.34^{c}	0.002 ± 0.0001^{C}	$13.0 \pm 1.21^{e,*}$	$0.005 \pm 0.0005^{F,*}$
RB	3.61 ± 0.83^{c}	0.002 ± 0.0004^{C}	$13.2 \pm 1.57^{e,*}$	$0.006 \pm 0.0006^{F,*}$
RBP	0.57 ± 0.05^{f}	0.001 ± 0.0000^{G}	$6.09 \pm 2.56^{c,*}$	$0.003 \pm 0.0010^{C,*}$

Abbreviations: CB, chokeberry berries; CBP, chokeberry pomace; BC, blackcurrant berries, BCP, blackcurrant pomace; RB, rowan berries; RBP, rowan berry pomace. Columns with the same letters ($^{a–f}$ in the Trolox and $^{A–G}$ in the GAE columns) do not differ significantly ($p \geq 0.05$). * $p < 0.05$, ** $p < 0.01$, *** $p > 0.001$ compared to H_2O extract.

3.3. Minimal Inhibitory Concentrations

In Table 3, the minimum inhibitory concentration (MIC) values of the tested berries and berry pomace extracts are presented for Gram-positive (G+) bacteria such as *L. monocytogenes* and *S. aureus*, as well as Gram-negative (G−) bacteria such as *E. coli* and *C. jejuni*. MIC results were expressed as gallic acid equivalents per mg/mL dry matter and gallic acid was positive control. The MIC value of pure gallic acid against the tested G+ bacteria was 0.02 mg GAE/mL (at a dilution of 1:64 of the extracts) and 0.03 mg GAE/mL (at a dilution of 1:32 of the extracts) for G− bacteria.

Table 3. Minimum inhibitory concentration (MIC) (mg GAE/mL) of aqueous (A) and 30% ethanolic (B) extracts of berries and their pomaces against selected bacteria.

Extracts		*L. monocytogenes* G+	*S. aureus* G+	*E. coli* G−	*C. jejuni* G−	Min and Max MIC Values *
CB	A	0.41	0.21	-	-	0.001–0.82
	B	0.37	0.19	0.75	0.75	0.003–1.49
CBP	A	-	0.32	-	-	0.001–0.63
	B	0.28	0.28	0.57	0.57	0.002–1.13
BC	A	0.34	0.17	-	-	0.001–0.68
	B	0.22	0.22	0.22	0.22	0.002–0.87
BCP	A	0.16	0.16	-	-	0.001–0.31
	B	0.14	0.27	0.14	0.27	0.001–0.54
RB	A	0.22	0.11	-	-	0.001–0.44
	B	0.13	0.13	0.13	0.13	0.001–0.50
RBC	A	-	0.10	-	-	0.001–0.20
	B	0.15	0.15	0.15	0.15	0.001–0.29

* The highest and lowest possible MIC values of the tested extracts were determined at a dilution of 1:1 and 1:512, respectively. -, no antibacterial effect. Abbreviations: CB, chokeberry berries; CBP, chokeberry pomace; BC, blackcurrant berries, BCP, blackcurrant pomace; RB, rowan berries; RBP, rowan berry pomace.

The sensitivity of tested microbes to the extracts of berries and berry pomaces starting from more sensitive bacteria was as follows: *S. aureus* > *L. monocytogenes* > *E. coli* ≈ *C. jejuni*. All tested berry extracts inhibited the growth of G+ bacteria such as *S. aureus* and *L. monocytogenes*. Among the berry pomaces, only BCP extracts had a similar effect against these G+ bacteria. G− bacteria such as *E. coli* and *C. jejuni* were inhibited by ethanolic extracts of berries and berry pomaces but not by aqueous extracts. Among all tested bacteria, *S. aureus* was the most sensitive to all tested aqueous and ethanolic extracts of berries and berry pomaces. CB showed the strongest antibacterial activity against *S. aureus* both in aqueous (at a dilution of 1:8) and ethanolic extracts (at a dilution of 1:16) with MIC values of 0.19–0.21 mg GAE/mL followed by BC with MIC values of 0.17–0.22 mg GAE/mL and CBP with MIC values of 0.28–0.32 mg GAE/mL. The remaining tested extracts showed an antibacterial effect only at the dilution of 1:2 or 1:4, where MIC values were 0.10–0.41 mg GAE/mL in aqueous extracts and 0.13–0.75 mg GAE/mL in ethanolic extracts depending on the total content of polyphenols (Figure 1) in the berries and their pomaces.

4. Discussion

The polyphenolic profile of berries and berry pomaces and their antibacterial effect have been investigated in several studies [10,18–20,24]. Chokeberry, blackcurrant, and rowan berries are rich sources of polyphenolic compounds [6–8]. In this study, the polyphenolic profile screening showed that tested berries and berry pomaces contained different polyphenols, e.g., hydroxycinnamic acids, flavanols, flavonols, dihydrochalcones, and anthocyanins. A total of 19 different polyphenolic compounds were detected in BC and its pomace extracts and 16 different compounds were found in both CB and RB and their pomace extracts. Therefore, BC and BCP were distinguished from other berries due to their more diverse composition of polyphenols. However, CB and RB and their pomace extracts had a similar polyphenolic profile and the proportion of chlorogenic acids was the highest in all these extracts. Polyphenolic profiles of CB and RB were similar in the study of Määttä-Riihinen et al. [25]. In this study, however, the EIC peak area of total anthocyanins was the largest in CB and BC samples, but the EIC peak area of total quercetin derivatives was the largest in CB and RB samples. The content and proportion of bioactive compounds in berries and berry pomaces are quite variable and they can vary depending on the berry variety, growing conditions, maturity and harvest time, and processing methods as well as climatic conditions [13,26].

In this study, both investigated extracts of chokeberry pomace contained approximately 77% of the total polyphenol content compared to the content in berries. Ethanolic extracts of blackcurrant and rowan berry pomaces contained 60% of the total polyphenol content of the respective berries. Therefore, according to our study, berry pomaces are a valuable source of polyphenols which has also been confirmed in several other studies [14,17]. Khanal et al. [27] found that after blueberry juice extraction, blueberry pomace contained 25% to 50% of the procyanidins compared to the content in fresh berries. In any case, the use of berry pomaces in food production deserves consideration for the development of new health-beneficial functional foods due to the valuable phenolic compounds [18]. Rowan berry extracts differed from other berries in this study, containing 1.5–3 times lower polyphenol levels than chokeberry or blackcurrant. This may be explained by the peculiarity of the rowan berry varieties and their low anthocyanin content.

In this study, both water and 30% aq. ethanol (30% EtOH) extraction were used to obtain the best possible yield of soluble polyphenolic compounds in berries and berry pomaces. The optimal concentration of ethanol in solvent was established with preliminary tests. Other types of possible solvents were not used to have only a food-grade solvent system. Salaheen et al. [19] found that total phenolic content was higher in a 10% ethanolic extract compared to 10% methanol or water extracts. Bobinaitė et al. [17] found that the aqueous extract had the highest polyphenolic compound diversity compared to ethanolic or acetone extracts obtained from rowan berry pomace. In this study, the aqueous and ethanolic extracts generally had a similar diversity of polyphenolic compounds with some minor differences. Compared to other aqueous extracts, the aqueous extracts of BC and BCP had a more diverse composition of polyphenols, including caffeic acid, (epi)gallocatechins, delphinidin rutinoside, and cyanidin hexoside 2, which were not detected in other aqueous extracts. Protocatechuic acid and caffeic acid were present in BC and CB ethanolic extracts but were lacking in RB and RBP ethanolic extracts. The differences in the polyphenolic profiles can be explained by the fact that more hydrophobic compounds are extracted with a higher % of ethanol [7].

In this study, 30% ethanolic extracts showed better antioxidative properties for all tested berries and pomaces compared to water extracts, indicating that the addition of ethanol in the extraction solvent facilitates the extraction of compounds with better AO properties. The tested berries and pomaces, especially in the case of chokeberry and blackcurrant, had good antioxidant properties and this result is consistent with other similar studies [8,10,26]. Antioxidative properties had the strongest correlation with the total content of polyphenols (TPC) both in aqueous and 30% ethanolic 1:20 (w/v) extracts. Within the group of polyphenolic compounds, anthocyanins also had a strong correlation in aqueous (r = 0.84) extracts or a moderate positive correlation in 30% ethanolic (r = 0.56) extracts with AO. That result may be due to the fact that anthocyanins are semipolar compounds and thus also extractable with hydrophilic solvent, whereas in 30% ethanolic extracts the AO results were also due to more nonpolar compounds. Interestingly, quercetin derivates had a very weak correlation with AO properties, which can be explained by the fact that in plant matrices there is usually no or very little free quercetin, which is known as a very good antioxidant, whereas quercetin derivatives also present in the extracts in this study exhibit lower antioxidant properties as the free OH groups of the aglycone are substituted with glycosyl groups. This result is in accordance with the study of Lesjak et al. [28].

Berry additives in food can play a role in inhibiting the growth of microorganisms due to their potential antibacterial properties. In this study, MIC values were generally high, ranging from 0.10 to 0.75 mg GAE/mL, although concentrations as low as 0.001 to 0.09 mg GAE/mL were also found. The sensitivity of tested bacteria to the berry and berry pomace extracts was as follows: *S. aureus* > *L. monocytogenes* > *E. coli* ≈ *C. jejuni*. The antibacterial activity of the extracts against tested bacteria did not depend only on the TPC but probably also on other factors such as the concentration of organic acids in the extract. This was also confirmed in the study of Adamczak et al. [29]. The MIC tests showed that all the tested berry extracts of CB, BC, and RB inhibited the growth of *S. aureus* and *L. monocytogenes*, but only BCP had the same effect among pomaces against the above-mentioned G+ bacteria. Contrary to our results, Bobinaitė et al. [17] showed that the growth of G+ bacteria was also effectively suppressed by acetone, ethanolic, and aqueous extracts of rowan berry pomace. LaPlante et al. [24] found that cranberry extracts inhibited the growth of G+ bacteria, such as *Staphylococcus* spp., but not G– bacteria, such as *E. coli*, with MIC in the range of 0.02–5 mg/mL. In this study, CB and BC had strong antibacterial activity against *S. aureus*, which was inhibited by 8- and 16-fold dilution of aqueous and ethanolic extracts, respectively. All ethanolic extracts also inhibited the growth of G– bacteria, but the antibacterial effect only appeared when the extracts were diluted up to two or four times. Unlike the above, the growth of *E. coli* and *C. jejuni* was not inhibited by aqueous extracts of tested berries and berry pomaces. Similarly, Shirzadi Karamolah et al. [30] found that, in contrast to the alcohol extract, the aqueous extracts had no effect on the tested bacteria. In the present study, the antibacterial effect on the growth of the tested bacteria was also confirmed mainly by the ethanolic extracts.

As previously known, the bioactive compounds found in berries are mainly phenolic compounds (phenolic acids, flavonoids, e.g., anthocyanins and flavonols, and tannins) [13] and they are associated with antibacterial properties of berries and berry pomaces [20]. Our previous studies [3,31] have shown good potential in using powdered plant material (including plant pomaces) as an antimicrobial agent in different products, ensuring the high microbiological quality of raw and cooked minced pork and marinated rainbow trout. Adamczak et al. [29] found that all tested individual pure compounds among flavonoids and organic acids showed antimicrobial properties, but their biological activity was moderate or relatively low against G+ and G− bacteria. Georgescu et al. [20] found that the extracted polyphenolic compounds from different berries showed an antibacterial effect on *E. coli*, *Bacillus subtilis*, and *S. aureus*. Furthermore, several studies have shown that procyanidins have an antibacterial effect against both G+ and G− bacteria [24,32]. The antibacterial mechanism of polyphenolic compounds is associated with the increased permeability of the pathogen's outer membrane and plasma membrane, facilitating the entry of hydrophobic compounds into the bacterial cell by breaking the lipopolysaccharide layer [33]. However, G+ bacteria are more sensitive to flavonoids than G− bacteria due to differences in cell wall structure [34]. The target of flavonoids is the cell membrane of G+ bacteria by damaging its phospholipid bilayer structure and by inhibiting the bacterial respiratory chain and ATP synthesis [35].

In this study, the ethanolic extract of CB had the strongest antibacterial activity against *S. aureus* which could be explained with the highest content of total polyphenols found in this extract. However, the effect of BC and its pomace can be highlighted because both their aqueous and ethanolic extracts had antibacterial activity against *S. aureus* and *L. monocytogenes*, while no similar effect was detected for other pomaces against *L. monocytogenes*. This may be related to blackcurrant and its pomace having a more diverse polyphenolic composition and higher content of anthocyanins (e.g. delphinidin rutinoside, cyanidin hexoside 2, and cyanidin rutinoside) compared to other berries and their pomaces. In addition, the antibacterial effect may be related to the effect of organic acids and their derivatives on G+ and G− bacteria [36].

A study by Ma et al. [37] concluded that anthocyanins exhibit antibacterial properties by destroying the cell wall of foodborne pathogens. Furthermore, caffeic acid and (epi)gallocatechins were identified only in the extracts of BC and BCP in this study. Perumal et al. [33] found that caffeic acid and epicatechin 3-gallate from *Euphorbia hirta* have remarkable bactericidal effects against *Pseudomonas aeruginosa* by increasing the permeability of the bacterial outer membrane and plasma membrane. Catechins may also have antibacterial properties by reducing the antioxidant capacity of foodborne pathogens [37].

The studied berries are also a rich source of hydroxycinnamic acids. The most abundant hydroxycinnamate was chlorogenic acid, which is a complex of caffeic acid linked to quinic acid through an ester bond [26]. In this study, the highest proportion of chlorogenic acids was found in the extracts of CB and RB and their pomaces. Adamczak et al. [29] showed that chlorogenic acid had moderate activity against *E. coli*, but no activity against *Pseudomonas aeruginosa*. In this study, the antibacterial activity of chokeberry and rowan berries against *S. aureus* and *L. monocytogenes* was shown in both aqueous and ethanolic extracts, but not their pomaces, where the proportions of chlorogenic acids were lower.

In the polyphenolic profile, the highest proportion of quercetin derivatives was found in CB and RB ethanolic extracts and these extracts had an antibacterial effect on all the tested G+ and G− bacteria. The inhibition of pathogenic bacteria by quercetin is associated with damage to peptidoglycan structures [34]. No free quercetin was detected in the aqueous extracts of RBP, and these extracts showed no antibacterial activity against *L. monocytogenes*, *E. coli*, and *C. jejuni*.

5. Conclusions

The study demonstrated a diverse profile of polyphenols in the extracts of chokeberry, blackcurrant, and rowan berry and their berry pomaces. Since the extracts of chokeberry and blackcurrant berries and berry pomaces contained a higher proportion of polyphenolic compounds, they also had better antibacterial and antioxidative properties compared to rowan berry and its pomace. In vitro antibacterial activity of tested extracts depended on the extraction solvent and mainly occurred for the ethanolic extracts. Findings suggest that chokeberry and blackcurrant berries and their pomaces are good sources of polyphenols with antioxidative properties and compounds that have antibacterial activity against selected bacteria. In practical applications, it is important to extract the valuable compounds from juice press residues for further usage in the composition of foods, especially for the enrichment of animal-origin foods. This kind of food valorization is in compliance with the zero-waste concept and goals of a circular bioeconomy.

Author Contributions: Conceptualization, K.M. and P.R.; methodology, K.M., P.R. and L.R.; validation, K.M., P.R. and L.R.; formal analysis, K.M., P.R. and L.R.; investigation, K.M., P.R. and L.R.; resources, M.R., D.A. and U.B.; data curation, K.M., P.R., and L.R.; writing—original draft preparation, K.M. and P.R.; writing—review and editing, K.M., P.R., L.R., D.A., M.R. and U.B.; visualization, K.M. and P.R.; supervision, M.R.; project administration, M.R.; funding acquisition, M.R. All authors have read and agreed to the published version of the manuscript.

Funding: This work was supported by the Estonian Research Council grant (PRG 1441) project "Effects and mechanisms of plant bioactive substances in foods of animal origin".

Institutional Review Board Statement: Not applicable.

Informed Consent Statement: Not applicable.

Data Availability Statement: Data are contained within the article.

Conflicts of Interest: The authors declare no conflicts of interest.

References

1. Koskar, J.; Meremäe, K.; Püssa, T.; Anton, D.; Elias, T.; Rätsep, R.; Mäesaar, M.; Kapp, K.; Roasto, M. Microbial growth dynamics in minced meat enriched with plant powders. *Appl. Sci.* **2022**, *12*, 11292. [CrossRef]
2. Kerner, K.; Kazernavičiūtė, R.; Jõudu, I.; Rocchetti, G.; Lucini, L.; Tänavots, A.; Hussain, S.; Venskutonis, P.R. Evaluation of different blackcurrant seed ingredients in meatballs by using conventional quality assessment and untargeted metabolomics. *Meat Sci.* **2023**, *200*, 109160. [CrossRef] [PubMed]
3. Roasto, M.; Mäesaar, M.; Püssa, T.; Anton, D.; Rätsep, R.; Elias, T.; Jortikka, S.; Pärna, M.; Kapp, K.; Tepper, M.; et al. The effect of fruit and berry pomaces on the growth dynamics of microorganisms and sensory properties of marinated rainbow trout. *Microorganisms* **2023**, *11*, 2960. [CrossRef] [PubMed]
4. Samtiya, M.; Aluko, R.E.; Dhewa, T.; Moreno-Rojas, J.M. Potential health benefits of plant food-derived bioactive components: An overview. *Foods* **2021**, *10*, 839. [CrossRef] [PubMed]
5. Liu, H.; Qin, S.; Sirohi, R.; Ahluwalia, V.; Zhou, Y.; Sindhu, R.; Binod, P.; Singhnia, R.R.; Patel, A.K.; Juneja, A.; et al. Sustainable blueberry waste recycling towards biorefinery strategy and circular bioeconomy: A review. *Bioresour. Technol.* **2021**, *332*, 125181. [CrossRef] [PubMed]
6. Borowska, S.; Brzóska, M. Chokeberries (*Aronia melanocarpa*) and their products as a possible means for the prevention and treatment of noncommunicable diseases and unfavorable health effects due to exposure to xenobiotics. *Compr. Rev. Food Sci. Food Saf.* **2016**, *15*, 982–1017. [CrossRef] [PubMed]
7. Raudsepp, P.; Koskar, J.; Anton, D.; Meremäe, K.; Kapp, K.; Laurson, P.; Bleive, U.; Kaldmäe, H.; Roasto, M.; Püssa, T. Antibacterial and antioxidative properties of different parts of garden rhubarb, blackcurrant, chokeberry and blue honeysuckle. *J. Sci. Food Agric.* **2018**, *99*, 2311–2320. [CrossRef]
8. Sarv, V.; Venskutonis, P.R.; Rätsep, R.; Aluvee, A.; Kazernavičiūtė, R.; Bhat, R. Antioxidants characterization of the fruit, juice, and pomace of sweet rowanberry (*Sorbus aucuparia* L.) cultivated in Estonia. *Antioxidants* **2021**, *10*, 1779. [CrossRef]
9. Zhang, H.; Tsao, R. Dietary polyphenols, oxidative stress and antioxidant and anti-inflammatory effects. *Curr. Opin. Food Sci.* **2016**, *8*, 33–42. [CrossRef]
10. Paunović, S.M.; Mašković, P.; Milinković, M. Phytochemical and antimicrobial profile of black currant berries and leaves. *Acta Agric. Serbica* **2022**, *27*, 25–29. [CrossRef]
11. Daglia, M. Polyphenols as antimicrobial agents. *Curr. Opin. Biotechnol.* **2012**, *23*, 174–181. [CrossRef]
12. Yahfoufi, N.; Alsadi, N.; Jambi, M.; Matar, C. The immunomodulatory and anti-inflammatory role of polyphenols. *Nutrients* **2018**, *10*, 1618. [CrossRef]
13. Skrovankova, S.; Sumczynski, D.; Mlcek, J.; Jurikova, T.; Sochor, J. Bioactive compounds and antioxidant activity in different types of berries. *Int. J. Mol. Sci.* **2015**, *16*, 24673–24706. [CrossRef] [PubMed]
14. Klavins, L.; Kviesis, J.; Nakurte, I.; Klavins, M. Berry press residues as a valuable source of polyphenolics: Extraction optimisation and analysis. *LWT Food Sci. Technol.* **2018**, *93*, 583–591. [CrossRef]
15. Fierascu, R.C.; Sieniawska, E.; Ortan, A.; Fierascu, I.; Xiao, J. Fruits by-products—A source of valuable active principles. A short review. *Front. Bioeng. Biotechnol.* **2020**, *8*, 319. [CrossRef] [PubMed]
16. Heinonen, M. Antioxidant activity and antimicrobial effect of berry phenolics—A Finnish perspective. *Mol. Nutr. Food Res.* **2007**, *51*, 684–691. [CrossRef] [PubMed]

17. Bobinaitė, R.; Grootaert, C.; Van Camp, J.; Šarkinas, A.; Liaudanskas, M.; Žvikas, V.; Viškelis, P.; Venskutonis, P.R. Chemical composition, antioxidant, antimicrobial and antiproliferative activities of the extracts isolated from the pomace of rowanberry (*Sorbus aucuparia* L.). *Food Res. Int.* **2020**, *136*, 109310. [CrossRef] [PubMed]
18. Garzón, G.A.; Soto, C.Y.; López-R, M.; Riedl, K.M.; Browmiller, C.R.; Howard, L. Phenolic profile, *in vitro* antimicrobial activity and antioxidant capacity of *Vaccinium meridionale* Swartz pomace. *Heliyon* **2020**, *6*, e03845. [CrossRef] [PubMed]
19. Salaheen, S.; Nguyen, C.; Hewes, D.; Biswas, D. Cheap extraction of antibacterial compounds of berry pomace and their mode of action against the pathogen *Campylobacter jejuni*. *Food Control* **2014**, *46*, 174–181. [CrossRef]
20. Georgescu, C.; Frum, A.; Virchea, L.-I.; Sumacheva, A.; Shamtsyan, M.; Gligor, F.-G.; Olah, N.K.; Mathe, E.; Mironescu, M. Geographic variability of berry phytochemicals with antioxidant and antimicrobial properties. *Molecules* **2022**, *27*, 4986. [CrossRef]
21. *EVS-EN ISO 20776-1:2020*; Susceptibility Testing of Infectious Agents and Evaluation of Performance of Antimicrobial Susceptibility Test Devices—Part 1: Broth Micro-Dilution Reference Method for Testing the In Vitro Activity of Antimicrobial Agents against Rapidly Growing Aerobic Bacteria Involved in Infectious Diseases. ISO: Tallinn, Estonia, 2020.
22. Akoglu, H. User's guide to correlation coefficients. *Turk. J. Emerg. Med.* **2018**, *18*, 91–93. [CrossRef] [PubMed]
23. Ratner, B. The correlation coefficient: Its values range between +1/−1, or do they? *J. Target. Meas. Anal. Mark.* **2009**, *17*, 139–142. [CrossRef]
24. LaPlante, K.L.; Sarkisian, S.A.; Woodmansee, S.; Rowley, D.C.; Seeram, N.B. Effects of cranberry extracts on growth and biofilm production of *Escherichia coli* and *Staphylococcus* species. *Phytother. Res.* **2012**, *26*, 1371–1374. [CrossRef] [PubMed]
25. Määttä-Riihinen, K.R.; Kamal-Eldin, A.; Mattila, P.H.; González-Paramás, A.M.; Törrönen, A.R. Distribution and contents of phenolic compounds in eighteen Scandinavian berry species. *J. Agric. Food Chem.* **2004**, *52*, 4477–4486. [CrossRef] [PubMed]
26. Jurikova, T.; Mlcek, J.; Skrovankova, S.; Sumczynski, D.; Sochor, J.; Hlavacova, I.; Snopek, L.; Orsavova, J. Fruits of black chokeberry *Aronia melanocarpa* in the prevention of chronic diseases. *Molecules* **2017**, *22*, 944. [CrossRef] [PubMed]
27. Khanal, R.C.; Howard, L.R.; Brownmiller, C.R.; Prior, R.L. Influence of extrusion processing on procyanidin composition and total anthocyanin contents of blueberry pomace. *J. Food Sci.* **2009**, *74*, 52–58. [CrossRef]
28. Lesjak, M.; Beara, I.; Simin, N.; Pintać, D.; Majkić, T.; Bekvalac, K.; Orčić, D.; Mimica-Dukić, N. Antioxidant and anti-inflammatory activities of quercetin and its derivatives. *J. Funct. Foods* **2018**, *40*, 68–75. [CrossRef]
29. Adamczak, A.; Ożarowski, M.; Karpiński, T.M. Antibacterial activity of some flavonoids and organic acids widely distributed in plants. *J. Clin. Med.* **2020**, *9*, 109. [CrossRef]
30. Shirzadi Karamolah, K.; Mousavi, F.; Mahmoudi, H. Antimicrobial inhibitory activity of aqueous, hydroalcoholic and alcoholic extracts of leaves and stem of *Daphne mucronata* on growth of oral bacteria. *GMS Hyg. Infect. Control* **2017**, *12*, Doc16. [CrossRef]
31. Anton, D.; Koskar, J.; Raudsepp, P.; Meremäe, K.; Kaart, T.; Püssa, T.; Roasto, M. Antimicrobial and antioxidative effects of plant powders in raw and cooked minced pork. *Foods* **2019**, *8*, 661. [CrossRef]
32. Leitão, D.P.; Polizello, A.C.; Ito, I.Y.; Spadaro, A.C. Antibacterial screening of anthocyanic and proanthocyanic fractions from cranberry juice. *J. Med. Food* **2005**, *8*, 36–40. [CrossRef] [PubMed]
33. Perumal, S.; Mahmud, R.; Ismail, S. Mechanism of action of isolated caffeic acid and epicatechin 3-gallate from *Euphorbia hirta* against *Pseudomonas aeruginosa*. *Pharmacogn. Mag.* **2017**, *13*, 311–315. [CrossRef] [PubMed]
34. Osonga, F.J.; Akgul, A.; Miller, R.M.; Eshun, G.B.; Yazgan, I.; Akgul, A.; Sadik, O.A. Antimicrobial activity of a new class of phosphorylated and modified flavonoids. *ACS Omega* **2019**, *4*, 12865–12871. [CrossRef]
35. Yuan, G.; Guan, Y.; Yi, H.; Lai, S.; Sun, Y.; Cao, S. Antibacterial activity and mechanism of plant flavonoids to gram-positive bacteria predicted from their lipophilicities. *Sci. Rep.* **2021**, *11*, 10471. [CrossRef] [PubMed]
36. Kovanda, L.; Zhang, W.; Wei, X.; Luo, J.; Wu, X.; Atwill, E.R.; Vaessen, S.; Li, X.; Liu, Y. In vitro antimicrobial activities of organic acids and their derivatives on several species of gram-negative and gram-positive bacteria. *Molecules* **2019**, *24*, 3770. [CrossRef]
37. Ma, Y.; Ding, S.; Fei, Y.; Liu, G.; Jang, H.; Fang, J. Antimicrobial activity of anthocyanins and catechins against foodborne pathogens *Escherichia coli* and Salmonella. *Food Control* **2019**, *106*, 106712. [CrossRef]

Disclaimer/Publisher's Note: The statements, opinions and data contained in all publications are solely those of the individual author(s) and contributor(s) and not of MDPI and/or the editor(s). MDPI and/or the editor(s) disclaim responsibility for any injury to people or property resulting from any ideas, methods, instructions or products referred to in the content.

foods

MDPI

Article

Elderberry Lipophilic and Hydrophilic Bioactive Compounds: Characterization and Extract Encapsulation

Rubén Domínguez-Valencia [1], Aurora Cittadini [1,2], Mirian Pateiro [1,*], Paulo E. S. Munekata [1] and José M. Lorenzo [1,3]

[1] Centro Tecnológico de la Carne de Galicia, Avd. Galicia nº 4, Parque Tecnológico de Galicia, San Cibrao das Viñas, 32900 Ourense, Spain; rubendominguez@ceteca.net (R.D.-V.); aurora.cittadini@unavarra.es (A.C.); paulosichetti@ceteca.net (P.E.S.M.); jmlorenzo@ceteca.net (J.M.L.)

[2] Instituto de Innovación y Sostenibilidad en la Cadena Agroalimentaria (IS-FOOD), Universidad Pública de Navarra (UPNA), Campus de Arrosadia, 31006 Pamplona, Spain

[3] Área de Tecnoloxía dos Alimentos, Facultade de Ciencias, Universidade de Vigo, 32004 Ourense, Spain

* Correspondence: mirianpateiro@ceteca.net

Abstract: There are few studies on the use of elderberry in the food industry, and its form of application differs between the different studies. Therefore, the objective of this study is to describe a procedure for obtaining a stabilized product with a high content of hydrophilic bioactive compounds (encapsulated elderberry extract). Moreover, the solid residue resulting from the extraction of the polyphenols was characterized, and the lipophilic compounds retained in this residue were analyzed. The results show an important antioxidant activity of the extracts obtained, mainly linked to the high content of anthocyanins, hydroxycinnamic acids, and flavonols. The lipophilic bioactive compounds were characterized by a high content of essential fatty acids and high proportions of tocopherols. The information and results of the present study provide novel information about both lipophilic and hydrophilic compounds for the integral valorization of elderberries to promote a circular economy strategy.

Keywords: *Sambucus nigra* L.; plant extracts; antioxidants; encapsulation; natural ingredients; functional food

1. Introduction

Elderberry is a robust and underutilized fruit used for value-added products [1]. The elderberry crop has shown a great capacity to adapt to many climate conditions [2]. For this reason, there is an increasing interest in wild-growing plant species rich in bioactive compounds [3]. In line with this, high amounts of elderberries from wild-growing shrubs are used for industrial purposes [4]. The black elderberry (*Sambucus nigra* L.) is widely distributed throughout Europe, with the majority of its market localized in Austria, the Czech Republic, Denmark, Germany, Italy, and Poland [5,6], although their production is small in comparison with other berries.

Elderberry is still rarely applied in food formulation, although it has been used in the food industry, mainly for the production of beverages (juices, tea, and alcoholic beverages) [7], as well as coloring in candies, drinks, jam, and confectionaries [8,9]. Elderberry contains one of the highest amounts of anthocyanins in comparison with other berries [10–12]; thus, there is an increased interest in this natural source to develop food colorants. It also contains significant amounts of other polyphenols with potential antioxidant activity. In addition to these hydrophilic compounds, other studies have provided convincing evidence that lipophilic substances also have a vital biological impact on human health and positive implications in foods. These substances include essential fatty acids (C18:2 *n*-6 and C18:3 *n*-3) as well as other compounds such as tocopherols. Therefore, the high amount of essential fatty acids, the specific ratio n-6/n-3, and the important content of tocopherols in the elderberry are also vital for the valorization of this berry [3,13].

However, due to their high water content, they are prone to microbial, chemical, and enzymatic degradation, which makes the marketing of fresh elderberries extremely difficult, and are generally marketed after processing [13]. The heat treatments not only considerably reduce cyanogenic glycosides concentration (up to 96%) in elderberry products [5] but also induce the degradation of elderberry beneficial compounds [14]. Thus, through processing, the stability of the bioactive compounds of elderberries is very low. In fact, various studies have shown that factors such as increased temperature and exposure to light or oxygen dramatically reduce the content of bioactive compounds in elderberry-derived products [15,16].

In order to increase the stability and bioavailability of these compounds, drying and encapsulation procedures have been proposed. To stabilize elderberry products, it is essential to reduce their water activity and moisture [15]. Moreover, several encapsulation techniques, which achieved a double effect (drying and encapsulation), including freeze- and spray-drying procedures, have been developed in recent decades [17,18]. Among all procedures, encapsulation by spray-drying continues to be one of the most widely used [19]. It is characterized by a minimal process duration, simplicity, flexibility, energy efficiency, low operating cost, larger-scale productions, scaling-up technique, high encapsulation efficiency, good stability of encapsulated powder, and available equipment in the market [15,20]. It has been described as an effective method to maintain the stability of anthocyanins throughout storage [21,22] and processing. Another important aspect is the correct selection of an appropriate encapsulant agent to obtain physically stable powders [19]. In this regard, maltodextrin is probably the polymer most used as an encapsulant agent for microencapsulation purposes [18,19]. This carrier produces powders with excellent physical properties and preserves the bioactive compounds [18].

Additionally, the studies published showed several ways to add elderberries to foods (in the form of juice, powder, extract, concentrate, vinegar, etc.); thus, the results and elderberry activity among studies are only somewhat comparable [9]. Consequently, the standardization and complete explanation of extraction and encapsulation conditions described in the present manuscript are vital to producing a reproducible product (encapsulated elderberry extract) to be used in different foods. Furthermore, although there are multiple studies that propose the use of anthocyanin-rich extracts as protective agents for the meat industry due to their antioxidant and colorant properties, such as the use of spray-dried raspberry powder [21], there are no studies on the use of encapsulated elderberry extracts in this industry. In addition, the high concentration of bioactive compounds in the encapsulated elderberry extract also allows food to be enriched with bioactive compounds, thus meeting the growing consumer demand for healthy and functional foods [9,17]. Recent research reported that anthocyanin-rich elderberry extract has a favorable toxicological profile, decreases liver lesions, and exerts a potent defense against oxidative DNA damage [23]. These authors also conclude this natural extract can be used as a food colorant without any negative effects on human health, and might even have some positive effects.

On the other hand, there are multiple studies on the characterization of elderberries, as well as their antioxidant capacity and the optimization of extraction processes [2–4,11,18]. However, practically all the studies are focused on phenolic compounds. Nevertheless, there are other important lipophilic bioactive compounds [5,9] that are retained in the residues resulting from the extraction. In addition, dried elderberry pomace has a high content of protein; thus, it could be an alternative vegetal protein source [2]. These residues are mainly composed of peels, pulp residues, and seeds, and they have important compounds that are generally discarded despite their great potential and biological activity [5]. Therefore, in order to promote a circular economy strategy, the integral valorization of elderberries is vital.

Therefore, complete information about the bioactive and valuable compounds present in elderberries is necessary for the rational and integral use of elderberries. The present study proposes not only the characterization of elderberry extracts but also to evaluate the potential of the solid residue resulting from the extraction of phenolic compounds and to fully value all parts of the berries. Currently, there is no study that comprehensively evaluates all potentially extractable bioactive compounds from elderberries, including those retained once the anthocyanins and other polyphenols are extracted. Consequently, the present work has several specific objectives, which include the complete characterization of the hydrophilic compounds extracted from elderberries, the characterization of the lyophilized and encapsulated extracts, the recovery of the resulting solid residue and its chemical analysis, as well as the characterization of the main lipophilic bioactive compounds (fatty acids and tocopherols) retained in the solid residue with the potential to be extracted and used in the food industry. Moreover, the results presented in this manuscript can be used in the development of new and innovative functional ingredients for food applications. In summary, the main objective of the present study was the complete characterization of the main lipophilic and hydrophilic bioactive compounds that can be obtained from elderberries, as well as the encapsulation of hydrophilic compounds.

2. Materials and Methods

2.1. Invention Information

The technological development and the product (encapsulated elderberry extract) presented in the following sections are protected (ES 1 300 302 U) by a utility model (Application number: U202232132/expedition date: 21 August 2023). The protection includes the process, the final product obtained (encapsulated elderberry extract), and its use as an additive in the meat industry. Therefore, the steps described below are included in the invention protection but do not limit its scope.

2.2. Chemicals and Reagents

All reagents were of analytical-grade purity, or HPLC/LC-MS-grade, for chromatographic analysis. 2-Propanol (gradient HPLC grade) (Scharlau; Barcelona, Spain), 2,4,6-tripyridyl-s-triazine (TPTZ) (ThermoScientific Chemicals; Madrid, Spain), 2,2-azinobis-(3-ethyl-benzothiazoline-6-sulphonate) (ThermoScientific Chemicals; Madrid, Spain), 2,20-azobis (2-methylpropionamidine) dihydrochloride (AAPH) (ThermoScientific Chemicals; Madrid, Spain), 2,2-diphenyl-1-picrylhydrazyl (DPPH) (TCL; Paris, France), 37 Component FAME (Supelco; Madrid, Spain), Acetone (VWR Chemicals; Galicia, Spain), acetonitrile (LC-MS grade) (Scharlau; Barcelona, Spain), ammonium sulfate (Scharlau; Barcelona, Spain), boric acid (Carlo Erba Reagents; Barcelona, Spain), chloroform (Scharlau; Barcelona, Spain), DL-α-Tocoferol (Supelco; Madrid, Spain), diethyl ether (Scharlau; Barcelona, Spain), dipotassium phosphate (Scharlau; Barcelona, Spain), ethanol (VWR Chemicals; Galicia, Spain), fluorescein (VWR Chemicals; Galicia, Spain), Folin–Ciocalteu (VWR Chemicals; Galicia, Spain), formic acid (LC-MS grade) (Scharlau; Barcelona, Spain), gallic acid (Scharlau; Barcelona, Spain), hydrochloric acid 37% (Carlo Erba Reagents; Barcelona, Spain), iron (III) chloride hexahydrate (VWR Chemicals; Galicia, Spain), iron (II) sulfate heptahydrate (VWR Chemicals; Galicia, Spain), Kjelcat Cu (Gerhardt; Barcelona, Spain), L(+)-ascorbic acid (Scharlau; Barcelona, Spain), maltodextrin (Ceylan; Valencia, Spain), methanol (Scharlau; Barcelona, Spain), n-Hexane (gradient HPLC grade) (Scharlau; Barcelona, Spain), potassium dihydrogen phosphate (Scharlau; Barcelona, Spain), sodium acetate (VWR Chemicals; Galicia, Spain), sodium carbonate (Scharlau; Barcelona, Spain), Sodium hydroxide (85%) (VWR Chemicals; Galicia, Spain), sodium metal (Acros Organics; Madrid, Spain), sulfuric acid (98%) (Carlo Erba Reagents; Barcelona, Spain), and Trolox (Acros Organics; Madrid, Spain).

2.3. Elderberry Material and Experimental Design

Wild elderberry samples were collected in quadruplicate (on different days/weeks) at the end of August 2021 in Casas da Veiga (GPS coordinates 42.0716854, −7.8093116) and Trasmiras (GPS coordinates 42.0228502, −7.6176959) (Ourense, Spain). Elderberries (only fully ripe fruits) from each collection (about 500 g of fresh elderberries) were treated as independent samples. They were washed, frozen (−80 °C), lyophilized (Lyovapor L300, Büchi; Barcelona, Spain; primary drying pressure limit: 0.500 mbar, time: 240 min; secondary drying pressure limit: 0.400 mbar, time: 120 min), and ground following the procedure previously described by Domínguez et al. [3]. The lyophilized elderberries (four different samples (approx. 110 g each), depending on the collection) were the material used for the extraction of bioactive compounds, according to the experimental design (Figure 1).

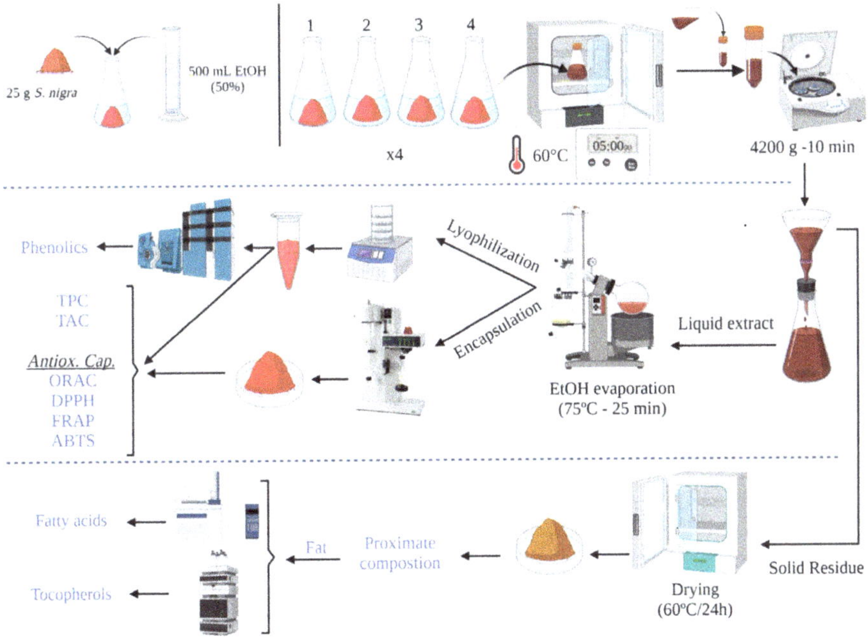

Figure 1. Experimental design for the characterization of bioactive compounds of elderberry.

The extraction parameters and procedure were based on the optimum conditions previously described by Domínguez et al. [3], with modifications. A total of 16 (n = 16) extractions (4 independent extractions per day (for each collection sample) × 4 different days) were carried out. The lyophilized material (25 g) was mixed with 500 mL of ethanolic solution (50%, *v*/*v*) and kept at 60 °C for 5 h (shaking manually every hour). After this time, the samples were centrifuged (4200× *g* for 10 min) and filtered (Filter-lab 1238, Barcelona, Spain) to separate the solid residue from the liquid extract. Ethanol was removed from the liquid extract via rotary evaporation (75 °C for 25 min under vacuum conditions), and the resulting aqueous extract was subdivided into two parts. Half of the extract was freeze-dried following the previously described conditions, while the other half was encapsulated using maltodextrin as the encapsulant agent (13.5 g maltodextrin/100 mL of extract). The encapsulation procedure was carried out in a spray-dryer (Mini Spray-Dryer B-290; Büchi, Barcelona, Spain) with a 145 °C inlet temperature, 28 m^3/h of aspirator flow (80%), and 600 g/h extract/maltodextrin feed flow [24]. The main groups of bioactive compounds (total phenolic compounds and total anthocyanins) (see Section 2.5.1) and the antioxidant capacity (see Section 2.5.3) were analyzed for both extracts. Additionally, on the lyophilized extracts, the phenolic compounds profile (anthocyanins and non-anthocyanins; see Section 2.5.2) responsible for their antioxidant capacity and bioactive character were analyzed and quantified via liquid chromatography coupled with mass spectrometry. On the other hand, the solid residue obtained after filtration was dried in an oven (60 °C for 24 h) and ground. In order to evaluate the potential of bioactive compounds that are still present in the residue after extraction, the chemical composition (see Section 2.4.1) and the lipid fraction (fatty acids and tocopherols; see Sections 2.4.2 and 2.4.3, respectively) were analyzed. Figure 2 shows the visual appearance of the elderberries (fresh and lyophilized), the solid residue, and the encapsulated extract obtained during this research.

Figure 2. Visual appearance of fresh elderberries (**a**), freeze-dried elderberries (**b**), solid residue after extraction (**c**), and encapsulated extract (**d**).

2.4. *Solid Residue Characterization*

2.4.1. Proximate Composition

The moisture [25], protein [26], and ash [27] content of the solid residue was measured according to ISO procedures, while total fat was quantified following the American Oil Chemists' Society method after acid hydrolysis [28]. The carbohydrate content was calculated by difference (carbohydrate content = 100—% protein (dry matter)—% fat (dry matter)—ash (dry matter).

2.4.2. Fatty Acid Composition

The oil extraction was carried out using solvents (methanol and chloroform) under the procedure and proportions described by Barros et al. [29]. After that, 20 mg of oil was transesterified with sodium methoxide and sulfuric-methanol solutions, extracted with n-hexane, and fatty acids were separated, identified, and quantified via gas chromatography according to the method and chromatographic conditions previously described [29]. The results were expressed as g/100 g of oil.

2.4.3. Tocopherol Composition

For the quantification of tocopherols, 100 mg of oil extracted, as indicated in the previous section, was re-dissolved in 3 mL of n-hexane, filtered through nylon syringe filters (0.45 μm), and injected into a liquid chromatograph (HPLC). For the analysis, 10 μL of each sample and standards (α-, β-, δ- and γ-tocopherol) were injected in isocratic mode (2-propanol/n-hexane, 2 and 98 v/v) at 1 mL/min flow in an HPLC system equipped with Alliance 2695 and 2475 fluorescence detector (Waters, Milford, CT, USA). The tocopherol separation was carried out in a column SunFire™ Prep Silica (4.6 mm ID × 250 mm, 5 μm particle size, Waters, Milford, MA, USA) at 35 °C, and was detected with a fluorescence detector (290 nm λ excitation and 330 nm λ emission). Results were expressed as μg/g oil.

2.5. Encapsulated and/or Lyophilized Extract Characterization

2.5.1. Determination of Total Phenolic Compounds (TPC) and Anthocyanins (TAC)

The determination of the main groups of bioactive compounds (TPC and TAC) was carried out using spectrophotometric techniques according to the procedures described by Domínguez et al. [3].

2.5.2. Characterization of Phenolic Compounds

The phenolic profile was determined in the lyophilized extracts re-dissolved in a methanol:water (80:20, v/v) solution and filtered through a 0.22 μm disposable filter. The chromatographic analysis was performed by LC-DAD-ESI/MS^n (Dionex Ultimate 3000 UPLC, Thermo Scientific, San Jose, CA, USA) using the conditions previously described by Bessada et al. [30]. Briefly, the compounds were separated with a reversed-phase column Spherisorb S3 ODS-2C18 (3 μm particle size, 4.6 mm I.D., 15 mm length). The mobile phase (A) was 0.1% formic acid in miliQ water, and phase (B) was acetonitrile. The elution was carried out at 0.5 mL/min flow and in gradient mode: isocratic 15% (B) (5 min); 15–20% (B) (5 min); 20–25% (B) (10 min); 25–35% (B) (10 min); 35–50% (B) (10 min); re-equilibration. Double online detection at 280 and 370 nm wavelengths were used (DAD detector). MS detection was performed in negative mode using a Linear Ion Trap LTQ XL mass spectrometer (Thermo Finnigan, San Jose, CA, USA) equipped with an ESI source. Sheat gas (N_2) was set at 50 psi; the spray voltage was 5 kV; source temperature was 325 °C; capillary voltage was −20 V; collision energy was 35. The full scan covered the mass range from 100 to 1500 m/z. Data acquisition was carried out with Xcalibur® data system (ThermoFinnigan, San Jose, CA, USA). The identification of the phenolic compounds was performed using standard compounds, when available, via comparison of their retention times, UV-vis, and mass spectra, and comparing the obtained information with available data reported in the literature, giving a tentative identification. For quantitative analysis, a calibration curve for each available phenolic standard was constructed based on the UV signal. For the identified phenolic compounds for which a commercial standard was not available, the quantification was performed through the calibration curve of the most similar available standard. The results were expressed as mg/100 g of lyophilized extract.

2.5.3. Determination of In Vitro Antioxidant Capacity

Four different techniques were used for the determination of in vitro antioxidant capacity, including oxygen radical absorbance capacity (ORAC), 2,2-diphenyl-1-picrylhydrazyl radical-scavenging activity (DPPH), ferric-reducing antioxidant power assay (FRAP), and 2,2-azinobis-(3-ethyl-benzothiazoline-6-sulphonate) radical (ABTS) according to the procedures described by Franco et al. [31].

2.6. Statistical Analysis

A total of 48 samples were analyzed, distributed into 16 samples of each type (lyophilized extract, encapsulated extract and solid residue) (n = 16 each) (4 independent extractions/day × 4 different extraction days). Experimental data were reported as mean values and standard deviations. Analysis of variance (one-way-ANOVA) for comparison of encapsulated and lyophilized extracts was performed using SPSS software (version 25). Differences were considered significant when $p < 0.05$. Pearson's correlation test was used to verify the correlations between the main groups of bioactive compounds, individual compounds, and the antioxidant capacity of the extracts.

3. Results and Discussion

3.1. Proximate Composition and Lipophilic Bioactive Compounds of the Solid Residue

After the extraction of the hydrophilic bioactive compounds (mainly anthocyanins and other polyphenols), it is evident that a large mass of material remains in the form of residue. In the present research, the dry solid residue after the extraction of the phenolic compounds represented 46.82 ± 4.96% of the initial mass of lyophilized elderberry. Therefore, it can be affirmed that this residue represents almost half of the initial mass, so there is enormous potential for the extraction of bioactive compounds that are still present in it, and thus achieve the integral use of the elderberry. For this, it is essential to know what types of compounds are present [13], so it was decided to study their chemical composition (Table 1).

Table 1. Proximate composition (g/100 g), tocopherols (μg/g of oil), and fatty acids (g/100 g of oil) of the solid residue obtained after extraction (n = 16).

	Solid Residue		
	Values	**Min**	**Max**
Proximate Composition (g/100 g)			
Moisture	18.34 ± 6.10	10.31	23.93
Fat †	17.07 ± 1.25	15.28	18.75
Protein †	15.66 ± 0.55	14.78	16.43
Ash †	1.36 ± 0.10	1.23	1.51
Carbohydrates †	65.91 ± 1.25	64.42	68.32
Tocopherols (μg/g of oil)			
α-Tocopherol	317 ± 37.5	260.53	360.30
β-Tocopherol	6.24 ± 0.72	5.25	7.44
δ-Tocopherol	463 ± 48.2	396.36	518.96
γ-Tocopherol	4.61 ± 0.49	3.99	5.29
Total Tocopherols	791 ± 81.4	667.06	877.70
Fatty acids * (g/100 g of oil)			
C16:0 (palmitic acid)	6.12 ± 0.25	5.70	6.43
C18:0 (stearic acid)	1.56 ± 0.03	1.52	1.61
C18:1 *n*-9 (oleic acid)	10.72 ± 0.21	10.33	11.07
C18:1 *n*-7 (vaccenic acid)	0.68 ± 0.03	0.63	0.71
C18:2 *n*-6 (linoleic acid)	32.88 ± 0.43	32.04	33.46
C18:3 *n*-3 (α-linolenic acid)	32.56 ± 0.47	31.68	33.19
9*c*,11*t*-C18:2 (conjugated linoleic acid)	0.17 ± 0.02	0.13	0.21
C20:0 (arachidic acid)	0.17 ± 0.01	0.15	0.18
C20:1 *n*-9 (*cis*-11-eicosenoic acid)	0.13 ± 0.003	0.13	0.14
C22:0 (behenic acid)	0.15 ± 0.01	0.13	0.18
SFA	8.50 ± 0.32	7.92	8.85
MUFA	11.80 ± 0.27	11.30	12.20
PUFA	65.75 ± 0.91	63.98	66.98
Σ *n*-3	32.59 ± 0.47	31.72	33.23
Σ *n*-6	32.98 ± 0.43	32.14	33.57
n-6/*n*-3	1.01 ± 0.00	1.01	1.02
Total fatty acids	86.05 ± 1.45	83.21	88.02

† Results of the chemical composition are expressed as g/100 g of dry matter. * In the table, only the fatty acids that represent more than 0.1% are presented, but for the sum (SFA, MUFA, PUFA, Σ *n*-3, Σ *n*-6, and total fatty acids), total identified and quantified fatty acids were taken into account; SFA: saturated fatty acids; MUFA: monounsaturated fatty acids; PUFA: polyunsaturated fatty acids.

Carbohydrates were the main macronutrient of the solid residue (65.91 g/100 g dry matter). As indicated, this solid residue is formed by the insoluble elderberry parts during extraction, so its composition is mainly skin, seeds, and exhausted pulp. With this in mind, it is reasonable that carbohydrates are the major components of this residue. In this case, sugars are soluble in the hydroalcoholic solution used during extraction; thus, the carbohydrates of solid residue are mainly composed of fiber. The same trend was observed by Costa et al. [2], who found that carbohydrates are the main macronutrient (82.4 g/100 g of dry matter) in elderberry dry pomace, while fiber represented 22.4 g/100 g of dry matter. Also, Domínguez et al. [3] observed that the total carbohydrates in elderberries were 79.42 g/100 g dry matter. The higher content of carbohydrates in these studies than in the present work is due to the fact that these authors used complete elderberry or elderberry pomace, which means that they have a high sugar content, while in our case, these simple sugars were removed in the extraction phase.

Another important macronutrient is protein, and solid residue contained important proportions (15.66 g/100 g of dry matter). This value was very similar to those reported by Domínguez et al. [3], who observed 14.08 g protein/100 g of dry matter in elderberries. In contrast, lower protein amounts were found in pomace powder (5.9 g/100 g of dry matter) [2]. Other studies also found lower protein values (2.7–2.9%) in elderberries [5,32]. A possible explanation for this high proportion of protein in our study may be that the removal of important compounds (all hydrophilic compounds) during the extraction process implies that the compounds that are not extracted and remain in the solid residue are concentrated in percentage terms. In addition, it is important to highlight that elderberry protein has a good nutritional value since it contains significant amounts of essential amino acids. In fact, a study reported isoleucine, lysine, and methionine as the limiting amino acids, but elderberry protein may still be considered as a satisfactory source for these amino acids [33]. The other essential amino acids' chemical scores ranged from 0.7 to 1.4, except for tyrosine (2.03). Moreover, these authors reported a chemical score of 0.94 for total essential amino acids, which indicates that elderberries are a food source of high biological value protein. Thus, taking into account the high proportion of protein in the solid residue, it could be an alternative vegetal high-quality protein source.

In elderberries, lipids are located primarily in the seeds [5]. Thus, it is expected that solid residue contains important non-polar constituents. In our research, total lipids in solid residue were the second most important macronutrient, and represented 17.07 g/100 g of dry matter. Therefore, an important proportion of lipophilic bioactive compounds could be extracted from this fraction. Our value was similar to those reported in elderberry seeds (10.5 g/100 g of dry matter) by Fazio et al. [13]. In contrast, the lipid content of elderberry pomace (2.5 g/100 g of dry matter) [2] and elderberries (1.6 g/100 g of dry matter) [3] was lower than those observed by us.

In addition to the high oil content observed in the present study, the elderberry fatty acids profile was characterized by a high amount of polyunsaturated fatty acids (PUFA; 65.75 g/100 g of oil), followed by monounsaturated fatty acids (MUFA; 11.8 g/100 g of oil) and very low proportions of saturated fatty acids (SFA; 8.50 g/100 g of oil). Fatty acids represent 86.05 g/100 g of total lipids (Table 1). The major individual fatty acids were C18:2 *n*-6 and C18:3 *n*-3 (about 32 g/100 g of oil each), with important amounts of C18:1 *n*-9 (10.72 g/100 g of oil). This profile was reported by other authors, who observe linoleic and α-linolenic fatty acids as the most important, and significant contents of oleic acid [3,5,13]. In fact, it has been concluded that elderberry oil is an adequate source of healthy fatty acids for a balanced diet [9]. The practically coincident value of the content of C18:2 *n*-6 and C18:3 *n*-3 affects the fact that this oil presents a *n*-6/*n*-3 ratio of 1.01. Following the recommendations of international health institutions, a healthy diet must present a high PUFA proportion, and these must be balanced. The recommended *n*-6/*n*-3 ratio is 4 or less, while the optimal value is 1. In this sense, it is clear that elderberry oil is not only within the recommended value but also has an optimal value of this ratio and is a balanced source of essential PUFA, which agrees with the findings observed in previous studies [3,5,13,32].

According to several epidemiological studies, the intake of PUFA, especially the essential PUFA, like C18:2 *n*-6 and C18:3 *n*-3, contributed to the reduced prevalence of chronic diseases such as cardiovascular diseases, hypercholesterolemia, inflammatory and autoimmune diseases, cancer, and diabetes [5,34]. Thus, both the high amounts of these two essential fatty acids (76% of total fatty acids) and the exceptional *n*-6/*n*-3 ratio indicate that elderberry oil is of high quality, and the oil extraction from solid residue is an opportunity to obtain valuable compounds to produce enriched and functional foods with added value to human health protection.

Regarding other lipophilic bioactive compounds, several antioxidant compounds belonging to tocopherols are present in elderberry solid residue. In our case, the total tocopherol content was 791 μg/g of oil, where the most important tocopherols were δ-tocopherol (463 μg/g of oil) and α- tocopherol (317 μg/g of oil), while we detected minor contents of β- and γ-tocopherol (Table 1). As occurs in the present research, some authors also reported abundant contents of α-tocopherol and γ-tocopherol in elderberry [13,22]. In contrast to our findings, Fazio et al. [13] did not detect β- and δ-tocopherol in elderberry seed oil, while the content of α-tocopherol (0.49 μg/g of oil) and γ-tocopherol (2.63 μg/g of oil) were significantly lower than those observed by us. As occurs with other bioactive compounds, there are many factors that can influence their content, and this can partially explain the differences between the two studies. In addition, Fazio et al. [13] indicated that the seeds were subjected to several stages to obtain the seed flour, including a phase of drying at high temperatures (110 °C for 2 h), which could also explain why these authors have a lower tocopherol content (highly susceptible to heat degradation) than those observed in the present study.

From the nutritional point of view, it is well known that α-tocopherol has the highest vitamin E bioactivity [13]; thus, the elderberry oil extracted from solid residue could be an important source of this vitamin. Additionally, all tocopherols show important antioxidant ability because they transfer hydrogen atoms to radicals, delay hydroperoxide decomposition, and scavenge singlet oxygen [9]. Thus, elderberry oil is also a source of strong antioxidants.

In conclusion, the solid residue, which represents a great proportion (approx. 47%) of initial elderberry mass, contained significant amounts of important macronutrients (lipids and proteins) and also high contents of lipophilic antioxidants and active compounds, such as essential fatty acids and tocopherols. Therefore, the results obtained in this study show the possibility of using the residue after extracting the hydrophilic compounds to obtain important substances while achieving integral use of elderberries, valuing the by-products that are generated during their processing, which are otherwise generally managed as waste and used as animal feed or discarded.

3.2. Main Bioactive Compounds and Antioxidant Capacity of Lyophilized and Encapsulated Elderberry Extracts

The extraction conditions of the phenolic compounds significantly impact their activity (antioxidant, colorant, and antimicrobial) as well as the recovery yields [22]. In the present study, the conditions applied for the extraction of phenolic compounds have been based on previous research [3] in which our research group has optimized these conditions to obtain high yields of extracts with excellent antioxidant properties. As specified in the experimental design (Figure 1), part of each hydroalcoholic extract, after removing ethanol content, was directly lyophilized (without carrier), while another part was microencapsulated (spray-drying) using maltodextrin. In this last case, the ratio of the encapsulant agent to the solids of the extract was 2.72 g of maltodextrin/g solids, and the content of solids in the liquid extract after ethanol removal was 4.27 ± 0.19 g solids/100 mL of extract. The encapsulated product yield (mass of powder recovered, considering total solids in the feed solution) obtained by the spray-drying procedure was 69.2 ± 3.07%. This value was lower than those obtained by Murugesan and Orsat [15] using 5:1 ratio proportions (80.1%) and similar to a 1:1 ratio (72.8%) but higher than those reported by Ribeiro et al. [20], who described that the product yield ranged from 25 to 41%. The differences could be attributed to the variations in encapsulant agent proportions and spray-drying conditions.

It is important to highlight that the lyophilized extract presented exceptional hygroscopicity, which makes its handling and storage difficult, limiting its potential use at an industrial level. In contrast, the encapsulated extract remained unaltered and was easily manipulated. Moreover, dehydration allows for obtaining powders rich in bioactive compounds due to the concentration of these compounds [17]. Additionally, reducing water activity is vital for preserving bioactive compounds during storage since dramatic degradative consequences in anthocyanin content were reported as increasing the water activity [10]. In the present study, the main groups of bioactive compounds (total phenolic and anthocyanin contents) and the antioxidant capacity were measured in each extract (lyophilized and encapsulated). These data are presented in Table 2. It is important to note that total phenolic compounds (TPC) assay could be easily affected by other interfering substances, such as sugars, proteins, or ascorbic acid, and chromatographic determinations report more reliable results, but TPC and total anthocyanin content (TAC) can be useful for making general comparisons with other studies [1].

Table 2. Quantification of bioactive compounds and in vitro antioxidant capacity of the lyophilized (n = 16) and encapsulated (n = 16) elderberry extracts.

	Extracts		Sig.
	Lyophilized	**Encapsulated**	
Bioactive compounds			
Total phenolic compounds (mg GAE/100 g)	7486 ± 395	1132 ± 168	***
Anthocyanins (mg/100 g)	520 ± 39.2	60.0 ± 7.55	***
Antioxidant Capacity			
DPPH (mg Trolox/g)	39.92 ± 1.73	3.84 ± 0.47	***
ABTS (mg ascorbic acid/g)	145.9 ± 7.4	12.91 ± 1.66	***
FRAP (mmol Fe^{2+}/100 g)	177.7 ± 8.76	12.27 ± 1.98	***
ORAC (mg Trolox/g)	286 ± 23.3	46.0 ± 13.1	***
IC_{50} (mg/mL)	4.24 ± 0.44	34.5 ± 7.81	***

*** Significance ($p < 0.001$).

The TPC value in the lyophilized extract was 7486 mg of GAE/100 g, while in the encapsulated extract, this value decreased and was 1132 mg of GAE/100 g. Similarly, TAC also decreases from 520 mg/100 g in the lyophilized extract to 60 mg/100 g in the encapsulated extract. Our TPC and TAC values in the lyophilized extract were higher than those reported in elderberry freeze-dried powders (1242 mg of GAE/100 and 171.8 mg/100 g for TPC and TAC, respectively) by Busso Casati et al. [17], but in our spray-dried powders, these values were similar to those obtained in the other study. These authors use a similar encapsulant/solid ratio (2.12) than in the present research, and both encapsulated powders presented similar results. In contrast, the use of a lower encapsulant/solid ratio (0.41) in elderberry freeze-dried powders resulted in higher total monomeric anthocyanin contents (between 1059 and 1363 mg/100 g) [10]. The present results in lyophilized extracts were higher than those described in our previous study [3], which obtained 2857 mg of GAE/100 g for TPC and 473 mg/100 g for TAC, and similar to those reported in ethanolic extracts obtained by Duymuş et al. [35] (7594 mg of GAE/100 g for TPC and 1066 mg/100 g for TAC). Moreover, in recent research, the authors reported similar values of TPC (1277 mg of GAE/100 g dry matter) in frozen elderberries and higher TAC content (4061 mg/100 g dry matter) [19] than in the present research, while Mattson et al. [18] found 2791 mg of GAE/100 g dry matter for TPC and 1514 mg/100 g dry matter for TAC in elderberry fruit. However, when these authors encapsulated (spray-drying) this extract using maltodextrin as an encapsulant agent (ratio of maltodextrin/solids of 11.2), these values decreased (457 mg of GAE/100 g and 239 mg/100 g for TPC and TAC, respectively). Finally, other authors reported lower TPC values in wild fruit elderberry extract (5678 mg of GAE/100 g dry matter) [4] and elderberries (4802 mg of GAE/100 g dry matter) [36] and higher TAC content (3071 mg/100 g dry matter [4]; 3278 mg/100 g dry matter [36]) than those obtained by us in the lyophilized extract.

In our research, after encapsulation, the TPC was 6.61 times lower, while the TAC was 8.67 times lower than that of the lyophilized extract. This fact agrees with those reported by other authors, who described that both TPC and TAC were considerably lower in spray-dried raspberry powder than in freeze-dried powders [21]. The differences between extracts could be attributed to the presence of an encapsulant agent in the spray-dried powders, while the lyophilized extract did not have any carrier, which creates a "dilution" effect in the spray-dried powders [21]. That is, for the same amount of extract, the lyophilized extract only contains the elderberry, while the encapsulated one has extract and maltodextrin, which determines that the proportions of TPC and TAC are lower since these compounds are only present in the elderberry. The use of encapsulant agents is vital to obtain high-quality powders, but the proportion between encapsulant and solids determines the final characteristics of the powders [10]. In fact, Baeza et al. [10] proposed the use of low encapsulant proportions to obtain elderberry powders with high amounts of anthocyanins and concluded that encapsulated powders with higher anthocyanins content were those obtained using the lowest encapsulant proportion. In any case, as aforementioned, in the present study, the proportion of maltodextrin with respect to solids is 2.72, so it also seems that the drying method, as well as the high temperature applied in the spray-drying, can cause changes (degradation) in the content of these bioactive compounds [15]. This explanation was also supported by Busso Casati et al. [17], who reported that the freeze-drying process produces better conservation of heat-sensitive bioactive compounds and antioxidant capacity than spray-drying.

According to the antioxidant and antiradical activities, the results show a similar trend to that of the TPC and TAC. The values of DPPH decreased from 39.92 mg of Trolox/g in the lyophilized extract to 3.84 mg of Trolox/g in spray-dried powders. ABTS and FRAP values also decreased from 145.9 mg of ascorbic acid/g and 177.7 mmol of Fe^{2+}/100 g, respectively, to about 12 in both cases, while ORAC decreased from 286 mg of Trolox/g in lyophilized powders to 46.0 mg of Trolox/g in encapsulated powders. Similarly, the values of IC_{50}, which represents the concentration required to scavenge 50% of the free radicals, increased from 4.24 to 34.5 mg/mL when using an encapsulant agent. A lower IC_{50} value refers to a higher antioxidant activity [13]. Therefore, the antioxidant capacity in the encapsulated powders was between 6.2 and 14.5 times lower than the lyophilized powders, depending on the determination method (DPPH: 10.4 times; ABTS: 11.3 times; FRAP: 14.5 times; ORAC: 6.2 times). As has been discussed, the decrease is due to both the dilution effect exerted by the encapsulant agent and possible degradation during the encapsulation process via spray-drying. In a previous study [3], the analysis revealed higher antioxidant activities for DPPH and ABTS but lower for ORAC and FRAP procedures than those reported in the lyophilized extract of this research. It is important to highlight that in the previous study, the results are expressed in dry elderberry, while in this study, the results are expressed in extract, so the differences between both studies could be partially related to this fact. Moreover, there are several other parameters (edaphic, climatic, cultivar, etc.) that can cause variation in the main bioactive compound content and, thus, the antioxidant activity of elderberries. Additionally, in a recent study, the authors demonstrated that wild-growing plants have lower antioxidant capacity than cultivated plants [4], which can also be related to the variations among different studies.

In our case, the antioxidant and antiradical activities were strongly, positively, and significantly correlated with TPC (r = 0.993 ($p < 0.001$) DPPH; r = 0.989 ($p < 0.001$) ABTS; r = 0.990 ($p < 0.001$) FRAP; r = 0.973 ($p < 0.001$) ORAC) and also with TAC (r = 0.994 ($p < 0.001$) DPPH; r = 0.997 ($p < 0.001$) ABTS; r = 0.996 ($p < 0.001$) FRAP; r = 0.990 ($p < 0.001$) ORAC). This fact was also observed in other studies [4,11,13,17,21,35], where positive and strong correlations between phenol and anthocyanin contents and antioxidant and radical-scavenging capacity were described. This high correlation is to be expected since phenols and anthocyanins are strong antioxidants and act as metal chelators, quenching single oxygen and limiting initiation and propagation reactions by scavenging reactive substances [35]. Additionally, the combination of phenolics and anthocyanins exerts a synergistic effect, resulting in an increase in antioxidant capacity.

In general conclusion, encapsulation has led to a significant reduction in the TPC and TAC content, as well as the antioxidant capacity of the extract (mainly due to the dilution effect exerted by the wall material used in encapsulation). However, the encapsulated extract presents significant benefits since the material is much more manipulable and more stable than the lyophilized extract, which would be practically impossible for industry to use. Although it has not been raised in this study, it would be necessary to expand the information on the stability of the encapsulated elderberry extract.

3.3. Phenolic Compounds of Lyophilized Elderberry Extracts

The lyophilized elderberry extracts analyzed using the LC-DAD-ESI/MS^n technique showed the presence of seven important polyphenolics, which include three non-anthocyanin and four anthocyanin compounds (Table 3). In the non-anthocyanin group, peaks at 5.83 and 6.63 min from the molecular $[M\text{-}H]^-$ ion at m/z 353 yielded deprotonated quinic acid (m/z at 191) as the base peak, and supplementary ions at m/z 179 (caffeic acid moiety) and at m/z 173 [37] were identified as "*cis*" and "*trans*" 5-*O*-caffeoylquinic acid, respectively. A peak detected at 16.52 min, with a parent ion $[M\text{-}H]^-$ ion at m/z 609 and loss of quercetin at m/z 301, was identified as quercetin-3-*O*-rutinoside (rutin) [37,38]. Additionally, the maximum absorption wavelengths for these compounds agree with those described by other authors [16,36–38], which confirms their identification.

Table 3. Identification and quantification (mg/100 g of extract) of the non-anthocyanin and anthocyanin phenolic compounds present in elderberry lyophilized extract (n = 16).

Peak No.	Rt (min)	λ Max (nm)	$[M\text{-}H]^-$ (m/z)	MS^2	Tentative Identification	Quantification (mg/100 g of Extract)
Non-Anthocyanin						
1	5.83	325	353	191(100), 179(9), 173(19), 161(5)	*cis* 5-*O*-Caffeoylquinic acid	218 ± 8.30
2	6.63	327	353	191(100), 179(6), 173(21), 161(9)	*trans* 5-*O*-Caffeoylquinic acid	286 ± 17.0
3	16.52	353	609	301(100)	Quercetin-3-*O*-rutinoside	218 ± 14.5
					Total non-anthocyanin	722 ± 31.5
Anthocyanin						
4	27.44	515	743	581(72), 449(100), 287(78)	Cyanidin-*O*-sambubioside-*O*-hexoside	525 ± 57.1
5	27.89	515	743	581(72), 449(100), 287(78)	Cyanidin-3-*O*-sambubioside-5-*O*-glucoside	943 ± 53.8
6	30.97	518	581	287(100)	Cyanidin-3-*O*-sambubioside	4827 ± 546
7	31.92	517	449	287(100)	Cyanidin-3-*O*-glucoside	4926 ± 756
					Total anthocyanin	11,221 ± 1143

Standard calibration curves used: chlorogenic acid (y = 168,823x − 161,172, R^2 = 0.9999, LOD = 0.20 μg/mL; LOQ = 0.68 μg/mL; peaks 1 and 2), quercetin-3-*O*-glucoside (y = 34,843x − 160,173, R2 = 0.9998, LOD = 0.21 μg/mL; LOQ = 0.71 μg/mL, peak 3), and cyanidin-3-*O*-glucoside (y = 105,078x − 12,437; R2: 0.9993; LOD = 0.28 μg/mL; LOQ = 0.84 μg/mL, peaks 4, 5, 6 and 7). Results are expressed as mean ± SD.

With this in mind, our results showed that 5-*O*-caffeoylquinic acids were the most representative hydroxycinnamic acids, while quercetin-3-*O*-rutinoside was the characteristic flavonol in elderberries. The amount of each compound in the non-anthocyanin group was very similar (between 218 and 286 mg/100 g of extract). Our findings agree with several authors [11,16,38,39], who conclude that quercetin-*O*-rutinoside and 5-caffeoylquinic acids are the major non-anthocyanin phenolic compounds in elderberry fruits. Important contents of 5-*O*-caffeoylquinic acid were also reported by other authors in several elderberry fruit cultivars [11,16,38,39].

On the other hand, four different anthocyanins have been detected in the lyophilized elderberry extracts. In this case, the characteristic fragment at m/z 287 corresponds to the cyanidin aglycone [13]; thus, it is expected that all anthocyanins found in our extracts are cyanidin-derivative anthocyanins. The peak detected at 27.89 min had an ion at $[H]^+$

at m/z 743, which yielded MS^2 fragments at m/z 581, 449, and 287. This characteristic fragmentation corresponds to cyanidin-3-O-sambubioside-5-O-glucoside [12,35].

The peak at 30.97 min was fragmented from a parent ion $[H]^+$ at m/z 581 to 287, indicating a loss of a sambubiosyl moiety. The peak was therefore identified as cyanidin-3-O-sambubioside. Moreover, the peak at 31.92 min, with a molecular ion $[H]^+$ at m/z 449 and a major fragment at m/z 287, was identified as cyanidin-3-O-glucoside [12,35,37].

In addition to the characteristic mass fragments, cyanidin-derived anthocyanins showed maximum absorption at the wavelengths of 515–518 nm, which agrees with the wavelengths reported in the literature for these compounds (between 515 and 525 nm) [16,18,35–37].

Our results indicated that the cyanidin-based anthocyanins were the major phenolic compounds detected in lyophilized elderberry extracts, which agrees with data reported by several authors [9,12]. Additionally, in accordance with the results of Mattson et al. [18], the only group of anthocyanins in elderberry were cyanidin-based anthocyanins. Among them, the highest contents were observed for cyanidin-3-O-sambubioside and cyanidin-3-O-glucoside, with very close values (about 4900 mg/100 g of extract each). In agreement with our findings, several authors reported both cyanidin-3-O-sambubioside and cyanidin-3-O-glucoside as the two major anthocyanins in elderberry [2,10,11,14,16–18,23,39]. In the present study, both compounds represented 88% of the total anthocyanins (44.5% for cyanidin-3-O-glucoside and 43.6% for cyanidin-3-O-sambubioside), and this value was within the range proposed by Domínguez et al. [9], Młynarczyk et al. [6], and Senica et al. [39], who reported that both anthocyanins contribute together around 85–90% of total anthocyanins present in elderberries. Similarly, Busso Casati et al. [17] reported that cyanidin-3-O-glucoside was the major anthocyanin in elderberry freeze-dried powders and that it represents 42% of total anthocyanins. Moreover, important amounts of cyanidin-3-O-sambubioside-5-O-glucoside were also reported by other studies in elderberries [2,11,16,39,40].

Similar to the results reported by Duymuş et al. [35], in the present study, significant correlations were observed between the content of cyanidin-3-O-glucoside and the antioxidant activity measured with the FRAP procedure (r = 0.642; p = 0.013). Moreover, the content of *trans*-5-O-caffeoylquinic acid (r = 0.753; $p < 0.001$) and total non-anthocyanin compounds (r = 0.568; p = 0.022) were strongly correlated with the ORAC values, while quercetin-3-O-rutinoside content was correlated with FRAP values (r = 0.563; p = 0.023). In view of the results, it seems evident that several compounds, including hydroxycinnamic acids, flavonols, and anthocyanins, determine the antioxidant and colorant properties of elderberry extracts.

4. Conclusions

In the present study, the effect of encapsulation on the concentration of the hydrophilic bioactive compounds and their antioxidant capacity was investigated. Encapsulant agents exert a dilution effect, which causes a decrease in their antioxidant capacity. However, it should be noted that the encapsulated elderberry powder is more manageable and stable than the lyophilized extract. Additionally, the complete description encapsulant process allows homogenizing the way of obtaining a functional and novel elderberry-derived product that is not present on the market.

The polyphenol profile indicated that the most important hydrophilic compounds in elderberry extract were cyanidin-derived anthocyanins, characterized by high amounts of cyanidin-3-O-glucoside and cyanidin-3-O-sambubioside. Moreover, significant amounts of non-anthocyanins compounds, such as hydroxycinnamic acids (5-O-caffeoylquinic acids) and flavonols (rutin), were also detected. All these compounds determined the high antioxidant properties of elderberry extract.

On the other hand, the solid residue resulting from the extraction of the hydrophilic bioactive compounds represents an important part of the initial mass of elderberry (over 47%). It has high amounts of important macronutrients (oil and protein). Among lipophilic compounds, oil extracted from solid residue has an exceptional fatty acid profile, characterized by its significant amount of PUFA and essential fatty acids (C18:2 n-6 and C18:3 n-3) and optimal ratio n-6/n-3, which indicate that this oil is a potential source for the obtention of important lipophilic substances and their use for the enrichment or manufacture of functional foods. Taking into account the increase in demand for natural food ingredients, elderberry may be an additional source of lipophilic and hydrophilic bioactive compounds.

Therefore, as a general conclusion, the entire process of obtaining a novel, stable, and versatile food ingredient (encapsulated elderberry extract) was completely explained and characterized. The strategy applied in this study provides important information about the use of elderberry as a potential source for the obtention of valuable compounds for the food industry. This strategy includes the extraction and stabilization (encapsulation) of hydrophilic compounds (mainly phenolics and anthocyanins), the integral use of solid residue for the extraction of lipophilic bioactive compounds (essential fatty acids and tocopherols), and the possibility of producing protein isolates. However, this study is a starting point and should be complemented with additional studies that demonstrate the possibility of obtaining lipophilic compounds and protein isolates from the solid residue, in addition to characterizing them and studying important aspects, such as their stability or possible toxicity.

5. Patents

The technological development and the product (encapsulated extract) are protected (ES 1 300 302 U) by a utility model (application number: U202232132/expedition date: 21 August 2023). The protection includes (i) the stages of the production process for obtaining and stabilizing (encapsulation) bioactive compounds from elderberry, (ii) the final product obtained (encapsulated elderberry extract), and (iii) its application in the meat industry. Therefore, the steps described in the M&M section are included in the invention protection but do not limit its scope.

Author Contributions: Conceptualization, R.D.-V. and J.M.L.; Investigation, R.D.-V.; Writing—Original Draft, R.D.-V.; Visualization, R.D.-V., M.P., P.E.S.M. and J.M.L.; Supervision, R.D.-V. and J.M.L.; Formal Analysis, A.C., M.P. and P.E.S.M.; Writing—Review and Editing, M.P., P.E.S.M. and J.M.L. All authors have read and agreed to the published version of the manuscript.

Funding: This research was supported by GAIN (Axencia Galega de Innovación) (grant number IN607A2023/01). The authors (R.D., M.P., P.E.S.M. and J.M.L.) are members of the Healthy Meat network, funded by CYTED (ref: 119RT0568).

Data Availability Statement: The data used to support the findings of this study can be made available by the corresponding author upon request.

Conflicts of Interest: The authors declare no conflict of interest.

References

1. Uhl, K.R.; Fyhrie, K.J.; Brodt, S.B.; Mitchell, A.E. Blue Elderberry (*Sambucus nigra* ssp. cerulea): A Robust and Underutilized Fruit for Value-Added Products. *ACS Food Sci. Technol.* **2022**, *2*, 347–358. [CrossRef]
2. Costa, C.P.; Patinha, S.; Rudnitskaya, A.; Santos, S.A.O.; Silvestre, A.J.D.; Rocha, S.M. Sustainable Valorization of *Sambucus nigra* L. Berries: From Crop Biodiversity to Nutritional Value of Juice and Pomace. *Foods* **2022**, *11*, 104. [CrossRef]
3. Domínguez, R.; Zhang, L.; Rocchetti, G.; Lucini, L.; Pateiro, M.; Munekata, P.E.S.; Lorenzo, J.M. Elderberry (*Sambucus nigra* L.) as potential source of antioxidants. Characterization, optimization of extraction parameters and bioactive properties. *Food Chem.* **2020**, *330*, 127266. [CrossRef] [PubMed]
4. Młynarczyk, K.; Walkowiak-Tomczak, D.; Staniek, H.; Kidoń, M.; Łysiak, G.P. The Content of Selected Minerals, Bioactive Compounds, and the Antioxidant Properties of the Flowers and Fruit of Selected Cultivars and Wildly Growing Plants of *Sambucus nigra* L. *Molecules* **2020**, *25*, 876. [CrossRef] [PubMed]
5. Kitryt, V.; Syrpas, M.; Pukalskas, A. Modeling and optimization of supercritical carbon dioxide extraction for isolation of valuable lipophilic constituents from elderberry (*Sambucus nigra* L.) pomace. *J. CO2 Util.* **2020**, *35*, 225–235. [CrossRef]
6. Młynarczyk, K.; Walkowiak-Tomczak, D.; Łysiak, G.P. Bioactive properties of *Sambucus nigra* L. As a functional ingredient for food and pharmaceutical industry. *J. Funct. Foods* **2018**, *40*, 377–390. [CrossRef]

7. Waswa, E.N.; Li, J.; Mkala, E.M.; Wanga, V.O.; Mutinda, E.S.; Nanjala, C.; Odago, W.O.; Katumo, D.M.; Gichua, M.K.; Gituru, R.W.; et al. Ethnobotany, phytochemistry, pharmacology, and toxicology of the genus *Sambucus* L. (Viburnaceae). *J. Ethnopharmacol.* **2022**, *292*, 115102. [CrossRef] [PubMed]
8. Gull, A.; Sheikh, M.A.; Kour, J.; Zehra, B.; Zargar, I.A.; Wani, A.A.; Bhatia, S.; Lone, M.A. Anthocyanins. In *Nutraceuticals and Health Care*; Kour, J., Nayik, G.A., Eds.; Academic Press: Cambridge, MA, USA, 2022; Volume 1, pp. 317–329. ISBN 9780323897792.
9. Domínguez, R.; Pateiro, M.; Munekata, P.E.S.; López, E.M.S.; Rodríguez, J.A.; Barros, L.; Lorenzo, J.M. Potential Use of Elderberry (*Sambucus nigra* L.) as Natural Colorant and Antioxidant in the Food Industry. A Review. *Foods* **2021**, *10*, 2713. [CrossRef]
10. Baeza, R.; Sánchez, V.; Salierno, G.; Molinari, F.; López, P.; Chirife, J. Storage stability of anthocyanins in freeze-dried elderberry pulp using low proportions of encapsulating agents. *Food Sci. Technol. Int.* **2021**, *27*, 135–144. [CrossRef]
11. Ferreira, S.S.; Silva, P.; Silva, A.M.; Nunes, F.M. Effect of harvesting year and elderberry cultivar on the chemical composition and potential bioactivity: A three-year study. *Food Chem.* **2020**, *302*, 125366. [CrossRef]
12. Mikulic-Petkovsek, M.; Schmitzer, V.; Slatnar, A.; Todorovic, B.; Veberic, R.; Stampar, F.; Ivancic, A. Investigation of anthocyanin profile of four elderberry species and interspecific hybrids. *J. Agric. Food Chem.* **2014**, *62*, 5573–5580. [CrossRef] [PubMed]
13. Fazio, A.; Plastina, P.; Meijerink, J.; Witkamp, R.F.; Gabriele, B. Comparative analyses of seeds of wild fruits of *Rubus* and *Sambucus* species from Southern Italy: Fatty acid composition of the oil, total phenolic content, antioxidant and anti-inflammatory properties of the methanolic extracts. *Food Chem.* **2013**, *140*, 817–824. [CrossRef] [PubMed]
14. Liu, D.; He, X.Q.; Wu, D.T.; Li, H.B.; Feng, Y.B.; Zou, L.; Gan, R.Y. Elderberry (*Sambucus nigra* L.): Bioactive Compounds, Health Functions, and Applications. *J. Agric. Food Chem.* **2022**, *70*, 4202–4220. [CrossRef] [PubMed]
15. Murugesan, R.; Orsat, V. Spray Drying of Elderberry (*Sambucus nigra* L.) Juice to Maintain Its Phenolic Content. *Dry. Technol.* **2011**, *29*, 1729–1740. [CrossRef]
16. Neves, C.M.B.; Pinto, A.; Gonçalves, F.; Wessel, D.F. Changes in Elderberry (*Sambucus nigra* L.) Juice Concentrate Polyphenols during Storage. *Appl. Sci.* **2021**, *11*, 6941. [CrossRef]
17. Busso Casati, C.; Baeza, R.; Sánchez, V. Physicochemical properties and bioactive compounds content in encapsulated freeze-dried powders obtained from blueberry, elderberry, blackcurrant and maqui berry. *J. Berry Res.* **2019**, *9*, 431–447. [CrossRef]
18. Mattson, M.L.G.; Corfield, R.; Bajda, L.; Pérez, O.E.; Schebor, C.; Salvatori, D. Potential bioactive ingredient from elderberry fruit: Process optimization for a maximum phenolic recovery, physicochemical characterization, and bioaccesibility. *J. Berry Res.* **2021**, *11*, 51–68. [CrossRef]
19. Gagneten, M.; Corfield, R.; Mattson, M.G.; Sozzi, A.; Leiva, G.; Salvatori, D.; Schebor, C. Spray-dried powders from berries extracts obtained upon several processing steps to improve the bioactive components content. *Powder Technol.* **2019**, *342*, 1008–1015. [CrossRef]
20. Ribeiro, A.M.; Estevinho, B.N.; Rocha, F. Spray Drying Encapsulation of Elderberry Extract and Evaluating the Release and Stability of Phenolic Compounds in Encapsulated Powders. *Food Bioprocess Technol.* **2019**, *12*, 1381–1394. [CrossRef]
21. Aksu, M.İ.; Turan, E.; Gülbandılar, A.; Tamtürk, F. Utilization of spray-dried raspberry powder as a natural additive to improve oxidative stability, microbial quality and overcome the perception of discoloration in vacuum-packed ground beef during chilled storage. *Meat Sci.* **2023**, *197*, 109072. [CrossRef]
22. Anusha Siddiqui, S.; Redha, A.A.; Esmaeili, Y.; Mehdizadeh, M. Novel insights on extraction and encapsulation techniques of elderberry bioactive compounds. *Crit. Rev. Food Sci. Nutr.* **2023**, *63*, 5937–5952. [CrossRef] [PubMed]
23. Azevedo, T.; Ferreira, T.; Ferreira, J.; Teixeira, F.; Ferreira, D.; Silva-Reis, R.; Neuparth, M.J.; Pires, M.J.; Pinto, M.d.L.; Gil da Costa, R.M.; et al. Supplementation of an Anthocyanin-Rich Elderberry (*Sambucus nigra* L.) Extract in FVB/n Mice: A Healthier Alternative to Synthetic Colorants. *Appl. Sci.* **2022**, *12*, 11928. [CrossRef]
24. Rocchetti, G.; Becchi, P.P.; Lucini, L.; Cittadini, A.; Munekata, P.E.S.; Pateiro, M.; Domínguez, R.; Lorenzo, J.M. Elderberry (*Sambucus nigra* L.) Encapsulated Extracts as Meat Extenders against Lipid and Protein Oxidation during the Shelf-Life of Beef Burgers. *Antioxidants* **2022**, *11*, 2130. [CrossRef]
25. *ISO 1442*; International Standards Meat and Meat Products—Determination of Moisture Content. International Organization for Standarization: Geneva, Switzerland, 1997.
26. *ISO 937*; International Standards Meat and Meat Products—Determination of Nitrogen Content. International Organization for Standarization: Geneva, Switzerland, 1978.
27. *ISO 936*; International Standards Meat and Meat Products—Determination of Ash Content. International Organization for Standarization: Geneva, Switzerland, 1998.
28. AOCS. *AOCS Official Procedure Am5-04. Rapid Determination of Oil/Fat Utilizing High Temperature Solvent Extraction*; American Oil Chemists Society: Urbana, IL, USA, 2005.
29. Barros, J.C.; Munekata, P.E.S.; De Carvalho, F.A.L.; Pateiro, M.; Barba, F.J.; Domínguez, R.; Trindade, M.A.; Lorenzo, J.M. Use of tiger nut (*Cyperus esculentus* L.) oil emulsion as animal fat replacement in beef burgers. *Foods* **2020**, *9*, 44. [CrossRef] [PubMed]
30. Bessada, S.M.F.; Barreira, J.C.M.; Barros, L.; Ferreira, I.C.F.R.; Oliveira, M.B.P.P. Phenolic profile and antioxidant activity of *Coleostephus myconis* (L.) Rchb.f.: An underexploited and highly disseminated species. *Ind. Crops Prod.* **2016**, *89*, 45–51. [CrossRef]
31. Franco, D.; Munekata, P.E.S.; Agregán, R.; Bermúdez, R.; López-Pedrouso, M.; Pateiro, M.; Lorenzo, J.M. Application of pulsed electric fields for obtaining antioxidant extracts from fish residues. *Antioxidants* **2020**, *9*, 90. [CrossRef]
32. Sidor, A.; Gramza-Michałowska, A. Advanced research on the antioxidant and health benefit of elderberry (*Sambucus nigra*) in food—A review. *J. Funct. Foods* **2015**, *18*, 941–958. [CrossRef]

33. Künsch, U.; Temperli, A. Changes in free and protein-bound amino acids in elderberry fruit (*Sambucus nigra*) during maturation. *J. Sci. Food Agric.* **1978**, *29*, 1037–1040. [CrossRef]
34. Câmara, J.S.; Albuquerque, B.R.; Aguiar, J.; Corrêa, R.C.G.; Gonçalves, J.L.; Granato, D.; Pereira, J.A.M.; Barros, L.; Ferreira, I.C.F.R. Food Bioactive Compounds and Emerging Techniques for Their Extraction: Polyphenols as a Case Study. *Foods* **2021**, *10*, 37. [CrossRef]
35. Duymuş, H.G.; Göger, F.; Başer, K.H.C. In vitro antioxidant properties and anthocyanin compositions of elderberry extracts. *Food Chem.* **2014**, *155*, 112–119. [CrossRef]
36. Silva, P.; Ferreira, S.; Nunes, F.M. Elderberry (*Sambucus nigra* L.) by-products a source of anthocyanins and antioxidant polyphenols. *Ind. Crops Prod.* **2017**, *95*, 227–234. [CrossRef]
37. Ferreira, S.S.; Domingues, M.R.; Barros, C.; Santos, S.A.O.; Silvestre, A.J.D.; Silva, A.M.; Nunes, F.M. Major anthocyanins in elderberry effectively trap methylglyoxal and reduce cytotoxicity of methylglyoxal in HepG2 cell line. *Food Chem. X* **2022**, *16*, 100468. [CrossRef] [PubMed]
38. Mikulic-Petkovsek, M.; Ivancic, A.; Todorovic, B.; Veberic, R.; Stampar, F. Fruit Phenolic Composition of Different Elderberry Species and Hybrids. *J. Food Sci.* **2015**, *80*, C2180–C2190. [CrossRef] [PubMed]
39. Senica, M.; Stampar, F.; Veberic, R.; Mikulic-Petkovsek, M. Processed elderberry (*Sambucus nigra* L.) products: A beneficial or harmful food alternative? *LWT Food Sci. Technol.* **2016**, *72*, 182–188. [CrossRef]
40. Veberic, R.; Jakopic, J.; Stampar, F.; Schmitzer, V. European elderberry (*Sambucus nigra* L.) rich in sugars, organic acids, anthocyanins and selected polyphenols. *Food Chem.* **2009**, *114*, 511–515. [CrossRef]

Disclaimer/Publisher's Note: The statements, opinions and data contained in all publications are solely those of the individual author(s) and contributor(s) and not of MDPI and/or the editor(s). MDPI and/or the editor(s) disclaim responsibility for any injury to people or property resulting from any ideas, methods, instructions or products referred to in the content.

foods

MDPI

Article

Effects of *Allium macrostemon* Bunge Extract on Adipose Tissue Inflammation and Hepatic Endoplasmic Reticulum Stress in High-Fat Diet-Fed and Bisphenol A-Treated C57BL/6N Mice

Juhae Kim [1], Na-Hyung Kim [1], Isoo Youn [2], Eun Kyoung Seo [2] and Choon Young Kim [1,3,*]

[1] Research Institute of Human Ecology, Yeungnam University, Gyeongsan 38541, Republic of Korea; kimjh825@yu.ac.kr (J.K.); knahyu@yu.ac.kr (N.-H.K.)
[2] Graduate School of Pharmaceutical Sciences, College of Pharmacy, Ewha Womans University, Seoul 03760, Republic of Korea; isooyoun87@gmail.com (I.Y.); yuny@ewha.ac.kr (E.K.S.)
[3] Department of Food and Nutrition, Yeungnam University, Gyeongsan 38541, Republic of Korea
* Correspondence: cykim@yu.ac.kr; Tel.: +82-53-810-2871

Abstract: The simultaneous exposure to a high-fat (HF) diet and to bisphenol A (BPA) from delivered foods and food-delivery containers is on the rise in humans, according to the increased frequency of food delivery during the COVID-19 pandemic. This co-exposure could cause harmful tissue toxicity in the human body. Here, the preventive effect of *Allium macrostemon* Bunge (AM) extract against dysfunction in adipose tissue and the liver under co-exposure to BPA and an HF diet was examined using mice. C57BL/6N mice were divided into four groups (n = 6 or 7/group) according to diet and treatment: control diet with vehicle (CON), HF diet with vehicle (HF), HF diet with an oral injection of BPA (HF + BP), and HF diet with an oral injection of BPA and AM extract (HF + BP + AM). HF feeding increased body weight gain compared to CON feeding, while BP + HF and BP + HF + AM feeding suppressed body weight gain compared with HF feeding. The BP + HF group had lower body weight than the HF group, but the two groups had similar epididymal fat mass. The HF + BP + AM group showed lower pro-inflammatory gene expression levels in adipose tissue and epididymal fat mass compared to the HF + BP group. Altered endoplasmic reticulum (ER) stress response in the liver was partly observed in the HF + BP group, as shown by increased total phosphorylated Jun N-terminal kinase protein levels compared to those in the HF group. In addition, ecdysterone 25-O-β-D-glucopyranoside and 6-gingerol were identified in AM extract by mass spectrometry and molecular networking analysis. In summary, the AM extract diminished adipose tissue inflammation and hepatic ER stress in an HF diet and BPA co-exposure condition. To utilize AM as a potential food component to alleviate the harmful effect of an HF diet and BPA exposure, further research investigating the specific impact of AM extract supplementation using additional experimental groups or various treatment doses is warranted.

Keywords: Bisphenol A; high-fat diet; *Allium macrostemon* Bunge; adipose tissue; inflammation; liver

1. Introduction

Food delivery services surged due to social distancing during the coronavirus disease 2019 (COVID-19) pandemic [1]. Representative delivery menus include foods containing high amounts of fat, which are frequently consumed in westernized diets. The fat in foods reportedly eases the migration of bisphenol A (4,4′-isopropylidenediphenol, BPA) particles to food, resulting in increased exposure of humans to BPA [2]. BPA is a frequently used compound in the production of polycarbonate plastics and epoxy resin. Due to its structural similarity to estrogen, BPA is classified as an endocrine-disrupting chemical with estrogenic activity. BPA is known to have negative health effects related to reproductive function, immune function, and metabolic function [2,3]. Notably, these health conditions can also be influenced by the consumption of a high-fat (HF) diet. In keeping with this trend, several studies on co-exposure have demonstrated the negative effects of a combined HF diet and BPA exposure on metabolic health.

The elevation of inflammation in adipose tissue is reported to be a consequence of BPA exposure when added to HF conditions [4]. The release of inflammatory adipokines and excessive amounts of free fatty acids (FFAs) are characteristics of dysfunctional adipose tissue [5]. Interleukin (IL)-1β, tumor necrosis factor-α (TNF-α), F4/80, and monocyte chemoattractant protein-1 (MCP-1) are examples of pro-inflammatory adipokines. Immune cell infiltration, histologically detected by a crown-like structure (CLS), is also a prominent feature of dysfunctional inflamed adipose tissue [6]. FFAs promote ectopic fat deposition and lipotoxicity, of which a representative molecular mechanism is endoplasmic reticulum (ER) stress in the liver. In addition to this secretory function, adipose tissue has been suggested as a potential site of toxicant accumulation [7]. Consistently, susceptibility of adipose tissue to BPA has resulted in its being reported as the site containing the highest amount of free BPA concentrations, followed by the liver and brain, as per a human study [8].

The synergistic harmful effects of BPA and an HF diet on the liver have also been reported as chronic ER stress-mediated steatosis progression [9] and immune-metabolic dysfunction [10]. The role of ER stress in hepatic dysfunction has been well demonstrated, as BPA exposure increased ER stress-associated apoptosis or hepatocellular injury in in vitro as well as in vivo models [9,11]. During ER stress, the unfolded protein response involves activation of transmembrane ER-resident stress sensors, which involves sequestering and binding of the chaperone, binding immunoglobulin protein (BiP), to unfolded proteins [12]. One sensor is inositol-requiring kinase 1 (IRE1), which subsequently activates Jun N-terminal kinase (JNK), an inducer of cell apoptosis and inflammation [12]. Another transducer, protein kinase RNA-like endoplasmic reticulum kinase (PERK), is activated by the phosphorylation of eukaryotic initiation factor 2 alpha (eIF2α) and translation of activating transcription factor 4 (ATF4). The transcription of C/EBP homologous protein (CHOP), known as the downstream target of ATF4, is a well-characterized proapoptotic pathway derived from the stressed ER [12]. The eIF2α/ATF4 pathway has also been reported to induce the transcription of genes involved in autophagy, such as *Becn 1* and *Ma1lc3b*, which encode autophagy marker proteins beclin 1 and microtubule-associated protein light chain 3 (LC3) [13].

The therapeutic properties of many nutraceuticals against the adverse health effects of BPA per se have been investigated [14]. Resveratrol [15], lycopene [16], curcumin [17], and thymoquinone [18] ameliorate BPA-induced metabolic disorders, including hyperlipidemia, upregulated inflammation, and insulin resistance. In addition to single compounds, food extracts from olive leaves, green tea, grape seeds, and ginger have been reported to possess a protective role against BPA-induced damage, including liver injury and lipid metabolism disturbance [19], vascular toxicity [20,21], and metabolic syndrome [15]. In line with these findings, it has been reported that plants have an intrinsic detoxification potential for reducing BPA content through oxygenation to quinones using polyphenol oxidase [22].

Allium macrostemon Bunge (AM), one of the plant species included in the Chinese Pharmacopoeia [23], is commonly grown in East Asia [24] and has shown a variety of beneficial effects, including lipid-lowering, antioxidant, and anti-obesity activities, in various study models, including in vitro, in vivo, and clinical systems [25–29]. Previously, we reported that AM whole extract (200 mg/kg BW daily) treatment ameliorates inflammation and endoplasmic reticulum stress in adipose tissue of HF diet-induced C57BL/6N mice [29]. Here, we hypothesized that the simultaneous exposure to HF and BPA from delivery containers and delivery foods in humans would be higher than before, as the delivery food culture increased rapidly as a result of COVID-19. Since we assumed that main targeting organs for BPA and HF co-exposure are the adipose tissue and liver, this study aimed to evaluate the effect of AM on adipose tissue and liver dysregulation, focusing on inflammation and ER stress, respectively.

2. Materials and Methods

The Flow chart representing the whole experimental design is presented in Figure 1.

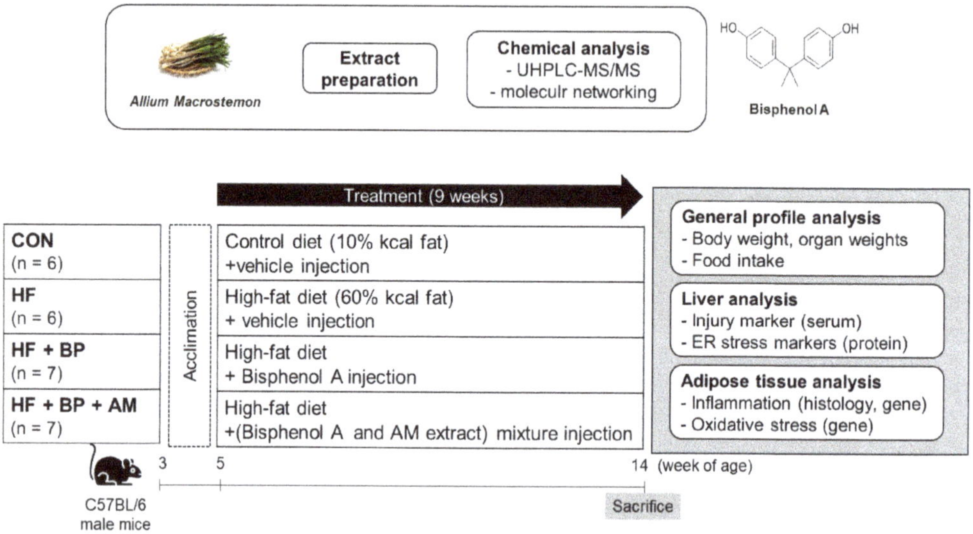

Figure 1. Flow chart for experiments. Information for treatment compounds (preparation and analysis for *Allium macrostemon* extract and chemical structure of bisphenol A) are indicated in the upper figure. In the bottom figure, the animal experimental scheme is presented. The specific analysis topic is presented in the gray box. CON, control diet group; HF, high-fat diet group; HF + BP, high-fat diet and bisphenol A–treated group; HF + BP + AM, high-fat diet and bisphenol A plus *Allium macrostemon* extract–treated group.

2.1. Allium macrostemon Bunge (AM) Preparation and Chemical Analysis

2.1.1. Preparation of AM Extract

AM was purchased from a commercial market (Sacheon, Gyeongnam, Republic of Korea) in April 2021. As previously reported [28], the dried part of the AM was extracted using a high-pressure/high-temperature reactor at 118 kPa/121 °C for 20 min. After filtration, the solvent was removed, and the extracts were concentrated using a rotary evaporator and lyophilized. The yield was 54.8%. The obtained AM extract was stored at −20 °C until use.

2.1.2. Mass Spectrometry Data Acquisition for Chemical Analysis of AM Extract

Ultra-high-performance liquid chromatography tandem mass spectrometry (UHPLC-MS/MS) data acquisition was conducted on a Waters SYNAPT G2-Si quadrupole time-of-flight mass spectrometer with MassLynx 4.1 software (Waters, MA, USA). Separation of the AM extract was achieved on an Agilent Zorbax Eclipse Plus C18 column (2.1 × 150 mm, 1.8 µm, Agilent Technologies, Santa Clara, CA, USA). The mobile phase was water with 0.1% formic acid (A) and acetonitrile with 0.1% formic acid (B). The gradient elution was as follows: 0–2.5 min, 5% B; 2.6–21 min, 5–100% B; 21–26 min, 100% B; 26.1–30 min, 5% B. The column oven was set to 40 °C, and the autosampler temperature was 15 °C. The flow rate was 0.4 mL/min, and the injection volume was 1 µL. Comprehensive mass spectra information was acquired using a negative MS^E mode. The parameters of the MS^E mode were set as follows: mass range, 100–1200 Da; capillary voltage, 2.2 kV; sampling cone voltage, 50 eV; source offset voltage, 30 eV; source temperature, 120 °C; desolvation temperature, 450 °C; cone gas flow rate, 50 L/h; and desolvation gas flow rate, 800 L/h. Nitrogen and argon were applied as cone and collision gases, respectively. The collision energy was 20–40 eV for high energy function, and the scan time was 0.5 s. The data were calibrated in real time using a leucine-enkephalin solution (m/z 554.2615 [M-H]$^-$) as an external reference (LockSpray™) at a flow rate of 5 µL/min.

2.1.3. Feature-Based Molecular Networking Analysis of AM Extract

The acquired data from UHPLC coupled with a quadrupole time-of-flight mass spectrometer was analyzed for feature-based molecular networking (FBMN) based on the online FBMN-Progenesis QI workflow (https://ccms-ucsd.github.io/GNPSDocumentation/featurebasedmolecularnetworking-with-progenesisQI/ (accessed on 1 May 2023) [30]. The obtained MS data were processed using Progenesis QI software (v3.0) for alignment and peak-picking (intensity > 10,000; spectral width > 0.1 min). Both the feature quantification table (CSV file) and MS/MS spectral fragment summary (MSP file) of the raw LC-MS/MS data were exported from the Progenesis QI and submitted to the GNPS platform (https://gnps.ucsd.edu (accessed on 1 May 2023)) using WinSCP (https://winscp.net (accessed on 1 May 2023)). The configurations of network parameters were as follows: precursor ion mass tolerance and fragment ion mass tolerance, 0.02; minimum cosine score, 0.70; minimum matched fragment ions, 6; maximum number of neighbor nodes for one single node, 10; and maximum size of a spectral family, 100. Cytoscape (ver. 3.9.1) was used for network visualization and analysis.

2.2. *Animal Experiment*

2.2.1. Animals and Experimental Design

Three-week-old male C57BL/6N mice were obtained from Central Lab Animal (SLC, Osaka, Japan) and housed under optimal environmental conditions at a temperature of 21 ± 2 °C and a humidity of 50 ± 5% under a 12:12 light:dark cycle. After acclimation for two weeks, the mice were randomly divided into four groups, as follows:

(i) the control (CON) group fed a control diet (n = 6);
(ii) the high-fat (HF) group fed an HF diet (n = 6);
(iii) the HF + BP group fed an HF diet and treated with BPA (n = 7);
(iv) the HF + BP + AM group fed an HF diet and treated with BPA and AM extract (n = 7).

The control diet (containing 10 kcal% fat, 20 kcal% protein, and 70 kcal% carbohydrate, D12450B) and the HF diet (containing 60 kcal% fat, 20 kcal% protein, and 20 kcal% carbohydrate, D12452) were purchased from Research Diets, Inc. (New Brunswick, NJ, USA) and provided ad libitum. BPA and/or AM extracts were dissolved in distilled water containing 0.01% dimethyl sulfoxide (Sigma-Aldrich Co., St. Louis, MO, USA). The mice in the CON and HF groups were administered a vehicle solution by oral gavage. The HF + BP group was treated with 50 µg/kg/day of BPA. The HF + BP + AM group was treated with 50 µg/kg/day of BPA plus 400 mg/kg/day of AM extract simultaneously. The treatment or vehicle solutions were administered daily via oral gavage (10 µL/g body weight) based on the measured body weight, neglecting daily body weight changes. During the experimental period of 9 weeks, body weight and food intake were measured twice or three times per week. Energy intake was calculated by multiplying food intake by the respective energy content of each diet (3.85 kcal/g for the CON group and 5.24 kcal/g for other groups fed an HF diet). All animal procedures were performed according to protocols approved by the Institutional Animal Care and Use Committee of Yeungnam University (approval number: #2021-033).

2.2.2. Serum and Tissue Isolation

After the experimental treatment for 9 weeks, all mice were fasted for 16 h and anesthetized with isoflurane (Isotroy 250; Troikaa Pharmaceuticals, Gujarat, India). Before tissue harvesting, blood was collected by cardiac puncture, transferred into serum separation tubes (BD Microtainer ® tube, #365967, BD Biosciences, Franklin Lakes, NJ, USA), clotted for 30 min at RT, and centrifuged at 15,000× g rpm at 4 °C for 90 s. Isolated serum was collected and analyzed for liver injury biomarkers (glutamic oxaloacetic transaminase (GOT) and glutamic pyruvic transaminase (GPT)) and kidney injury biomarkers (blood urea nitrogen (BUN)) (Asan Pharmaceutical Co., Seoul, Republic of Korea), using commercial colorimetric kits according to the manufacturer's instructions. Absorbance was measured at 505 nm for GOT and GPT and 580 nm for BUN using a SpectraMax ID3 microplate reader (Molecular Devices, San Jose, CA, USA). Epididymal, perirenal, and retroperitoneal fat pads as well as liver tissue were rapidly removed from the mice and weighed immediately. The tissues were snap-frozen in liquid nitrogen and stored at −70 °C until analysis.

2.2.3. Histological Analysis for Adipose Tissue

Epididymal adipose tissues were fixed in 10% neutral formalin solution (#HT501128; Sigma-Aldrich Co.) and embedded in paraffin. The tissues were cut to a thickness of 6 μm and stained with hematoxylin and eosin (H&E). Using a light microscope (Eclipse Ni-U, Nikon, Tokyo, Japan), the number of CLSs in each stained section from four different mice in each group was manually counted and recorded. For each slide, the analysis was performed in four randomly selected fields (100× magnification). The total number of CLSs was expressed as the mean number per field.

2.2.4. RNA Extraction, cDNA Synthesis, and Real-Time Polymerase Chain Reaction for Adipose Tissue

Epididymal adipose tissues were homogenized using the TissueLyser system (#85300, Qiagen, Venlo, The Netherlands), and total RNA was isolated using TRIzol reagent (#15596018, Invitrogen Life Technologies, Carlsbad, CA, USA) according to the manufacturer's instructions. RNA content and purity were measured using a NanoDrop spectrophotometer (Thermo Fisher Scientific, Waltham, MA, USA). One microgram of total RNA was reverse-transcribed into cDNA using the AMPIGENE® cDNA synthesis kit (#END-KIT106, Enzo Life Sciences, Farmingdale, NY, USA) and a SimpliAmp Thermal Cycler (Applied Biosystems, Waltham, MA, USA) according to the manufacturer's instructions. Real-time PCR was performed using AMPIGENE® qPCR Green Mix Hi-ROX (#ENZ-NUC104, Enzo Life Sciences) on a StepOne Plus real-time PCR system (Applied Biosystems). The thermal cycles were as follows: 94 °C for 3 min, followed by 40 cycles at 95 °C for 10 s, 60 °C for 15 s, and 72 °C for 20 s. Mouse 18s rRNA was used as a reference gene, and relative gene expression levels were analyzed using the $2^{-\Delta\Delta Ct}$ method. The primers used for the real-time PCR analysis are listed in Supplementary Materials Table S1.

2.2.5. Protein Extraction and Western Blotting Analysis for Liver Tissue

Liver tissues were homogenized in ice-cold lysis buffer (100 mM Tris-HCl, pH 7.6, 100 mM NaCl, and 0.5% Triton X-100) containing 1 mM sodium orthovanadate, 10 mM sodium fluoride, and a protease inhibitor cocktail (#P3100, GenDEPOT, Katy, TX, USA) using the TissueLyser system (#85300, Qiagen). The liver homogenates were incubated at 4 °C for 20 min with gentle rotation and centrifuged at 10,000× g for 30 min at 4 °C. The epididymal fat homogenates were incubated at 4 °C for 1 h with gentle rotation and centrifuged three times at 20,000× g for 15 min at 4 °C to remove excess lipids. The final supernatant from each tissue sample was collected and stored at −70 °C until further analysis. The protein content of the lysates was determined using the Bradford protein assay kit (Bio-Rad, Hercules, CA, USA). Equal amounts of protein were loaded onto the lanes of a sodium dodecyl sulfate-polyacrylamide gel electrophoresis gel, separated, and transferred onto a polyvinylidene difluoride membrane (Millipore, Bedford, MA, USA). After blocking with 5% non-fat milk (BD Biosciences) or bovine serum albumin (Bovostar, Bovogen, Victoria, Australia) in a Tris-buffered saline solution containing 0.05% Tween-20 (pH 7.5), the membrane was probed with a specific primary antibody at 1:1000 dilution of one of the following: ATF4 (#11815, Cell Signaling Technology, Danvers, MA, USA), beclin-1 (#3495, Cell Signaling Technology), phosphorylated eIF2α (#3398, Cell Signaling Technology), eIF2α (#9722, Cell Signaling Technology), heat shock cognate 70 (HSC70; #sc-7298, Santa Cruz Biotechnology, Dallas, TX, USA), phosphorylated IRE1α (#NB-100-2323, Novus Biologicals, Centennial, CO, USA), IRE1α (sc-390960, Santa Cruz Biotechnology), phosphorylated JNK (#4668, Cell Signaling Technology), JNK (#9252, Cell Signaling Technology), LC3B (#2775, Cell Signaling Technology), and SQSTM1/p62 (#5114, Cell Signaling Technology). The membrane was then incubated with anti-mouse or anti-rabbit horseradish peroxidase-linked secondary antibody (Jackson ImmunoResearch, West Grove, PA, USA). The protein expression signal was detected using the Amersham ECL western blotting Detection Reagent (GE Healthcare, Piscataway, NJ, USA). Immunoreactive bands were visualized and analyzed using Amersham ImageQuant 800 (Cytiva, Marlborough, MA, USA). The expression of HSC70 was used as a control to monitor equal protein loading in each lane.

2.3. Statistical Analysis

Data are presented as mean ± standard error of the mean (SEM). For serum parameters, outliers identified using Grubbs' outlier test were excluded from the analyses, as previously reported [29]. The statistical analysis was conducted with one-way analysis of variance (ANOVA) followed by a Tukey post hoc test for multiple comparison. Significance was determined at a p-value of less than 0.05. All statistical analyses were carried out using SPSS 25 (IBM, Chicago, IL, USA). A minimum sample size of six mice per group was based on the ability to detect a minimal meaningful difference to provide an $\alpha = 0.05$ and a power of 80%. Calculation of the sample size was performed using G*Power Software (version 3.0.10, Germany).

3. Results

3.1. Effects of AM Supplementation on Body Weight Gain, Energy Intake, Organ Weights, and Serum Analysis

During the 9-week experimental period, HF feeding induced a significant increase in body weight compared to that in the CON diet feeding (Table 1). The body weight gain in the HF + BP group was significantly lower by 1.3-fold than that in the HF group. No further effect of AM supplementation on body weight gain was observed. Energy intake was 1.3-fold higher in both the HF group and the HF + BP group than that in the CON group (Table 1). Similar to the body weight changes, the weights of the epididymal fat, kidneys, and spleen were 3.6-fold, 1.2-fold, and 1.2-fold higher, respectively, than those of the CON group (Table 1). Notably, the epididymal fat weight of the HF + BP + AM group was significantly decreased by 1.3-fold compared to that of the HF + BP group. A similar pattern was observed in the case of the sum of the perirenal and retroperitoneal fat weights. The spleen weight was lower in the HF + BP group than in the HF group, with no further effect of AM supplementation. Liver and kidney injury biomarkers were also evaluated as indicated in Table 1. Although there were no significant alterations in biomarker levels among all groups, we observed a 1.51-fold increase in GPT levels in the HF + BP group compared to that in the HF group.

Table 1. Effects of *Allium macrostemon* (AM) extract on body weight gain, energy intake, organ weights, and serum analysis.

Parameters	CON	HF	HF + BP	HF + BP + AM
Body weight gain (g)	6.3 ± 0.5	16.6 ± 0.9 $^{*,\#}$	12.8 ± 0.7 $^{*,\#}$	12.9 ± 0.7 $^{*,\#}$
Energy intake (kcal/d/mouse)	8.94 ± 0.00	11.76 ± 0.72 *	11.47 ± 0.46 *	11.06 ± 0.21
Epididymal fat (g)	0.58 ± 0.06	2.09 ± 0.11 *	2.00 ± 0.06 *	1.57 ± 0.11 $^{*,\#,\dagger}$
Perirenal and retroperitoneal fat (g)	0.29 ± 0.04	1.12 ± 0.07 *	0.87 ± 0.06 *	0.70 ± 0.11 $^{*,\#}$
Liver (g)	0.74 ± 0.03	0.80 ± 0.02	0.77 ± 0.02	0.74 ± 0.01
Kidneys (g)	0.26 ± 0.01	0.31 ± 0.02	0.27 ± 0.01	0.28 ± 0.01
Spleen (g)	0.06 ± 0.00	0.07 ± 0.00 *	0.06 ± 0.00 #	0.06 ± 0.00 #
Serum GOT (IU/L)	31.30 ± 7.08	21.78 ± 3.61	34.81 ± 7.39	46.74 ± 10.33
Serum GPT (IU/L)	24.50 ± 1.92	21.03 ± 0.72	31.71 ± 5.90	26.83 ± 3.08
Serum BUN (mg/mL)	18.22 ± 0.92	18.03 ± 0.44	17.03 ± 1.70	16.61 ± 0.95

Data are expressed as mean ± SEM (n = 6–7). * $p < 0.05$ versus CON; # $p < 0.05$ versus HF; † $p < 0.05$ versus HF + BP. CON, control diet group; HF, high-fat diet group; HF + BP, high-fat diet and bisphenol A–treated group; HF + BP + AM, high-fat diet and bisphenol A plus *Allium macrostemon* extract–treated group; GOT, glutamic oxaloacetic transaminase; GPT; glutamic pyruvic transaminase; BUN, blood urea nitrogen.

3.2. Effects of AM Supplementation on Adipose Tissue Inflammation and Oxidative Stress

Since there were significant changes in adipose tissue weight among the groups (Table 1), histological and molecular changes were examined by H&E staining and qRT-PCR analysis in adipose tissue. The number of CLSs, a hallmark of macrophage infiltration, was markedly increased in the HF group as well as in the HF + BP group by 13.5-fold and 10.0-fold, respectively, compared to that in the CON group (Figure 2a,b). Notably, the HF + BP + AM group showed significantly lower levels of CLS than the HF + BP group, by 2.86-fold. Similar to histological observations, the mRNA expression levels of pro-inflammatory markers, including IL-1β and TNF-α, and macrophage markers, including F4/80 and MCP-1, were higher in the HF group than in the CON group. Similarly, slightly lower levels of these pro-inflammatory markers were observed in the HF + BP group than in the HF group (Figure 2c). Significantly lowered levels of inflammatory gene expression were detected in the HF + BP + AM group compared to the HF group. Although there was a decreasing tendency, no significant differences between the HF + BP + AM group and HF + BP group were determined. Changes in the expression of antioxidative enzymes were also evaluated, as shown in Figure 2d. The gene expression levels of catalase (CAT), superoxide dismutase 2 (SOD2), and thiredoxin2 (TRX2) were higher in the HF group than in the CON group. The HF + BP group showed a similar or lower tendency of expression levels compared to those in the HF group. Significantly lowered levels of three of antioxidative enzymes' gene expression (CAT, SOD2, TRX2) were detected in the HF + BP + AM group compared to the HF group. AM supplementation lowered the gene expression levels of SOD2 and TRX2 by 1.56-fold and 1.95-fold, respectively, compared to those of the HF + BP group.

3.3. Effects of AM Supplementation on Hepatic ER Stress and Autophagy

As increased levels of serum FFA from triacylglycerol induce hepatic toxicity by upregulating ER stress [31], we further examined ER stress-related protein and gene expressions. Phosphorylated JNK and eIF2α levels were significantly increased by the HF diet. The protein levels of activated JNK and its upstream transmembrane ER-residence stress sensor IRE1α were higher in the HF + BP group than in the HF group by 1.36-fold and 1.95-fold, respectively, although only JNK levels showed significant differences. Downstream members of other transmembrane ER-residence stress sensors, eIF2α and ATF4, showed lower expression levels in the HF + BP group than in the HF group. Notably, AM supplementation resulted in lower levels of t-IRE1α/HSC70, p-JNK/t-JNK, p-eIF2α/t-eIF2α, and ATF4/HSC70 by 2.43-fold, 1.85-fold, 1.61-fold, and 1.41-fold, respectively, compared to those of the HF + BP group (Figure 3a). In gene expression analysis, the levels of BiP and CHOP, which are mediators of ER stress induction, were higher in the HF + BP group than in the HF group by 1.63-fold and 1.65-fold, respectively. These increased expressions of BPA-aggravated ER stress induction were reduced by AM treatment, resulting in 0.58-fold and 0.67-fold reduction, respectively (Figure 3b). In addition, the expression of autophagy-related proteins was examined. Although HF feeding did not alter autophagy-related protein expression, 1.30-fold higher levels of p62, as well as 0.69-fold and 0.82-fold levels of LC3-II/LC3-I ratio and beclin-1, respectively, were detected in the HF + BP group compared to those in the HF group. In response to AM supplementation, p62 expression levels were decreased compared to those in the HF + BP group, but further effects on other proteins were not observed (Figure 3c).

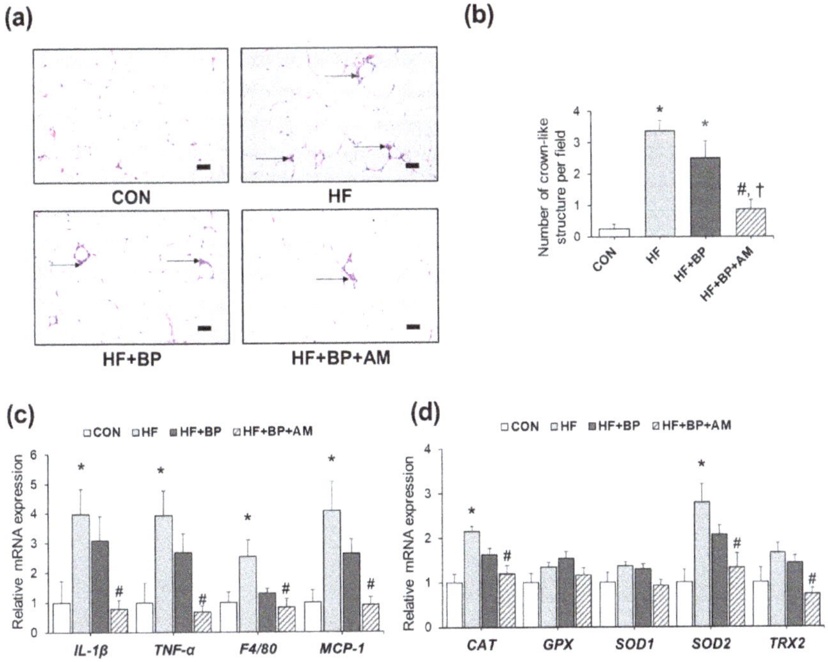

Figure 2. Effects of *Allium macrostemon* (AM) extract supplementation on adipose tissue inflammation and oxidative stress in high-fat diet-treated and BPA-treated mice. (**a**) Representative hematoxylin and eosin (H&E) staining of epididymal adipose tissue (200× magnification, scale bar = 50 μm) is shown. Crown-like structures (CLSs) are depicted with arrows. (**b**) The number of CLSs was counted under one field in 100× magnification. (**c**) Relative mRNA levels of pro-inflammatory genes were measured by real-time PCR. (**d**) Relative mRNA levels of antioxidant enzyme genes were measured by real-time PCR. Data are expressed as mean ± SEM (n = 4 for histological analysis, n = 5–6 for gene expression analysis). * $p < 0.05$ versus CON; # $p < 0.05$ versus HF; † $p < 0.05$ versus HF + BP. CON, control diet group; HF, high-fat diet group; HF + BP, high-fat diet and bisphenol A–treated group; HF + BP + AM, high-fat diet and bisphenol A plus *Allium macrostemon* extract–treated group.

3.4. Identification of Compounds in AM Extract by UHPLC-MS/MS and Molecular Network Analysis

As illustrated in Figure 4, 162 precursor ions were organized into a molecular network, which included 12 clusters (nodes ≥ 2) and 162 single nodes. Two compounds from the AM extract were identified as ecdysterone 25-O-β-D-glucopyranoside (1) in Cluster 2 and 6-gingerol (2) in Cluster 11 by the FBMN analysis; additional details can be found on the GNPS website (https://gnps.ucsd.edu/ProteoSAFe/status.jsp?task=d827b40931454be6bb67f9521e2a64c8 (accessed on 1 May 2023)). Compound **1** is identified as a new metabolite of ecdysone in the nematode Parascaris equorum [32], and 2 is known as a major bioactive component in ginger [33]. Compounds **1** and **2** have never been identified in *Allium macrostemon*.

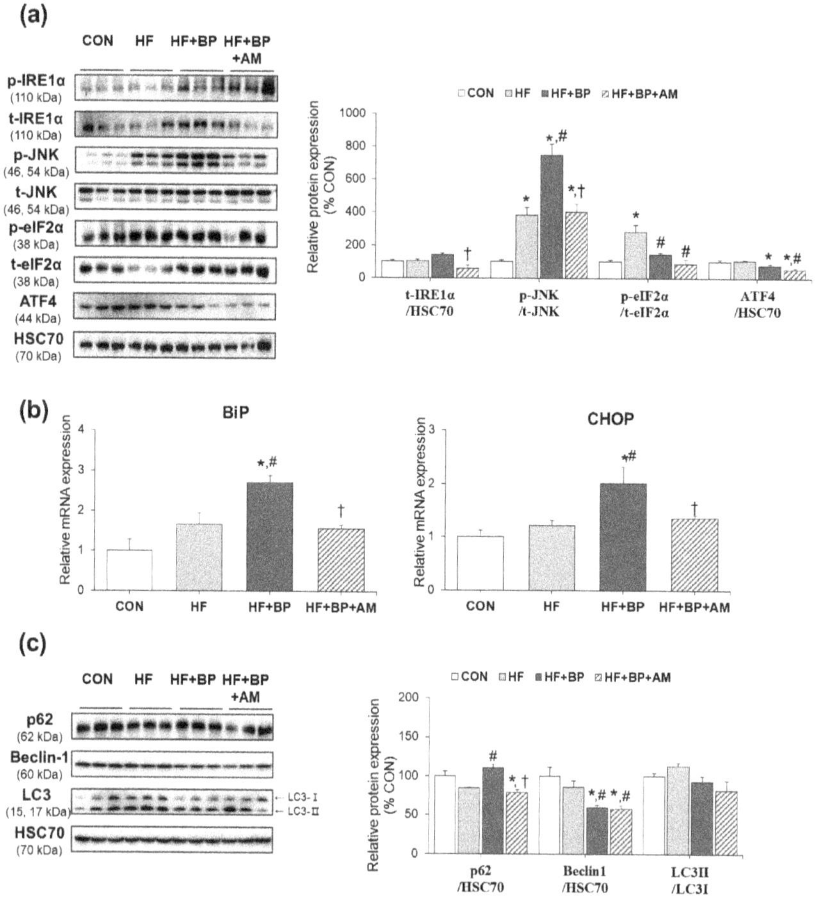

Figure 3. Effects of *Allium macrostemon* (AM) extract supplementation on hepatic endoplasmic reticulum (ER) stress and autophagy in high-fat diet-treated and BPA-treated mice. (**a**) Protein expression of ER stress-related mediators, including total IRE1α, phosphorylated and total JNK, phosphorylated and total eIF2α, and ATF4 was measured by Western blot analysis. (**b**) Relative mRNA levels of ER stress-related mediators, including BiP and CHOP, were measured by real-time PCR. (**c**) Protein expression of autophagy-related mediators, including p62, beclin-1, and LC3, were determined by Western blot analysis. Data are expressed as mean ± SEM (n = 3 for protein analysis, n = 4–5 for gene expression analysis). * $p < 0.05$ versus CON; # $p < 0.05$ versus HF; † $p < 0.05$ versus HF + BP.

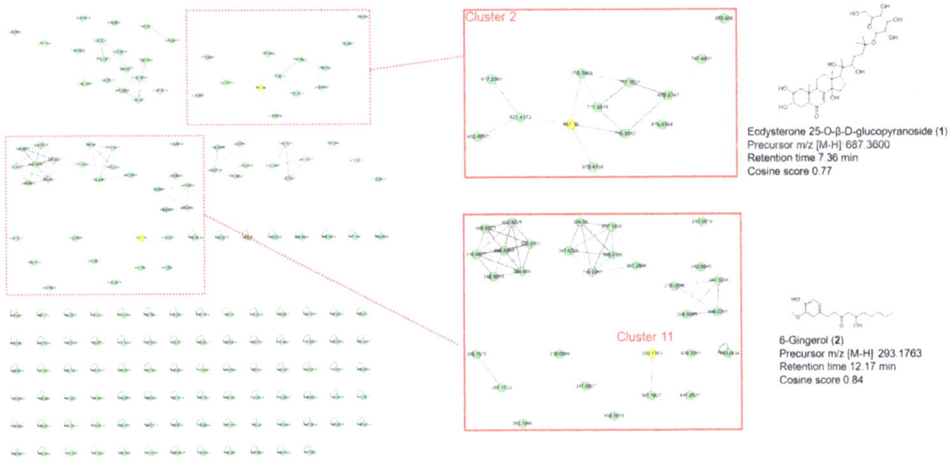

Figure 4. Feature-based molecular networking of the *Allium macrostemon* extract. A total of 162 nodes and 12 clusters (nodes $\geq$ 2) were organized into a network, and ecdysterone 25-O-β-D-glucopyranoside (1, Cluster 2) and 6-gingerol (2, Cluster 11) were identified.

4. Discussion

The primary aim of the present study was to evaluate the preventive effect of AM supplementation against the endocrine-disrupting chemical BPA in combination with HF-diet-altered adverse health effects. Although no difference in epididymal fat mass between the HF + BP group and HF group was observed, BPA suppressed HF diet-induced body weight gain. These results indicated that BPA may exert toxicity by inhibiting body weight gain. This result agrees with the results of a previous study, in which 8-week-old HF/high-cholesterol/high-cholic-acid diet (HFCCD)-fed C57BL/6 male mice were treated with 50 mg/kg/day BPA for 8 weeks [32]. BPA coupled with HFCCD increased body weight loss as a result of severe liver damage, as shown by the significantly increased level of liver injury serum marker, GPT, whose level was increased 2.23-fold compared to that in the HFCCD group. Here, we also determined serum liver injury marker levels, revealing modest increases in GOT and GPT levels in the HF + BP group compared to those in the HF group. In addition, no additional impact of BPA administration on epididymal fat mass in HF diet-fed mice indicates that lower body weight gain is not derived from lower adipose tissue weight. Previous studies have demonstrated that BPA-induced adverse health outcomes include bone loss [33]. It would be interesting to clarify the mechanism by which body weight loss occurs in future studies.

Similar to comparable weights, we observed no significant alterations of inflammation in the adipose tissue between the HF group and HF + BP group in our 9-week exposure period. This implies that there was no additional effect of BPA exposure in HF-induced adipose tissue inflammation. Similarly, previous studies also reported no adipose tissue mass difference between HF only and HF + BPA treatments in 4-week [34] or 12-week exposure schemes [35,36]. These previous studies proposed tissue-specific impacts of the HFD and BPA co-exposure condition. Although no mass alteration was seen in adipose tissue, brain inflammation [36] and skeletal muscle insulin signaling deterioration [35] were exaggerated in the co-exposure condition compared to only HF exposure. In line with previous studies, we observed an exaggerated harmful effect of additional bisphenol A exposure in terms of ER stress induction in liver tissue (Figure 3) as well as in testis tissue.

Activation of ER stress is a key pathological mechanism of steatosis [37]. Our data showed that HF + BP activated one ER branch, the IRE/JNK pathway, and inhibited the other branch, at the p-eIF2α/ATF4 level. Interestingly, ATF4 expression was significantly positively correlated with autophagy marker protein beclin-1 (r = 0.762, p = 0.004) and LC3-

II/LC3-I (r = 0.611, $p = 0.035$) levels. These associations suggest that decreased autophagy in the HF + BP group may be linked to changes in the eIF2α/ATF4 pathway [13]. In a previous study using Neuro-2a cells, autophagy was decreased by BPA, as determined by the reduction in beclin-1 and LC3-II/LC3-I levels, suggesting BPA-induced neurotoxicity [38]. A macrophage study also showed a BPA-induced reduction in LC3-II/LC3-I levels and an increase in p62 levels, contributing to BPA-induced immune cell injury [39].

BPA's exacerbation of dysfunction in the liver of mice fed with an HF diet was evident, as shown by the expression of certain ER stress markers, including p-JNK, CHOP, and BiP. Therefore, we propose that BPA's exacerbation of dysfunction under the HF diet condition may be tissue-specific in our experimental model. Similar to our results, the aggravation effect of BPA on the HF diet in hepatic ER stress induction was reported in a rodent study of 3 months of BPA and HF diet co-exposure [32]. The differences in the experimental design, including a shorter treatment period (2 months in our study), sex, and species, may explain the inconsistent expression patterns of several other ER stress induction markers in our study compared to the previous report.

In this study, significant reduction in epididymal fat mass by AM supplementation against HF and BPA co-exposure was demonstrated. These results are consistent with those of a previous AM study [28]. In HF-diet-fed mice, daily administration of a steroidal saponin, macrostemonoside A, isolated from the bulbs of AM, at a concentration of 4 mg/kg for 30 days showed a decrease in serum TC levels and visceral fat accumulation [28]. In the present study, we used a dose of BPA of 50 μg/kg/day; this dose was set based on the dose considered safe for tolerable daily intake by the U.S. Environmental Protection Agency, and the dose was also tested in previous mouse studies [34,40]. The AM extract dose was also tested in a previous mouse study, with no side effects [41]. In human subjects with hyperglyceridemia, daily administration of a Chinese medicine extract of AM Xuezhitong at 2700 mg for 12 weeks induced a superior reduction in TC levels by 14.18% compared to 3.89% in the placebo group [28]. In addition, similar to the present results, a previous study demonstrated that administration of AM-derived saponins reduced visceral fat accumulation but was not accompanied by changes in body weight [28]. The lack of body weight reduction by AM may be explained by changes in body composition, including other fat depot alterations, as well as muscle weight, which warrants further study.

AM supplementation also improved adipose tissue dysfunction in the HF diet and BPA co-exposure condition. Similar to decreased epididymal fat pad weight, the reduction of inflammation, as determined by CLS and pro-inflammatory gene expression, was observed in the HF + BP + AM group compared to the HF group. In line with our data, a previous study investigating lycopene supplementation in a BPA-treated rats showed loss of adipose tissue weight and reduction of pro-inflammatory cytokine levels, explaining its antioxidant potential [16]. Similarly, lowered antioxidant enzyme expression levels were also observed in the HF + BP + AM group compared to the HF group. Previous studies have suggested that AM exhibits antioxidant and anti-inflammatory properties [24,26]. Nevertheless, since there was no difference in expression levels of inflammation and antioxidant markers between the HF group and the HF + BP group, the possibility of lowering adipose tissue dysfunction with regard to HF per se, rather than HF and BPA co-exposure, cannot be precluded.

We also demonstrated that AM reduced hepatic ER stress levels, although not all markers exhibited consistent alterations. The ameliorating effects of supplementation with stronger antioxidants, such as n-acetylcysteine, on HF plus BPA co-exposure has been reported previously [32]. To the best of our knowledge, amelioration of hepatic ER stress induction by AM extract was firstly reported in our study, with the possible mediator being the JNK pathway. The JNK pathway is known to mediate the induction of pancreatic ER stress by a major active liver metabolite of BPA, 4-methyl-2,4-bis(4-hydroxyphenyl)pent-1-ene [42]. Suppression of the JNK pathway by food components such as curcumin has been reported in a BPA-exposed liver cell model [5]. Still, the decrease in ER stress induction in the liver was not significant when compared to the control group. In summary, our study proved that AM supplementation can modify BPA-exaggerated ER stress induction in the liver, possibly through the JNK pathway, but the lowering effect is insufficient to completely mitigate the ER stress. Also, since the AM supplementation to the HF group was not tested, further study would be warranted to specify the AM's ameliorating effect on the HF as well as the BPA exposure condition.

It is believed that the polyphenols in AM extract are responsible for its biological function. We previously reported the results of HPLC analysis showing that AM contains several phenolic compounds, including ellagic acid, protocatechuic acid, catechin, ferulic acid, chlorogenic, *p*-coumaric acid, and caffeic acid, which have anti-obesity, antioxidant, and anti-inflammatory activities [29]. Other major bioactive compounds, including steroidal saponins, flavonoids, phenylpropanoids, alkaloids, volatile oils, polysaccharides, and amino acids [23], were also reported from AM. In this study, we further analyzed AM using UHPLC-MS/MS and molecular networking analysis. The results suggest that AM supplementation can exert a preventative effect on the HF and BPA exposure condition. The analysis identified two compounds, named ecdysterone 25-O-β-D-glucopyranoside and 6-gingerol. Our results proposed identification of minor active compounds using nontargeted profiling analysis rather than NMR analysis, which mainly identifies major compounds that have been previously reported. Taken together, the various phytochemicals in AM extract, identified by two different analysis methods, might explain the beneficial effects, including anti-ER stress, antioxidant, and anti-inflammatory properties.

One of the limitations of our study was its relatively short duration of 9 weeks, which may have limited detection of significant negative effects on the liver and adipose tissue parameters induced by co-exposure of HF and BPA. Previous research reported that co-treatment of BPA for 3 months worsened hepatic steatosis in HF diet-fed mice [9], suggesting the need for longer testing periods in future studies to explore the role of supplementation in co-exposure scenarios. Another limitation of our study is that the number of animals used in this research was relatively small, although it was sufficient to demonstrate statistical differences between the groups. Nonetheless, the use of the minimum number of mice aligns with the ethical principles of the "Three Rs (Replacement, Reduction, and Refinement)" [43], which aims to minimize unnecessary pain and suffering to animals. Additionally, we did not include groups exposed to BPA only or BPA with AM supplementation. Despite this, we observed a significant ameliorating effect of AM in the HF + BP + AM group. The possibility of AM's protective role against HF diet-induced effects cannot be precluded. Future research incorporating more groups or multiple dose provision could further elucidate the role of AM in countering the effects of an HF diet alone or in combination with BPA exposure.

5. Conclusions

Co-exposure to a high-fat diet and bisphenol A suppressed high-fat diet-induced body weight gain and hepatic endoplasmic reticulum stress induction. Although not all perturbations caused by co-exposure were recovered by *Allium macrostemon* extract supplementation, the *Allium macrostemon* extract–supplemented group showed reduced adipose tissue weight, adipose tissue inflammation, and hepatic endoplasmic reticulum stress levels compared to the non-supplemented high-fat + bisphenol A group. In the context of bisphenol A–exaggerated metabolic perturbations with a high-fat diet condition, the antioxidant potential of *Allium macrostemon* may play an important role in counteracting high-fat diet and bisphenol A–induced metabolic dysfunction in adipose tissue and liver. Further studies are needed to extend the evaluation of the effect of *Allium macrostemon* t against high-fat diet and bisphenol A co-exposure scenarios, which would highlight the necessity to develop a potential functional ingredient with *Allium macrostemon* to counteract the dietary issues in modern life.

Supplementary Materials: The following supporting information can be downloaded at: https://www.mdpi.com/article/10.3390/foods12203777/s1. Table S1, Primer sequences used in quantitative real-time PCR.

Author Contributions: Conceptualization, C.Y.K.; methodology, software, J.K. and I.Y.; validation, E.K.S.; formal analysis, data curation, writing—original draft preparation, J.K.; writing—review and editing, N.-H.K. and C.Y.K.; supervision, project administration, funding acquisition, C.Y.K. All authors have read and agreed to the published version of the manuscript.

Funding: This research was supported by the Basic Science Research Program through the National Research Foundation of Korea (NRF) funded by the Ministry of Education (2021R1A6A1A03040177).

Institutional Review Board Statement: All animal procedures were performed according to protocols approved by the Institutional Animal Care and Use Committee of Yeungnam University (approval number: #2021-033).

Data Availability Statement: The data presented in this study are available on request from the corresponding author.

Conflicts of Interest: The authors declare no conflict of interest.

References

1. Nguyen, T.H.D.; Vu, D.C. Food Delivery Service During Social Distancing: Proactively Preventing or Potentially Spreading Coronavirus Disease-2019? *Disaster Med. Public Health Prep.* **2020**, *14*, e9–e10. [CrossRef]
2. Manzoor, M.F.; Tariq, T.; Fatima, B.; Sahar, A.; Tariq, F.; Munir, S.; Khan, S.; Nawaz Ranjha, M.M.A.; Sameen, A.; Zeng, X.A.; et al. An insight into bisphenol A, food exposure and its adverse effects on health: A review. *Front. Nutr.* **2022**, *9*, 1047827. [CrossRef]
3. Rochester, J.R. Bisphenol A and human health: A review of the literature. *Reprod. Toxicol.* **2013**, *42*, 132–155. [CrossRef]
4. Luo, S.; Li, Y.; Lu, X.; Jiang, J.; Zhu, Q.; Shen, T. Effects of bisphenol A on inflammation and Th17 cells in adipose tissue of high-fat fed mice. *Wei Sheng Yan Jiu* **2017**, *46*, 7–12. [PubMed]
5. Cusi, K. The role of adipose tissue and lipotoxicity in the pathogenesis of type 2 diabetes. *Curr. Diabetes Rep.* **2010**, *10*, 306–315. [CrossRef] [PubMed]
6. Guzik, T.J.; Skiba, D.S.; Touyz, R.M.; Harrison, D.G. The role of infiltrating immune cells in dysfunctional adipose tissue. *Cardiovasc. Res.* **2017**, *113*, 1009–1023. [CrossRef]
7. Jackson, E.; Shoemaker, R.; Larian, N.; Cassis, L. Adipose Tissue as a Site of Toxin Accumulation. *Compr. Physiol.* **2017**, *7*, 1085–1135. [CrossRef] [PubMed]
8. Geens, T.; Neels, H.; Covaci, A. Distribution of bisphenol-A, triclosan and n-nonylphenol in human adipose tissue, liver and brain. *Chemosphere* **2012**, *87*, 796–802. [CrossRef]
9. Figueiredo, L.S.; Oliveira, K.M.; Freitas, I.N.; Silva, J.A., Jr.; Silva, J.N.; Favero-Santos, B.C.; Bonfleur, M.L.; Carneiro, E.M.; Ribeiro, R.A. Bisphenol-A exposure worsens hepatic steatosis in ovariectomized mice fed on a high-fat diet: Role of endoplasmic reticulum stress and fibrogenic pathways. *Life Sci.* **2020**, *256*, 118012. [CrossRef]
10. Pirozzi, C.; Lama, A.; Annunziata, C.; Cavaliere, G.; Ruiz-Fernandez, C.; Monnolo, A.; Comella, F.; Gualillo, O.; Stornaiuolo, M.; Mollica, M.P.; et al. Oral Bisphenol A worsens liver immune-metabolic and mitochondrial dysfunction induced by high-fat diet in adult mice: Cross-talk between oxidative stress and inflammasome pathway. *Antioxidants* **2020**, *9*, 1201. [CrossRef]
11. Asahi, J.; Kamo, H.; Baba, R.; Doi, Y.; Yamashita, A.; Murakami, D.; Hanada, A.; Hirano, T. Bisphenol A induces endoplasmic reticulum stress-associated apoptosis in mouse non-parenchymal hepatocytes. *Life Sci.* **2010**, *87*, 431–438. [CrossRef] [PubMed]
12. Malhi, H.; Kaufman, R.J. Endoplasmic reticulum stress in liver disease. *J. Hepatol.* **2011**, *54*, 795–809. [CrossRef] [PubMed]
13. B'Chir, W.; Maurin, A.C.; Carraro, V.; Averous, J.; Jousse, C.; Muranishi, Y.; Parry, L.; Stepien, G.; Fafournoux, P.; Bruhat, A. The eIF2alpha/ATF4 pathway is essential for stress-induced autophagy gene expression. *Nucleic Acids Res.* **2013**, *41*, 7683–7699. [CrossRef] [PubMed]
14. D'Angelo, S.; Scafuro, M.; Meccariello, R. BPA and Nutraceuticals, Simultaneous Effects on Endocrine Functions. *Endocr. Metab. Immune Disord. Drug Targets* **2019**, *19*, 594–604. [CrossRef]
15. Rameshrad, M.; Razavi, B.M.; Imenshahidi, M.; Hosseinzadeh, H. *Vitis vinifera* (grape) seed extract and resveratrol alleviate bisphenol-A-induced metabolic syndrome: Biochemical and molecular evidences. *Phytother. Res.* **2019**, *33*, 832–844. [CrossRef]
16. Elgawish, R.A.; El-Beltagy, M.A.; El-Sayed, R.M.; Gaber, A.A.; Abdelrazek, H.M.A. Protective role of lycopene against metabolic disorders induced by chronic bisphenol A exposure in rats. *Environ. Sci. Pollut. Res. Int.* **2020**, *27*, 9192–9201. [CrossRef]
17. Geng, S.; Wang, S.; Zhu, W.; Xie, C.; Li, X.; Wu, J.; Zhu, J.; Jiang, Y.; Yang, X.; Li, Y.; et al. Curcumin suppresses JNK pathway to attenuate BPA-induced insulin resistance in LO_2 cells. *Biomed. Pharmacother.* **2018**, *97*, 1538–1543. [CrossRef]

18. Fadishei, M.; Ghasemzadeh Rahbardar, M.; Imenshahidi, M.; Mohajeri, A.; Razavi, B.M.; Hosseinzadeh, H. Effects of Nigella sativa oil and thymoquinone against bisphenol A-induced metabolic disorder in rats. *Phytother. Res.* **2021**, *35*, 2005–2024. [CrossRef]
19. Mahmoudi, A.; Hadrich, F.; Feki, I.; Ghorbel, H.; Bouallagui, Z.; Marrekchi, R.; Fourati, H.; Sayadi, S. Oleuropein and hydroxytyrosol rich extracts from olive leaves attenuate liver injury and lipid metabolism disturbance in bisphenol A-treated rats. *Food Funct.* **2018**, *9*, 3220–3234. [CrossRef]
20. Mohsenzadeh, M.S.; Razavi, B.M.; Imenshahidi, M.; Mohajeri, S.A.; Rameshrad, M.; Hosseinzadeh, H. Evaluation of green tea extract and epigallocatechin gallate effects on bisphenol A-induced vascular toxicity in isolated rat aorta and cytotoxicity in human umbilical vein endothelial cells. *Phytother. Res.* **2021**, *35*, 996–1009. [CrossRef]
21. Kazemi, S.; Feizi, F.; Aghapour, F.; Joorsaraee, G.A.; Moghadamnia, A.A. Histopathology and Histomorphometric Investigation of Bisphenol A and Nonylphenol on the Male Rat Reproductive System. *N. Am. J. Med. Sci.* **2016**, *8*, 215–221. [CrossRef] [PubMed]
22. Yoshida, M.; Ono, H.; Mori, Y.; Chuda, Y.; Mori, M. Oxygenation of bisphenol A to quinones by polyphenol oxidase in vegetables. *J. Agric. Food Chem.* **2002**, *50*, 4377–4381. [CrossRef] [PubMed]
23. Wu, J.; Wang, L.; Cui, Y.; Liu, F.; Zhang, J. *Allii macrostemonis* Bulbus: A Comprehensive Review of Ethnopharmacology, Phytochemistry and Pharmacology. *Molecules* **2023**, *28*, 2485. [CrossRef] [PubMed]
24. Yao, Z.H.; Qin, Z.F.; Dai, Y.; Yao, X.S. Phytochemistry and pharmacology of Allii Macrostemonis Bulbus, a traditional Chinese medicine. *Chin. J. Nat. Med.* **2016**, *14*, 481–498. [CrossRef]
25. Jia, W.; Li, Y.; Wan, J.; Cui, X.; Lu, J.; Liu, J.; Li, D.; Li, L.; Zou, T.; Ding, J.; et al. Effects of Xuezhitong in Patients with Hypertriglyceridemia: A Multicentre, Randomized, Double-Blind, Double Simulation, Positive Drug and Placebo Parallel Control Study. *Cardiovasc. Drugs Ther.* **2020**, *34*, 525–534. [CrossRef]
26. Wu, Z.Q.; Li, K.; Ma, J.K.; Huang, Q.; Tian, X.; Li, Z.J. Antioxidant activity of organic sulfides from fresh *Allium macrostemon* Bunge and their protective effects against oxidative stress in *Caenorhabditis elegans*. *J. Food Biochem.* **2020**, *44*, e13447. [CrossRef]
27. Lee, J.-Y.; Jeong, Y.; Kim, J.; Kim, C.Y. Inhibitory effect of *Allium macrostemon* extracts on adipogenesis of 3T3-L1 preadipocytes. *Korean J. Food Sci. Technol.* **2020**, *52*, 441–449. [CrossRef]
28. Xie, W.; Zhang, Y.; Wang, N.; Zhou, H.; Du, L.; Ma, X.; Shi, X.; Cai, G. Novel effects of macrostemonoside A, a compound from *Allium macrostemon* Bung, on hyperglycemia, hyperlipidemia, and visceral obesity in high-fat diet-fed C57BL/6 mice. *Eur. J. Pharmacol.* **2008**, *599*, 159–165. [CrossRef]
29. Kim, J.; Lee, J.Y.; Kim, C.Y. *Allium macrostemon* whole extract ameliorates obesity-induced inflammation and endoplasmic reticulum stress in adipose tissue of high-fat diet-fed C57BL/6N mice. *Food Nutr. Res.* **2023**, *67*, 9256. [CrossRef]
30. Nothias, L.-F.; Petras, D.; Schmid, R.; Dührkop, K.; Rainer, J.; Sarvepalli, A.; Protsyuk, I.; Ernst, M.; Tsugawa, H.; Fleischauer, M. Feature-based molecular networking in the GNPS analysis environment. *Nat. Methods* **2020**, *17*, 905–908. [CrossRef]
31. Estadella, D.; da Penha Oller, do.; Nascimento, C.M.; Oyama, L.M.; Ribeiro, E.B.; Damaso, A.R.; de Piano, A. Lipotoxicity: Effects of dietary saturated and transfatty acids. *Mediat. Inflamm* **2013**, 137579. [CrossRef] [PubMed]
32. Lee, J.L.; Wang, Y.C.; Hsu, Y.A.; Chen, C.S.; Weng, R.C.; Lu, Y.P.; Chuang, C.Y.; Wan, L. Bisphenol A coupled with a high-fat diet promotes hepatosteatosis through reactive-oxygen-species-Induced CD36 overexpression. *Toxics* **2022**, *10*, 208. [CrossRef] [PubMed]
33. Thent, Z.C.; Froemming, G.R.A.; Muid, S. Bisphenol A exposure disturbs the bone metabolism: An evolving interest towards an old culprit. *Life Sci.* **2018**, *198*, 1–7. [CrossRef]
34. Yang, M.; Chen, M.; Wang, J.; Xu, M.; Sun, J.; Ding, L.; Lv, X.; Ma, Q.; Bi, Y.; Liu, R.; et al. Bisphenol A Promotes Adiposity and Inflammation in a Nonmonotonic Dose-response Way in 5-week-old Male and Female C57BL/6J Mice Fed a Low-calorie Diet. *Endocrinology* **2016**, *157*, 2333–2345. [CrossRef]
35. Moon, M.K.; Jeong, I.K.; Jung Oh, T.; Ahn, H.Y.; Kim, H.H.; Park, Y.J.; Jang, H.C.; Park, K.S. Long-term oral exposure to bisphenol A induces glucose intolerance and insulin resistance. *J. Endocrinol.* **2015**, *226*, 35–42. [CrossRef]
36. Ma, Q.; Deng, P.; Lin, M.; Yang, L.; Li, L.; Guo, L.; Zhang, L.; He, M.; Lu, Y.; Pi, H.; et al. Long-term bisphenol A exposure exacerbates diet-induced prediabetes via TLR4-dependent hypothalamic inflammation. *J. Hazard Mater.* **2021**, *402*, 123926. [CrossRef] [PubMed]
37. Zhang, X.Q.; Xu, C.F.; Yu, C.H.; Chen, W.X.; Li, Y.M. Role of endoplasmic reticulum stress in the pathogenesis of nonalcoholic fatty liver disease. *World J. Gastroenterol.* **2014**, *20*, 1768–1776. [CrossRef]
38. Li, J.; Yin, Z.; Hua, L.; Wang, X.; Ren, F.; Ge, Y. Evaluation of BPA effects on autophagy in Neuro-2a cells. *Toxicol. Ind. Health* **2022**, *38*, 151–161. [CrossRef]
39. Wu, M.; Cong, Y.; Wang, K.; Yu, H.; Zhang, X.; Ma, M.; Duan, Z.; Pei, X. Bisphenol A impairs macrophages through inhibiting autophagy via AMPK/mTOR signaling pathway and inducing apoptosis. *Ecotoxicol. Environ. Saf.* **2022**, *234*, 113395. [CrossRef]
40. Ding, S.; Zuo, X.; Fan, Y.; Li, H.; Zhao, N.; Yang, H.; Ye, X.; He, D.; Yang, H.; Jin, X.; et al. Environmentally Relevant Dose of Bisphenol A Does Not Affect Lipid Metabolism and Has No Synergetic or Antagonistic Effects on Genistein's Beneficial Roles on Lipid Metabolism. *PLoS ONE* **2016**, *11*, e0155352. [CrossRef]
41. Lee, S.; Kim, D.H.; Lee, C.H.; Jung, J.W.; Seo, Y.T.; Jang, Y.P.; Ryu, J.H. Antidepressant-like activity of the aqueous extract of *Allium macrostemon* in mice. *J. Ethnopharmacol.* **2010**, *131*, 386–395. [CrossRef] [PubMed]

42. Huang, C.C.; Yang, C.Y.; Su, C.C.; Fang, K.M.; Yen, C.C.; Lin, C.T.; Liu, J.M.; Lee, K.I.; Chen, Y.W.; Liu, S.H.; et al. 4-Methyl-2,4-bis(4-hydroxyphenyl)pent-1-ene, a Major Active Metabolite of Bisphenol A, Triggers Pancreatic beta-Cell Death via a JNK/AMPKalpha Activation-Regulated Endoplasmic Reticulum Stress-Mediated Apoptotic Pathway. *Int. J. Mol. Sci.* **2021**, *22*, 4379. [CrossRef] [PubMed]
43. Russell, W.M.S.; Burch, R.L. *The Principles of Humane Experimental Technique*; Methuen: London, UK, 1959.

Disclaimer/Publisher's Note: The statements, opinions and data contained in all publications are solely those of the individual author(s) and contributor(s) and not of MDPI and/or the editor(s). MDPI and/or the editor(s) disclaim responsibility for any injury to people or property resulting from any ideas, methods, instructions or products referred to in the content.

foods

MDPI

Article

Digestion-Related Enzyme Inhibition Potential of Selected Mexican Medicinal Plants (*Ludwigia octovalvis* (Jacq.) P.H.Raven, *Cnidoscolus aconitifolius* and *Crotalaria longirostrata*)

Kimberly Calonico * and Julian De La Rosa-Millan

Tecnologico de Monterrey, Escuela de Ingenieria y Ciencias, BioFoods Research Lab, Av. Eugenio Garza Sada 2501 Sur, Tecnologico, Monterrey 64849, Nuevo Leon, Mexico; juliandlrm@tec.mx
* Correspondence: a01203981@tec.mx

Abstract: Medicinal plants offer a valuable source of natural compounds with specific and selective bioactivity. These compounds have been isolated since the mid-nineteenth century and are now commonly used in modern medications. *L. octovalvis* (Jacq.) P.H.Raven, *C. aconitifolius*, and *C. longirostrata* are Mexican medicinal plants consumed regularly, and research has shown that they contain bioactive compounds capable of promoting the inhibition of digestive enzymes. This is noteworthy since enzyme inhibitors are bioactive substances that interact with enzymes, diminishing their activity and thereby contributing to the management of diseases and metabolic disturbances. To investigate the activity of these plants, individual analyses were conducted, assessing their proximal composition, bioactive compounds, and inhibition of α-Amylase, α-Glucosidase, lipase, and pepsin. The results revealed that all three plants exhibited enzymatic inhibition. When comparing the plants, it was determined that *C. aconitifolius* had the lowest concentration required for a 50% inhibition in α-Amylase, α-Glucosidase, and lipase, as indicated by the IC_{50} values. For pepsin, *C. longirostrata* demonstrated the lowest IC_{50} value. By understanding the bioactive compounds present in these plants, we can establish the relationship they have with enzymatic inhibition, which can be utilized for future investigations.

Keywords: enzymatic inhibition; *Ludwigia octovalvis*; *Cnidoscolus aconitifolius*; *Crotalaria longirostrata*; antidiabetic properties; digestive enzymes; α-Amylase; α-Glucosidase; lipase; pepsin; enzyme inhibitors

1. Introduction

Medicinal plants are a valuable source of natural compounds with precise and selective bioactivity. Since the mid-nineteenth century, many of these have been extracted, isolated, purified, and used in modern medications [1]. The WHO reports that 80% of the world's population uses medicinal plants to satisfy their health needs, and in Mexico alone, more than 4000 species with medicinal attributes have been identified, representing 15% of the country's total flora [2]. Medicinal plants can be an excellent source of secondary metabolites, some of which are valuable due to their enzyme-inhibiting properties. In many cases, enzymes serve as the molecular targets in various diseases, as they play a crucial role in regulating metabolic pathways. The relative ease in controlling and modifying them via different strategies makes them attractive targets for drug treatments. One of the interesting features related with enzymes is their inhibition characteristics via different mechanisms, related to the molecular characteristics of plant bioactive compounds that reduce their activity and conversion capacity, which could be useful in treating diseases or correcting metabolic imbalances [1].

Inhibitors targeting the enzymes α-glucosidase and α-Amylase have demonstrated efficacy in reducing postprandial hyperglycemia and are commonly used in the treatment of diabetes [3]; as an example, active compounds such as corilagin (1), repandusinic acid A (2), and mallotinin (3) found in *Phyllanthus urinaria* and *Phyllanthus amarus* have been demonstrated to have inhibited α-glucosidase, particularly through 1, which is a mixed-type mode of inhibition, whereas 2 and 3 competitively inhibit α-glucosidase [4].

Research on phenolic compounds derived from various plants has demonstrated their ability to inhibit activities and pancreatic lipase. These lipase inhibitors have the capacity to reduce the absorption of triglycerides [5]. According to reports, plants such as *Aleurites moluccana* (L.), *Acer mono*, and *Baccharis trimera* Less have demonstrated inhibitory activity against pancreatic lipase [6]. Pepsin, an aspartic protease [7], plays a crucial role in the release of amino acids and short peptides from proteins and peptides, and it is highly expressed in the human gastrointestinal tract, which is why it is essential to maintain some of this activity for nutritional purposes. Certain protease inhibitors derived from plants may hinder the digestive function of dietary proteins by inhibiting trypsin and chymotrypsin activities in the human gastrointestinal system. However, scientific evidence also supports their positive effects, such as anti-inflammatory actions in the gut, and these are believed to be achieved through the inhibition of oxygen free radicals released by impaired cells, as proposed in previous studies in vitro and in vivo [8].

Studies investigating the inhibitory effects on α-glucosidase, α-Amylase, and lipase enzymes are commonly conducted to identify substances suitable for addressing and controlling metabolic disorders. Plant extracts and natural products are screened to identify potential substances with inhibitory activity against these enzymes [3]. The search for new, high-quality, affordable, and easily accessible natural compound enzyme inhibitors is a significant focus for drug discovery and research organizations worldwide [1]. Considering the prevalence of type II diabetes mellitus (TII-DM) as a major health concern in the twenty-first century, allopathic medicines remain the primary option for initial management. However, herbal remedies have gained widespread acceptance as alternative treatments, particularly in countries such as Mexico, where medicinal plants hold deep cultural significance and are highly valued [9].

As per the existing literature, *L. octovalvis* (Jacq.) P.H.Raven and *C. aconitifolius* have historically been employed in Mexico and Central America to mitigate and manage a wide array of conditions linked to metabolic disturbances. Regarding *C. longirostrata*, its selection is attributed to its inclusion in Mexican cuisine; while possessing acknowledged bioactive compounds, it has remained relatively underexplored in the scientific literature. This is significant considering the growing prevalence of metabolic disorders in developing nations, with Mexico being prominently affected by disturbances in lipid and glucose metabolism, particularly diabetes. With over 10% of the Mexican population aged 20 years and older impacted by diabetes, it represents a substantial public health concern [10].

L. octovalvis (Jacq.) P.H.Raven, commonly known as "clavillo" [10], is a flowering plant that is widespread in tropical regions across the globe. It has a historical use as a beverage for promoting health and treating conditions such as edema, nephritis, hypotension, and diabetes [11]. *C. aconitifolius* belongs to the shrub family and can reach a height of up to 6 m. It features palmate lobed leaves with milky sap and small flowers arranged in dichotomously branched cymes. Known as "chaya" [12], this plant was cultivated as a leafy green vegetable during the pre-Cambrian era in the Maya region of Guatemala, Belize, and Southeast Mexico. It continues to be valued today for its ease of cultivation, high productivity potential, and rich nutritional content, and is used as a food, medicine, and ornamental plant. *C. longirostrata*, also known as "chipilín", is native to Central America and Mexico, thriving in diverse environments. The tender shoots, leaves, and young stems of this plant have been utilized since pre-Hispanic times for food preparation, beverages, herbal medicine, and as fodder due to their nutritional composition, which includes carotenoids, vitamin C, iron, calcium, proteins, and various phenolic compounds such as flavonoids, saponins, coumarins, tannins, anthraquinones, anthrones, and alkaloids [13].

The aim of this study was to assess the role of bioactive compounds in *Ludwigia octovalvis*, *Cnidoscolus aconitifolius*, and *Crotalaria longirostrata* against some digestive enzymes related to metabolic diseases. These plants were examined because enzyme inhibitors act as bioactive entities that interact with enzymes. This interaction results in a decrease in the functional activity of enzymes, which can contribute to the management of diseases and metabolic imbalances.

2. Materials and Methods

2.1. General

For all the methods applied in this research, samples of *C. aconitifolius* and *C. longirostrata* were purchased from a local vendor in Monterrey, Mexico. The samples of *L. octovalvis* (Jacq.) P.H.Raven were germinated and cultivated from *L. octovalvis* seeds donated by Dulce Lourdes Morales-Ferra, who has done previous work on *L. octovalvis* [14]. Every material and method will be described in each section below.

2.2. Proximal Analysis

For the proximal analysis, the aerial parts of *L. octovalvis* (Jacq.) P.H.Raven, *C. aconitifolius*, and *C. longirostrata* were used. The plant material was milled in a Wiley knife mill (Thomas Scientific, Swedesboro, NJ, USA). The AACC methods 44-15.02, 08-17.01, and 32-10.01 were utilized for the moisture, ash, and ether extract of the plant flour, which were all calculated [15]. protein content was determined using the AOAC method 960.52 [16].

2.2.1. Dietary Fiber

The Megazyme K-TDFR Kit was used to measure the total dietary fiber (TDF), insoluble dietary fiber (FDI), and soluble dietary fiber (SDF) (Megazyme, Wicklow, Ireland). The kit was also used for the AOAC 991.43 [16] and AACC 32-07.01 methods [15].

2.2.2. Soluble Sugars (Sucrose, Glucose, and Fructose)

Following the procedure developed by Karkacier et al. in 2003 [17], soluble sugars were measured. Briefly, liquid nitrogen was used in a mortar to grind 50 mg of leaf sample. After adding one milliliter of nanopure water, the mixture was centrifuged at 10,000 rpm for 15 min. A 0.45 μm membrane filter (VertiPureTM, Vertical R) was used to filter the supernatant after it had been collected. A Waters HPLC with a MetaCarb 87C column and a guard column was used to analyze 20 microliters of the filtrate. At a flow rate of 0.5 mL per min, deionized water was employed as the mobile phase. Using a Waters 410 differential refractometer detector, online detection was carried out, and Empower 3.7 software was used to analyze the results. Analytical standards for fructose (Fluka, Charlotte, Greenwood Village, NC, USA), glucose, and sucrose were employed.

2.3. Bioactive Compounds

2.3.1. Preparation of Leaves Extract

The hot water extract of the entire plant of *L. octovalvis* (Jacq.) P.H.Raven, *C. aconitifolius*, and *C. longirostrata* was prepared as follows: (a) Each plant was separated and cleaned with distilled water, then dried with clean paper towels. (b) The leaves and small branches were separated, and 5 g of each plant were weighed. (c) Three separate beakers of potable water were heated to 90 °C. (d) Then, 5 g of *L. octovalvis* (Jacq.) P.H.Raven, *C. aconitifolius*, and *C. longirostrata* were added to each beaker and allowed to soak for 5 min at 90 °C. (e) The extract was filtered using Whatman n.1 filter paper and stored in amber-colored bottles at 4 °C until use.

2.3.2. Free Amino Nitrogen (FAN)

The ninhydrin colorimetric method was used to determine the FAN content. The standard solution was a glycine solution in water (MB013–500 G). Next, 0.1085 g of pure glycine ($C_2H_5NO_2$; MW, 75.07; purity, 99%) was dissolved in 100 mL of deionized water in a volumetric flask in accordance with the European Brewery Convention Protocol (Analytica-EBC, 2000). FAN content in this solution was 200 ppm. The 100 ppm, 50 ppm, and

25 ppm solutions were diluted from this stock solution, and a standard curve was created. Following the addition of 1 mL of ninhydrin solution to 5 mL of sample, the mixture was gently stirred for 4–7 min at 80–100 °C. The sample's amino acids and ninhydrin combined created Ruhemann's complex, which took on a purple hue. The absorbance at 570 nm was measured after cooling to room temperature in a cold-water bath, and the concentrations of the samples were determined from the standard curve (Lie, 1973) [18].

2.3.3. Total Phenolics

The total phenolic content (TPC) was determined using the Folin–Ciocalteu method. Briefly, 200 μL of each plant infusion was mixed with 50 μL of Folin–Ciocalteu reagent (2 M) and 750 μL of sodium carbonate solution (7.5% w/v) in a 1.5 mL Eppendorf tube. The mixture was vortexed for 15 s using a Standard Heavy-Duty Vortex Mixer. The tubes were then kept in complete darkness for 2 h and centrifuged at 2500 rpm for 15 min. Next, 200 μL of each sample reaction was transferred to a 96-well microplate, and the absorbance was measured at 620 nm using a microplate reader (Epoch 2, BioTek Instruments Inc., Charlotte, VT, USA). A standard curve was generated using gallic acid (GA) concentrations ranging from 0 to 600 mg/L [19].

2.3.4. Total Flavonoid Content

The total flavonoid content of the plant material was quantified using a colorimetric assay. Specifically, 100 mg of the dried plant material was mixed with 4 mL of distilled water. To this mixture, 0.3 mL of 5% sodium nitrite was added, and after 5 min, 0.3 mL of 10% aluminum chloride was introduced. Following a 6-min incubation period, 2 mL of 1 M sodium hydroxide was incorporated into the solution. Immediately, the mixture was diluted with 3.3 mL of distilled water and thoroughly mixed.

Subsequently, the sample was subjected to centrifugation at 1000× g and 4 °C, and 2 mL of the resulting supernatant was collected. The absorbance of the supernatant was measured at 510 nm, using a blank for comparison. Catechin was utilized as a standard to construct the calibration curve. The total flavonoid content of the plant extract was then expressed in milligrams of catechin equivalents per gram of the sample (mg/g) [20].

2.3.5. DPPH Trolox Equivalent

DPPH free radical scavenging activity was determined following the methodology described by García-Becerra et al., 2010 [21]. To estimate the antioxidant capacity of plant infusions, a mixture of 60 μM of a methanolic solution of DPPH (2940 μL) with 60 μL of extract in a polystyrene cuvette was used to measure the absorbance at 515 nm on two occasions: before adding the extract, and 60 min after adding it with a spectrophotometer (Genesys 10S, Annapolis, MD, USA). The standard curve was determined with Trolox, and the results were expressed as a Trolox equivalent in mmol.

2.3.6. Reducing Power μg Ascorbic Acid Equivalents

The method used to determine the reducing power was based on the procedure described by Juntachote et al., 2006 [22], with modifications from Terpinc et al., 2012 [23]. The results are presented as the coefficient of the reducing power (CR). The ability of a compound to act as a reducing agent can indicate its potential antioxidant activity. To conduct the assay, aliquots of the standard and test sample extracts at various concentrations (ranging from 10 to 100 μg/mL) were mixed with 1.0 mL of deionized water, 2.5 mL of phosphate buffer (pH 6.6), and 2.5 mL of potassium ferricyanide (1%). The mixture was then incubated at 50 °C in a water bath for 20 min, followed by cooling. Next, 2.5 mL of trichloroacetic acid (10%) was added, and the mixture was centrifuged at 3000 rpm for 10 min. The upper layer (2.5 mL) of the solution was combined with 2.5 mL of distilled water and 0.5 mL of freshly prepared ferric chloride solution (0.1%). The absorbance was measured at 700 nm using a UV spectrometer (Sytronic double beam-UV-2201). A blank sample without the addition of the extract was prepared, and ascorbic acid at various concentrations (ranging from 10 to 100 µg/mL) was used as the standard [24].

2.3.7. Determination of Antinutrients (Tanins, Alkaloids, Saponins, Oxalate)

The amount of oxalate present in the avocado extract was measured using a modified technique based on Libert and Franceschi's method from 1987 [25]. A sodium oxalate curve (Sigma Aldrich, 71804, St. Louis, MO, USA) was used as a reference to compare the plant extract. To measure the total tannin content, the spectrophotometric method described by Griffiths and Jones in 1977 was used, utilizing a Tannic acid solution (Sigma Aldrich, 403040, St. Louis, MO, USA) [26]. The saponin content was determined following the approach outlined by Makkar and Becker in 1996, employing a saponin standard (Sigma Aldrich, SAE0073, St. Louis, MO, USA) [27]. Lastly, the alkaloid content was estimated using the spectrometric method developed by Shamsa et al. in 2008, which involves the use of the bromocresol green method [28].

2.4. Determination the Inhibition of α-Amylase and α-Glucosidase

The inhibitory activity for α-Amylase was obtained following the method of Kazeem et al. [29]. Data were compared by using different extract dilutions and using acarbose (200, 400, 600, 800, 1000, and 1200 µg mL^{-1}). Briefly, tubes containing 250 µL acarbose (1 mg mL^{-1}) or plant extract along with a sodium phosphate buffer (0.02 M pH 6.9) or 500 µL α-Amylase solution in a phosphate buffer (with 0.006 M de NaCl, 13 U mL^{-1}) were preincubated at room temperature for 10 min with a 250 µL 1% starch solution (sodium phosphate buffer). Tubes were then boiled for 5 min after adding 1 mL of DNS (method of Miller, 1959) [30] to stop the reaction. As the control blank, a tube without acarbose or plant extract was used. Samples were diluted in 10 mL of distilled water, and absorbance was measured at 540 nm (UNICO S1200, Dayton, NJ, USA). α-Amylase inhibition was calculated using Equation (1):

$$\text{Inhibition}(\%) = \frac{K - (S1 - S0)}{K} * 100 \tag{1}$$

where K = absorbance of control blank, $S1$ = absorbance of sample with enzyme, and $S0$ = absorbance of sample without enzyme. Inhibitory activity was expressed as the concentration, allowing half of the maximal inhibitory activity (IC_{50}). The IC_{50} value of each extract was determined from the plots of percent inhibition versus inhibitor concentration.

The α-Glucosidase test was performed using different extract concentrations as described by Mayur et al. [31]. Rat intestinal acetone powder was mixed with distilled water (10 mg mL^{-1}) and centrifuged for 10 min at 10,000 rpm. The assay buffer was 100 mM HEPES at pH 6.8, and the substrate was 2 mM 4-nitrophenyl-α-d-glucopyranoside. The assay constituents were added to 96-well microplates in the following order: 100 µL of enzymatic solution (10 mg rat intestinal acetone powder per 1 mL distilled water), then 50 µL of acarbose or seaweed extract or buffer, and finally 50 µL of substrate. Samples were incubated at 25 °C for 2 h. Absorbance was recorded at the beginning and end of the incubation time at 405 nm. α-Glucosidase inhibition was calculated using Equation (1). Acarbose (1 mg mL^{-1}) was used as a positive control. The inhibitory activity was expressed as the concentration required to obtain half of the maximal inhibitory activity (IC_{50}). The extracts' IC_{50} values were determined from the plots of percent inhibition versus inhibitor concentration.

2.5. Determination the Inhibition of Lipase and Pepsin

To conduct the experiment, stock solutions from each herbal product and orlistat were prepared. Orlistat is the only commercial drug known to effectively inhibit pancreatic lipase. The contents of each capsule were dissolved in a 100 mL flask, and these solutions were then diluted to obtain concentrations of 50, 100, 200, 300, and 400 µg/mL. Next, 0.2 mL of each diluted solution was added to separate test tubes. Before proceeding, a stock solution of porcine pancreatic lipase enzyme was prepared at a concentration of 1 mg/mL in a 10% DMSO solution. Subsequently, 0.7 mL of Tris-HCl buffer (pH 7.4, 0.1 M) and 0.1 mL of pancreatic lipase enzyme were added to each test tube. A blank tube was also prepared, containing only 0.9 mL of Tris-HCl and 0.1 mL of pancreatic lipase enzyme without any inhibitor. All the test tubes, including the blank, that contained p-nitrophenyl-butyrate (PNPB) were incubated at 37 °C for 15 min. After the initial incubation, 0.1 mL of PNPB (100 mM in acetonitrile) was added to all the test tubes, including the blank. Another incubation period of 30 min followed. The activity of pancreatic lipase was assessed by measuring the hydrolysis of PNPB to p-nitrophenol at a wavelength of 410 nm using a UV-visible spectrophotometer (Shimadzu-UV-1800, Kyoto, Japan) [32].

Later, the absolute inhibitory activity was calculated using the following equation:

$$\textbf{Inhibition}\ (\%) = \frac{(B - S)}{B} * \textbf{100\%} \quad (2)$$

where S refers to the sample absorbance and B to the blank absorbance.

2.6. Fourier Transformed Infrared Spectroscopy (FTIR)

After being dried at 40 °C for six hours, the sample was crushed in a coffee grinder. The sample underwent ATR-FTIR analysis (Spectrum 1, Perkin Elmer, Norwalk, VA, USA). Data on molecular spectrum emissions were gathered, adjusted for the air background, and then examined using the Spectrum program (ver. 5.3.0). The spectra were produced using the mid-IR (about 4000–8000) in absorption mode with a resolution of 4 cm^{-1} by 20 scans and a half-band width of 15 cm^{-1}. Using published studies (De La Rosa-Millán, 2017; Chávez-Murillo et al., 2018) [33,34] all chemical functional groups were identified.

2.7. Statistical Analysis

The obtained results underwent a one-way ANOVA followed by the Tukey multiple comparison test to determine the presence of statistically significant differences. Additionally, a hierarchical cluster analysis was conducted to identify meaningful correlations among the data collected: protein, total flavonoids, free amino nitrogen, IC50 α-Amylase, IC50 α-Glucosidase, lipids, soluble fiber, insoluble fiber, IC50 lipase, ash, reducing power, oxalates, IC50 pepsin, free sugars, DPPH, total phenolics, alkaloids, tanins, and saponins. A hierarchical clustering was applied to the principal components analysis to create a dendrogram.

3. Results

3.1. Proximal Composition

The proximal composition of *L. octovalvis*, *C. aconitifolius*, and *C. longirostrata* is presented in Table 1, where the major component of *L. octovalvis* are carbohydrates that correspond to the total dietary fiber, with a 65.30 ± 1.83% from which 28.99 ± 0.75% is soluble fiber, 36.31 ± 1.13% is insoluble fiber, and 13.99 ± 0.55% correspond to free sugars. Another major component is protein with 12.70 ± 1.54%, where the free amino nitrogen was also determined. In the case of *C. aconitifolius*, the major component was also carbohydrates, with a 70.27 ± 1.02% from which 25.35 ± 1.38% is soluble fiber, 28.98 ± 1.78 is insoluble fiber, and 15.94 ± 0.80% correspond to free sugars. Comparing all plants, *C. longirostrata* had the highest percentage of carbohydrates (73.70 ± 2.3%), with soluble fiber accounting for 26.34 ± 1.05% of those carbohydrates, 31.25 ± 1.24% representing insoluble fiber, and 16.10 ± 0.20% corresponding to free sugars. Regarding protein, *C. longirostrata* has the largest proportion of protein (14.51 ± 2.04) compared to *L. octovalvis* and *C. aconitifolius*. *C. longirostrata* has the highest concentration of free amino nitrogen (FAN), which is 330.33 ± 12.22 mg/100 g^{-1}, followed by *L. octovalvis* with 259.67 ± 4.16 mg/100 g^{-1}, and *C. aconitifolius* with 225.83 ± 5.55 mg/100 g^{-1}.

Table 1. Proximal chemical composition of *L. octovalvis*, *C. aconitifolius*, and *C. longirostrata* plant (% dry matter).

Sample	*L. octovalvis*	*C. aconitifolius*	*C. longirostrata*
Moisture (%)	5.53 ± 0.21	4.88 ± 0.34	4.83 ± 0.16
Protein (%)	12.70 ± 1.54	12.52 ± 1.18	14.51 ± 2.04
Lipids (%)	6.67 ± 0.66	5.67 ± 0.50	6.26 ± 0.67
Ash (%)	6.01 ± 0.57	6.66 ± 0.61	0.71 ± 0.14
Carbohydrates (%)	65.30 ± 1.83	70.27 ± 1.02	73.70 ± 2.31
Free sugars (%)	13.99 ± 0.55	15.94 ± 0.80	16.10 ± 0.20
Soluble fiber (%)	28.99 ± 0.75	25.35 ± 1.38	26.34 ± 1.05
Insoluble fiber (%)	36.31 ± 1.13	28.98 ± 1.78	31.25 ± 1.24
Total dietary Fiber (%)	65.30 ± 1.83	54.33 ± 0.42	57.59 ± 2.29

3.2. Bioactive Compounds

The bioactive compounds of *L. octovalvis*, *C. aconitifolius*, and *C. longirostrata* are presented in Table 2.

Table 2. Bioactive compounds of *L. octovalvis*, *C. aconitifolius*, and *C. longirostrata* plant (% dry matter).

Sample	*L. octovalvis*	*C. aconitifolius*	*C. longirostrata*
Free amino nitrogen (mg/100 g^{-1})	259.67 ± 4.16	225.83 ± 5.55	330.33 ± 12.22
Total phenolics (mg GAE/100 g)	652.96 ± 10.58	740.57 ± 9.43	889.66 ± 2.82
Total flavonoid content (mg of chatequin equivalents/100 g)	337.00 ± 3.61	346.03 ± 4.28	531.53 ± 10.07
DPPH Trolox equivalent (μMTE/g)	2.35 ± 0.06	4.34 ± 0.12	4.52 ± 0.22
Reducing power μg ascorbic acid equivalents	443.36 ± 2.14	572.41 ± 4.93	411.26 ± 3.92

C. longirostrata, which has the greatest concentration of phenolic compounds (889.66 ± 2.82 mg GAE/100 g), is followed by *C. aconitifolius*, which has a concentration of 740.57 ± 9.43 mg GAE/100 g, and *L. octovalvis*, which has the lowest concentration (652.96 ± 10.58 mg GAE/100 g). In terms of total flavonoid content, *C. longirostrata* similarly has the largest quantity of flavonoids, with 531.53 ± 10.07 mg of chatequin equivalents/100 g, followed by *C. aconitifolius* and *L. octovalvis*, both of which have 346.03 ± 4.28 and 337.00 ± 3.61 mg of chatequin equivalents/100 g, respectively.

The plant with the largest level of DPPH Trolox equivalent was *C. longirostrata* with 4.52 ± 0.22 μMTE/g, followed by *C. aconitifolius* with 4.34 ± 0.12 μMTE/g, and *L. octovalvis* with 2.35 ± 0.06 μMTE/g. Regarding the reducing power in μg ascorbic acid equivalents, *C. aconitifolius* had the largest quantity at 572.41 ± 4.92, followed by *L. octovalvis* at 443.36 ± 2.14, and *C. longirostrata* at 411.26 ± 3.92.

3.3. Antinutrients

Plants also contain a category of compounds called antinutrients, which they use for their own defense [35]. These compounds can have antinutrients that can disrupt nutrient absorption, reduce nutrient intake, hinder digestion and utilization, and potentially lead to other negative consequences [35], but they can also have positive effects on health [36]; for example, at lower concentrations, phytic acid, lectins, phenolic compounds, enzyme inhibitors, and saponins have demonstrated the ability to lower blood glucose and/or plasma cholesterol and triacylglycerol levels. Additionally, saponins have been reported to effectively maintain liver function, prevent osteoporosis, and inhibit platelet agglutination [35].

Analyzing compounds such as tannins, alkaloids, saponins, and oxalates reveals that *C. aconitifolius* has the largest quantity of tannins per unit weight (17.58 ± 0.63 mg/g), followed by *C. longirostrata* (15.34 ± 0.14 mg/g), and *L. octovalvis* (14.13 ± 0.14 mg/g). Similar results were found for the alkaloids in terms of the amount of tannins: *C. aconitifolius* had 2.37 ± 0.04 mg/g, *C. longirostrata* had 1.74 ± 0.11 mg/g, and *L. octovalvis* had 1.12 ± 0.14 mg/g. *C. aconitifolius* (5.49 ± 0.21 mg/g) had the greatest concentration of saponins, followed by *C. longirostrata* (4.42 ± 0.23 mg/g) and *L. octovalvis* (2.26 ± 0.30 mg/g). *C. aconitifolius* had the greatest concentration of oxalates (8.91 ± 0.48 mg/g), followed by *L. octovalvis* (4.41 ± 0.22 mg/g) and *C. longirostrata* (3.32 ± 0.12 mg/g). The information is presented in Table 3.

Table 3. Antinutrients of *L. octovalvis*, *C. aconitifolius*, and *C. longirostrata* plants (% dry matter).

Sample	*L. octovalvis*	*C. aconitifolius*	*C. longirostrata*
Tannin (mg/g)	14.13 ± 0.14	17.58 ± 0.63	15.34 ± 0.14
Alkaloid (mg/g)	1.12 ± 0.14	1.74 ± 0.11	2.37 ± 0.04
Saponins (mg/g)	2.26 ± 0.30	5.49 ± 0.21	4.42 ± 0.23
Oxalate (mg/g)	4.41 ± 0.22	8.91 ± 0.48	3.32 ± 0.12

3.4. α-Amylase Inhibition

The ability to inhibit α-Amylase was found to vary among *L. octovalvis*, *C. aconitifolius*, and *C. longirostrata* against the positive control, acarbose. Table 4 presents the results of the α-Amylase inhibition assay conducted on three plant samples (*L. octovalvis*, *C. aconitifolius*, and *C. longirostrata*) and Acarbose at different concentrations. The inhibitory effects of the plants and drug were expressed as percentages of dry matter. The results demonstrate that the IC_{50} values for *L. octovalvis*, *C. aconitifolius*, and *C. longirostrata* were 519.40 mg/mL, 471.39 mg/mL, and 585.12 mg/mL, respectively, while Acarbose had an IC_{50} value of 338.49 mg/mL. The IC_{50} value refers to the concentration at which a 50% inhibition of α-Amylase activity is achieved.

Table 4. α-Amylase inhibition of *L. octovalvis*, *C. aconitifolius*, and *C. longirostrata* plants (% dry matter).

Sample	*L. octovalvis*	*C. aconitifolius*	*C. longirostrata*	*Acarbose*
100 mg/mL	5.64 ± 0.37	7.18 ± 0.36	8.78 ± 0.48	24.30 ± 0.98
200 mg/mL	22.70 ± 0.47	27.43 ± 1.06	13.60 ± 0.48	36.78 ± 0.42
400 mg/mL	35.77 ± 1.68	44.10 ± 2.19	28.50 ± 0.57	47.79 ± 0.57
600 mg/mL	63.93 ± 0.71	71.89 ± 1.54	47.50 ± 0.80	84.07 ± 0.64
800 mg/mL	86.85 ± 2.25	87.59 ± 1.10	72.53 ± 1.55	95.46 ± 1.15
1000 mg/mL	94.39 ± 0.86	98.37 ± 1.17	85.28 ± 2.76	100.00 ± 0.00
1200 mg/mL	100.00 ± 0.00	100.00 ± 0.00	93.22 ± 0.95	1200.00
IC_{50} mg/mL	519.40	471.39	585.12	338.49

It is evident that the inhibition of α-Amylase activity becomes more pronounced as the concentration of the samples increases. Among the tested samples, *L. octovalvis* exhibits the lowest inhibition activity, with an IC_{50} value of 519.40 mg/mL, while *C. aconitifolius* demonstrates the highest inhibition activity, with an IC_{50} value of 471.39 mg/mL. These findings indicate that *C. aconitifolius* is the most potent inhibitor of α-Amylase activity among the samples examined.

Figure 1 depicts the percentage of α-Amylase inhibition of the *L. octovalvis*, *C. aconitifolius*, and *C. longirostrata* plants (dm). The graph shows an increasing trend in inhibition percentage with increasing extract concentration. It is evident that there is a significant difference between the groups, from concentrations ranging from 100 mg/mL to 600 mg/mL, as indicated by the non-overlapping error bars. At a concentration of 800 mg/mL, there is no significant difference between *L. octovalvis* and *C. aconitifolius*, while there is a significant difference between *C. longirostrata* and acarbose. From 1000 mg/mL, there is no significant difference between *L. octovalvis*, *C. aconitifolius*, and acarbose, but there is a significant difference with *C. longirostrata*. In Figure A1, the data are presented in line charts with trendlines.

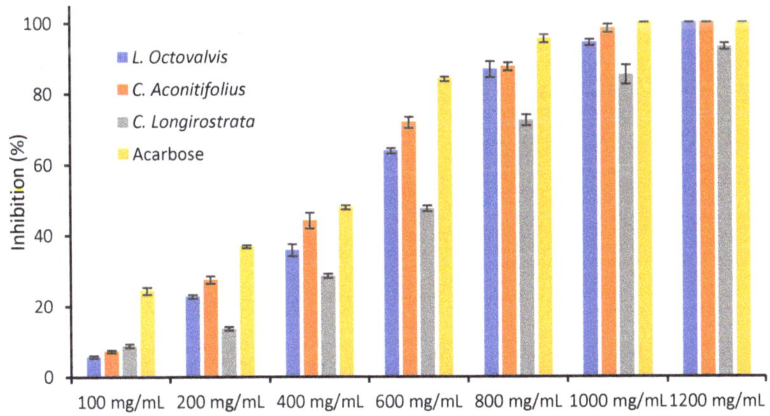

Figure 1. α-Amylase inhibition of *L. octovalvis*, *C. aconitifolius*, and *C. longirostrata* plant extracts.

3.5. Glucosidase Inhibition

Table 5 shows the α-Glucosidase inhibition of three different plants (*L. octovalvis*, *C. aconitifolius*, and *C. longirostrata*) at various concentrations of plant extract (100–1200 mg/mL) and acarbose. As can be seen, if the concentration of the plant extract increased, the percentage of α-Glucosidase inhibition also increased for all three plant species. Finally, the IC_{50} values for each plant extract and Acarbose were also reported, indicating the concentration required to inhibit 50% of the α-Glucosidase activity.

Table 5. α-Glucosidase inhibition of *L. octovalvis*, *C. aconitifolius*, and *C. longirostrata* plants (% dry matter).

Sample	*L. octovalvis*	*C. aconitifolius*	*C. longirostrata*	Acarbose
100 mg/mL	5.05 ± 0.48	6.36 ± 0.22	8.36 ± 0.73	19.61 ± 1.45
200 mg/mL	27.71 ± 1.04	32.54 ± 1.42	18.54 ± 0.59	23.92 ± 0.51
400 mg/mL	40.78 ± 2.14	49.21 ± 2.67	33.44 ± 1.04	35.78 ± 0.52
600 mg/mL	68.93 ± 0.15	77.00 ± 1.89	52.44 ± 0.87	55.70 ± 0.70
800 mg/mL	91.86 ± 2.65	92.70 ± 0.63	77.47 ± 1.35	65.55 ± 1.56
1000 mg/mL	99.39 ± 0.96	100.00 ± 0.00	86.54 ± 2.83	74.59 ± 2.29
1200 mg/mL	100.00 ± 0.00	100.00 ± 0.00	93.17 ± 0.97	94.84 ± 0.72
IC_{50} mg/mL	480.50	433.78	579.71	571.61

Figure 2 shows the concentrations between 100–600 mg/mL, and there was a significant difference in α-Glucosidase inhibition between the three plant species, as indicated by the non-overlapping error bars.

At concentrations above 800 mg/mL, there was no significant difference in α-Glucosidase inhibition between *L. octovalvis* and *C. aconitifolius*, but there was a significant difference between both of these plants and *C. longirostrata*. At 1000 mg/mL, there was no significant difference in α-Glucosidase inhibition between *L. octovalvis*, *C. aconitifolius*, and acarbose, but there was a significant difference with *C. longirostrata* (Figure 2).

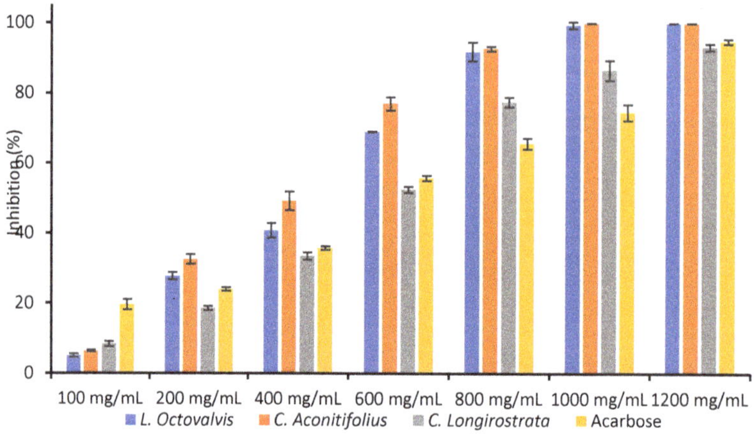

Figure 2. α-Glucosidase inhibition of *L. octovalvis*, *C. aconitifolius*, and *C. longirostrata* plant extracts.

3.6. Lipase Inhibition

In Table 6 it is shown the Lipase inhibition of three different plants (*L. octovalvis*, *C. aconitifolius*, and *C. longirostrata*) at various concentrations of plant extract (100–1200 mg/mL) and Orlistat as control. In the observed data, it is evident that elevating the concentration of the plant extract resulted in a proportional increase in the inhibition of Lipase activity across all three plant species.

Table 6. Lipase inhibition of *L. octovalvis*, *C. aconitifolius*, and *C. longirostrata* plants (% dry matter).

Sample	*L. octovalvis*	*C. aconitifolius*	*C. longirostrata*	Orlistat
100 mg/mL	4.34 ± 0.29	5.53 ± 0.27	6.76 ± 0.37	35.36 ± 0.09
200 mg/mL	17.48 ± 0.36	21.12 ± 0.81	10.47 ± 0.37	67.13 ± 0.78
400 mg/mL	27.54 ± 1.29	33.96 ± 1.68	21.95 ± 0.44	100.00 ± 0.00
600 mg/mL	49.22 ± 0.55	55.36 ± 1.18	36.58 ± 0.61	-
800 mg/mL	66.87 ± 1.74	67.45 ± 0.84	55.85 ± 1.19	-
1000 mg/mL	72.68 ± 0.66	75.74 ± 0.90	65.67 ± 2.13	-
1200 mg/mL	73.29 ± 0.90	84.27 ± 0.94	71.78 ± 0.73	-
IC_{50} mg/mL	697.17	627.84	677.27	-

In Figure 3 it is displayed the percentage of lipase inhibition for the *L. octovalvis*, *C. aconitifolius*, and *C. longirostrata* plants at various concentrations, as well as for the positive control Orlistat. As the concentration of the plant samples increases, so does the percentage of lipase inhibition. At the highest concentration of 1200 mg/mL, all three plant samples exhibit substantial lipase inhibition, with *C. aconitifolius* showing the highest percentage (84.27 ± 0.94). Orlistat, the positive control, completely inhibits lipase activity at all concentrations. In general, the plant samples differ significantly in their lipase inhibitory activity, with *C. aconitifolius* consistently displaying greater inhibition than the other two plant samples.

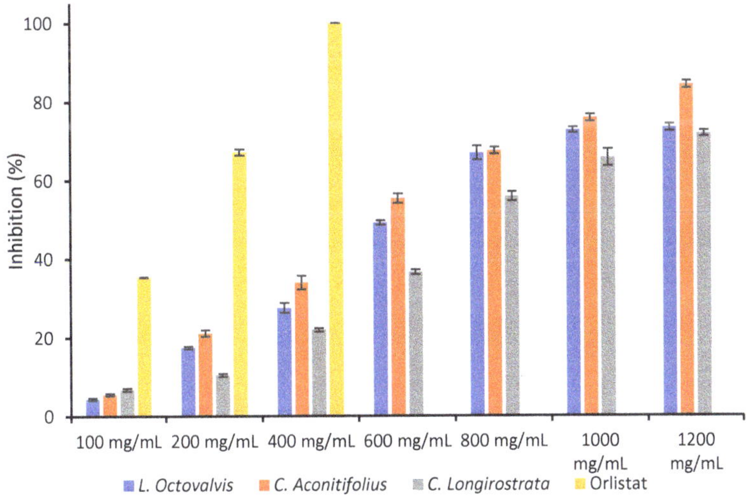

Figure 3. Lipase inhibition of *L. octovalvis*, *C. aconitifolius*, and *C. longirostrata* plant extracts.

3.7. Pepsin Inhibition

Table 7 displays the outcomes of the pepsin inhibition test performed on the three plant specimens (*L. octovalvis*, *C. aconitifolius*, and *C. longirostrata*) and a positive control (protease inhibitor cocktail) at varying concentrations (ranging from 100 to 1200 mg/mL) and the inhibitory effects were expressed as percentages of dry matter. The findings indicate that the higher the concentration of the plant extract, the greater the percentage of pepsin inhibition. The IC_{50} values for *L. octovalvis*, *C. aconitifolius*, and *C. longirostrata* were 910.94 mg/mL, 1039.93 mg/mL, and 823.02 mg/mL, respectively, which refers to the concentration at which a 50% inhibition of pepsin activity occurs. At 200 mg/mL, *C. aconitifolius* and the protease inhibitor cocktail showed complete inhibition of pepsin activity, while *L. octovalvis* and *C. longirostrata* did not achieve full inhibition at any of the tested concentrations. It is crucial to note that the protease inhibitor cocktail was employed as a positive control and not as a sample for comparison with the plant extracts.

Table 7. Pepsin inhibition of *L. octovalvis*, *C. aconitifolius*, and *C. longirostrata* plants (% dry matter).

Sample	*L. octovalvis*	*C. aconitifolius*	*C. longirostrata*	Protease Inhibitor Cocktail
100 mg/mL	2.58 ± 0.25	3.32 ± 0.42	4.29 ± 0.21	86.60 ± 2.19
200 mg/mL	7.14 ± 1.29	9.86 ± 1.20	2.92 ± 0.40	100.00 ± 0.00
400 mg/mL	14.79 ± 1.72	19.31 ± 1.35	11.42 ± 0.44	-
600 mg/mL	25.00 ± 1.13	29.46 ± 1.75	17.27 ± 0.09	-
800 mg/mL	39.63 ± 1.85	40.31 ± 1.03	32.76 ± 0.58	
1000 mg/mL	46.39 ± 0.82	46.88 ± 0.36	46.52 ± 0.33	-
1200 mg/mL	57.85 ± 0.30	56.68 ± 0.59	59.21 ± 0.12	-
IC_{50} mg/mL	910.94	1039.93	823.02	−173.19

The plant samples show significant differences in their inhibitory effects at various concentrations. For instance, *C. longirostrata* has a higher inhibitory effect than *L. octovalvis* and *C. aconitifolius* at 100 mg/mL. However, at 200 mg/mL, *C. aconitifolius* exhibits significantly higher inhibition than the other two plant samples, achieving complete inhibition. Neither *L. octovalvis* nor *C. longirostrata* show complete inhibition at any tested concentration.

Furthermore, it is important to note that the protease inhibitor cocktail has a much stronger inhibitory effect on pepsin than the plant samples at all concentrations, indicating its high effectiveness as an inhibitor (Figure 4).

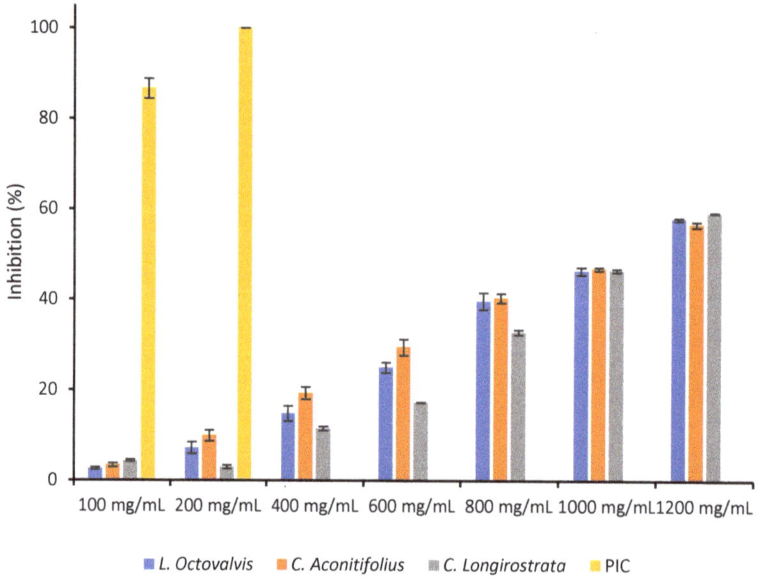

Figure 4. Pepsin inhibition of *L. octovalvis*, *C. aconitifolius*, and *C. longirostrata* plant extracts.

3.8. FTIR

Figure 5 shows the spectrum determined by the FTIR of *L. octovalvis*, *C. aconitifolius*, and *C. longirostrata*.

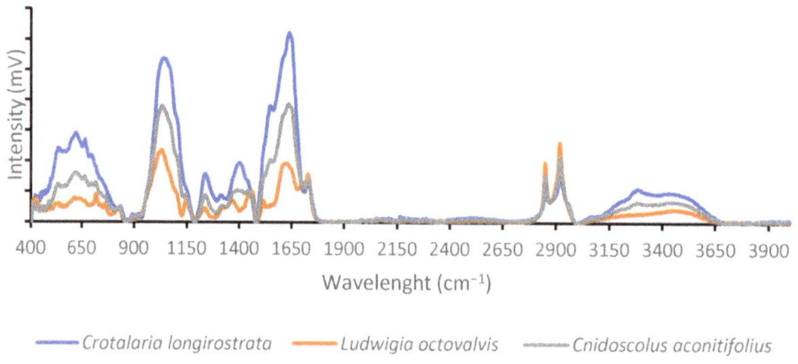

Figure 5. *L. octovalvis*, *C. aconitifolius*, and *C. longirostrata* FTIR spectrum.

3.9. Statiscal Analysis

With the objective of comprehensively analyzing the various evaluated parameters in this research, a hierarchical clustering analysis was done. Significant correlations are illustrated in Figure 6. The analysis resulted in the identification of five distinct clusters, each closely associated with the IC_{50} values of enzyme inhibition. The first cluster comprises proteins, flavonoid content, FAN, IC_{50} α-Amylase, and IC_{50} α-Glucosidase. The second cluster includes lipids, soluble fiber, insoluble fiber, and IC_{50} lipase. In the third cluster, we find Ash, FRAP, oxalates, and IC_{50} pepsin. The fourth cluster contains free sugars, DDPH, total phenolics, and alkaloids, while the fifth cluster consists of tannins and saponins.

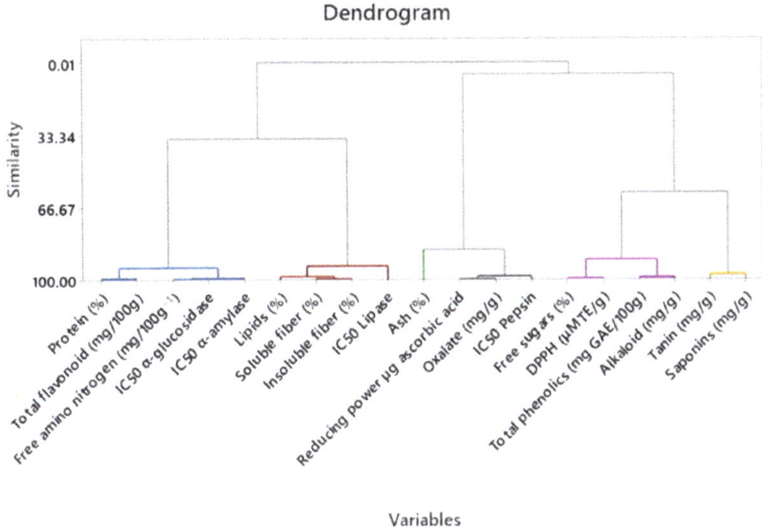

Figure 6. Hierarchical cluster analysis of *L. octovalvis*, *C. aconitifolius*, and *C. longirostrata* components.

4. Discussion

4.1. Proximal Composition

One of the objectives of this study was to determine the chemical composition of *L. octovalvis* (Jacq.) P.H.Raven, *C. aconitifolius*, and *C. longirostrata* to gain a better understanding of the nutritional properties of each plant. Oyagbemi et al. (2011) conducted a proximate analysis of *C. aconitifolius*, revealing protein content at 14.61%, fat at 2.58%, ash at 9.98%, and fiber at 9.88%. When comparing our results solely with the parameters examined by Oyagbemi et al., we found that the protein content reported in our study (12.52%) was lower than their finding (14.61%). Conversely, our data showed higher fat content (5.67%) compared to theirs (2.58%), and lower ash content (6.66%) in contrast to their value of 9.98%. Notably, our fiber content (54.33%) was substantially higher than their reported value (9.885%). These differences might be influenced by factors such as sample variability, analytical methodologies, environmental conditions, or genetic diversity among the studied *C. aconitifolius* specimens. Hence, when comparing nutritional data from different studies, it is crucial to consider these factors for accurate interpretation and understanding of the observed variations [37].

For *C. longirostrata*, a comparative analysis was conducted with a plant of the same genus, *Crotalaria retusa*, as reported in the study by Alalade in 2019. The findings revealed notable differences between the two datasets. Specifically, the protein content reported in Alalade et al. (2019) (18.00%) was higher than that observed in our research (14.51%). Likewise, the ash content reported by Alalade et al. (2019) (7.00%) significantly exceeded the value found in our research (0.71%). Furthermore, the insoluble fiber content in our research (31.25%) was substantially higher than the value reported in Alalade et al. (2019) (16.75%) [38].

These observed variations in protein, ash, and insoluble fiber content may be attributed to several factors, such as differences in plant samples, analytical methods, environmental conditions, or genetic factors that influence the nutritional composition of *C. longirostrata*. Therefore, it is imperative to consider these factors when comparing nutritional data from various studies to ensure accurate interpretation and comprehension of the discrepancies observed.

A comparison was conducted between *L. octovalvis* (Jacq.) P.H.Raven and *Ludwigia grandiflora* in a study carried out by Oku et al. in 2020. The choice of Ludwigia grandiflora was due to the unavailability of data for *L. octovalvis* in the existing literature. Our experimental findings indicate that *L. octovalvis* exhibits a higher protein content (12.70%) compared to the value reported by Oku et al. (10.31%). Additionally, our results show that *L. octovalvis* has a lower ash content (6.01%) in contrast to the value presented by Oku et al. (7.89%). Moreover, our data demonstrate a significantly higher level of insoluble fiber in *L. octovalvis* (36.31%) compared to Oku et al.'s findings (19.94%) [39]. These disparities between the datasets for *L. octovalvis* should be interpreted with consideration of potential factors such as the comparison with different species that have the same genus, sample source, analytical methods, and experimental procedures, that could contribute to the observed differences.

4.2. Bioactive Compounds

It is important to emphasize that the plant extracts in this study were prepared using hot water instead of an alcoholic solvent. While many research studies commonly employ alcoholic extracts to comprehensively analyze plant composition, the objective of our investigation is to gain insights into the plants from a more realistic perspective, particularly when considering edible plants that are naturally consumed without the extraction of compounds through alcoholic solvents such as methanol or ethanol. The selection of an appropriate solvent and extraction technique is crucial for effectively extracting biologically active compounds from medicinal plants. The polarity of solvents plays a critical role in separating specific compounds with diverse structures and physicochemical properties. Water, ethanol, and glycerol are well-established solvents that are recognized and approved for use in pharmaceutical formulations [40].

According to Shopska et al. (2019), "free amino nitrogen" (FAN) refers to nitrogen-containing molecules that are readily available, such as amino acids, peptides, and ammonium ions [41]. The aim of this study was to determine the form in which protein is present, and the results suggest that it contains free amino acids. Amino acids are crucial constituents of plants and serve as the basic building blocks for proteins, the main nitrogen carriers, and signaling molecules. Plant amino acids are produced through three different processes: absorption by roots, reduction in nitrates, and breakdown of ammonium, as explained by Guo et al. (2021) [42]. Due to the scarce information on the free amino nitrogen content in *L. octovalvis*, *C. aconitifolius*, and *C. longirostrata*, an analysis of FAN content was carried out. When comparing the free amino nitrogen content of *L. octovalvis*, *C. aconitifolius*, and *C. longirostrata*, it can be observed that *C. longirostrata* has the highest content, with a mean value of 330.33 ± 12.22 mg/100 g^{-1}, followed by *L. octovalvis*, with a mean value of 259.67 ± 4.16 mg/100 g^{-1}, and finally, *C. aconitifolius*, with a mean value of 225.83 ± 5.55 mg/100 g^{-1}. Free amino nitrogen includes amino acids, peptides, and ammonium ions and is an important component of plant protein.

Phenolics represent the most extensive category of phytochemicals, playing a pivotal role in the majority of antioxidant activity found within plants and plant-derived products [43]. The quantification of total phenolics is of significant importance as phenolic compounds serve as vital constituents of plants, exhibiting redox properties that contribute to their antioxidant activity [44]. Compared to a study made by Yakob et al. in 2012, the total phenolic compounds found in *L. octovalvis* using the same methodology was 264.76 ± 0.23 µg GAE/g [45]. Another study conducted by Lin et al. in 2017 found that the polyphenol content was 146.3 ± 3.1 mg GAE/g [11], which is lower than the quantity obtained in this research, 652.96 ± 10.58 mg GAE/g. This could be because of a difference in the solvents used (methanol vs. warm water). In the case of *C. aconitifolius*, there are few research studies reporting the total phenolic content. In a study conducted by Godínez-Santillán et al. in 2019 [46], it was found that the total phenolic content was 59.2 ± 2.3 mg CE/g in a methanol:water (50:50) extract. In comparison, in our research, we found a content of 740.57 ± 9.43 mg GAE/100 g, which differs significantly; this was also possibly due to the different solvents used. A study conducted by Jiménez Aguilar and Grusak in 2015 [47] on *C. longirostrata* revealed that the total phenolic compounds ranged from 2.68 ± 0.14 mg GAE/g FW to 3.38 ± 0.24. In comparison, our research found that the total phenolics were 889.66 ± 2.82 mg GAE/100 g. This differs because the result of Aguilar and Grusak's study is expressed on a fresh weight basis, and ours is expressed on a dry weight basis. Additionally, this difference in results could be attributed to variations in the extraction method employed, as well as the solvent; the previous study utilized 2.5 mL of 90% methanol at 90 °C for 2 h in a water bath, with intermittent vortexing every 30 min. On the other hand, we used hot water at 90 °C for 5 min.

Flavonoids, the largest subgroup among naturally occurring phenolic compounds, are present in various plant components, existing in their unbound state or as glycosides [43]. Flavonoids, which are widely distributed in plants, have been extensively utilized in traditional herbal remedies. They constitute essential constituents of our diet and are predominantly present in edible plant parts. Flavonoids can be found in various sources such as fruits, vegetables, grains, tree bark, stems, tea, and wine. When addressing complex chronic ailments, conventional treatment methods often rely on polypharmacy, involving the use of multiple medications. It is crucial to recognize that herbal medicines are intricate mixtures comprising diverse elements, both major and minor, and exhibiting multiple targets and processes, thus possessing complex chemical characteristics [48].

In our investigation, the total flavonoid content in *L. octovalvis* was determined to be 337.00 ± 3.61 mg of catechin equivalents per 100 g. However, another study conducted by Pandey et al. in 2023 reported a relatively lower total flavonoid content of 43.9 mg of catechin equivalents per gram. Pandey et al., 2023, employed an extraction method utilizing a mixture of ethanol and water (70:30 *v*/*v*) [49]. In the case of *C. aconitifolius*, the total flavonoid content was determined to be 346.03 ± 4.28 mg of catechin equivalents per 100 g, as indicated in a study conducted by Padilla-Camberos. Another investigation by Padilla-Camberos revealed a flavonoid content of 154.23 ± 3.35 mg of catechin equivalents per milliliter, utilizing an ethyl acetate extract [50]. The observed significant difference in results could potentially be attributed to the use of different solvents. For *C. longirostrata*, we determined that the total flavonoid content was 531.53 ± 10.07 mg of catechin equivalents per 100 g. However, Mendez-Lopez et al. (2022) reported a flavonoid content of 20.8–35.3 mg/g [51], demonstrating significant discrepancies. These variations can also be attributed to the choice of solvents, specifically hexane in this case.

Trolox, a synthetic analog of α-tocopherol that can dissolve in water, is commonly used as a standard compound to evaluate the Trolox equivalent antioxidant capacity (TEAC) of various food samples. α-tocopherol is widely recognized as the most potent form of vitamin E. The TEAC comparison is based on Trolox's ability to demonstrate antioxidant activity similar to other natural substances found in food. With its high solubility of approximately 3 mg/mL at neutral pH levels, Trolox is suitable for conducting various antioxidant assays. This allows for the measurement of the antioxidant capacity of food samples by comparing their activity to the known antioxidant capacity of Trolox. Among the most commonly employed methods, a 1,1-diphenyl-2-picrylhydrazyl (DPPH) assay is frequently performed, which involves evaluating the scavenging ability of antioxidants against the DPPH radical [52].

To validate the findings of this study, a comprehensive review of the literature was conducted to examine previous research on the utilization of 1,1-diphenyl-2-picrylhydrazyl (DPPH) in *L. octovalvis*. Although no studies utilizing water extracts were discovered, a study conducted by Yakob et al. in 2012 investigated the application of DPPH using various alcoholic extracts. The outcomes indicated that the methanol extract exhibited the highest DPPH content (1080.84 ± 6.07 μM TE/mg d.w), followed by the ethyl acetate extract (301.48 ± 8.92 μM TE/mg d.w). Conversely, the chloroform extract displayed a lower DPPH content (113.04 ± 1.58 μM TE/mg d.w), while the n-hexane extract exhibited the lowest DPPH content (91.00 ± 0.31 μM TE/mg d.w) [45]. These findings underscore the significant influence of solvent selection on the DPPH content present in the extracts. In our investigation, a DPPH value of 2.35 ± 0.06 μg ascorbic acid equivalents were determined, thus confirming the substantial impact of solvent choice on the DPPH content.

In the case of *C. aconitifolius*, an investigation conducted by Godínez-Santillán et al. in 2019 was discovered, wherein this plant and its DPPH content were examined. It was observed that *C. aconitifolius* exhibited a content of 140 ± 6 in the methanol–water (50:50) extract, while the ethanol–water (50:50) extract displayed a content of 102 ± 6 [46]. These findings diverge significantly from our own research, similar to the DPPH content of *L. octovalvis*. Our investigation revealed a DPPH content of 4.34 ± 0.12, representing a substantial difference. This discrepancy may be ascribed to the variances in methodology and the differing solvents employed (methanol, ethanol vs. warm water).

In the case of *C. longirostrata*, a comparison was conducted with a previous investigation carried out on *Crotalaria pallida* by Alam et al. in 2014. Since no specific studies were found on the same plant species, *Crotalaria pallida* was used as it belongs to the same genus [53]. Comparing the DPPH content between *Crotalaria pallida* and *C. longirostrata* reveals a significant difference in their antioxidant capacities. The study conducted on *Crotalaria pallida* reported a DPPH content of 37.60 μg/mL. In contrast, our investigation on *C. longirostrata* found a considerably lower DPPH content of 4.52 ± 0.22 μM TE/g. This discrepancy suggests that *Crotalaria pallida* exhibits a higher antioxidant capacity, as indicated by its higher DPPH content, compared to *C. longirostrata*. The variation in DPPH content could be attributed to factors such as plant species, extraction methods, or other environmental factors. Further research is needed to elucidate the underlying reasons for this divergence.

Reducing power (FRAP) assay is an efficient and cost-effective technique for directly assessing the overall antioxidant activity of electron-donating antioxidants in a given sample. It involves the reduction of ferric ions (Fe^{3+}) to ferrous ions (Fe^{2+}), which triggers a color change, serving as an indicator of the reaction. The method is known for its simplicity, rapidity, and affordability [47]. Based on our investigation we can conclude that *C. aconitifolius* has the highest FRAP value (572.41 ± 4.93 μg ascorbic acid equivalents), indicating stronger total antioxidant activity. *L. octovalvis* has a lower FRAP value (443.36 ± 2.14 μg ascorbic acid equivalents), suggesting a moderate level of antioxidant activity. Finally, *C. longirostrata* exhibits the lowest FRAP value (411.26 ± 3.92 μg ascorbic acid equivalents), indicating relatively weaker antioxidant activity compared to the other two plants.

To validate the accuracy of our obtained results, a comprehensive literature search was conducted to identify relevant studies that investigated the FRAP (Ferric Reducing Antioxidant Power) content of the same plant species. Although no studies utilizing water extracts were found, a notable study by Yakob et al. in 2012 provided insights into the FRAP content of *L. octovalvis* using various alcoholic extracts. Their findings indicated that the methanol extract exhibited the highest FRAP content (1256.88 ± 5.38 μM TE/mg d.w), followed by the ethyl acetate extract (253.45 ± 5.97 μM TE/mg d.w). In contrast, the chloroform extract displayed a relatively lower FRAP content (136.83 ± 3.48 μM TE/mg d.w), while the n-hexane extract demonstrated a similar FRAP content (142.04 ± 3.53 μM TE/mg d.w) [45]. These results strongly suggest that the choice of solvent for extraction significantly influences the FRAP content of the extracts, with methanol exhibiting the highest antioxidant capacity.

Upon comparing our investigation's results (443.36 ± 2.14 μg ascorbic acid equivalents) with the findings from the study conducted by Yakob et al. (2012), a notable discrepancy in the FRAP content of the plant extracts became evident. Our investigation reported a significantly higher FRAP content of 443.36 ± 2.14, which contrasts with the values obtained in the Yakob et al. study [45]. This difference may be attributed to various factors, such as variations in plant samples, extraction methods, or discrepancies in laboratory techniques employed. Further analysis and exploration are required to determine the specific reasons underlying this discrepancy.

In the case of *C. aconitifolius*, research conducted by Godínez-Santillán et al. in 2019 was discovered [46]. In this investigation, the FRAP content of *C. aconitifolius* was determined, revealing a content of 133 ± 1 μmol TE/g DW in the methanol–water (50:50) extract, and 86 ± 2 μmol TE/g DW in the ethanol–water (50:50) extract. However, in our own research, we found a content of 572.41 ± 4.93 μg ascorbic acid equivalents. These discrepancies can primarily be attributed to the use of water-only extraction in our study, while their investigation involved a combination with an alcohol extract, leading to potential differences in the measured quantities.

In the case of *C. longirostrata*, no FRAP content reports were available. However, in a study conducted by Govindappa et al. in 2011, the FRAP content of the water extract from *Crotalaria pallida* was reported as 1323.08 ± 0.03 μmol/L [54]. In contrast, our research on *C. longirostrata* determined a FRAP value of 411.26 ± 3.92 μg ascorbic acid equivalents. These results indicate a significant difference in the FRAP content between the two studies. Various factors could contribute to this disparity, such as variations in the specific plant species, differences in methodology, variations in the growth conditions or harvesting period, and the choice of solvent for extracting and quantifying the FRAP content.

5. Antinutrients

Tannins, which are bitter and astringent plant compounds classified as polyphenols, possess the capability to bind or cause proteins, as well as other organic substances such as amino acids and alkaloids, to form precipitates. These compounds are frequently present in food items and have been observed to inhibit the functions of various enzymes, including trypsin, chymotrypsin, Amylase, and lipase. Additionally, tannins can diminish the protein quality of foods and disrupt the absorption of dietary iron. It has been discovered that tannins can impede the digestion process by exhibiting anti-Amylase activity [55].

According to our research, *C. aconitifolius* has the highest tannin content, with 17.58 ± 0.63 mg/g. This is followed by *C. longirostrata* with 15.34 ± 0.14 mg/g, and *L. octovalvis* with the lowest tannin content of 14.13 ± 0.14 mg/g. However, it is important to note that the values we provide are approximate and include a margin of error (indicated by the ± symbol). These values suggest the average tannin content per gram of dried plant material, but variations may occur due to factors such as growing conditions, plant age, and analytical methods used.

Alkaloids are diverse chemical compounds synthesized by plants, often found as the salts of plant acids. They are present in approximately 15 to 20 percent of vascular plants and consist of small organic molecules with carbon rings and side chains, frequently containing nitrogen atoms [55]. Alkaloids have vital roles in both human medicine and the natural defense mechanisms of organisms. They contribute to about 20% of known plant secondary metabolites, providing protection against predators and regulating plant growth. Moreover, alkaloids possess therapeutic properties and are recognized for their anesthetic, cardioprotective, and anti-inflammatory effects [56]. It is important to note that alkaloids are classified as anti-nutrients as they can interfere with the nervous system, leading to disruptions or excessive enhancement of electrochemical transmission [55] From the research conducted, it was found that *C. longirostrata* has the highest alkaloid content with 2.37 ± 0.04 mg/g. *C. aconitifolius* has a higher alkaloid content compared to *L. octovalvis*, with values of 1.74 ± 0.11 mg/g and 1.12 ± 0.14 mg/g, respectively.

Saponins are a diverse group of natural compounds found in various plants that have the ability to produce foam. They are classified as secondary compounds and are widely present in the plant kingdom. Saponins are typically non-volatile and have surfactant properties. They have been found to reduce the availability of nutrients and decrease enzyme activity, affecting the digestion of proteins by inhibiting enzymes such as trypsin and chymotrypsin. However, saponins have also gained attention due to their potential positive effects on human health. Recent evidence suggests that saponins possess properties such as hypocholesterolemia (lowering cholesterol levels), immunostimulatory (boosting the immune system), and anticarcinogenic properties (preventing or inhibiting the growth of cancer cells) [55]. Thanks to our research, it was found that when comparing the saponin levels, *C. aconitifolius* has the highest saponin content at 5.49 ± 0.21 mg/g. *C. longirostrata* has a slightly lower saponin content of 4.42 ± 0.23 mg/g, while *L. octovalvis* has the lowest saponin content among the three plants at 2.26 ± 0.30 mg/g.

Oxalates are natural compounds found in certain vegetables, including spinach, chard, beets, and rhubarb. They are considered antinutrients and have strong acidic properties. Oxalates can form soluble salts with minerals such as sodium and potassium, as well as insoluble salts with calcium, iron, or zinc. Consuming foods high in oxalates has traditionally been associated with an increased risk of developing kidney stones, based on human studies. When assessing oxalate content in foods, it is important to recognize that soluble oxalates have a greater impact on bioavailability and the risk of stone formation compared to insoluble oxalates [36]. Comparing the oxalate levels obtained, *C. aconitifolius* has the highest oxalate content at 8.91 ± 0.48 mg/g. *L. octovalvis* has a lower oxalate content of 4.41 ± 0.22 mg/g, while *C. longirostrata* has the lowest oxalate content among the three plants at 3.32 ± 0.12 mg/g.

5.1. α-Amylase Inhibition

α-Amylase is an essential enzyme synthesized by the pancreas and salivary glands that plays a vital role in the digestion of starch and glycogen. It is found in various organisms, including microorganisms, plants, and higher organisms. This enzyme is responsible for the initial breakdown of starch into smaller oligosaccharides, such as maltose, maltotriose, and branched α-(1–6) and α-(1–4) oligoglucans. These smaller molecules are subsequently further degraded into glucose by α-Glucosidases, facilitating their absorption into the bloodstream. By inhibiting the activity of α-Amylase, the increase in postprandial glucose levels can be controlled by slowing down the hydrolysis and absorption of carbohydrates. Furthermore, the antioxidant properties of plants are often highlighted for their beneficial effects in preventing diabetes and other chronic diseases [57].

A literature review was conducted to explore previous studies on α-Amylase inhibition in *L. octovalvis*. One study by Binh et al., 2016, was identified, in which they investigated α-Amylase inhibition but did not observe any inhibitory effects, despite conducting tests with ethanol and water for the plant extraction [4]. In contrast, our own study found a significant inhibition of 519.40. These disparities could be attributed to variations in the experimental methodologies employed in both studies, as well as the inherent variability in the composition of the parts of the plant used. Factors such as harvesting season, and geographical region of cultivation can influence the chemical profile of the plant, thus potentially impacting its α-Amylase inhibitory activity.

C. aconitifolius was studied by Ramos-Gómez, where an inhibition of 0.094 IC_{50} in mg/mL was found for the α-Amylase [58]. However, in our research, we found an IC_{50} of 471.39 mg/mL, which can be attributed to differences in the methodology employed. Both studies used the same water extraction method, but they differed in the boiling time. While Ramos-Gómez boiled the sample for 20 min, we only performed a 5 min infusion.

Due to *C. longirostrata* being a poorly studied plant, there is limited information available regarding studies on its enzymatic inhibition. However, a comparison was made with research conducted by Anwar et al. in 2022 [59], where a plant called *Crotalaria burhia* Buch.-Ham, belonging to the same genus as *C. longirostrata*, was studied. In this study, the inhibitory activity of α-Amylase was measured, and it was found that with methanol extracts, an IC_{50} inhibition of 0.63 mmol/g was observed. This may suggest that the enzymatic inhibition is different to that found in our study, where we found that *C. longirostrata* exhibited an IC_{50} inhibition of 585.12 mg/mL. Although the studied plants are not the same, it can help guide us in assessing the accuracy of our results because both plants belong to the same genus. However, there is no available information with which we can directly compare the results obtained in our research.

5.2. α-Glucosidase Inhibition

Slowing down the increase in blood sugar after consuming carbohydrates can be achieved by inhibiting a specific enzyme. This enzyme is located in the membrane of the small intestine's epithelium. Its main function is to assist in the absorption of glucose by catalyzing the breakdown of oligosaccharides into easily absorbable monosaccharides.

When α-Glucosidase in the intestine is inhibited, the hydrolysis of oligosaccharides is reduced, causing the process of carbohydrate digestion to occur in the lower portion of the small intestine. As a result, the overall absorption rate of glucose into the bloodstream is delayed. This approach has proven to be highly effective in reducing the rise in blood glucose levels after a meal and can help prevent the development of complications associated with diabetes [57].

Based on these values, it can be observed that *C. aconitifolius* has the lowest IC_{50} value (433.78 mg/mL), indicating a stronger inhibition of α-Glucosidase activity. *L. octovalvis* has a slightly higher IC_{50} value (480.50 mg/mL), suggesting a moderate level of α-Glucosidase inhibition. On the other hand, *C. longirostrata* (579.71 mg/mL) and Acarbose (571.61 mg/mL) have higher IC_{50} values, indicating weaker inhibition compared to the other samples. It is important to note that these values represent the inhibitory concentration required, with lower values indicating stronger inhibition.

Regarding *L. octovalvis*, Ramirez et al. (2012) found an inhibition with an IC_{50} value of 0.202 mg/mL [10], while Morales et al. (2018) found an inhibition with an IC_{50} value of 0.700 mg/mL [14]. In our study, we found an inhibition with an IC_{50} value of 480.50 mg/mL. These differences are attributed to the use of different hydroalcoholic extracts, as we did not use alcohol in our research. *C. aconitifolius* was studied by Ramos-Gómez, where an inhibition of 0.089 ± 0.7 IC_{50} in mg/mL was found for the α-Glucosidase [58]. However, in our research, we found an IC_{50} of 433.78 mg/mL, which can be attributed to differences in the methodologies employed. Both studies used the same water extraction method, but they differed in the boiling time. While Ramos-Gómez boiled the sample for 20 min, whereas we only performed a 5 min infusion.

To make a comparison in the literature on *C. longirostrata*, a study conducted by Sut et al. in 2020 was used [60]. In this investigation, *Crotalaria retusa* and its inhibitory power were studied. The comparison was made with this plant because it belongs to the same genus as *C. longirostrata*. The study concluded that α-Glucosidase inhibition was not active, which differs from our findings. In our research, we found an inhibition of 579.71 mg/mL. This difference may be due to the fact that it is not exactly the same plant. Although they belong to the same genus, their chemical composition is different, which may affect α-Glucosidase inhibition and explain the contrasting results.

5.3. Lipase Inhibition

By binding to the active part of lipase in the stomach and small intestine, lipase inhibitors alter the structure of both stomach and trypsin enzymes, leading to the inhibition of their catalytic activity. Consequently, the breakdown of lipids, such as triglycerides, is diminished. This reduction in hydrolysis hampers the digestion and absorption of dietary lipids, ultimately resulting in decreased accumulation of adipose tissue. These effects contribute to the control and treatment of obesity [61].

L. octovalvis was studied by Morales et al. in 2018, and the dry extract displayed an IC_{50} value of 0.480 mg/mL [14]. In our study it was found that *L. octovalvis* has an IC_{50} value of 697.17 mg/mL, and this difference can be attributed to the extraction method and the solution, which was 60% ethanol. It is also important to mention that these differences can be due to the parts of the plant used or the plant itself, which is related to the harvest season and its growing conditions.

C. aconitifolius was studied by Ramos-Gómez, where an inhibition of 0.065 ± 0.000 IC_{50} in mg/mL was found [58]. However, in our research, we found an IC_{50} of 627.84 mg/mL, which can be attributed to differences in the methodology employed. Both studies used an aqueous extraction, but they differed in the boiling time. While Ramos-Gómez boiled the sample for 20 min, we conducted a 5 min infusion. Due to the lack of information on *C. longirostrata*, a comparison will be made with *Crotalaria pumila* Ort., which belongs to the same Fabaceae family. In an investigation conducted by Villa-Ruano et al. in 2013, it was found that *Crotalaria pumila* has an IC_{50} inhibition of 0.075 ± 0.002 μg/mL [61]. In order to determine the lipase inhibition, an ethanol extraction was performed for 15 days. This differs from our research because we found that *C. longirostrata* has an IC_{50} inhibition of 677.27 mg/mL. This difference could be due to the applied methodology as well as the type of extraction used.

5.4. Pepsin Inhibition

Pepsin, which is an enzyme categorized as an aspartic protease [7], facilitates the breakdown of proteins and peptides by releasing amino acids and small peptides. It is predominantly found in significant amounts in the digestive tracts of humans. Certain protease inhibitors derived from plants have the potential to hinder the digestive process of dietary proteins in the human gastrointestinal system by inhibiting trypsin and chymotrypsin. However, scientific evidence also suggests that these inhibitors offer favorable effects in the gut, including anti-inflammatory and chemo-preventive actions, both in laboratory studies and in living organisms [8]. Considering that the inhibition of pepsin has not been thoroughly studied, there are not many studies in the literature to compare the plants studied in this paper. Therefore, a comparison between them will be made based on the inhibition found in the IC_{50} values. The values found represent the concentration of a substance required to inhibit the activity of the protease enzyme by 50%. The lower the IC_{50} value, the more potent the inhibitory effect of the sample. Based on the given data, the protease inhibitor cocktail exhibits the most significant inhibitory effect with a negative IC_{50} value (−173.19 mg/mL), suggesting it has a highly potent inhibitory activity against protease enzymes. Among the plant samples, *C. longirostrata* has the lowest IC_{50} value (823.02 mg/mL), indicating that it also possesses a relatively strong inhibitory effect. *L. octovalvis* (910.94 mg/mL) and *C. aconitifolius* (1039.93 mg/mL) have higher IC_{50} values, suggesting they have relatively weaker inhibitory activity compared to the other samples.

5.5. FTIR

With the FTIR spectrum, it was found that the most prominent peaks were the same for all three samples, only with different intensities. The most prominent peaks were observed around 650, 1000, and 1650 cm^{-1}. This indicates that the composition is similar in all three plants, but the quantity differs. Therefore, a literature search was conducted to identify the major compounds present.

For *C. longirostrata*, the most prominent peak is located at approximately 1650 cm^{-1}, followed by a peak at approximately 1600 cm^{-1}. According to a study published by Nandiyanto et al. in 2019, this may indicate the presence of an ether and oxy compound, specifically a primary amine NH bend, which typically occurs in the 1650–1590 wavenumber (cm^{-1}) range. The second highest peak is observed at 1070 cm^{-1}, which, according to the same article, suggests the possibility of an alcohol and hydroxy compound, specifically a primary alcohol C-O stretch, as these compounds typically exhibit a wavenumber of approximately 1050 cm^{-1} [62]. The third most prominent peak, which exhibits a broad shape, reaches its highest point at approximately 640 cm^{-1}. According to Coates in 2000, this peak may be attributed to an alkyne C-H bend, as alkyne compounds typically exhibit a wavenumber around 630 cm^{-1} [63]. There is also a broad peak observed between 3000 and 3650 according to the literature, which could correspond to a hydrogen-bonded alcohol OH stretch. This is supported by the wavenumber range of 3550–3200 and the broad nature of the peak [64].

For *C. aconitifolius* and *L. octovalvis*, we observed that the peaks occur in the same locations as *C. longirostrata*, leading us to conclude that they belong to the same functional groups. However, unlike *C. longirostrata*, they also exhibit two weak peaks around 2850–2900, which could be attributed to the presence of C-H stretch alkenes. These alkenes are typically found in the range of 2990–2850 [64].

5.6. Statistical Analysis

The primary aim of this analysis was to explore the correlation between the compounds under investigation and enzyme inhibition in plants. The results and interpretation of the dendrogram indicate an initial association between the content of free amino nitrogen and the inhibition of α-Amylase and α-Glucosidase. This suggests that higher levels of free amino nitrogen may lead to greater inhibition of these two enzymes. However, this relationship might not be direct due to the potential involvement of other factors. The observed effects on free amino nitrogen content would likely be secondary effects resulting from changes in nutrient availability or metabolic responses, rather than a direct consequence of enzyme inhibition.

Free amino nitrogen (FAN) refers to nitrogen-containing molecules readily available in the system, such as amino acids, peptides, and ammonium ions [41]. It is primarily derived from protein hydrolysis. The clustering analysis reveals that proteins and FAN are grouped in the same cluster, suggesting a possible association between these two variables; since FAN is predominantly derived from protein hydrolysis this is a reasonable relationship [65]. A more direct relationship may be observed with the content of flavonoids, which are widely occurring polyphenolic compounds [66]. Polyphenols are known to hinder glucose absorption by inhibiting α-Amylase and α-Glucosidase, which are crucial enzymes in carbohydrate digestion [67]. This association can be observed in the first cluster of the dendrogram.

In the second cluster, a relationship between lipase inhibition and fiber content can be observed. Based on the literature, there have only been a limited number of studies investigating the mechanism of how dietary fiber regulates lipid metabolism, with a particular focus on its impact on pancreatic lipase (PL) activity, which is the primary lipolytic enzyme synthesized by the pancreas. One such study conducted by Yu et al. in 2023 revealed that insoluble dietary fiber from citrus peels (CIDF) exhibited both adsorption and inhibitory effects on PL activity [68]. Another study demonstrated that soluble dietary fiber (SDF) also had a noticeable effect on PL activity. Specifically, SDF derived from *K. alvarezii* and *E. denticulatum* showed a reduction in PL activity by 60% and 57%, respectively [69].

In the third cluster, a noticeable relationship can be observed regarding the inhibition of pepsin in connection with the levels of oxalates and FRAP. As we understand it, the FRAP content represents the overall antioxidant activity of electron-donating antioxidants present in a given sample. A higher quantity of antioxidants in a sample corresponds to a greater FRAP content [70]. A study conducted by Salma et al. in 2021 demonstrated that the extract of *E. alte* possesses inhibitory effects on pepsin enzyme, as well as antibacterial and antioxidant activities. However, this alone does not suffice to conclude the existence of a direct relationship between antioxidant content and pepsin inhibition [71]. Our current investigation allows us to potentially confirm this relationship.

Regarding the content of oxalates and enzymatic inhibition, a correlation could plausibly exist, despite the absence of reported evidence in the literature suggesting a direct association. This potential relationship might stem from the involvement of other factors which have yet to be documented.

Based on the information presented earlier and the literature review, this research has the potential to be further developed in the future. Investigating the synergistic mechanisms of herbal ingredients not only has the potential to assist researchers in discovering new phytomedicines or drug combinations but also has the potential to mitigate the risk of negative synergies. Additional clinical research will be essential to validate the reported drug combinations and the elucidated synergistic mechanisms [72]. It is also important to highlight that there are Glycosidase inhibition studies in which it has been found that iminosugars with the gauche–gauche side chain conformations exhibit 6- to 10-fold greater potency compared to isosteric compounds with the gauche–trans conformation. A manno-configured iminosugar with the gauche–gauche conformation is a 27-fold more effective inhibitor than 1-deoxymannojirimycin, which would be worth exploring in future investigations [73].

6. Conclusions

The results obtained from the present study suggest the positive therapeutic effects of *L. octovalvis* (Jacq.) P.H.Raven, *C. aconitifolius*, and *C. longirostrata*. The analysis conducted revealed that all three plants exhibited enzymatic inhibition. Upon comparing the plants, it was determined that *C. aconitifolius* had the lowest concentration required for a 50% inhibition in α-Amylase, α-Glucosidase, and lipase, as indicated by the IC_{50} values. In the case of pepsin, *C. longirostrata* demonstrated the lowest IC_{50} value. After analyzing the components of each plant, it can be concluded that there is a correlation between certain bioactive components and enzyme inhibition. Based on our findings, it was determined that certain Polyphenols, particularly flavonoids, possess the ability to impede the activity of α-Amylase and α-Glucosidase, pivotal enzymes involved in the breakdown of carbohydrates. Furthermore, a correlation between lipase inhibition and dietary fiber content was established. Lastly, our study provides support for the presence of a direct association between antioxidant levels and the inhibition of pepsin enzyme.

Author Contributions: Conceptualization, methodology, software, validation, formal analysis, resources, data curation, visualization, supervision, project administration, J.D.L.R.-M.; Conceptualization, software, investigation, writing—original draft preparation, writing—review and editing, K.C. All authors have read and agreed to the published version of the manuscript.

Funding: This research received no external funding.

Data Availability Statement: The data used to support the findings of this study can be made available by the corresponding author upon request. Additional graphics from enzyme inhibition are shown in the Appendix A.

Acknowledgments: Kimberly Calónico acknowledges the support of the doctoral grant (CVU No. 992151) from the Consejo Nacional de Ciencia y Tecnología (Conacyt), as well as Julián de la Rosa for his guidance throughout the research.

Conflicts of Interest: The authors declare no conflict of interest.

Appendix A

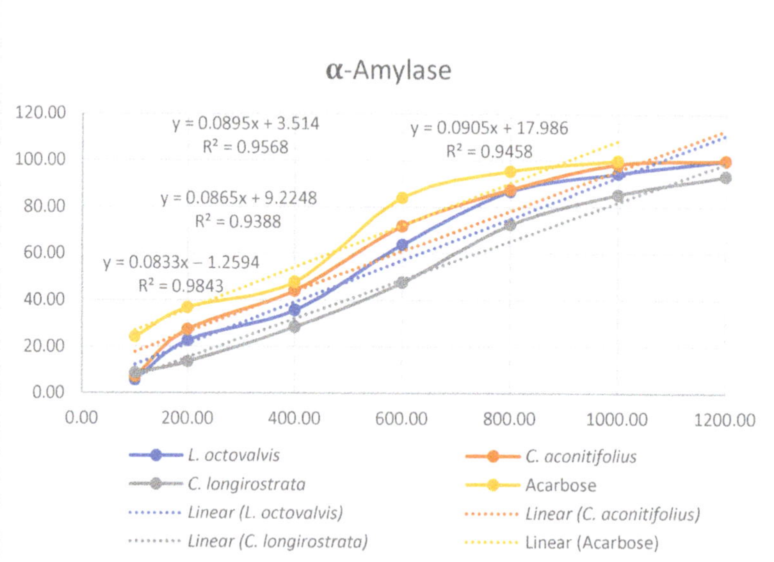

Figure A1. The α-Amylase inhibition of *L. octovalvis*, *C. aconitifolius*, and *C. longirostrata* plants expressed as a percentage of dry matter. The data is presented in line charts with trendlines.

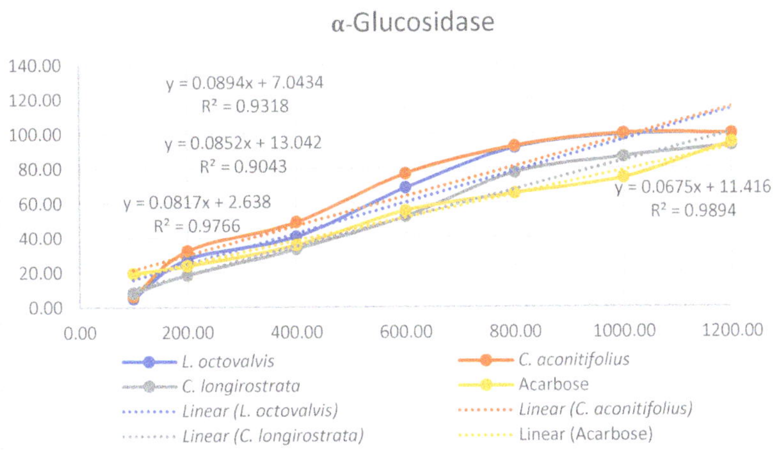

Figure A2. The α-Glucosidase inhibition of *L. octovalvis*, *C. aconitifolius*, and *C. longirostrata* plants expressed as a percentage of dry matter. The data is presented in line charts with trendlines.

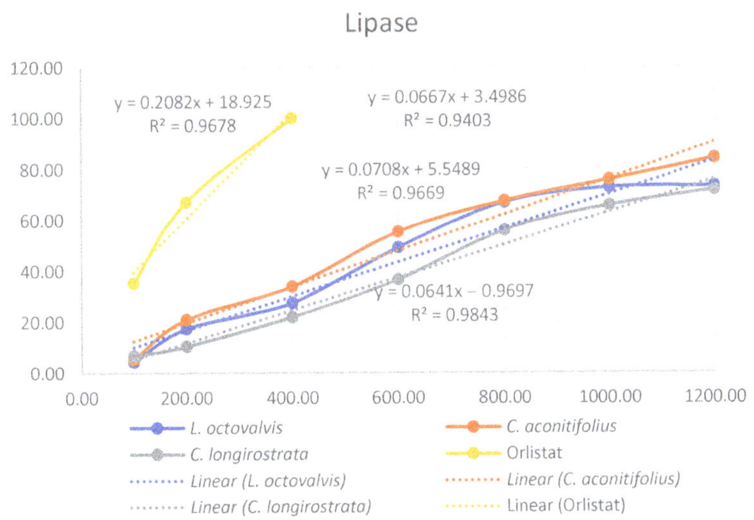

Figure A3. The lipase inhibition of *L. octovalvis*, *C. aconitifolius*, and *C. longirostrata* plants expressed as a percentage of dry matter. The data is presented in line charts with trendlines.

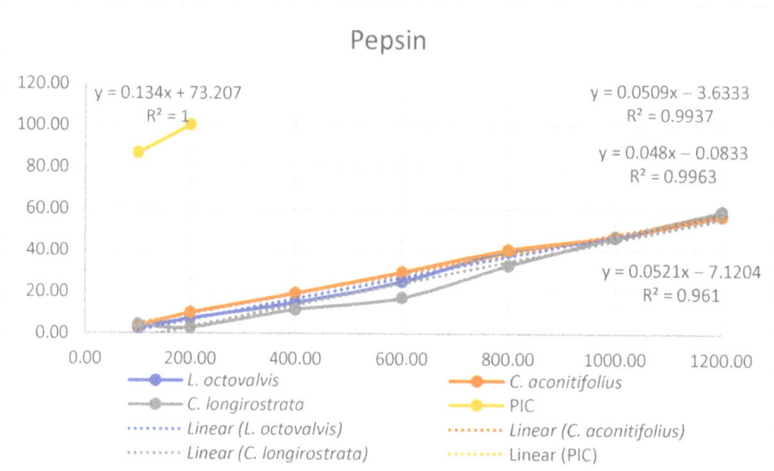

Figure A4. The pepsin inhibition of *L. octovalvis*, *C. aconitifolius*, and *C. longirostrata* plants expressed as a percentage of dry matter. The data is presented in line charts with trendlines.

Figure A5. *L. octovalvis* in its reproductive stage, the inflorescence begins to open, and the flowers completely extend.

References

1. Senturk, M. *Enzyme Inhibitors and Activators*; IntechOpen: Rijeka, Croatia, 2017. [CrossRef]
2. Gobierno de México. Plantas Medicinales: Nuestro Patrimonio. Available online: https://www.gob.mx/bienestar/sembrandovida/articulos/plantas-medicinales-nuestro-patrimonio (accessed on 13 April 2023).
3. Lankatillake, C.; Luo, S.; Flavel, M.; Lenon, G.B.; Gill, H.; Huynh, T.; Dias, D.A. Screening natural product extracts for potential enzyme inhibitors: Protocols, and the standardisation of the usage of blanks in α-Amylase, α-Glucosidase and lipase assays. *Plant Methods* **2021**, *17*, 3. [CrossRef] [PubMed]
4. Trinh, B.T.D.; Staerk, D.; Jäger, A.K. Screening for potential α-Glucosidase and α-Amylase inhibitory constituents from selected Vietnamese plants used to treat type 2 diabetes. *J. Ethnopharmacol.* **2016**, *186*, 189–195. [CrossRef] [PubMed]
5. Tang, Y.; Zhang, B.; Li, X.; Chen, P.X.; Zhang, H.; Liu, R.; Tsao, R. Bound Phenolics of Quinoa Seeds Released by Acid, Alkaline, and Enzymatic Treatments and Their Antioxidant and α-Glucosidase and Pancreatic Lipase Inhibitory Effects. *J. Agric. Food Chem.* **2016**, *64*, 1712–1719. [CrossRef]
6. Seyedan, A.; Alshawsh, M.A.; Alshagga, M.A.; Koosha, S.; Mohamed, Z. Medicinal Plants and Their Inhibitory Activities against Pancreatic Lipase: A Review. In *Evidence-Based Complementary and Alternative Medicine;* Hindawi Publishing Corporation: London, UK, 2015; Volume 2015. [CrossRef]
7. Moradi, S.; Ahmadi, P.; Karami, C.; Farhadian, N.; Balaei, F.; Ansari, M.; Shahlaei, M. Evaluation of the effects of isoniazid and rifampin on the structure and activity of pepsin enzyme by multi spectroscopy and molecular modeling methods. *Spectrochim. Acta A Mol. Biomol. Spectrosc.* **2021**, *253*, 119523. [CrossRef] [PubMed]
8. Kårlund, A.; Paukkonen, I.; Gómez-Gallego, C.; Kolehmainen, M. Intestinal exposure to food-derived protease inhibitors: Digestion physiology-and gut health-related effects. *Healthcare* **2021**, *9*, 1002. [CrossRef]
9. Mata, R.; Cristians, S.; Escandón-Rivera, S.; Juárez-Reyes, K.; Rivero-Cruz, I. Mexican antidiabetic herbs: Valuable sources of inhibitors of α-Glucosidases. *J. Nat. Prod.* **2013**, *76*, 468–483. [CrossRef]
10. Ramírez, G.; Zavala, M.; Pérez, J.; Zamilpa, A. In vitro screening of medicinal plants used in Mexico as antidiabetics with Glucosidase and lipase inhibitory activities. *Evid.-Based Complement. Altern. Med.* **2012**, *2015*, 701261. [CrossRef]
11. Lin, W.S.; Lo, J.H.; Yang, J.H.; Wang, H.W.; Fan, S.Z.; Yen, J.H.; Wang, P.Y. *L. octovalvis* extract improves glycemic control and memory performance in diabetic mice. *J. Eth-Nopharmacol.* **2017**, *207*, 211–219. [CrossRef]
12. A. Awoyinka and Ogunnowo, "Phytochemical screening and in vitro bioactivity of *C. aconitifolius* (Euphorbia-ceae). *J. Med. Plants Res.* **2007**, *1*, 63–065.
13. Peña-Valdivia, C.; Salinas Morales, J.; Vázquez Sánchez, M.; Trejo, C.; López Palacios, C.; Padilla Chacón, D.D. Componentes del rendimiento de *C. longirostrata* Hook. & Arn. en Guerrero, México. *Polibotanica* **2022**, *1*, 101–121. [CrossRef]
14. Morales, D.; Ramirez, G.; Herrera-Arellano, A.; Tortoriello, J.; Zavala, M.; Zamilpa, A. Identification of Digestive Enzyme Inhibitors from *L. octovalvis* (Jacq.) P.H.Raven. *Evid.-Based Complement. Altern. Med.* **2018**, *2018*, 8781352. [CrossRef] [PubMed]
15. AACC. *Approved Methods of Analysis*, 11th ed.; American Association of Cereal Chemists: St. Paul, MN, USA, 2000.
16. AOAC. *Official Methods of Analysis*, 17th ed.; The Association of Official Analytical Chemists: Gaithersburg, MD, USA, 2000.
17. Karkacier, M.; Ebras, M.; Uslu, M.; Aksu, M. Comparison of different extraction and detection methods for sugars using amino-bonded phase HPLC. *J. Chromatogr. Sci.* **2011**, *41*, 331–333. [CrossRef] [PubMed]
18. Lie, S. The EBC-Ninhydrin method for determination of free Alpha amino nitrogen. *J. Inst. Brew.* **1973**, *79*, 37–41. [CrossRef]
19. Cabello-Pasini, A.; Macías-Carranza, V.; Abdala, R.; Korbee, N.; Figueroa, F.L. Effect of nitrate concentration and UVR on photosynthesis, respiration, nitrate reductase activity, and phenolic compounds in Ulva rigida (Chlorophyta). *J. Appl. Phycol.* **2011**, *23*, 363–369. [CrossRef]
20. Jia, Z.; Tang, M.; Wu, J. The determination of flavonoid contents in mulberry and their scavenging effects on superoxide radicals. *Food Chem.* **1999**, *64*, 555–559.
21. García, L.; Verde, J.S.; Castro, R.R.; Chávez, M.A.; Oranday, C.A.; Núñez, G.A.; Rivas, M.C. Biological activity of mexican grape pomace extract. *Rev. Mex. Cienc. Farm.* **2010**, *41*, 28–36.
22. Juntachote, T.; Berghofer, E.; Siebenhandl, S.; Bauer, F. The antioxidative properties of Holy basil and Galangal in cooked ground pork. *Meat Sci.* **2006**, *72*, 446–456. [CrossRef]
23. Terpinc, P.; Polak, T.; Makuc, D.; Ulrih, N.P.; Abramovič, H. The occurrence and characterization of phenolic compounds in *Camelina sativa* seed, cake and oil. *Food Chem.* **2012**, *131*, 580–589. [CrossRef]
24. Oyaizu, M. Studies on product of browning reaction prepared from glucose amine. *JPN J. Nutr.* **1986**, *44*, 307–315. [CrossRef]
25. Libert, B.; Franceschi, V.R. Oxalate in crop plants. *J. Agric. Food Chem.* **1987**, *35*, 926–938. [CrossRef]
26. Griffiths, D.W.; Jones, D.I. Cellulase inhibition by tannins in the testa of field beans (*Vicia faba*). *J. Sci. Food Agric.* **1977**, *28*, 938–989. [CrossRef] [PubMed]
27. Makkar, H.P.S.; Becker, K. Nutritional value and antinutritional components of whole and ethanol extracted *Moringa oleifera* leaves. *Anim. Feed Sci. Technol.* **1996**, *63*, 211–228. [CrossRef]
28. Shamsa, F.; Monsef, H.; Ghamooshi, R.; Verdian-Rizi, M. Spectrophotometric determination of total alkaloids in some Iranian medicinal plants. *Thai J. Pharm. Sci.* **2008**, *32*, 17–20.
29. Makkar, H.P.S.; Becker, K. Effect of Quillaja Saponins on in Vitro Rumen Fermentation. In *Saponins Used in Food and Agriculture. Advances in Experimental Medicine and Biology*; Waller, G.R., Yamasaki, K., Eds.; Springer: Boston, MA, USA, 1996; Volume 405. [CrossRef]

30. Miller, G.L. Use of Dinitrosalicylic Acid Reagent for Determination of Reducing Sugar. *J. Anal. Chem.* **1959**, *31*, 426–428. [CrossRef]
31. Mayur, B.; Sandesh, S.; Shruti, S.; Sung-Yum, S. Antioxidant and α-Glucosidase inhibitory properties of *Carpesium abrotanoides* L. *J. Med. Plants Res.* **2010**, *4*, 1547–1553. [CrossRef]
32. Jaradat, N.; Khasati, A.; Hawi, M.; Hawash, M.; Shekfeh, S.; Qneibi, M.; Eid, A.M.; Arar, M.; Qaoud, M.T. Antidiabetic, antioxidant, and anti-obesity effects of phenylthio-ethyl benzoate derivatives, and molecular docking study regarding α-Amylase enzyme. *Sci. Rep.* **2022**, *12*, 3108. [CrossRef]
33. De La Rosa-Millán, J. Physicochemical, Molecular, and Digestion Characteristics of Annealed and Heat–Moisture Treated Starches Under Acidic, Neutral, or Alkaline pH. *Cereal Chem.* **2017**, *94*, 770–779. [CrossRef]
34. Chávez-Murillo, C.; Veyna-Torres, J.; Cavazos-Tamez, L.; de la Rosa-Millán, J.; Serna-Saldívar, S. Physicochemical charac-teristics, ATR-FTIR molecular interactions and in vitro starch and protein digestion of thermally-treated whole pulse flours. *Food Res. Int.* **2018**, *105*, 371–383. [CrossRef]
35. Popova, A.; Mihaylova, D. Antinutrients in Plant-based Foods: A Review. *Open Biotechnol. J.* **2019**, *13*, 68–76. [CrossRef]
36. López-Moreno, M.; Garcés-Rimón, M.; Miguel, M. Antinutrients: Lectins, goitrogens, phytates and oxalates, friends or foe? *J. Funct. Foods* **2022**, *89*, 104938. [CrossRef]
37. Oyagbemi, A.A.; Odetola, A.A.; Azeez, O.I. Phytochemical Investigation and Proximate Analysis on the Leaves of *Cnidoscolus aconitifolius*. *J. Med. Food* **2011**, *14*, 322–324. [CrossRef] [PubMed]
38. Alalade, J.A.; Akinlade, J.A.; Akingbade, A.A.; Emiola, C.B.; Adebisi, I.A. Proximate Composition and Phytochemical Screenings of *Crotalaria retusa* Leaves and Seeds. *Open Access Libr. J.* **2019**, *6*, 1–9. [CrossRef]
39. Oku, M.; Inoue, C.; Hieda, S.; Noma, N.; Nakagawa, T. Chemical composition and in vitro ruminal digestibility of *Ludwigia grandiflora*. *Anim. Sci. J.* **2021**, *92*, e13509. [CrossRef] [PubMed]
40. Kaczorová, D.; Karalija, E.; Dahija, S.; Bešta-Gajević, R.; Parić, A.; Zeljković, S.Ć. Influence of extraction solvent on the phenolic profile and bioactivity of two achillea species. *Molecules* **2021**, *26*, 1601. [CrossRef] [PubMed]
41. Shopska, V.; Denkova, R.; Lyubenova, V.; Kostov, G. Kinetic characteristics of alcohol fermentation in brewing: State of art and control of the fermentation process. *Fermented Beverages* **2019**, *5*, 529–575. [CrossRef]
42. Guo, N.; Zhang, S.; Gu, M.; Xu, G. Function, transport, and regulation of amino acids: What is missing in rice? *Crop J.* **2021**, *9*, 530–542. [CrossRef]
43. Sulaiman, C.T.; Balachandran, I. Total phenolics and total flavonoids in selected Indian medicinal plants. *Indian J. Pharm. Sci.* **2012**, *74*, 258–260. [CrossRef]
44. Aryal, S.; Baniya, M.K.; Danekhu, K.; Kunwar, P.; Gurung, R.; Koirala, N. Total Phenolic content, Flavonoid content and antioxidant potential of wild vegetables from western Nepal. *Plants* **2019**, *8*, 96. [CrossRef]
45. Al-Salamn, H.K.Y.; Sulaiman, S.F.; Uyub, A.M. Antioxidant and Antibacterial Activity of *L. octovalvis* on *Escherichia coli* O157:H7 and Some Pathogenic Bacteria. *World Appl. Sci. J.* **2012**, *16*, 22–29.
46. Godínez-Santillán, R.I.; Chávez-Servín, J.L.; García-Gasca, T.; Guzmán-Maldonado, S.H. Caracterización fenólica y capacidad antioxidante de extractos alcohólicos de hojas crudas y hervidas de *C. aconitifolius* (Euphorbia-ceae). *Acta Bot. Mex.* **2019**, *126*, 1–15. [CrossRef]
47. Jiménez-Aguilar, D.M.; Grusak, M.A. Evaluation of Minerals, Phytochemical Compounds and Antioxidant Activity of Mexican, Central American, and African Green Leafy Vegetables. *Plant Foods Hum. Nutr.* **2015**, *70*, 357–364. [CrossRef] [PubMed]
48. Roy, A.; Khan, A.; Ahmad, I.; Alghamdi, S.; Rajab, B.S.; Babalghith, A.O.; Alshahrani, M.Y.; Islam, S.; Islam, M.R. Flavonoids a Bioactive Compound from Medicinal Plants and Its Therapeutic Applications. *BioMed Res. Int.* **2022**, *2022*, 5445291. [CrossRef] [PubMed]
49. Pandey, S.; Kumar, P.; Niranjan, A.; Verma, H.; Kumar, M. Targeting biosignatures of hyperglycemia and oxidative stress in diabetes comorbid depressive rats: Effectiveness of hydroethanolic extract of the whole plant of *Ludwigia octovalvis*. *Tradit. Med. Res.* **2023**, *8*, 9. [CrossRef]
50. Padilla-Camberos, E.; Torres-Gonzalez, O.R.; Sanchez-Hernandez, I.M.; Diaz-Martinez, N.E.; Hernandez-Perez, O.R.; Flores-Fernandez, J.M. Anti-inflammatory activity of *Cnidoscolus aconitifolius* (Mill.) ethyl acetate extract on croton oil-induced mouse ear edema. *Appl. Sci.* **2021**, *11*, 9697. [CrossRef]
51. Mendez-Lopez, A.Y.; del Carmen Lagunes-Espinoza, L.; González-Esquinca, A.R.; Hernández-Nataren, E.; Ortiz-García, C.F. Phenological characterization of chipilín (*C. longirostrata* Hook. & Arn.) and relationship between the phenological stage and chemical composition of leaves. *South Afr. J. Bot.* **2023**, *154*, 140–148. [CrossRef]
52. Frangu, A.; Ashrafi, A.M.; Sýs, M.; Arbneshi, T.; Metelka, R.; Adam, V.; Vlček, M.; Richtera, L. Determination of trolox equivalent antioxidant capacity in berries using amperometric tyrosinase biosensor based on multi-walled carbon nanotubes. *Appl. Sci.* **2020**, *10*, 2497. [CrossRef]
53. Alam, M.H.; Rahman, M.M.; Sathi, S.I.; Bellah, S.F.; Mazid, M.A.; Murshid, G.M. Assessment of Antioxidant, Antibacterial and Anthelmintic Activities of Ethanol Extract of Leaves of *Crotalaria pallida* (Aiton). *Int. J. Pharmacogn.* **2014**, *1*, 438–444. [CrossRef]
54. Govindappa, M.; Bharath, N.; Shruthi, H.B.; Sadananda, T.S.; Sharanappa, P. Antimicrobial, antioxidant and in vitro anti-inflammatory activity and phytochemical screening of *Crotalaria pallida* Aiton. *Afr. J. Pharm. Pharmacol.* **2011**, *5*, 2359–2371. [CrossRef]
55. Thakur, A.; Sharma, V.; Thakur, A.; Sharma, C.V. An overview of anti-nutritional factors in food. *Int. J. Chem. Stud.* **2019**, *7*, 2472–2479.

56. Heinrich, M.; Mah, J.; Amirkia, V. Alkaloids used as medicines: Structural phytochemistry meets biodiversity—An update and forward look. *Molecules* **2021**, *26*, 1836. [CrossRef]
57. Kamtekar, S.; Keer, V.; Patil, V. Estimation of phenolic content, flavonoid content, antioxidant and alpha amylase inhibitory activity of marketed polyherbal formulation. *J. Appl. Pharm. Sci.* **2014**, *4*, 61–65. [CrossRef]
58. Ramos-Gomez, M.; Figueroa-Pérez, M.G.; Guzman-Maldonado, H.; Loarca-Piña, G.; Mendoza, S.; Quezada-Tristán, T.; Reynoso-Camacho, R. Phytochemical Profile, Antioxidant Properties and Hypoglycemic Effect of Chaya (*Cnidoscolus Chayamansa*) in STZ-Induced Diabetic Rats. *J. Food Biochem.* **2017**, *41*, e12281. [CrossRef]
59. Anwar, S.; Faisal Nadeem, M.; Pervaiz, I.; Khurshid, U.; Akmal, N.; Aamir, K.; Haseeb ur Rehman, M.; Almansour, K.; Alshammari, F.; Shaikh, M.F.; et al. A comprehensive phytochemical, biological, and toxicological studies of roots and aerial parts of *Crotalaria burhia* Buch.-Ham: An important medicinal plant. *Front. Plant Sci.* **2022**, *13*, 988352. [CrossRef] [PubMed]
60. Sut, S.; Dall'Acqua, S.; Sinan, K.I.; Bene, K.; Kumar, G.; Mahomoodally, M.F.; Picot-Allain, C.; Zengin, G. *Cola caricifolia* (G.Don) K. Schum and *Crotalaria letusa* L. from Ivory Coast as sources of bioactive constituents. *Ind. Crops Prod.* **2020**, *147*, 112246. [CrossRef]
61. Villa-Ruano, N.; Zurita-Vásquez, G.; Pacheco-Hernández, Y.; Betancourt-Jiménez, M.; Cruz-Durán, R.; Duque-Bautista, H. Anti-lipase and antioxidant properties of 30 medicinal plants used in Oaxaca, México. *Biol. Res.* **2013**, *46*, 153–160. [CrossRef]
62. Nandiyanto, A.B.D.; Oktiani, R.; Ragadhita, R. How to read and interpret ftir spectroscope of organic material. *Indones. J. Sci. Technol.* **2019**, *4*, 97–118. [CrossRef]
63. Coates, J. *Interpretation of Infrared Spectra, A Practical Approach*; Wiley: Hoboken, NJ, USA, 2006.
64. Mohring, J.; Noring, C.; Schatz, P. *Techniques in Organic Chemistry*; W. H. Freeman: New York, NY, USA, 2010.
65. Hill, A.E.; Stewart, G.G. Free amino nitrogen in brewing. *Fermentation* **2019**, *5*, 22. [CrossRef]
66. Mutha, R.E.; Tatiya, A.U.; Surana, S.J. Flavonoids as natural phenolic compounds and their role in therapeutics: An 1231 overview. *Futur. J. Pharm. Sci.* **2021**, *7*, 25. [CrossRef]
67. Ćorković, I.; Gas, D.; Pichler, A.; Šimunović, J.; Kopjar, M. Dietary Polyphenols as Natural Inhibitors of α-Amylase 1233 and α-Glucosidase. *Life* **2022**, *12*, 1692. [CrossRef]
68. Yu, B.; Tang, Q.; Fu, C.; Regenstein, J.; Huang, J.; Wang, L. Effects of different particle-sized insoluble dietary fibre from citrus peel on adsorption and activity inhibition of pancreatic lipase. *Food Chem.* **2022**, *398*, 133834. [CrossRef]
69. Chen, H.; Li, J.; Yao, R.; Yan, S.; Wang, Q. Mechanism of lipid metabolism regulation by soluble dietary fibre from micronized and non-micronized powders of lotus root nodes as revealed by their adsorption and activity inhibition of pancreatic lipase. *Food Chem.* **2020**, *305*, 125435. [CrossRef] [PubMed]
70. Benzie, I.F.F.; Devaki, M. The ferric reducing/antioxidant power (FRAP) assay for non-enzymatic antioxidant capacity: Concepts, procedures, limitations and applications. In *Measurement of Antioxidant Activity & Capacity: Recent Trends and Applications*; Wiley: Hoboken, NJ, USA, 2018.
71. Salman, H.A.; Yaakop, A.S.; Al-Mustafa, A.; Tarawneh, K.; Aladaileh, S.; Al-Rimawi, F.; Alakhras, F.; Abu-Lafi, S.; Al Zarzour, R.; Wahab, H. The dual impact of Jordanian *Ephedra alte* for inhibiting pepsin and treating microbial infections. *Saudi J. Biol. Sci.* **2021**, *28*, 6245–6253. [CrossRef] [PubMed]
72. Yang, Y.; Zhang, Z.; Li, S.; Ye, X.; Li, X.; He, K. Synergy effects of herb extracts: Pharmacokinetics and pharmacodynamic basis. *Fitoterapia* **2014**, *92*, 133–147. [CrossRef] [PubMed]
73. Tseng, P.S.; Ande, C.; Moremen, K.W.; Crich, D. Influence of Side Chain Conformation on the Activity of Glycosidase Inhibitors. *Angew. Chem.-Int. Ed.* **2023**, *62*, e202217809. [CrossRef] [PubMed]

Disclaimer/Publisher's Note: The statements, opinions and data contained in all publications are solely those of the individual author(s) and contributor(s) and not of MDPI and/or the editor(s). MDPI and/or the editor(s) disclaim responsibility for any injury to people or property resulting from any ideas, methods, instructions or products referred to in the content.

foods

MDPI

Article

Phytochemical Composition and Bioactive Potential of *Melissa officinalis* L., *Salvia officinalis* L. and *Mentha spicata* L. Extracts

Beatriz Nunes Silva [1,2,3,4], Vasco Cadavez [1,2], Cristina Caleja [1,2], Eliana Pereira [1,2], Ricardo C. Calhelha [1,2], Mikel Añibarro-Ortega [1,2], Tiane Finimundy [1,2], Marina Kostić [5], Marina Soković [5], José António Teixeira [3,4], Lillian Barros [1,2] and Ursula Gonzales-Barron [1,2,*]

1 Centro de Investigação de Montanha (CIMO), Instituto Politécnico de Bragança, Campus de Santa Apolónia, 5300-253 Bragança, Portugal
2 Laboratório Associado para a Sustentabilidade e Tecnologia em Regiões de Montanha (SusTEC), Instituto Politécnico de Bragança, Campus de Santa Apolónia, 5300-253 Bragança, Portugal
3 CEB—Centre of Biological Engineering, University of Minho, Campus de Gualtar, 4710-057 Braga, Portugal
4 LABBELS—Associate Laboratory, 4710-057 Braga, Portugal
5 Institute for Biological Research "Siniša Stanković"—National Institute of Republic of Serbia, University of Belgrade, Bulevar Despota Stefana 142, 11000 Belgrade, Serbia
* Correspondence: ubarron@ipb.pt; Tel.: +35-12-7330-3325

Abstract: Plants are rich in bioactive phytochemicals that often display medicinal properties. These can play an important role in the production of health-promoting food additives and the replacement of artificial ones. In this sense, this study aimed to characterise the polyphenolic profile and bioactive properties of the decoctions, infusions and hydroethanolic extracts of three plants: lemon balm (*Melissa officinalis* L.), sage (*Salvia officinalis* L.) and spearmint (*Mentha spicata* L.). Total phenolic content ranged from 38.79 mg/g extract to 84.51 mg/g extract, depending on the extract. The main phenolic compound detected in all cases was rosmarinic acid. The results highlighted that some of these extracts may have the ability to prevent food spoilage (due to antibacterial and antifungal effects) and promote health benefits (due to anti-inflammatory and antioxidant capacities) while not displaying toxicity against healthy cells. Furthermore, although no anti-inflammatory capacity was observed from sage extracts, these stood out for often displaying the best outcomes in terms of other bioactivities. Overall, the results of our research provide insight into the potential of plant extracts as a source of active phytochemicals and as natural food additives. They also support the current trends in the food industry of replacing synthetic additives and developing foods with added beneficial health effects beyond basic nutrition.

Keywords: Lamiaceae; polyphenols; antimicrobials; antifungals; antioxidants; anti-inflammatory effect; antiproliferative effect; alternative preservatives; biopreservation

1. Introduction

Over recent years, with the increasingly negative perception of consumers towards artificial food additives [1] and the higher demand for nutritious foods with additional health benefits, two major trends in the food industry have been to replace synthetic additives, which may be harmful to human health [2], and to develop nutraceuticals/functional foods [3].

In line with these trends, modern science has shown that plant matrices are sources of valuable molecules (for example, phenolic compounds) with promising biological value (e.g., antioxidant, anti-inflammatory, antibacterial and antifungal), thus encouraging their use for the development of functional foods and nutraceuticals, and as possible substitutes for artificial additives in foods or their packaging [4–6]. However, it is necessary to guarantee that the herbal extracts are safe for human consumption, and, among other considerations, it is crucial that they are obtained: (i) using nontoxic solvents authorised for the industrial production of foodstuffs and food ingredients, which do not leave residues or derivatives in the product after removal (or leave them in technically unavoidable quantities that pose negligible risk to human health) [7,8]; and (ii) from herbs with documented traditional use, commonly used in cooking as aroma and/or flavour enhancers [6,8,9].

To this, lemon balm (*Melissa officinalis* L., Lamiaceae), sage (*Salvia officinalis* L., Lamiaceae) and spearmint (*Mentha spicata* L., Lamiaceae) are among the various plants widely used in traditional Mediterranean cuisine and medicine, and for which several researchers have reported health-promoting capacities and potential as natural food additives [10–12]. Lemon balm has many beneficial capacities, such as spasmolytic, sedative, antitumoral, antimicrobial and antioxidant effects [13]. Furthermore, this plant has shown therapeutic effects for the treatment of the cognitive disturbance of Alzheimer's disease, and has been traditionally used to reduce anxiety, sleep disturbance, depression and gastrointestinal disorders [13,14]. In relation to sage, this herb has been used as a gargle for throat inflammations, to reduce perspiration, improve regularity of menstrual cycle, decrease hot flashes in menopause, battle gastrointestinal problems, prevent neurodegenerative diseases and improve mental capacity [11,15]. Furthermore, sage has shown anti-inflammatory, antimicrobial, hypoglycemic, antidiabetic, antioxidant and antitumor activities [15]. As for spearmint, it is frequently used in folk medicine against gastrointestinal and respiratory complications, haemorrhoids, stomach ache, memory dysfunction, and can be used as a carminative, antispasmodic, diuretic, antibacterial, antifungal and antioxidant agent [12,16,17].

Considering the recognised beneficial effects for human health of lemon balm, sage and spearmint, the goal of this research was to chemically characterise and appraise the bioactivities of extracts from such plants, produced through different environmentally friendly extraction methods (decoction, infusion and maceration), using water and 80% ethanol (*v*/*v*) as solvents. More specifically, the extracts' cytotoxicity, antibacterial, antifungal, anti-inflammatory and antioxidant capacities were evaluated to assess their safety and preservative effects.

2. Materials and Methods

2.1. Plant Material and Extraction Procedures

Sage, lemon balm and spearmint dry aerial parts were supplied by Pragmático Aroma, Lda. ("Mais Ervas", Trás-os-Montes, Portugal), mechanically milled and submitted to the following extraction methods: infusion, decoction and dynamic maceration.

For the infusions, 2 g of plant material was mixed with 200 mL of boiling distilled water and left to rest for 5 min without additional heating. For the decoctions, 2 g of plant material was mixed with 200 mL of distilled water, heated to boiling and boiled for 5 min. Infusions and decoctions were then filtrated (7–10 µm), frozen and lyophilised (FreeZone 4.5, Labconco, Kansas City, MO, USA). To obtain hydroethanolic extracts, dynamic macerations were conducted by incorporating 1 g of plant material in 30 mL of ethanol at 80% (*v*/*v*) and stirring at room temperature for 1 h. The supernatants were filtrated (7–10 µm), another 30 mL of ethanol 80% (*v*/*v*) was added to the extraction residues, and the maceration was repeated for 1 h. Finally, the ethanolic portion was evaporated (Büchi R-210, Flawil, Switzerland) and the resulting extracts were frozen and lyophilised. The extractions were carried out in triplicate (n = 3).

2.2. Identification and Quantification of Individual Phenolic Compounds

Individual phenolic compounds were investigated using a previously validated method, as described by Restivo et al. [18]. First, the samples were dissolved in ethanol 20% (*v*/*v*) up to a final concentration of 10 mg/mL and filtered through disposable 0.22 µm filters. The phenolic profiles were then determined by a liquid chromatography system (Dionex Ultimate 3000 UPLC, Thermo Scientific, San Jose, CA, USA) equipped with a quaternary pump, an automatic injector at 5 °C, a degasser and a column compartment with an automated thermostat. Compound detection was carried out with a diode-array detector (at wavelengths of 280 nm, 330 nm and 370 nm), coupled to a mass spectrometry (MS) detector. Separation was performed on a reverse phase Waters Spherisorb S3 ODS-2 C18 column (4.6 mm × 150 mm, 3 µm) at 35 °C. The flow rate was 0.5 mL/min. The mobile phase used was water/formic acid 0.1% (A) and acetonitrile (B). The elution gradient for solvent B was as follows: 10–15% eluent B up to 5 min, 15–20% B up to 5 min, 20–25% B 10 min, 25–35% B 10 min, 35–50% B 10 min and column re-equilibration for 10 min. For MS detection, a Linear Ion Trap LTQ XL spectrophotometer equipped with an electrospray ionization source was used. Nitrogen (50 psi) was used as a carrier gas, and the system worked with an initial temperature of 325 °C, a spray voltage of 5 kV and a capillary voltage of −20 V. The tube lens offset voltage remained at −66 V. Spectra were recorded in negative ion mode 100–1500 m/z.

The phenolic compounds were identified through their chromatographic characteristics by comparison to the obtained standard compounds (4-hydroxybenzoic acid, apigenin-6-C-glucoside, apigenin-7-*O*-glucoside, caffeic acid, chlorogenic acid, naringenin and rosmarinic acid) and with the literature [19–21]. For quantitative analysis, calibration curves prepared with appropriate standards (between 100 and 2.5 mg/L) were used. Limits of detection and quantification were also calculated, and, in all cases, the coefficient of linear correlation was $R^2 > 0.99$ (supplementary materials, Table S1). All analyses were made in triplicate (n = 3). The results were expressed in mg per g of dry extract (mg/g).

2.3. Biological Evaluation

2.3.1. Antibacterial and Antifungal Activity

For the antibacterial and antifungal activity screening, six bacterial strains were used: *Escherichia coli* (ATCC 25922), *Salmonella enterica* serovar Typhimurium (ATCC 13311), *Enterobacter cloacae* (clinical isolate), *Staphylococcus aureus* (ATCC 11632), *Bacillus cereus* (food isolate) and *Listeria monocytogenes* (NCTC 7973), and six micromycetes: *Aspergillus fumigatus* (human isolate), *Aspergillus niger* (ATCC 6275), *Aspergillus versicolor* (ATCC 11730), *Penicillium funiculosum* (ATCC 36839), *Penicillium verrucosum* var. *cyclopium* (food isolate) and *Trichoderma viride* (IAM 5061).

Minimum inhibitory (MIC), minimum bactericidal (MBC) and minimum fungicidal (MFC) concentrations were determined using a broth microdilution method and 96-well microplates [22]. The streak plate culture method, conducted on tryptic soy agar (Torlak, Belgrade, Serbia) incubated at 37 °C for 24 h, was used to obtain bacterial cells in the exponential growth phase. Then, an adequate number of individual colonies were placed in tubes with sterile water to achieve bacterial suspensions with a concentration of approximately 1.0×10^5 CFU/well in the microplates. For the antifungal activity essay, fungal spores were washed from the surface of malt agar plates (Neogen, Heywood, UK) with sterile 0.85% saline added with 0.1% Tween 80 (v/v) (Zorka pharma, Šabac, Belgrade). Sterile saline was then used to adjust the spore suspension to a concentration of approximately 1.0×10^5 in a final volume of 100 µL per well.

For the antibacterial and antifungal essay, resuspended extracts were obtained by dissolving them in ethanol 30% (v/v) to obtain a final concentration of 10 mg/mL. The liquid media (90 µL) used in the microplate wells was tryptic soy broth (Torlak, Belgrade, Serbia) for the antibacterial essay, or malt extract broth (Neogen, Heywood, UK) in the case of the antifungal essay.

After placing the inoculum, resuspended extract and liquid media in the microplate wells as appropriate, the microdilution plates were incubated at 37 °C for 24 h for the determination of the antibacterial activity, or 28 °C for 72 h for the determination of the antifungal activity. After that, 40 µL of iodonitrotetrazolium (Sigma-Aldrich, St. Louis, MO, USA), at a concentration of 0.2 mg/mL, was added to each well, and the microplate incubated again at 37 °C for 1 h. Afterwards, the microplates were evaluated, and the lowest concentrations without visible growth were determined as the MICs. The MBCs were determined as the lowest concentration with no visible growth after serial sub-cultivation of 10 µL into microdilution plates containing 100 µL of tryptic soy broth per well and further incubated for 24 h at 37 °C.

For the antifungal essay, MICs were determined under binocular microscope using the same procedure as described above. After that, the MFC was determined by serial sub-cultivation of 2 µL of the content of the wells and further incubation at 28 °C for 72 h. The lowest concentration of this sub-culture with no visible growth was defined as the MFC.

Two commonly used artificial food preservatives, sodium benzoate (E211) and potassium metabisulfite (E224), were also tested to evaluate the sensitivity of the microorganisms to such additives. The MIC, MBC and MFC were expressed in mg/mL of the resuspended lyophilised extracts.

2.3.2. Antioxidant Activity

The antioxidant activity was evaluated through two in vitro essays, using previously described methodologies [23,24]: inhibition of lipid peroxidation by decrease in the formation of thiobarbituric acid reactive substances (TBARS), and the oxidative haemolysis inhibition essay (OxHLIA). The extracts were initially diluted in distilled water (for TBARS) or phosphate-buffered saline (PBS, pH 7.4) (for OxHLIA) to different concentrations. Trolox was used as a positive control in both essays.

For TBARS essay: the extracts were examined for their power to inhibit the ferrous sulphate-induced lipid peroxidation, using porcine brain cell homogenates, through monitorisation of the colour strength (at 532 nm) provided by malondialdehyde-thiobarbituric acid complexes. The results were expressed as the extract concentration (µg/mL) required to inhibit 50% of the TBARS formation (half-maximal inhibitory concentration, IC_{50}).

For OxHLIA essay: 200 µL of an erythrocyte solution at 2.8% prepared in PBS was added to 400 µL of either: extract solution (13–800 µg/mL in PBS), PBS solution (negative control), distilled water (baseline), or Trolox (7.81–250 µg/mL). After incubation for 10 min at 37 °C with agitation, 200 µL of 2,2′-azobis(2-methylpropionamidine) dihydrochloride (AAPH, 160 mM in PBS) was added, and the optical density (at 690 nm) was measured in a microplate reader (Bio-Tek Instruments, ELx800, Winooski, VT, USA) every 10 min until complete haemolysis. The percentage of the erythrocyte population that remained undamaged (P) was calculated using Equation (1), where S_t and S_0 are the optical density of the sample at t and 0 min, respectively, and CH_0 is the optical density of the complete haemolysis at 0 min.

$$P\% = 100 \times \left(\frac{S_t - CH_0}{S_0 - CH_0}\right) \quad (1)$$

The delayed time of haemolysis (Δt) was calculated using Equation (2), where the 50% haemolytic time (min) graphically obtained from the haemolysis curve of each sample concentration is represented by Ht_{50}:

$$\Delta t \text{ (min)} = Ht_{50} \text{ (sample)} - Ht_{50} \text{ (control)} \quad (2)$$

Lastly, the Δt values were correlated to the various sample concentrations. From that correlation, the concentrations able to promote Δt haemolysis delays of 60 min and 120 min were calculated. The results were expressed as IC_{50} values (µg/mL) at Δt = 60 min and Δt = 120 min, i.e., the sample concentration required to protect 50% of the erythrocyte population from the haemolytic action of AAPH for 60 min and 120 min, respectively.

2.3.3. Anti-Inflammatory Activity

The anti-inflammatory activity was evaluated using a previously described essay, with modifications [25]. First, cells from the mouse macrophage-like cell line RAW264.7 were seeded in plates of 96-wells, and their attachment was allowed overnight. Subsequently, cells were subjected to different extract concentrations (6.25–400 µg/mL) for 1 h, and then stimulated with lipopolysaccharides (1 µg/mL) for 18 h. This procedure enabled observation of the occurrence of induced changes in nitric oxide basal levels, using a Griess Reagent System kit (Promega, Madison, WI, USA). The nitrite level produced was determined in a microplate reader (Bio-Tek Instruments, ELx800, Winooski, VT, USA) by assessing the optical density at 540 nm and comparing it with the standard calibration curve. The positive control used was dexamethasone (50 μM). The results are stated as the sample concentration (μg/mL) necessary to inhibit 50% of the nitric oxide production (IC_{50}).

2.3.4. Cytotoxic Activity

The lyophilised extracts were dissolved in water and successively diluted to obtain the stock solutions. The cytotoxic activity was then assessed against six human tumour cell lines, namely AGS (gastric adenocarcinoma), CaCo-2 (colorectal adenocarcinoma), HeLa (cervical carcinoma), MCF-7 (breast adenocarcinoma), NCI-H460 (large cell lung carcinoma) and non-tumour hFOB (human foetal osteoblasts), using the previously described sulforhodamine B essay [25]. For this, each of the cell lines (190 μL, 10^4 cells/mL) was incubated with the plant extracts at various concentrations (6.25–400 μg/mL). Ellipticine was used as a positive control. The results were expressed as the extract concentration required to inhibit 50% of the net cell growth (half-maximal cell growth inhibitory concentration, GI_{50}).

2.4. Statistical Analysis

Data were presented as mean ± standard deviation (SD) values. One-way analysis of variance (ANOVA, $\alpha = 0.05$) was used to assess statistical differences between the means. Clustered heatmaps were generated using the pheatmap function from the pheatmap package [26]. Statistical analysis was conducted in R software (version 4.1.0, R Foundation for Statistical Computing, Vienna, Austria).

3. Results and Discussion

3.1. Phenolic Profile

The peak characteristics (retention time, wavelength of maximum absorption and mass spectral data), tentative identification and quantification of the phenolic compounds detected in the extracts produced are reported in Tables S2–S4 of the Supplementary Materials (sage, lemon balm and spearmint, respectively). Heatmaps for a fast visualisation of the phenolic compounds identified and their concentrations were produced and are shown in Figures 1–3 (sage, lemon balm and spearmint, respectively).

The dendrograms of each clustered heatmap arrange the information on phenolic composition in terms of similarities, where the lower the height at which any two objects are joined, the greater the similarity. In this sense, one dendrogram (left) offers insight regarding compounds detected in similar concentrations across extracts obtained through different methodologies (infusion, decoction and hydroethanolic maceration), whereas the other dendrogram (upper) informs about similar total phenolic compound content across the extracts produced, for each plant.

Twenty-four phenolic compounds were identified in all sage extracts. From Figure 1 and Table S2, sage decoction and infusion contained higher and similar total phenolic compounds content (84.07 and 77.67 mg/g extract, respectively), compared to the hydroethanolic extract (63.17 mg/g extract). In the case of lemon balm, a maximum of fourteen compounds were identified, depending on the extract type. Figure 2 and Table S3 indicate that its infusion and hydroethanolic extract showed comparable total phenolic compounds content (61.00 and 58.35 mg/g extract, respectively); however, lower than that of the decoction (84.51 mg/g extract). As for the spearmint extracts, a maximum of fourteen compounds were identified. Figure 3 and Table S4 reveal that spearmint infusion and hydroethanolic extract had closer total phenolic compounds concentration (38.79 and 57.92 mg/g extract, respectively) than spearmint decoction (77.20 mg/g extract). Considering these results, decoctions revealed the highest amount of total phenolic compounds when compared to infusions and hydroethanolic extracts, regardless of the plant (Tables S2–S4). Overall, sage and lemon balm decoctions stood out for their higher total phenolic content (84.07 and 84.51 mg/mL, respectively). Oppositely, spearmint infusion yielded the lowest total phenolic content among the nine extracts (38.79 mg/g extract, Table S4).

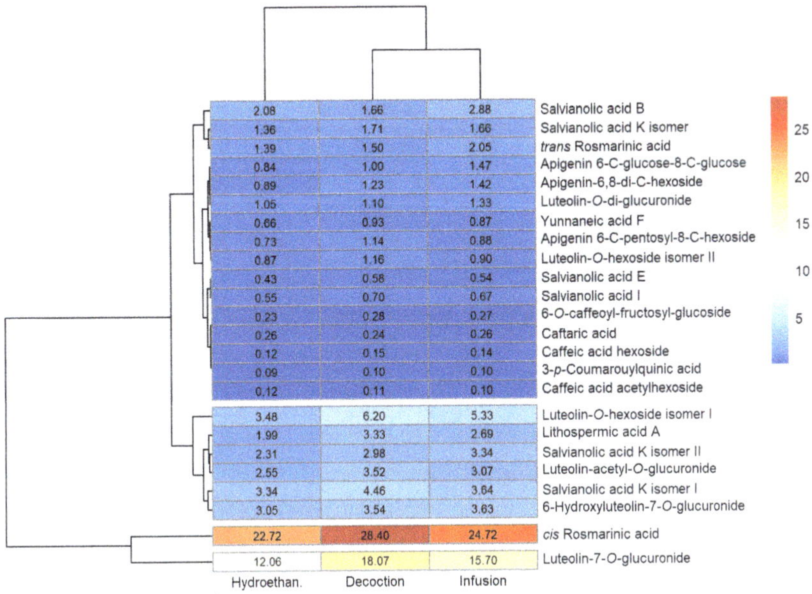

Figure 1. Clustered heatmap visualisation of phenolic compounds detected in sage infusion, decoction and hydroethanolic extract (units: mg/g).

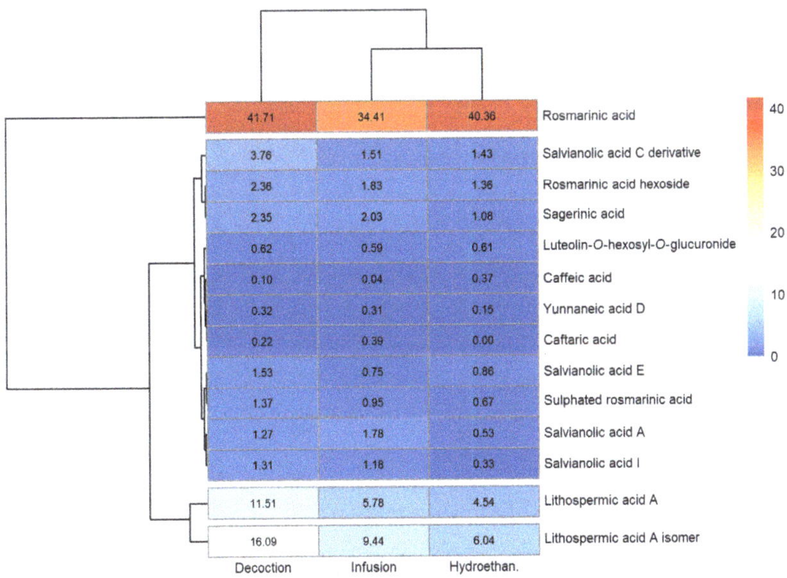

Figure 2. Clustered heatmap visualisation of phenolic compounds detected in lemon balm infusion, decoction and hydroethanolic extract (units: mg/g).

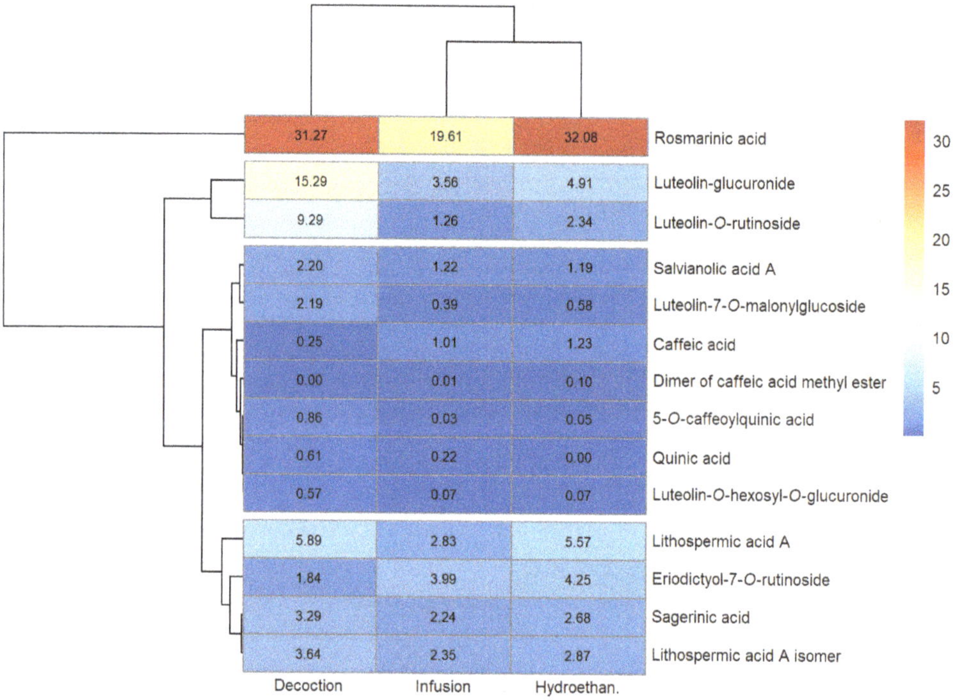

Figure 3. Clustered heatmap visualisation of phenolic compounds detected in spearmint infusion, decoction and hydroethanolic extract (units: mg/g).

In all cases, the plant extracts revealed a higher content of total phenolic acids compared with total flavonoids (Tables S2–S4). This was particularly noticeable in lemon balm extracts, which presented total flavonoid concentrations lower than 0.7 mg/g extract, in comparison with the total phenolic acids content, which ranged between 57.74 and 83.90 mg/g extract. In terms of qualitative profile, sage showed the highest variety of phenolic acids, with a total of fifteen different acids regardless of the type of extract. In comparison, in lemon balm, twelve or thirteen distinct phenolic acids were identified, depending on the extract type, whereas nine or ten acids were identified in spearmint extracts.

Some phenolic acids were found across all the evaluated extracts, namely, rosmarinic and salvianolic acids, as well as lithospermic acid A. Among these, the major compound in all the sage extracts was *cis*-rosmarinic acid (22.72 to 28.40 mg/g extract; Figure 1 and Table S2), followed by a derivative of luteolin, luteolin-7-*O*-glucuronide (12.06 to 18.07 mg/g extract; Figure 1 and Table S2). In lemon balm, rosmarinic acid was found in the greatest amount, irrespective of the type of extract, with concentrations between 34.40 and 41.71 mg/g extract (Figure 2 and Table S3). Similarly, the major phenolic compound in spearmint extracts was rosmarinic acid (19.61 to 32.08 mg/g extract; Figure 3, Table S4). Rosmarinic acid is known to possess extraordinary therapeutic potential, which includes antiviral, antibacterial, anticarcinogenic, antioxidant, anti-aging, antidiabetic, cardioprotective, hepatoprotective, nephroprotective, antidepressant, antiallergic and anti-inflammatory activities [27].

Sage extracts presented the highest number of different flavonoids (nine in total). These were derivates of apigenin and luteolin, with the most abundant compound being luteolin-7-O-glucuronide. Flavonoids were also detected in lemon balm and spearmint extracts, although in lesser variety (one and four in total, respectively), and these were also luteolin derivatives.

Given these results, sage, lemon balm and spearmint extracts appear to be valuable sources of valuable bioactive compounds, particularly of phenolic acids. Previous studies also investigated the phenolic profile of the plant materials used in this work. In sage and lemon balm hydroethanolic extracts, Spréa et al. [6] identified twenty-one and twelve phenolic compounds, respectively, several of which were also detected in the present study. Maliki et al. [28] studied the polyphenolic profile of a sage aqueous extract, identifying eighteen compounds, most of which belonged to hydroxycinnamic acid, rosmarinic acid and luteolin derivatives. Both the studies of Spréa et al. and Maliki et al. [6,28] found rosmarinic acid (51.00 mg/g and 2.192 mg/g, respectively) and luteolin-7-*O*-glucuronide (27.00 and 1.877 mg/g, respectively) to be the compounds of the highest concentrations in sage extracts, thus supporting the findings of our study. Also, in agreement with our results, Cirlini et al. [12] identified rosmarinic acid and its derivatives as the most prevalent polyphenolic compounds in an aqueous spearmint extract (230.5 mg/g), followed by salvianolic acids (14.70 mg/g) and caffeoylquinic acids (3.06 mg/g). Silva et al. [29] identified rosmarinic acid as the main compound in aqueous (204 mg/L) and hydroethanolic (333 mg/L) spearmint extracts; however, in lemon balm hydroethanolic extract, naringin was the principal compound (116 mg/L), and in sage aqueous and hydroethanolic extracts, hesperidin was present in the greatest amount (279 and 805 mg/L, respectively). This and other studies may have reported different phytochemical compositions [30,31], which however does not conflict with our results, since variations can be caused by different environmental factors during plant development, including soil type, change in season, salinity, light, altitude and humidity, as well as plant growth stage and extraction procedure [12,32]. Since the health-promoting properties of plants have been largely attributed to their phenolic compounds (among other secondary metabolites) [33,34], it is intuitive that differences in phenolic profile among extracts produced from the same plant matrix will also originate variations in their bioactivities (antimicrobial, antioxidant and anti-inflammatory, for example).

3.2. Antibacterial and Antifungal Activity

The results of the antibacterial and fungicidal activity are shown in Tables 1 and 2, respectively.

Table 1. Antibacterial activity of plant extracts expressed as minimum inhibitory concentration and minimum bactericidal concentration, MIC/MBC, respectively (mg/mL; mean ± SD, n = 3).

Extraction	Plant	SA [1]	BC [2]	LM [3]	EC [4]	ST [5]	EntC [6]
Infusion	Lemon balm	0.5/1	1/2	0.5/1	0.5/1	0.5/1	1/2
	Spearmint	0.5/1	0.5/1	0.5/1	0.5/1	0.5/1	1/2
	Sage	0.25/0.5	0.25/0.5	1/2	1/2	0.5/1	1/2
Decoction	Lemon balm	0.5/1	0.5/1	2/4	1/2	0.5/1	1/2
	Spearmint	0.5/1	0.5/1	1/2	0.5/1	0.5/1	1/2
	Sage	0.5/1	0.5/1	1/2	0.5/1	0.5/1	1/2
Hydroethanolic extraction	Lemon balm	0.5/1	0.5/1	1/2	0.5/1	0.5/1	0.5/1
	Spearmint	0.5/1	1/2	0.5/1	0.5/1	0.5/1	0.5/1
	Sage	0.5/1	1/2	0.5/1	0.5/1	0.5/1	0.5/1
E211 [7]		4/4	0.5/0.5	1/2	1/2	1/2	2/4
E224 [8]		1/1	2/4	0.5/1	0.5/1	1/1	0.5/0.5

Legend: [1] *S. aureus*, [2] *B. cereus*, [3] *L. monocytogenes*, [4] *E. coli*, [5] *Salmonella enterica* ser. Typhimurium, [6] *E. cloacae*, [7] Sodium benzoate, [8] Potassium metabisulfite.

Table 2. Antifungal activity of plant extracts expressed as minimum inhibitory and minimum fungicidal concentration, MIC/MFC, respectively (mg/mL; mean ± SD, n = 3).

Extraction	Plant	AF [1]	AN [2]	AV [3]	PF [4]	PVC [5]	TV [6]
Infusion	Lemon balm	0.125/0.25	0.125/0.25	0.25/0.5	0.5/1	0.5/1	0.25/0.5
	Spearmint	0.125/0.25	0.25/0.5	0.25/0.5	0.5/1	1/2	0.25/0.5
	Sage	0.25/0.5	0.25/0.5	0.25/0.5	0.25/0.5	0.25/0.5	0.125/0.25
Decoction	Lemon balm	0.25/0.5	>4/>4	0.25/0.5	0.5/1	0.25/0.5	0.125/0.25
	Spearmint	0.25/0.5	>4/>4	0.25/0.5	0.5/1	0.25/0.5	0.25/0.5
	Sage	0.25/0.5	0.5/1	0.5/1	0.5/1	0.5/1	0.25/0.5
Hydroethanolic extraction	Lemon balm	0.5/1	0.5/1	0.25/0.5	0.25/0.5	0.25/0.5	0.25/0.5
	Spearmint	0.25/0.5	0.25/0.5	0.25/0.5	0.25/0.5	0.25/0.5	0.125/0.25
	Sage	0.5/1	0.5/1	0.25/0.5	0.25/0.5	0.125/0.25	0.125/0.25
E211 [7]		1/2	1/2	2/2	1/2	2/4	1/2
E224 [8]		1/1	1/1	1/1	0.5/0.5	1/1	0.5/0.5

Legend: [1] *A. fumigatus*, [2] *A. niger*, [3] *A. versicolor*, [4] *P. funiculosum*, [5] *P. verrucosum* var. *cyclopium*, [6] *T. viride*, [7] Sodium benzoate, [8] Potassium metabisulfite.

Overall, the extracts revealed antimicrobial activity against all foodborne pathogens tested, namely *S. aureus*, *B. cereus*, *L. monocytogenes*, *E. coli*, *S.* Typhimurium and *E. cloacae* (MIC ≤ 2 mg/mL; MBC ≤ 4 mg/mL). Sage infusion presented the lowest MIC and MBC values of all extracts (i.e., the greatest antimicrobial potential), particularly against *S. aureus* and *B. cereus* (MIC = 0.25 and MBC = 0.5 mg/mL in both cases). On the other hand, lemon balm decoction displayed the highest MIC and MBC values, specifically against *L. monocytogenes* (MIC = 2 and MBC = 4 mg/mL). With a few exceptions, hydroethanolic extracts showed uniform activity (MIC = 0.5 and 1 mg/mL) for all tested bacteria.

In terms of antifungal capacity, all the infusions and hydroethanolic extracts were effective in inhibiting the six fungi tested, *A. fumigatus*, *A. niger*, *A. versicolor*, *P. funiculosum*, *P. verrucosum* and *T. viride* (MIC ≤ 1 mg/mL; MFC ≤ 2 mg/mL). Infusions demonstrated inhibition activity against the tested fungi with MIC values between 0.125 and 0.5 mg/mL, except for spearmint infusion against *P. verrucosum* var. *cyclopium* (MIC = 1 mg/mL). Hydroethanolic extracts stood out for inhibiting *T. viride* at a low concentration (MIC = 0.125 mg/mL for spearmint and sage extracts; MIC = 0.25 mg/mL for lemon balm extract), which demonstrates the susceptibility of this microorganism to such extracts. The three decoctions were also effective against all fungi (MIC ≤ 0.5 mg/mL; MFC ≤ 1 mg/mL) except *A. niger* (MIC > 4 mg/mL for lemon balm and spearmint).

In general, the infusions, decoctions and hydroethanolic extracts showed comparable or higher antimicrobial and fungicidal activities than those of the artificial food preservatives E211 and E224. In particular, the results of E211 against *S. aureus* (MIC and MBC = 4 mg/mL) and *P. verrucosum* (MIC = 2 and MFC = 4 mg/mL), and those of E224 against *B. cereus* (MIC = 2 and MBC = 4 mg/mL) differ noticeably from the lower concentration of plant extracts needed to prevent the growth of such microorganisms. These findings point out the potential of the extracts tested in this study as good candidates for applications in food and possible alternatives for replacing synthetic preservatives, aiming to delay the proliferation of food spoilage and pathogenic bacteria and fungi.

In line with our research, some previous studies have also reported on the antimicrobial and antifungal effects of these plants. The sage infusions of Abdel-Wahab et al. [35] showed MIC values of 50 mg/mL for *E. coli*, and 75 mg/mL for *Salmonella* spp., *S. aureus* and *B. cereus*. Hydroethanolic sage extracts produced by Hemeg et al. [36] revealed MIC values of 5 mg/mL for *S. aureus*, 0.625 mg/mL for *B. cereus* and 2.5 mg/mL for *E. coli* and *S.* Enteritidis. Silva et al. [29] hydroethanolic sage extracts revealed MIC values of 2.5–5 mg/mL for *L. monocytogenes*, 0.625 mg/mL for *S. aureus*, 10 mg/mL for *S.* Typhimurium and 1.25 mg/mL for *E. coli*. In turn, Ueda et al. [37] investigated hydroethanolic sage extracts obtained through ultrasound-assisted extraction, and MIC values were 1 mg/mL for *S. aureus*, *B. cereus*, *L. monocytogenes*, *E. coli*, *S.* Typhimurium and *E. cloacae*, 0.25 mg/mL for *A. fumigatus*, *A. versicolor*, *P. funiculosum* and *P. verrucosum* and 0.5 mg/mL for *A. niger* and *T. viride*.

Silva et al. [29] also tested the hydroethanolic extracts of spearmint and lemon balm, which revealed MIC values of 2.5 mg/mL for *L. monocytogenes*, 1.25 mg/mL for *S. aureus*, 20 mg/mL for *S.* Typhimurium and 1.25 mg/mL for *E. coli* for spearmint, and 5 mg/mL for *L. monocytogenes*, 2.5 mg/mL for *S. aureus*, 20 mg/mL for *S.* Typhimurium and 2.5 mg/mL for *E. coli* for lemon balm. Caleja et al. [38] analysed the antimicrobial activity of spearmint infusions, reporting MIC values of 0.5 mg/mL for *L. monocytogenes*, *B. cereus* and *E. coli* and 0.25 mg/mL for *S.* Typhimurium. The same study also determined the MIC of lemon balm infusions, which revealed values of 1 mg/mL for all bacteria mentioned before [38]. Furthermore, Caleja et al. [38] evaluated the MIC of said infusions against *A. niger*, *A. versicolor*, *P. funiculosum* and *P. verrucosum*, and the values ranged between 0.25 and 1 mg/mL.

3.3. Antioxidant Activity

The results of the TBARS and OxHLIA essays, which assess the ability of the plant extracts to inhibit lipid peroxidation and oxidative haemolysis in vitro, are presented in Table 3. The results are expressed as IC_{50} values, meaning that lower values correspond to greater antioxidant potential.

Table 3. Antioxidant activity of plant extracts expressed as half-maximal inhibitory concentration (IC_{50}, µg/mL) measured by the TBARS (mean ± SD, n = 9) and OxHLIA (mean ± SD, n = 3) essays.

Essay	Plant	Infusion	Decoction	Hydroethanolic Extract
TBARS [1]	Lemon balm	125 ± 2.08^{a}	204 ± 2.66^{b}	206 ± 8.99^{b}
	Spearmint	255 ± 11.0^{c}	197 ± 5.68^{a}	295 ± 9.77^{c}
	Sage	235 ± 6.43^{b}	196 ± 5.04^{a}	132 ± 5.07^{a}
OxHLIA [2] Δt = 60 min	Lemon balm	61.4 ± 1.31^{b}	27.0 ± 0.43^{b}	13.5 ± 0.38^{a}
	Spearmint	83.5 ± 1.84^{c}	42.2 ± 0.62^{c}	12.5 ± 0.17^{a}
	Sage	21.9 ± 0.77^{a}	8.93 ± 0.44^{a}	23.9 ± 0.94^{b}
OxHLIA [2] Δt = 120 min	Lemon balm	95.5 ± 2.16^{b}	41.6 ± 0.63^{b}	27.4 ± 0.85^{a}
	Spearmint	120 ± 1.84^{c}	66.8 ± 0.92^{c}	27.6 ± 1.28^{a}
	Sage	38.4 ± 0.89^{a}	23.5 ± 0.67^{a}	56.4 ± 1.51^{b}

Legend: [1] Thiobarbituric acid reactive substances, [2] Oxidative haemolysis inhibition essay. Trolox IC_{50} value: 5.4 ± 0.3 µg/mL (TBARS), 21.8 ± 0.25 µg/mL (OxHLIA Δt = 60 min), 43.5 ± 1.00 µg/mL (OxHLIA Δt = 120 min). For each essay, values with different superscript letters in a column mean significant differences (ANOVA, $p < 0.05$).

In both TBARS and OxHLIA essays, the antioxidant capacity of each plant infusion was significantly different from that of the other two ($p < 0.05$). Differences were also found among the decoctions, in both essays, depending on the plant species ($p < 0.05$). The antioxidant power of the hydroethanolic extracts also displayed differences depending on the plant used ($p < 0.05$), although not all of them were significant in the case of the OxHLIA essay. Moreover, in both essays, for each plant, different extraction methods yielded distinct antioxidant activities ($p < 0.05$). The exception was the decoction and hydroethanolic extract of lemon balm, which presented similar antioxidant potential in the TBARS essay ($p > 0.05$).

Overall, according to the statistical analysis, lemon balm infusion and sage hydroethanolic extract (125 µg/mL and 132 µg/mL, respectively) showed the best capacities to inhibit the formation of malondialdehyde and other reactive substances that are the result of the ex vivo decomposition of lipid peroxidation products (in the TBARS essay).

The results of the OxHLIA essay showed that the sage decoction (8.93 µg/mL and 23.5 µg/mL, for Δt = 60 min and 120 min) and the hydroethanolic extracts of spearmint (12.5 µg/mL and 27.6 µg/mL, for Δt = 60 min and 120 min) and of lemon balm (13.5 µg/mL and 27.4 µg/mL, for Δt = 60 min and 120 min) exhibited the greatest antioxidant protection for the erythrocyte membrane, even compared to the pure antioxidant compound used as a positive control, Trolox (21.8 µg/mL and 43.5 µg/mL, for Δt = 60 min and 120 min). These results suggest the potential of such extracts to be used against free radical-induced oxidative damage of biological membranes.

Furthermore, the OxHLIA essays allows us to distinguish between short-term and long-term antioxidant protection, as the antioxidant behaviour is monitored over time and the oxidative haemolysis assessed at two Δt. It was observed that all the infusions had anti-haemolytic activity for longer exposure times, as the concentration necessary to protect 50% of the red blood cells for 120 min was less than double the concentration necessary for this protection for 60 min. This also occurred in the case of spearmint and lemon balm decoctions, but not for the remaining extracts.

Our findings agree with other researchers that have also reported on the antioxidant capacities of lemon balm, spearmint and sage. Groupwise summary statistics calculated by Silva et al. [29] showed the high antioxidant power of these three plants, determined by the free radical scavenging (DPPH), radical cation decolorization (ABTS) and ferric reducing antioxidant power (FRAP) essays: the results were between 259 and 507 μmol Trolox Equivalent/g dry plant, for the DPPH and ABTS essays, and between 722 and 1013 μmol Fe^{2+}/g dry plant for the FRAP essay. Abdel-Wahab et al. [35] also evaluated a sage extract, using the DPPH method, and reported an IC_{50} of 13.34 μg/mL. Ueda et al. [37] reported an IC_{50} of 2.6 mg/g of sage extract, determined by the OxHLIA method, for the time period of 120 min. Caleja et al. [38] used two methods to assess the antioxidant power, reporting IC_{50} values of 6.6 μg/mL and 4.2 μg/mL for lemon balm and spearmint extracts, respectively (using the TBARS essay), and IC_{50} values of 24.8 μg/mL and 38.3 μg/mL for lemon balm and spearmint extracts, respectively (using the OxHLIA method for Δt = 60 min).

3.4. Anti-Inflammatory Activity

Table 4 presents the anti-inflammatory activity essay results. These are expressed as IC_{50} values, so lower values correspond to greater anti-inflammatory potential.

Table 4. Anti-inflammatory activity of plant extracts expressed as half-maximal inhibitory concentration (IC_{50}, μg/mL) measured by nitric oxide production inhibitory capacity (mean ± SD, n = 2).

Plant	Infusion	Decoction	Hydroethanolic Extract
Lemon balm	>400 ^{b}	>400 ^{b}	>400 ^{b}
Spearmint	44.4 ± 0.66 ^{a}	43.9 ± 4.26 ^{a}	26.6 ± 1.65 ^{a}
Sage	>400 ^{b}	>400 ^{b}	>400 ^{b}

Dexamethasone IC_{50} value: 6 ± 1 μg/mL. Values with different superscript letters in a column mean significant differences (ANOVA, $p < 0.05$).

The outcomes shown in Table 4 indicate that most extracts did not reveal anti-inflammatory action at the tested concentrations (IC_{50} > 400 μg/mL). Only those of spearmint showed this capability, regardless of the extraction method. Spearmint hydroethanolic extract showed the greatest anti-inflammatory capacity, considering its IC_{50} of 26.6 μg/mL.

In agreement with our results, the spearmint infusions of Caleja et al. [38] also displayed anti-inflammatory activity against the RAW 246.7 cell line (IC_{50} = 324 μg/mL), whereas those of lemon balm did not (IC_{50} > 400 μg/mL).

Nonetheless, and despite our results, some researchers have reported anti-inflammatory effects of sage and lemon balm extracts, meaning that these plants may be capable of offering such beneficial capacity under different circumstances [39,40].

It could be expected that extracts with high rosmarinic acid concentrations and promising antioxidant activity (low IC_{50} values in Table 3), such as sage or lemon balm infusions, for example, would also show anti-inflammatory potential, as antioxidants can reduce the inflammatory process caused by the overproduction of free radicals [25]. However, from the results in Table 4, it is noticeable that extracts presenting anti-inflammatory activity were not always the ones with the highest antioxidant capacity (except for spearmint hydroethanolic extract, which presented the lowest IC_{50} = 12.5 μg/mL in the OxHLIA essay among that type of extract). In this sense, it is important, when conducting analyses, to evaluate all bioactivities, and not to infer the results of one essay from the outcomes of another, to avoid arriving at wrongful conclusions, or even discarding plant extracts with substantial potential in terms of one particular bioactivity.

3.5. Cytotoxic Activity

The cytotoxicity essay results are shown in Table 5. These are expressed as GI_{50}, meaning that lower outcomes correspond to greater cytotoxic capacity.

Table 5. Cytotoxic activity of plant extracts expressed as half-maximal cell growth inhibitory concentration (GI_{50}, μg/mL) measured by the sulforhodamine B essay (mean ± SD, n = 3).

Extraction	Plant	AGS [1]	CaCo-2 [2]	HeLa [3]	MCF-7 [4]	NCI-H460 [5]	hFOB [6]
Infusion	Lemon balm	215 ± 6.22 ^{a}	290 ± 0.19 ^{b}	249 ± 11.5 ^{a}	239 ± 0.99 ^{b}	>400	>400
	Spearmint	196 ± 7.44 ^{a}	304 ± 0.55 ^{c}	229 ± 21.2 ^{a}	203 ± 1.50 ^{a}	>400	>400
	Sage	249 ± 8.68 ^{b}	242 ± 0.40 ^{a}	248 ± 25.6 ^{a}	198 ± 0.97 ^{a}	>400	>400
Decoction	Lemon balm	255 ± 7.45 ^{b}	>400 ^{c}	301 ± 10.9 ^{b}	>400	>400	>400
	Spearmint	258 ± 5.49 ^{b}	396 ± 0.05 ^{b}	289 ± 1.49 ^{b}	>400	>400	>400
	Sage	215 ± 6.25 ^{a}	269 ± 0.31 ^{a}	111 ± 2.14 ^{a}	320 ± 1.05 ^{a}	>400	350 ± 4.25 ^{a}
Hydroethanolic extract	Lemon balm	231 ± 2.75 ^{b}	351 ± 3.30 ^{c}	266 ± 11.5 ^{b}	180 ± 4.43 ^{a}	369 ± 3.37 ^{a}	271 ± 2.52 ^{a}
	Spearmint	162 ± 8.05 ^{a}	285 ± 0.43 ^{b}	215 ± 2.21 ^{a}	210 ± 2.20 ^{b}	381 ± 0.63 ^{b}	264 ± 2.29 ^{a}
	Sage	361 ± 3.74 ^{c}	272 ± 0.06 ^{a}	257 ± 1.17 ^{b}	206 ± 2.34 ^{b}	>400 ^{c}	>400 ^{b}

Legend: [1] Gastric adenocarcinoma, [2] Colorectal adenocarcinoma, [3] Cervical carcinoma, [4] Breast adenocarcinoma, [5] Large cell lung carcinoma, [6] non-tumour hFOB (human foetal osteoblasts). Ellipticine GI_{50} values: 1.23 ± 0.03 μg/mL (AGS), 1.21 ± 0.02 μg/mL (CaCo-2), 1.91 ± 0.12 μg/mL (HeLa), 1.02 ±0.02 μg/mL (MCF-7), 1.01 ± 0.01 μg/mL (NCI-H460) and 1.21 ± 0.08 μg/mL (hFOB). Values with different superscript letters in a column mean significant differences (ANOVA, $p < 0.05$).

All nine extracts produced revealed inhibitory potential ($GI_{50} < 400$ μg/mL) against at least one tumour cell line. Overall, the extracts were more active in tumour cells AGS, CaCo-2, HeLa and MCF-7 than NCI-H460. In fact, the cytotoxic capacity of the infusions and decoctions in the NCI-H460 tumour line was non-existent ($GI_{50} > 400$ μg/mL); however, some hydroethanolic extracts revealed activity.

The absence of toxicity ($GI_{50} > 400$ μg/mL) against non-tumour human foetal osteoblast cells, hFOB, was evident in the case of infusions and two decoctions (the exception was that of sage), which is a desirable outcome as extracts to be used in food products must be safe for consumption and cannot display toxicity against healthy cells. In contrast, the majority of hydroethanolic extracts (except that of sage, curiously) showed a cytotoxic effect towards hFOB cells, suggesting that this methodology may induce toxicity to the extracts, thus compromising their applicability as food additives.

From all the extracts, those that are non-toxic against hFOB and simultaneously present inhibitory potential against AGS, CaCo-2, HeLa and MCF-7 cells are: sage hydroethanolic extract and the infusions of lemon balm, spearmint and sage. These results point out the cytotoxic potential of the infusions produced in comparison to other extraction methods. The infusion of spearmint, specifically, showed overall greater antiproliferative capacity, with GI_{50} values of 196 μg/mL for the AGS cell line, 304 μg/mL for the CaCo-2 cell line, 229 μg/mL for the HeLa cell line and 203 μg/mL for the MCF-7 cell line.

The results obtained in this study agree, to some extent, with those of other researchers. Sage hydroethanolic extracts produced by Ueda et al. [37] did not show hepatotoxicity in PLP2 cells (non-tumour) at the maximum tested concentration of 400 μg/mL. Lemon balm and spearmint infusions of Caleja et al. [38] did not show toxicity for non-tumour cells PLP2 ($GI_{50} > 400$ μg/mL) and inhibited the growth of the HeLa cell line (GI_{50} = 241 μg/mL and GI_{50} = 251 μg/mL, respectively), in agreement with our results. Their spearmint infusion also inhibited MCF-7 growth (GI_{50} = 283 μg/mL), as found in our study. However, in contrast to our findings, lemon balm and spearmint infusions were able to inhibit NCI-H460 (GI_{50} = 290 μg/mL and GI_{50} = 322 μg/mL, respectively), and lemon balm infusion was incapable of affecting MCF-7 viability ($GI_{50} > 400$ μg/mL) [38].

Overall, these results indicate that extracts originating from any of the plants examined are potentially valuable for their cytotoxic impact on various tumour cell lines. However, it is crucial to further evaluate potential undesired effects against healthy cell lines, as even reduced concentrations may result in dangerous consequences for human health.

4. Conclusions

This work revealed the biological capacities of sage, spearmint and lemon balm extracts. Although only spearmint extracts showed anti-inflammatory potential, all infusions, decoctions and hydroethanolic extracts presented encouraging results in terms of antibacterial, antifungal and antioxidant capacities. Infusions revealed the most promising results, compared to decoctions and hydroethanolic extracts, as they yielded the best outcomes in each of the essays conducted (antimicrobial, antioxidant, anti-inflammatory and antiproliferative tests), while displaying an absence of toxicity against non-tumour cells, and even though infusions did not contain the highest total phenolic contents. Extracts from sage stood out from the remainder as they were often among those presenting the best capacities, both in terms of inhibiting the oxidation and growth of pathogenic bacteria and fungi, as well as impairing the viability of tumour cells. Nonetheless, no anti-inflammatory action was detected.

Overall, the results of this study emphasise the potential value of sage, spearmint and lemon balm extracts as natural food ingredients to prevent spoilage, provide beneficial health effects and potentially replace artificial additives, hence aligning with current trends in the food industry. However, further in vitro and in vivo studies must be conducted to verify the functionality of these extracts: for example, evaluating their pharmacokinetic parameters (bioavailability and bioaccessibility). It is also expected that the food matrix has some impact on the bioactivities of plant extracts, causing differences between the results observed in vitro and in vivo, which may limit the bio-functionalities of such extracts in food products. Another obstacle that must be investigated and that herbal extracts may face is related to their effect on the sensory characteristics of foods, since the concentrations necessary to provide the desired biological capacities can be very high and, therefore, negatively affect the aroma and taste of the products. In this sense, further research must be conducted to complement in vitro studies and address these and other limitations.

Supplementary Materials: The following supporting information can be downloaded at: https://www.mdpi.com/article/10.3390/foods12050947/s1, Table S1: Limit of detection (LOD), limit of quantification (LOQ) and coefficient of linear correlation (R^2) of the different standards used to obtain the calibration curves required for phenolic compound quantification; Table S2. Phenolic compound content (mg/g dry extract) of sage (*Salvia officinalis* L.) extracts; Table S3. Phenolic compound content (mg/g dry extract) of lemon balm (*Melissa officinalis* L.) extracts; Table S4. Phenolic compound content (mg/g dry extract) of spearmint (*Mentha spicata* L.) extracts.

Author Contributions: Conceptualization, V.C., J.A.T. and U.G.-B.; Data curation, B.N.S., C.C., E.P. and U.G.-B.; Formal analysis, B.N.S. and U.G.-B.; Funding acquisition, V.C., J.A.T. and U.G.-B.; Investigation, B.N.S., C.C., E.P., R.C.C., M.A.-O., T.F., M.K. and M.S.; Methodology, C.C., E.P., M.K., M.S., L.B. and U.G.-B.; Project administration, V.C., J.A.T. and U.G.-B.; Resources, V.C., L.B., J.A.T. and U.G.-B.; Software, B.N.S., V.C. and U.G.-B.; Supervision, V.C., L.B., J.A.T. and U.G.-B.; Validation, B.N.S., V.C., C.C., E.P. and U.G.-B.; Visualization, B.N.S., C.C., E.P. and U.G.-B.; Writing—original draft, B.N.S.; Writing—review and editing, C.C., E.P. and U.G.-B. All authors have read and agreed to the published version of the manuscript.

Funding: The authors are grateful to the EU PRIMA program and the Portuguese Foundation for Science and Technology (FCT) for funding the ArtiSaneFood project (PRIMA/0001/2018) and for financial support through national funds FCT/MCTES to CIMO (UIDB/00690/2020) and SusTEC (LA/P/0007/2021). This study was supported by FCT under the scope of the strategic funding of UIDB/04469/2020 unit and BioTecNorte operation (NORTE-01-0145-FEDER-000004) funded by the European Regional Development Fund under the scope of Norte2020—Programa Operacional Regional do Norte. This work has been supported by the Ministry of Science, Technological Development and Innovation of the Republic of Serbia (451-03-47/2023-01/200007). B.N. Silva and M. Añibarro-Ortega acknowledge the financial support provided by FCT through the Ph.D. grants SFRH/BD/137801/2018 and 2020.06297.BD, respectively. E. Pereira (2021.03908.CEECIND), R.C.

Calhelha, L. Barros and U. Gonzales-Barron acknowledge the national funding by FCT, P.I., through the Individual and Institutional Scientific Employment Program contract. C. Caleja is grateful for her contract through the project Healthy-PETFOOD (POCI-01-0247-FEDER-047073).

Data Availability Statement: Summary data available upon request.

Conflicts of Interest: The authors declare no conflict of interest.

References

1. Mesías, F.J.; Martín, A.; Hernández, A. Consumers' growing appetite for natural foods: Perceptions towards the use of natural preservatives in fresh fruit. *Food Res. Int.* **2021**, *150*, 110749. [CrossRef]
2. Abiega-Franyutti, P.; Freyre-Fonseca, V. Chronic consumption of food-additives lead to changes via microbiota gut-brain axis. *Toxicology* **2021**, *464*, 153001. [CrossRef]
3. Donato-Capel, L.; Garcia-Rodenas, C.L.; Pouteau, E.; Lehmann, U.; Srichuwong, S.; Erkner, A.; Kolodziejczyk, E.; Hughes, E.; Wooster, T.J.; Sagalowicz, L. Technological Means to Modulate Food Digestion and Physiological Response. In *Food Structures, Digestion and Health*; Boland, M., Golding, M., Singh, H., Eds.; Academic Press: Cambridge, MA, USA, 2014; pp. 389–422. [CrossRef]
4. Putnik, P.; Lorenzo, J.M.; Barba, F.J.; Roohinejad, S.; Režek Jambrak, A.; Granato, D.; Montesano, D.; Bursać Kovačević, D. Novel Food Processing and Extraction Technologies of High-Added Value Compounds from Plant Materials. *Foods* **2018**, *7*, 106. [CrossRef]
5. Vinceković, M.; Viskić, M.; Jurić, S.; Giacometti, J.; Kovačević, D.B.; Putnik, P.; Donsì, F.; Barba, F.J.; Jambrak, A.R. Innovative technologies for encapsulation of Mediterranean plants extracts. *Trends Food Sci. Technol.* **2017**, *69*, 1–12. [CrossRef]
6. Spréa, R.M.; Caleja, C.; Pinela, J.; Finimundy, T.C.; Calhelha, R.C.; Kostić, M.; Sokovic, M.; Prieto, M.A.; Pereira, E.; Amaral, J.S.; et al. Comparative study on the phenolic composition and in vitro bioactivity of medicinal and aromatic plants from the Lamiaceae family. *Food Res. Int.* **2022**, *161*, 111875. [CrossRef]
7. Union, Europäische. Directive 2009/32/EC of the European Parliament and of the Council of 23 April 2009 on the approximation of the laws of the Member States on extraction solvents used in the production of foodstuffs and food ingredients. *Off. J. Eur. Union L 141* **2009**, *5*, 3–9.
8. Granato, D.; Santos, J.S.; Salem, R.D.; Mortazavian, A.M.; Rocha, R.S.; Cruz, A.G. Effects of herbal extracts on quality traits of yogurts, cheeses, fermented milks, and ice creams: A technological perspective. *Curr. Opin. Food Sci.* **2018**, *19*, 1–7. [CrossRef]
9. Granato, D.; Nunes, D.S.; Barba, F.J. An integrated strategy between food chemistry, biology, nutrition, pharmacology, and statistics in the development of functional foods: A proposal. *Trends Food Sci. Technol.* **2017**, *62*, 13–22. [CrossRef]
10. Özogul, F.; Tugce Aksun, E.; Öztekin, R.; Lorenzo, J.M. Effect of lavender and lemon balm extracts on fatty acid profile, chemical quality parameters and sensory quality of vacuum packaged anchovy (*Engraulis encrasicolus*) fillets under refrigerated condition. *LWT* **2017**, *84*, 529–535. [CrossRef]
11. Jakovljević, M.; Jokić, S.; Molnar, M.; Jašić, M.; Babić, J.; Jukić, H.; Banjari, I. Bioactive Profile of Various *Salvia officinalis* L. Preparations. *Plants* **2019**, *8*, 55. [CrossRef]
12. Cirlini, M.; Mena, P.; Tassotti, M.; Herrlinger, K.A.; Nieman, K.M.; Dall'Asta, C.; Del Rio, D. Phenolic and Volatile Composition of a Dry Spearmint (*Mentha spicata* L.) Extract. *Molecules* **2016**, *21*, 1007. [CrossRef]
13. Miron, T.L.; Herrero, M.; Ibáñez, E. Enrichment of antioxidant compounds from lemon balm (*Melissa officinalis*) by pressurized liquid extraction and enzyme-assisted extraction. *J. Chromatogr. A* **2013**, *1288*, 1–9. [CrossRef]
14. Dos Santos-Neto, L.L.; De Vilhena Toledo, M.A.; Medeiros-Souza, P.; De Souza, G.A. The Use of Herbal Medicine in Alzheimer's Disease—A Systematic Review. *Evid.-based Complement. Altern. Med.* **2006**, *3*, 441. [CrossRef]
15. Garcia, C.S.C.; Menti, C.; Lambert, A.P.F.; Barcellos, T.; Moura, S.; Calloni, C.; Branco, C.S.; Salvador, M.; Roesch-Ely, M.; Henriques, J.A.P. Pharmacological perspectives from Brazilian *Salvia officinalis* (Lamiaceae): Antioxidant, and antitumor in mammalian cells. *An. Acad. Bras. Cienc.* **2016**, *88*, 281–292. [CrossRef]
16. Adsersen, A.; Gauguin, B.; Gudiksen, L.; Jäger, A.K. Screening of plants used in Danish folk medicine to treat memory dysfunction for acetylcholinesterase inhibitory activity. *J. Ethnopharmacol.* **2006**, *104*, 418–422. [CrossRef]
17. Snoussi, M.; Noumi, E.; Trabelsi, N.; Flamini, C.; Papetti, A.; De Feo, V. *Mentha spicata* Essential Oil: Chemical Composition, Antioxidant and Antibacterial Activities against Planktonic and Biofilm Cultures of *Vibrio* spp. Strains. *Molecules* **2015**, *20*, 14402–14424. [CrossRef]
18. Restivo, A.; Degano, I.; Ribechini, E.; Colombini, M.P. Development and Optimisation of an HPLC-DAD-ESI-QToF Method for the Determination of Phenolic Acids and Derivatives. *PLoS ONE* **2014**, *9*, e88762. [CrossRef]
19. Clifford, M.N.; Knight, S.; Kuhnert, N. Discriminating between the Six Isomers of Dicaffeoylquinic Acid by LC-MS^n. *J. Agric. Food Chem.* **2005**, *53*, 3821–3832. [CrossRef]
20. Clifford, M.N.; Johnston, K.L.; Knight, S.; Kuhnert, N. Hierarchical scheme for LC-MS^n identification of chlorogenic acids. *J Agric Food Chem.* **2003**, *51*, 2900–2911. [CrossRef]
21. Lee, J.S.; Kim, D.H.; Liu, K.H.; Oh, T.K.; Lee, C.H. Identification of flavonoids using liquid chromatography with electrospray ionization and ion trap tandem mass spectrometry with an MS/MS library. *Rapid Commun. Mass Spectrom.* **2005**, *19*, 3539–3548. [CrossRef]

22. Kostić, M.; Smiljković, M.; Petrović, J.; Glamočlija, J.; Barros, L.; Ferreira, I.C.F.R.; Ćirić, A.; Soković, M. Chemical, nutritive composition and a wide range of bioactive properties of honey mushroom *Armillaria mellea* (Vahl: Fr.) Kummer. *Food Funct.* **2017**, *8*, 3239–3249. [CrossRef]
23. Takebayashi, J.; Chen, J.; Tai, A. A method for evaluation of antioxidant activity based on inhibition of free radical-induced erythrocyte hemolysis. *Methods Mol. Biol.* **2010**, *594*, 287–296. [CrossRef]
24. Silva de Sá, I.; Peron, A.P.; Leimann, F.V.; Bressan, G.N.; Krum, B.N.; Fachinetto, R.; Pinela, J.; Calhelha, R.C.; Barreiro, M.F.; Ferreira, I.C.F.R.; et al. In vitro and in vivo evaluation of enzymatic and antioxidant activity, cytotoxicity and genotoxicity of curcumin-loaded solid dispersions. *Food Chem. Toxicol.* **2019**, *125*, 29–37. [CrossRef]
25. Jabeur, I.; Tobaldini, F.; Martins, N.; Barros, L.; Martins, I.; Calhelha, R.C.; Henriques, M.; Silva, S.; Achour, L.; Santos-Buelga, C.; et al. Bioactive properties and functional constituents of *Hypericum androsaemum* L.: A focus on the phenolic profile. *Food Res. Int.* **2016**, *89*, 422–431. [CrossRef]
26. Kolde, P. Pheatmap: Pretty Heatmaps. R Package Version 1.0.12. 2018. Available online: https://cran.r-project.org/web/packages/pheatmap/ (accessed on 25 November 2022).
27. Nadeem, M.; Imran, M.; Aslam Gondal, T.; Imran, A.; Shahbaz, M.; Muhammad Amir, R.; Wasim Sajid, M.; Batool Qaisrani, T.; Atif, M.; Hussain, G.; et al. Therapeutic Potential of Rosmarinic Acid: A Comprehensive Review. *Appl. Sci.* **2019**, *9*, 3139. [CrossRef]
28. Maliki, I.; Es-safi, I.; El Moussaoui, A.; Mechchate, H.; El Majdoub, Y.O.; Bouymajane, A.; Cacciola, F.; Mondello, L.; Elbadaouia, K. *Salvia officinalis* and *Lippia triphylla*: Chemical characterization and evaluation of antidepressant-like activity. *J. Pharm. Biomed. Anal.* **2021**, *203*, 114207. [CrossRef]
29. Silva, B.N.; Cadavez, V.; Ferreira-Santos, P.; Alves, M.J.; Ferreira, I.C.F.R.; Barros, L.; Teixeira, J.A.; Gonzales-Barron, U. Chemical Profile and Bioactivities of Extracts from Edible Plants Readily Available in Portugal. *Foods* **2021**, *10*, 673. [CrossRef]
30. Skendi, A.; Irakli, M.; Chatzopoulou, P. Analysis of phenolic compounds in Greek plants of Lamiaceae family by HPLC. *J. Appl. Res. Med. Aromat. Plants* **2017**, *6*, 62–69. [CrossRef]
31. Oliveira, A.S.; Ribeiro-Santos, R.; Ramos, F.; Castilho, M.C.; Sanches-Silva, A. UHPLC-DAD Multi-Method for Determination of Phenolics in Aromatic Plants. *Food Anal. Methods* **2018**, *11*, 440–450. [CrossRef]
32. Ramasar, R.; Naidoo, Y.; Dewir, Y.H.; El-Banna, A.N. Seasonal Change in Phytochemical Composition and Biological Activities of *Carissa macrocarpa* (Eckl.) A. DC. Leaf Extract. *Horticulturae* **2022**, *8*, 780. [CrossRef]
33. Guldiken, B.; Ozkan, G.; Catalkaya, G.; Ceylan, F.D.; Yalcinkaya, I.E.; Capanoglu, E. Phytochemicals of herbs and spices: Health versus toxicological effects. *Food Chem. Toxicol.* **2018**, *119*, 37–49. [CrossRef]
34. Panzella, L.; Moccia, F.; Nasti, R.; Marzorati, S.; Verotta, L.; Napolitano, A. Bioactive Phenolic Compounds From Agri-Food Wastes: An Update on Green and Sustainable Extraction Methodologies. *Front. Nutr.* **2020**, *7*, 60. [CrossRef]
35. Abdel-Wahab, M.; El-Sohaimy, S.A.; Ibrahim, H.A.; Abo El-Makarem, H.S. Evaluation the efficacy of clove, sage and kiwifruit peels extracts as natural preservatives for fish fingers. *Ann. Agric. Sci.* **2020**, *65*, 98–106. [CrossRef]
36. Hemeg, H.A.; Moussa, I.M.; Ibrahim, S.; Dawoud, T.M.; Alhaji, J.H.; Mubarak, A.S.; Kabli, S.A.; Alsubki, R.A.; Tawfik, A.M.; Marouf, S.A. Antimicrobial effect of different herbal plant extracts against different microbial population. *Saudi J. Biol. Sci.* **2020**, *27*, 3221–3227. [CrossRef]
37. Ueda, J.M.; Pedrosa, M.C.; Fernandes, F.A.; Rodrigues, P.; Melgar, B.; Dias, M.I.; Pinela, J.; Calhelha, R.C.; Ivanov, M.; Soković, M.; et al. Promising Preserving Agents from Sage and Basil: A Case Study with Yogurts. *Foods* **2021**, *10*, 676. [CrossRef]
38. Caleja, C.; Finimundy, T.; Pereira, C.; Barros, L.; Calhelha, R.C.; Sokovic, M.; Smiljkovic, M.; Carvalho, A.M.; Rosa, E.; Ferreira, I.C.F.R. Challenges of traditional herbal teas: Plant infusions and their mixtures with bioactive properties. *Food Funct.* **2019**, *10*, 5939–5951. [CrossRef]
39. Kolac, U.K.; Ustuner, M.C.; Tekin, N.; Ustuner, D.; Colak, E.; Entok, E. The Anti-Inflammatory and Antioxidant Effects of *Salvia officinalis* on Lipopolysaccharide-Induced Inflammation in Rats. *J. Med. Food* **2017**, *20*, 1193–1200. [CrossRef]
40. Birdane, Y.O.; Büyükokuroglu, M.E.; Birdane, F.M.; Cemek, M.; Yavuz, H. Anti-inflammatory and antinociceptive effects of *Melissa officinalis* L. in rodents. *Rev. Med. Vet.* **2007**, *158*, 75–81.

Disclaimer/Publisher's Note: The statements, opinions and data contained in all publications are solely those of the individual author(s) and contributor(s) and not of MDPI and/or the editor(s). MDPI and/or the editor(s) disclaim responsibility for any injury to people or property resulting from any ideas, methods, instructions or products referred to in the content.

foods

MDPI

Article

Antioxidant and Anti-Inflammatory Effect of Cinnamon (*Cinnamomum verum* J. Presl) Bark Extract after In Vitro Digestion Simulation

Stefania Pagliari [1], Matilde Forcella [2], Elena Lonati [3], Grazia Sacco [2], Francesco Romaniello [4], Pierangela Rovellini [4], Paola Fusi [2], Paola Palestini [3], Luca Campone [1], Massimo Labra [1], Alessandra Bulbarelli [3,*] and Ilaria Bruni [1,*]

[1] ZooPlantLab, Department of Biotechnology and Biosciences, University of Milano-Bicocca, Piazza della Scienza 2, 20126 Milan, Italy
[2] Department of Biotechnology and Biosciences, University of Milano-Bicocca, Piazza della Scienza 2, 20126 Milan, Italy
[3] School of Medicine and Surgery, University of Milano-Bicocca, Via Cadore 48, 20900 Monza, Italy
[4] Innovhub Stazioni Sperimentali per l'Industria S.r.l., Via Giuseppe Colombo 79, 20133 Milano, Italy
* Correspondence: alessandra.bulbarelli@unimib.it (A.B.); ilaria.bruni@unimib.it (I.B.)

Abstract: Cinnamon bark is widely used for its organoleptic features in the food context and growing evidence supports its beneficial effect on human health. The market offers an increasingly wide range of food products and supplements enriched with cinnamon extracts which are eliciting beneficial and health-promoting properties. Specifically, the extract of *Cinnamomum* spp. is rich in antioxidant, anti-inflammatory and anticancer biomolecules. These include widely reported cinnamic acid and some phenolic compounds, such asproanthocyanidins A and B, and kaempferol. These molecules are sensitive to physical-chemical properties (such as pH and temperature) and biological agents that act during gastric digestion, which could impair molecules' bioactivity. Therefore, in this study, the cinnamon's antioxidant and anti-inflammatory bioactivity after simulated digestion was evaluated by analyzing the chemical profile of the pure extract and digested one, as well as the cellular effect in vitro models, such as Caco2 and intestinal barrier. The results showed that the digestive process reduces the total content of polyphenols, especially tannins, while preserving other bioactive compounds such as cinnamic acid. At the functional level, the digested extract maintains an antioxidant and anti-inflammatory effect at the cellular level.

Keywords: cinnamon; antioxidant; oxidative stress; inflammation; in vitro digestion simulation; natural product; functional food

1. Introduction

Herbs and spices have been used since ancient times for nutritional and traditional medicine applications as they are rich in natural bioactive substances eliciting, for example, antioxidant and anti-inflammatory responses. Among these, cinnamon is widely used both as a bark extract and powder. There are over 300 species of cinnamon, but the most diffused as food/nutraceuticals are *Cinnamomum verum* J. Presl, *Cinnamomum aromaticum* or *Cinnamomum cassia* (L.) J. Presl, *Cinnamomum burmanni* (Nees and Th. Nees) Nees ex Blume, 2012 and *Cinnamomum loureiroi* Nees. These species differ in morphology and chemical composition. Specifically, at the phytochemical level, significant differences in the content of polyphenols and volatile phenols have been observed among various genotypes [1]. Moreover, the secondary metabolites profile also depends on plant growth conditions and the associated environmental pressures [2,3]. Among the most interesting compounds of bioactive Cinnamomum accessions, there are vanillic acid, caffeic acid, gallic acid, *p*-coumaric acid, ferulic acid, proanthocyanidins A and B, kaempferol, cinnamic acid and cinnamaldehyde, which exhibit several human beneficial effects, such as neuroprotective, hepatoprotective, cardioprotective and gastroprotective [1,4]. Most of these compounds are related to the antioxidant activity, enhancing the activities of catalase (CAT), superoxide dismutase (SOD) and glutathione peroxidase (GPx) [5,6]. Another important activity of cinnamon phyto-complexes is the anti-inflammatory one, which has been demonstrated in various cell and animal models and diseases, such as colitis, arthritis and diabetes [7–10].

Because of these healthful properties, the cinnamon extract and/or specific interesting compounds have been used as a bioactive ingredient in food products and supplements. However, it should be considered that the use of only isolated molecules is quite limited both because, in many cases, the beneficial effect is performed by a combination of molecules that act in synergy, and due to their poor systemic distribution and relative bioavailability [11]. For this reason, in recent years, the use of total cinnamon phytoextracts obtained by aqueous or hydroalcoholic extractions capable of optimizing the content of various bioactive molecules has become increasingly widespread. The extracts are often used to produce functional foods, such as yoghurt, creams and ice creams, rather than confectionery products up to food supplements and dietetic foods [12].

Starting from these premises, the present work was aimed at the study of the functional bioactivity of cinnamon extracts used in food supplements. Specifically, this paper analyzes the cinnamon bark, normally used in cooking, pulverized, to test the permanence of its antioxidant and anti-inflammatory properties after digestion processes. To better stimulate the real effect of a possible functional food based on cinnamon extract, the obtained aqueous extract was subjected to simulate gastrointestinal digestion (i.e., INFOGEST protocol [13]), with the aim of evaluating the stability of the bioactive molecules and therefore the functionality of the digested extract at the cellular level. A semi quantitative chemical analysis on total extract and digested ones was performed to define quantitative changes on phyto-complexes. Moreover, cellular tests were performed to evaluate the antioxidant and anti-inflammatory activity of the digested samples, with the final aim of supporting the production processes of functional foods and to confirm the beneficial properties of cinnamon-based extracts.

2. Materials and Methods

2.1. Samples Preparation and In Vitro Digestion

C. verum L. dried barks (batch number: C1900010, date: 16 January 2019) obtained by EPO s.r.l., was pulverized with an electric laboratory grinder. The method used for the preparation of cinnamon extracts has been described by De Giani et al., 2022 [14]. Briefly, 1 g of powder was extracted with hot water by 1 h incubation in a rotavapor (Strike-300, Steroglass Italia Srl, San Martino in Campo, Italy). At the end of the process, a centrifugation was carried out at 600 rpm for 15 min to recover the supernatant. Then, 75% ethanol was added in a 1:1 ratio to precipitate polysaccharides. To improve and accelerate the precipitation process, the mixture was stored 1 h at 4 °C. Then, the extract was filtered under vacuum using Whatman filters n.1. Finally, the solvent was evaporated with rotavapor and the dry fraction was resuspended in 5–10 mL of water. The samples were freeze-dried and stored at −20 °C.

2.2. Gastrointestinal Digestion Process In Vitro

The gastrointestinal digestion simulation was performed using INFOGEST protocol according to Minekeus et al., 2014 [13]. The process was simulated in 3 phases: oral phase, gastric phase and intestinal phase.

Oral phase: 1.4 mL of the oral saline phase (SSF) previously prepared as in Minekeus et al., 2014, 200 μL of α-amylase 75 U/mL, 10 μL of $CaCl_2$ and 390 μL of water were added to 2 mL of cinnamon water extract 10 mg/mL. The oral phase was set up at 37 °C for a duration of 2 min.

Gastric phase: 3 mL of the gastric saline phase as in Minekeus et al., 2014, 640 μL of swine pepsin 2000 U/mL, 2 μL of $CaCl_2$ and 278 μL of water were added to the 4 mL of bolus obtained from the oral phase. Finally, the pH of the gastric mixture was acidified with 160 μL of HCl 2 M to reach pH 3. The gastric phase was set up at 37 °C for a duration of 2 h.

Intestinal phase: 8 mL of gastric chyme was mixed with 4.4 mL of the intestinal saline phase previously prepared as in Minekeus et al., 2014. The pH was then measured to ensure that it was neutral or slightly alkaline. At this point 14 mL of porcine bile, 2 mL of porcine pancreatin to obtain trypsin with activity of 100 U/mL, 16 μL of $CaCl_2$ and 1.58 mL of water were added. The intestinal phase was set up at 37 °C for a duration of 2 h.

Upon completion of the intestinal phase, the digestion product was acidified to pH 2–3 to precipitate enzymes and to preserve the polyphenols as suggested by Pineda-Vadillo et al., 2016 [15]. Finally, to stop the enzymatic activity and to preserve the sample, protease inhibition was added.

2.3. Chemical Analysis of Polyphenols

The Folin–Ciocalteu assay was performed to determine the total polyphenol content of the samples as described in Pagliari et al., 2022 [16]. For the quantification of polyphenols, a calibration line was prepared using as a standard the gallic acid prepared at different operating concentrations, 100-75-50-25-12.5-6.25-0 mg/mL. Sample volume apart (i.e., 80 μL), the solvents and the respective volumes used in the assay were 400 μL of water, 40 μL of the Folin–Ciocalteu reagent and 480 μL of sodium carbonate at 10.75%.

The readings of the samples and the points of the line were taken at a wavelength of 760 nm after a half-hour wait, using a Cary 60 UV-Vis spectrophotometer (Agilent Technologies, Inc. Santa Clara, CA, USA). Results were expressed as mg [GAE]/g.

For the samples obtained from the in vitro digestion simulation, it was necessary to make some modifications to the Folin–Ciocalteu assay. The main reasons of this choice relied on the liquid physical state of samples and the need to reduce interference due to enzymes and salts used during the digestion process. For this purpose, the indications reported by Helal and Tagliazucchi, 2018, were followed [17]. Therefore, 150 μL of sodium carbonate 20% was used, while the volumes of water, sample and Folin–Ciocalteu reagent were 790 μL, 10 μL and 50 μL, respectively. The assay required an incubation of 2 h and was followed by a reading of the absorbance at a wavelength of 760 nm.

2.4. UHPLC-DAD-ESI-HRMS Profile of Cinnamon Extract and Digestion Product

The identification of target bioactive molecules was carried out using a Thermo Vanquish UHPLC system (Thermo Scientific, Rodano, Italy) coupled with a Thermo Orbitrap Exploris 120 mass spectrometer (Thermo Scientific, Rodano, Italy) and a Vanquish Diode Array Detector (Thermo Scientific, Rodano, Italy). The chromatographic separation was conducted on a Luna Omega Polar C18 (150 mm × 2.1 mm, 3 μm) (Phenomenex, Castelmaggiore, Italy).

The eluents, all MS grade from Fisher (Fisher scientific, Rodano, Italy), were (A) water and (B) acetonitrile/methanol (1:1), both containing 0.1% of formic acid. The gradient elution started from 2% of B, then raised at 40% in 40 min, at 50 min, the B concentration was 60% and at 55 min, 100% remaining constant for 10 min; finally, the mobile phase was brought back to the initial condition in 1 min. Moreover, 15 min of re-equilibration was necessary.

The flow rate was set at 300 μL/min, and the injection volume was 5 μL. The column was maintained at 30 °C. UV spectra were recorded between 210 and 600 nm, 280 and 340 nm wavelengths were employed, respectively, for the characterization of acids and flavonoids compounds. The UHPLC was coupled with the Orbitrap Exploris 120 through a HESI Optamax NG source. The source parameters were as follows: spray voltage 3500 V, ion transfer tube temperature 320 °C, vaporizer temperature 300 °C; sheath, auxiliary and sweep gas flow were 45, 10 and 0 arbitrary units, respectively. Spectra were recorded in full-mass mode within a range of 100–1500 m/z in positive and negative ionization, mild trapping condition was exploited. The Orbitrap resolution was set at 120,000 and the RF Lens (%) was 80. Data-dependent MS experiments were performed in stepped collision energy mode, the normalized collision energy was 20, 40 and 80%, HCD (higher energy collisional dissociation) fragmentation was achieved.

Phenolic compounds were characterized according to the corresponding spectral characteristics (UV and MS/MS spectra), accurate molecular mass, characteristic MS fragmentation pattern and libraries comparison in semi-automatic way through Compound Discoverer Software (2.1, Thermo Scientific, Rodano, Italy). Then, a semi-quantification was conducted in order to disentangle the phyto-complex composition change after digestion of the cinnamon extract. Briefly, all compounds were quantified, depending on their structural similarity, as trans-cinnamic acid (* in Table S1) or catechin (** in Table S1) at 280 nm by external calibration line. The quantification was performed in duplicate.

2.5. Caco-2 Cell Cultures

Cellular tests were performed on the Caco-2 (ATCC® HTB-37™) human colorectal cancer cell line, which is a continuous line of heterogeneous human epithelial colorectal adenocarcinoma cells that, upon reaching confluence, express the characteristics of enterocytic differentiation: tight junctions, microvilli and a number of enzymes and transporters that are characteristic of enterocytes [18]. Caco-2 cells were grown in EMEM medium supplemented with 10% heat-inactivated fetal bovine serum (FBS), 2 mM L-glutamine, 0.1 mM non-essential amino acids, 100 U/mL penicillin, 100 μg/mL streptomycin. The Caco-2 cell line was maintained at 37 °C in a humidified 5% CO_2 incubator. The ATCC cell line was validated by short tandem repeat profiles that are generated by simultaneous amplification of multiple short tandem repeat loci and amelogenin (for gender identification). All the reagents for cell cultures were supplied by Euroclone (Euroclone Spa, Milan, Italy). Caco-2 cells were seeded at a density of 1×10^4 cells/well in 96-well microtiter plates, at 1×10^6 cells/dish in 100 mm dish or at 90,000 cell/cm^2 onto collagen coated Transwell® polyester membrane inserts, based on the suitable conditions for oxidative stress and inflammation assays.

2.6. Intestinal Barrier In Vitro Model

The anti-inflammatory capacity experiments were performed in an in vitro intestinal barrier model since it represents the most suitable condition to study the gut maintenance and defense to inflammatory mediators and the modulation of intestinal epithelial permeability. Caco-2 cellswere seeded onto collagen coated Transwell® polyester membrane inserts (Corning 3640) and maintained at 37 °C in a 5% CO_2 atmosphere for 21 days, in order to reach the proper differentiation in a polarized epithelial cell monolayer. The basolateral and the apical sides of the insert represent the circulatory and luminal poles of the intestinal epithelium, respectively. Medium was replaced both in apical and basolateral chambers every 3 days, and Transendothelial Electrical Resistance (TEER) was measured every 7 days by means of EVOM EndOhm-12 chamber (World Precision Instruments, Sarasota, FL, USA), to check the formation of a reliable intestinal barrier.

TEER was evaluated as a parameter of barrier functionality after inflammation in cells pretreated or not with cinnamon digested extract. TEER measures were performed before and after the 24 h-inflammatory treatment for every insert, and the ΔTEER [Δ ($\Omega \cdot cm^2$)] between the two measures was calculated and analyzed within the groups.

2.7. Intestinal Barrier In Vitro Model

Caco-2 cells were treated with two different concentrations in polyphenols content (23 μg/mL or 46 μg/mL) of digested cinnamon extract 24 h before the induction of pro-inflammatory or pro-oxidative stimuli.

In order to mimic the acute phase of intestinal inflammation, Caco-2 cells cultured onto Transwell® inserts were exposed to a cocktail of inflammatory mediators (TNFα 10 ng/mL + IL-1β 5 ng/mL) applied at the basolateral side of the cells for 24 h, as described by Van de Walle et al. (2010) with minor modifications.

In order to induce oxidative stress, Caco-2 cells cultured in microplates or dishes were treated for 1 h with 1 mM H_2O_2 [19].

2.8. Cell Viability Assay

Cell viability was investigated using MTT-based in vitro toxicology assay kit (Sigma, St. Louis, MO, USA), according to manufacturer's protocols. Briefly, MTT [3-(4,5-Dimethythiazol-2-yl)-2,5-diphenyltetrazolium bromide] assay is based on the reduction of tetrazolium salts to colored formazan compounds that occurs in metabolically active cells. Optical density was measured using a multi-detection microplate reader at a wavelength of 570 nm. Cell viability was expressed as a percentage against untreated cell lines used as controls. MTT was performed in cells subjected to pro-inflammatory or pro-oxidative stimuli in the respective culture systems.

2.9. Detection of Reactive Oxygen Species (ROS)

The generation of intracellular reactive oxygen species (ROS) was detected by the oxidation of 2′,7′-dichlorofluorescin diacetate (H_2DCFDA) (Sigma Chemical Co., St. Louis, MO, USA), an indicator for both reactive oxygen species and nitric oxide (•NO). Caco-2 cell lines were seeded in 96-well black microtiter plates at a density of 1×10^4 cells/well, cultured in complete medium and treated with 23 and 46 µg/mL of polyphenols for 24 h and 1 mM H_2O_2 for 1 h for positive control. To evaluate the protective role of the enzymatic digestion, after 24 h of treatment, cells were washed in PBS and incubated with 5 µM H_2DCFDA in PBS for 30 min in the dark at 37 °C. After two washes in PBS, cells were treated with 1mM H_2O_2 for 1 h. The fluorescence (λ_{em} = 485 nm/λ_{ex} = 535 nm) was measured at 37 °C using a fluorescence microtiter plate reader (VICTOR X3) and analyzed by the PerkinElmer 2030 Manager software for Windows.

2.10. Enzyme Assay

To evaluate the effect of enzymatic digestion of cinnamon on enzyme activities, Caco-2 cell lines were seeded at 1×10^6 cells/100 mm dish and treated for 24 h with 23 and 46 µg/mL of polyphenols and with 1 mM H_2O_2 for 1 h. The cells were rinsed with ice-cold PBS and lysed in 50 mM Tris-HCl pH 7.4, 150 mM NaCl, 5 mM EDTA, 10% Glycerol, 1% NP-40, containing protease inhibitors (1 µM leupeptin, 2 µg/mL aprotinin, 1 µg/mL pepstatin and 1 mM PMSF). Homogenates were obtained by passing 5 times through a blunt 20-gauge needle fitted to a syringe and then centrifuged at 15,000× *g* for 30 min at 4 °C. Supernatants were used to measure enzyme activities: superoxide dismutase (SOD) was assayed as previously described by Vance et al., 1972 [20]; catalase (CAT) was assayed according to Oldani et al., 2020 [21]; glutathione reductase (GR) according to Wang et al., 2001 [22]; glutathione peroxidase (GPx) as reported in Nakamura et al., 1974 [23]; glutathione-S-Transferase (GST) as previously described by Habig et al., 1974 [24]. Enzyme activities were expressed in international units and referred to protein concentration as determined by the Bradford method [25]. All of the assays were performed in triplicate at 25 °C in a Jasco V-550 Spectrophotometer and analyzed by the Spectra Manager (version 1.33.02) software for Windows.

2.11. Electrophoresis and Immunoblotting

Equal amounts (as protein) of homogenate were analyzed by SDS-PAGE electrophoresis on 10% polyacrylamide tris-glycine gels. Proteins were transferred to a nitrocellulose membrane (Amersham, GE Healthcare Europe GmbH, Milano, Italy) and revealed by immunoblotting with specific antibodies. Immunoblotting was performed using rabbit polyclonal anti-COX2 (1:1000) (#12282 Cell signaling technology, Danvers, MA, USA), rabbit polyclonal phospho p65-NFκB (Ser536) (1:1000) and rabbit polyclonal p65-NF-κB (1:1000) (#93H1 and #4764 Cell signaling technology), rabbit polyclonal anti-β-actin (1:1500) (A2066 Sigma-Aldrich, St.Louis, MO, USA) antibodies. Immunoreactive proteins were revealed by enhanced chemiluminescence (ECL) and semi-quantitatively estimated by

LAS800 Image Station. Normalization in the same sample was carried out with respect to β-actin homogenate samples [26,27].

2.12. IL-8 ELISA Assay

In order to assess IL-8 secretion in the compartment mimicking blood circulation, extracellular media were collected from the basolateral chamber. IL-8 concentration was evaluated using commercially available ELISA assay (PeproTech EC, Ltd., London, UK) according to the manufacturer's instructions. Samples concentration was quantified in pg/mL using the standard provided by the kit, after reading absorbance with FLUOstar Omega (BMG Labtech, Ortenberg, Germany) multi-detection microplate reader at a wavelength of 405 nm and a reference wavelength of 690 nm. Then, to facilitate comparison between groups and to understand the anti-inflammatory potential of cinnamon, data normalized by cell protein concentration were expressed in relative terms to the inflammatory treatment.

2.13. Statistical Analysis

All data have been expressed as mean ± SEM (standard error of the means). Regarding the total content of phenols in different extraction condition, the water extraction method was compared with the other hydroalcoholic condition using Dunnet's test following one-way ANOVA calculation, a p-value < 0.05 was considered statistically significant, while for cell assay, the values were compared to the negative control (untreated cells) or positive control (oxidant or inflammatory stimulus) using the Dunn test following one-way ANOVA calculation. A p-value < 0.05 was considered to be statistically significant.

3. Results and Discussion

3.1. Samples Preparation and In Vitro Digestion

Several studies are available in the literature on the efficacy of aqueous and hydroalcoholic extracts of cinnamon, which can be correlated with the phenolic compound in the spice [7,8,10]. The first phase of the work involved the comparison of different extraction protocols to define the one most suitable for the study of antioxidant and anti-inflammatory activities in vitro after stimulation of the digestive process. The results reported in Table 1 show a higher % yield in hydroalcoholic extracts than in aqueous extracts. However, a comparison of total phenol content, using the Foli—Ciocalteu spectrophotometric assay, shows that the extract obtained with the aqueous solvent is comparable to a 30% and 70% hydroalcoholic extract (p-value < 0.05) with significant values (p-value > 0.05) only at ethanol percentages of 50%.

Table 1. Comparison between Folin–Ciocalteu assay among different *Cinnamomum* extract. The statistical analysis was carried out comparing water extraction conditions against other solvent ratios in the recovery of phenols compounds.

Sample	Solvent Condition	Yield (%)	Water Solubility	Folin–Ciocalteu Assay (mgGAE/gEXT)
CIN_{EtOH70}	EtOH-H2O 70:30	10.57 ± 2.04	≤50%	517.39 ± 30.15
CIN_{EtOH50}	EtOH-H2O 50:50	9.40 ± 1.64	≤50%	532.12 ± 20.61 **
CIN_{EtOH30}	EtOH-H2O 30:70	7.06 ± 1.03	≤50%	480.18 ± 36.92
CIN_{H2O}	H2O	5.44 ± 1.39	100%	482.87 ± 24.22

** p-value < 0.05.

Moreover all hydroalcoholic extracts are insoluble in the aqueous environment of the gastrointestinal tract, indicating that only the aqueous fraction is bioavailable for the cells [14]. Accordingly, for this study, it was decided to select the aqueous extract for the following steps due to its high total phenolic content, comparable to some hydroalcoholic extracts, and especially because it contains only the water-soluble fraction that is bioavailable to cells in the gastrointestinal tract.

3.2. UHPLC-DAD-HESI-HRMS Untargeted Analysis of Cinnamon Extracts

In order to quantify the active compounds in the aqueous extract of cinnamon, the complete characterization by UHPLC-DAD-HESI-HRMS was carried out, before and after the simulated digestion, in a completely non targeted way. The structural elucidation of the molecules was performed by using UV spectra, HRMS data (accurate molecular mass, isotopic pattern, MS/MS fragmentation spectra) and literature databases comparison. UHPLC-DAD (extracted at 280 nm) chromatograms between 0 and 40 min of cinnamon extract before (A) and after (B) digestion are reported in Figure 1.

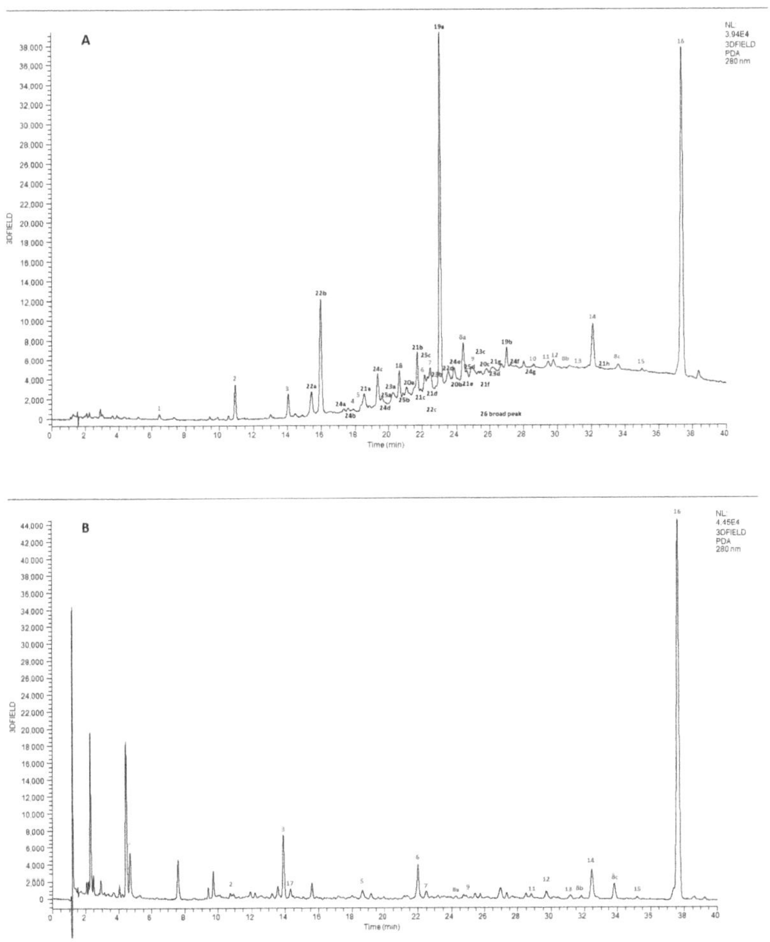

Figure 1. UHPLC-DAD chromatographic separation (280 nm) of cinnamon extract before (**A**) and after (**B**) digestion.

The chromatogram (A) is extremely complex and rich in different chromatographic peaks. All the identified molecules belong either to the class of procyanidins or they are derivatives of phenolic compounds. Procyanidins are oligomers of catechin, epicatechin and their gallic acid esters most frequently linked, either 4–6 or 4–8 (B-type procyanidins) [28]. Structural variations to procyanidins oligomers may also occur with formation of a second interflavonoid bond by C-O oxidative coupling to form A-type, as shown in additional Figure S1.

During the first ten minutes, small and hydrophilic molecules such as: xylitol, raffinose and gluconic acid, quinic acid, malic acid, citric acid were eluted. All of them were identified on the basis of HRMS spectra and Compound Discoverer libraries. Furthermore, the chromatographic trace was characterized by two very intense peaks at 22.99 min (number 29a on chromatogram) and 37.35 (number 16 on chromatogram), attributed respectively to cinnamtannin B1 and *trans*-cinnamic acid, and by a broad peak (number 26) starting at 18 min and ending at 40 min. This wide-ranging peak was ascribable to the procyanidins with different degrees of polymerization (from 2 to 8). After digestion (Figure 1B), the complete disappearance of all catechin polymers was observed, while the *trans*-cinnamic acid and other phenolic compounds, such as 2-hydoxybenzoic acid, coumarin and hydroxybenzaldehyde also present in the extract, became the principal components in the digested sample.

Table S1 shows the identification, also supported by reference [17,29–32], of all compounds individuated in *Cinnamomum verum* extract and the digested sample, their retention time, exact and observed molecular mass and their MS/MS principal fragments.

3.3. Gastrointestinal Digestion Process

The second important result obtained in this study is that the digestion process produced a significant reduction in the amount of catechin and its polymers (tannins). It is known that catechins can be easily degraded by the digestive process. Concerning the transformations of phenols along the digestive tract, it is known that these compounds are quite resistant to the acidic environment, while they can be easily degraded to the alkaline environment of the intestine and by the enzymes present. Catechins, for example, can easily undergo autoxidative processes due to the alkaline pH of the intestinal tract and by the presence of bile salts and pancreatin, explaining the disappearance in the extract cinnamon digested [33,34]. Other phenolic compounds, on the other hand, such as cinnamic acid or simple phenols, are less prone to oxidative processes preserving themselves more. As shown in Table S1, the content of *trans*-cinnamic acid, 2-hydroxybenzoic acid, coumarin and hydroxybenzaldehyde was very similar before and after the simulated digestion step. This reduction in total polyphenol content as a result of digestion is also confirmed by results obtained by spectrophotometric tests (Folin–Ciocalteu assay). Indeed, the total phenolic content observed before and after the digestion process showed that the aqueous cinnamon extract initially contained 1029.8 ± 24.1 μg of polyphenols, about 56 percent of the extract, while after digestion, in the gastrointestinal condition, the total phenol content decreased to 430.15 ± 12.5 μg, indicating a reduction of about 50 percent.

More in general, these data suggest that the occurrence of bioactive (i.e., the amount of water-soluble phytochemicals released at each of the stages of digestion from the complex food material that could be considered absorbable by the human body) may vary significantly due to digestion. Therefore, it becomes essential to evaluate the beneficial effects of food extracts following the digestive process, in order to assess their real potential for human health. For this purpose, some bioactivity in vitro experiments were conducted, choosing the human colorectal epithelial adenocarcinoma cell line Caco-2 because it represents an in vitro model of the human intestinal mucosa and these cells are widely used to study absorption of digested compounds [35].

3.4. Protective Effect of Digested Cinnamon Extracts Againstr Oxidative Stress

The effect of digested cinnamon extracts on Caco-2 cells was evaluated by determining cell viability using the MTT assay. As reported in Figure 2A, the digested cinnamon extract at 23 and 46 μg/mL polyphenol concentrations was not cytotoxic. The treatment with 1 mM H_2O_2 for 1 h reduced cell viability by 55% compared to the control; however, pre-treatment of Caco-2 cells with digested cinnamon extract at 23 and 46 μg/mL polyphenols for 24 h before adding H_2O_2 significantly increased cell viability compared to treatment with only

H_2O_2. Therefore, the digested cinnamon extract shows a protective effects against H_2O_2-induced damage in Caco-2 cells (Figure 2A).

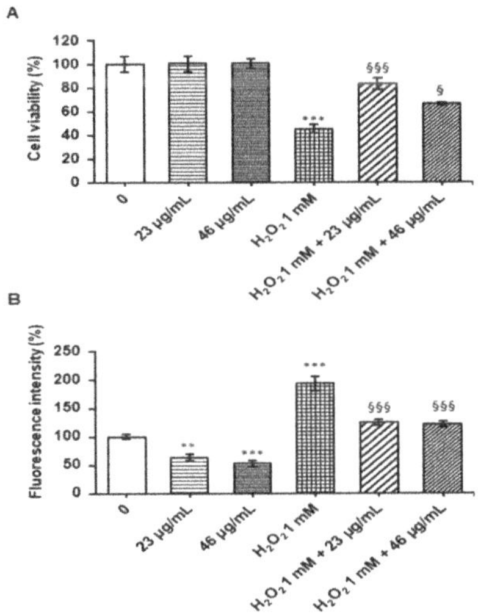

Figure 2. Viability assay and ROS evaluation. Caco-2 cells were treated with a digested cinnamon extract at 23 and 46 µg/mL polyphenols for 24 h, with 1 mM H_2O_2 for 1 h and with 1 mM H_2O_2 for 1 h after digested cinnamon extract pretreatment. In panel (**A**), the cell viability evaluated by MTT assay is reported; panel (**B**) shows the level of fluorescence after cells incubation with 5 µM H_2DCFDA. Data are expressed as a percentage against the untreated cell line used as a control and are shown as mean ± SEM from three independent experiments. Statistical significance: ** $p < 0.01$, *** $p < 0.001$ respect to untreated cell line; § $p < 0.05$, §§§ $p < 0.001$ respect to H_2O_2-treated cell line.

Total cytoplasmic ROS were evaluated using H_2DCFDA, which is oxidized to DCF fluorescent probe. After treatment with a digested cinnamon extract at 23 and 46 µg/mL polyphenols, a significant decrease in ROS production was measured, confirming the protective effect (Figure 2B). A significant ROS increase was observed after treatment with H_2O_2 compared to the control; however, a pre-treatment with digested cinnamon extract at 23 and 46 µg/mL polyphenols for 24 h significantly decreased ROS generation as compared to H_2O_2-stressed cells, bringing the level of ROS close to the control values (Figure 2B). The inability of digested cinnamon extract to lower ROS levels to exactly the control values suggests that, in the presence of 1 mM H_2O_2, extract concentration is limiting even at 46 µg/mL. Rather than testing higher extracts concentrations, further work will be aimed at identifying key compound(s) endowed with the highest protective effect against oxidative stress. The antioxidant properties of cinnamon extract after digestion are attributable to the ability of phenolic constituents to quench reactive oxygen species. Cinnamic acid and coumarin, which increase after digestion, can scavenge superoxide anions and hydroxyl radicals as well as other free radicals [36].

3.5. *Effect of Digested Cinnamon Extract on Antioxidant Enzyme Activity*

In order to evaluate the effect of digested cinnamon extract on antioxidant enzyme activities in the Caco-2 cell line, it was assessed the activity of superoxide dismutase (SOD), catalyzing the dismutation of the highly reactive superoxide anion to H_2O_2, catalase (CAT), catalyzing the decomposition of hydrogen peroxide to water and oxygen, glutathione reductase (GR), a flavoprotein that catalyzes the reduction of glutathione disulfide (GSSG) to glutathione (GSH) with the participation of NADPH as an electron donor, glutathione peroxidase (GPx), a cytosolic enzyme that catalyzes the reduction of hydrogen peroxide to water and oxygen as well as the reduction of peroxide radicals to alcohols and oxygen, and glutathione S-transferase (GST), involved in detoxification mechanisms, via conjugation of reduced glutathione with numerous substrates.

Treatment of Caco-2 cells with 1 mM H_2O_2 for 1 h induced a significant increase in the enzyme activity of superoxide dismutase and glutathione S-transferase and a significant decrease in glutathione peroxidase activity, compared to the control. However, SOD and GST increase was reduced by a pre-treatment with a digested cinnamon extract at 23 and 46 µg/mL polyphenols for 24 h before adding H_2O_2, while GPx activity returned close to the control level (Figure 3). On the other hand, both CAT and GR activities did not change upon H_2O_2 treatment, suggesting that these enzymes are already effective in managing oxidative stress; the decrease in GPx activity suggests the most H_2O_2 is detoxified within the cytosol and that GPx activity is not sufficient to fully inactivate 1 mM H_2O_2; pretreatment with digested cinnamon extract at both concentrations showed a protective effect against H_2O_2 damage, by increasing GPx activity (Figure 3).

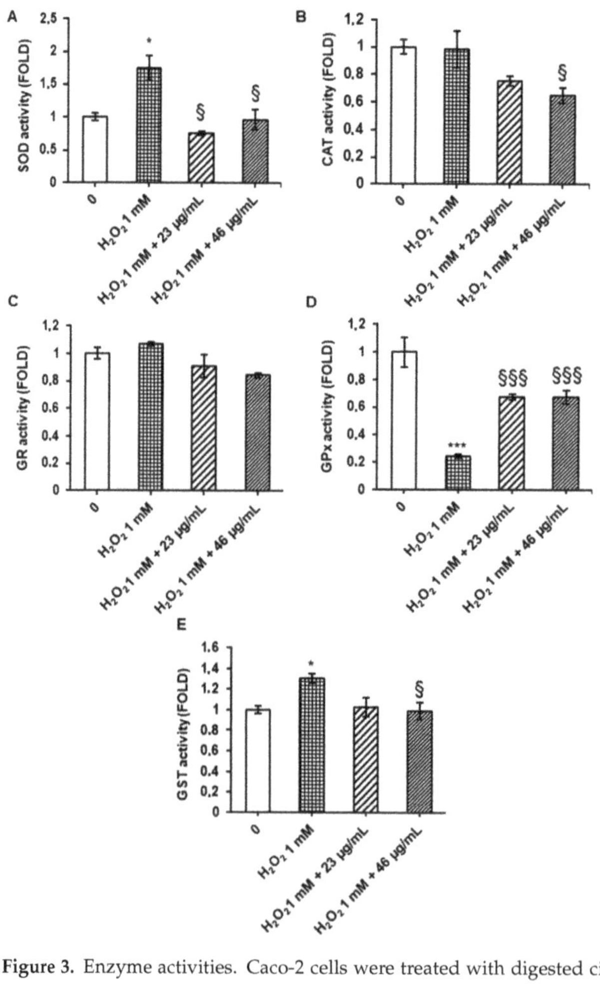

Figure 3. Enzyme activities. Caco-2 cells were treated with digested cinnamon extract at 23 and 46 µg/mL polyphenols for 24 h, with 1 mM H_2O_2 for 1 h and with 1 mM H_2O_2 for 1 h after digested cinnamon extract pretreatment. Results of SOD (**A**), CAT (**B**), GR (**C**), GPx (**D**) and GST (**E**) activities are expressed as folds with respect to untreated control and are shown as mean ± SEM from three independent experiments. Statistical significance: * $p < 0.05$, *** $p < 0.001$ respect to untreated cell line; § $p < 0.05$, §§§ $p < 0.001$ respect to H_2O_2-treated cell line.

Mammalian cells have evolved antioxidant enzymes to protect against oxidative stress. A change in the level of antioxidant enzyme activity can be considered a sensitive biomarker of cellular response to oxidative stress [37,38]. Therefore, these data confirm the protective effect of digested cinnamon extract in Caco-2 cells. Phenolic compounds in digested cinnamon extract, such as trans-cinnamic acid and coumarin, may be responsible for antioxidant activity through various mechanisms: removing free radicals, binding metal ions, inhibiting enzymatic systems that produce free radical forms, increasing the concentration of biologically important endogenous antioxidants and inducing the expression of a variety of genes responsible for the synthesis of enzymes that reduce oxidative stress, such as superoxide-dismutase, catalase, glutathione peroxidase, glutathione reductase, glutathione-S-transferase, which represent a primary line of antioxidant protection [39].

3.6. Effect of Digested Cinnamon Extracts on Anti-Inflammatory Processes

Acute inflammation in the gut leads to disruption of tight junctions anchoring the epithelial cells which are part of the intestinal barrier. The loss of integrity and the increase of barrier permeability have, as important consequences, the alteration of microbiota and immune system homeostasis [8]. Since several polyphenols have beneficial effects on epithelial barrier dysfunctions [40], the potential anti-inflammatory power of digested cinnamon was evaluated in an intestinal barrier in vitro model (see Material and Methods) exposed to pro-inflammatory cytokines stimulus.

First of all, after polarization in Transwell inserts, a MTT assay was performed also under inflammatory stimulus in cells pre-treated or not with digested cinnamon. Increasing concentrations in polyphenols of digested cinnamon (23 μg/mL and 46 μg/mL) were administered to Caco-2 cells for 24 h in the apical compartment, then, the inflammatory stimulus was induced by adding the cytokine cocktail (TNFα/IL-1β) in the basolateral compartment for other 24 h. Afterward, an MTT assay was performed in Caco-2 on Transwell filters to evaluate the effects of single or combined treatments. None of the performed treatments induced cytotoxicity (Figure 4A).

In parallel, the intestinal barrier integrity was investigated by evaluation of transepithelial resistance (TEER). According to the increment of barrier paracellular permeability observed in inflammatory diseases of the gastro-intestinal tract, TEER was reduced in Caco-2 under inflammatory conditions with respect to untreated cells. Interestingly, pretreatment with digested cinnamon extracts partially prevent the decrement of TEER in a dose-dependent manner (Figure 4B). Results are more evident when shown as intra-treatment ΔTEER and compared inter-groups (Figure 4C). Indeed, in the major concentration used (46 μg/mL), cinnamon can significantly prevent 25% of TEER loss during inflammatory treatment. No significant changes were observed after exposure to cinnamon alone. Taken together, these data suggest that this spice has a protective effect against the barrier disruption induced by inflammatory cytokines. The high amount of coumarin, cinnamic acid and derivatives, enriched after digestion, probably exerted the observed bioactivity. Indeed, rescue of barrier integrity was observed in presence of cinnamic acid when Ca-co-2 cells were treated with another inflammatory stimulus, such as LPS. In this case, cinnamic acid seems to act as a modulator of tight junction (TJ) protein expression, which is the major determinant of paracellular permeability [7]. On the other hand, it has been already demonstrated that several polyphenols and phenolic catabolites are able to partially reverse the TEER decrement induced by a pro-inflammatory cytokine cocktail in the Caco-2 monolayer [41]. Moreover, in a recent review, Bernardi and colleagues highlighted the role of polyphenols as intestinal permeability modulators both in vitro and in vivo studies [40].

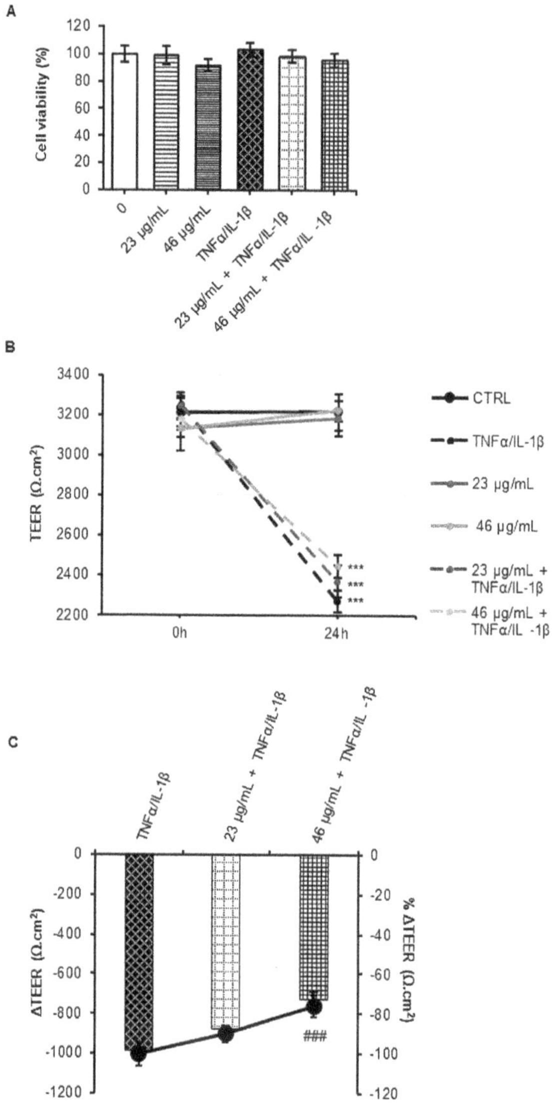

Figure 4. Cell viability assay and measurement of TEER under different experimental conditions. Caco-2 cells, cultured on Transwell inserts for 21 days were exposed to pro-inflammatory cytokines (TNFα 10 ng/mL + IL-1β 5 ng/ml) for 24 h after a 24 h pre-treatment with two different concentrations in polyphenols of digested cinnamon extract (23 and 46 μg/mL). Then, a MTT assay was performed, and results are reported in the histograms. Single conditions were also tested. Data are expressed as a percentage against the untreated cell line used as a control and are shown as mean ± SE from three independent experiments (**A**). TEER measures (´Ω·cm^2) were performed before (0 h) and after (24 h) cytokines treatment (**B**). The ΔTEER of cells exposed to cytokines was calculated and reported in the histograms as number (left Y-axes) and percentage (right Y-axes). Data represent the mean ± SEM from at least three independent experiments (**C**). Statistical significance: *** $p < 0.001$ vs. untreated cells, ### $p < 0.001$ vs. TNFα/IL-1β.

3.7. Effect of Digested Cinnamon Extract on NF-κB Pathway and Its Targets in Cells Exposed to Pro-Inflammatory Cytokines

The intestinal barrier alterations are sustained and amplified by an excessive production of pro-inflammatory cytokines and chemokines secreted locally by immune cells and enterocytes [42]. Most of them are highly produced by the transcriptional activity of the nuclear factor-kappa B (NF-κB) in a sort of vicious cycle sustained by pro-inflammatory cytokines themselves (e.g., TNF-α). The NF-κB is a complex constituted by homo-heterodimers of p50, p52, p65 (Rel-A), c-Rel and Rel-B proteins, sequestered in the cytosol by its inhibitor IκBα. Under an inflammatory stimulus, the activation of the canonical pathway passes through the phosphorylation and the following proteasome degradation of the inhibitor, allowing the NF-κB p65 migration to the nucleus. Here, when activated by phosphorylation on Ser536, NF-κB p65 promotes its target gene expression. Accordingly, in polarized Caco-2 on the Transwell system, the exposure to TNF-α/IL-1β induced an increment of phosphorylated NF-κB about three times versus untreated cells (Figure 5). Interestingly, a reduction of 30% in p65-NF-κB phosphorylation, after inflammation, was detected in cells pre-treated with the major dose in polyphenols (46 μg/mL) of cinnamon digested extract, indicating an anti-inflammatory capacity of the phyto-compound. Indeed, many of the flavonoids revealed by LC-MS mass analysis, such as quercetin, kaempferol, rutin (flavonols) and apigenin (flavone), as well as coumarin, concurred in down-regulation of the pro-inflammatory NF-κB pathway [43] and in overproduction of chemokines [44].

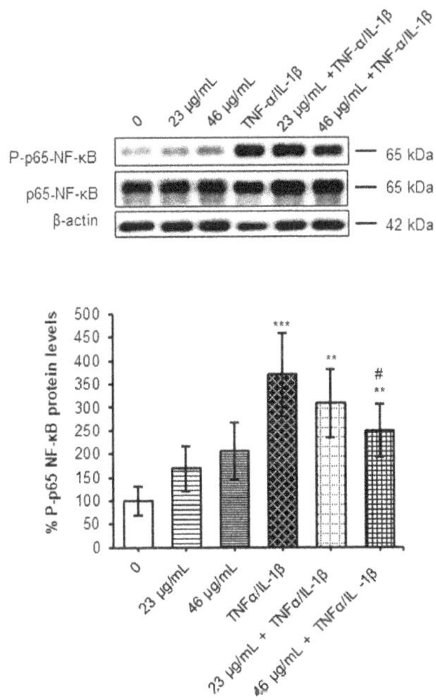

Figure 5. Evaluation of NF-κB phosphorylation and protein levels. Caco-2 cells, cultured on Transwell inserts for 21 days were exposed to pro-inflammatory cytokines (TNFα 10ng/mL + IL-1β 5 ng/mL) for 24 h after a 24 h pre-treatment with two different concentrations in polyphenols of digested cinnamon extract (23 and 46 μg/mL). Cell lysates were harvested, and samples analyzed for protein concentration by bicinchoninic acid (BCA) assay. Equal amounts of homogenate samples (as protein) were analyzed by SDS-PAGE electrophoresis and Western blotting. p-p65-NF-κB and NF-κB were detected with specific antibodies and revealed by enhanced chemiluminescence (ECL). Samples were normalized on β-actin immunoreactivity. Histograms, obtained from at least three distinct experiments, represent the percentage of protein levels with respect to untreated cells as mean ± S.E. Statistical significance: ** $p < 0.01$, *** $p < 0.001$ vs. untreated cells, # $p < 0.05$ vs. TNFα/IL-1β.

In several cellular trials, the modulation of NF-κB is strictly associated to reduction of gene targets expression. Thus, to follow this pathway and considering that IL-8 secretion increased under pro-inflammatory stimulus [19,45,46], this marker was chosen to be assessed under the different experimental conditions. IL-8 was measured in the basolateral compartment, which mimics the blood circulatory district, and reported as concentration (pg/mL) in Table 2. As expected, inflammatory treatment induced the IL-8 secretion of about six times higher than untreated cells. Then, to facilitate the comparison among groups under inflammation and to understand the effect of cinnamon treatment, data were normalized by cell protein concentration and expressed in relative terms to inflammatory treatment (Figure 6A).

Table 2. Concentration (pg/mL) of IL-8 secreted in the basolateral media at the different conditions.

	0	23 μg/mL	46 μg/mL	TNFα/IL-1β	23 μg/mL + TNFα/IL-1β	46 μg/mL + TNFα/IL-1β
pg/mL	48.76 ± 4.74	39.96 ± 2.42	42.87 ± 7.7	322 ± 16.96	275.52 ± 20.49	266.60 ± 13.85
Vs 0				***	***	***

Results reported are expressed in relative terms to inflammatory treatment and are shown as mean ± SEM from at least three independent experiments. Statistical significance: *** $p < 0.001$ vs. untreated cells.

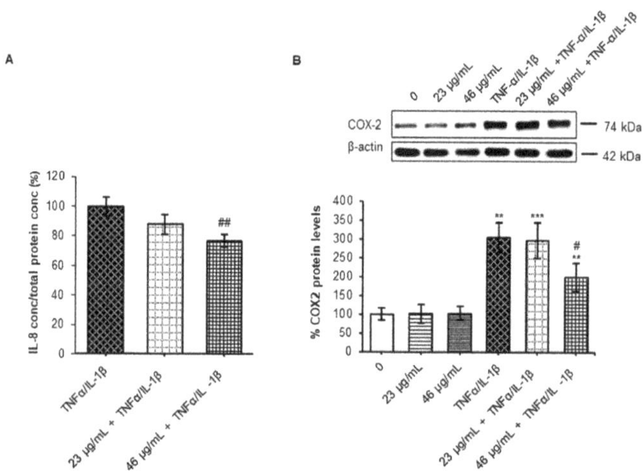

Figure 6. Evaluation of NFKB targets. Caco-2 cells, cultured on Transwell inserts for 21 days, were exposed to pro-inflammatory cytokines (TNFα 10 ng/mL + IL-1β 5 ng/mL) for 24 h after a 24 h pre-treatment with two different concentrations in polyphenols of digested cinnamon extract (23 and 46 μg/mL). Then, cell lysates and basolateral media were harvested. (**A**) IL-8 secretion was evaluated by ELISA assay. Data were normalized by cell total protein concentration. Results reported in histograms are expressed in relative terms to inflammatory treatment and are shown a mean ± SE from at least three independent experiments. Statistical significance: ## $p < 0.01$ vs. TNFα/IL-1β. (**B**) COX-2 protein levels were analyzed by SDS-PAGE electrophoresis and Western blotting. COX-2 was detected with a specific antibody and revealed by enhanced chemiluminescence (ECL). Samples were normalized on β-actin immunoreactivity. Histograms, obtained from at least three distinct experiments, represent the percentage of protein levels with respect to untreated cells as mean ± S.E. Statistical significance: ** $p < 0.01$ and *** $p < 0.001$ vs. CTRL and # $p < 0.05$ vs. TNFα/IL-1β.

The pre-treatment with cinnamon digested extract at the higher concentration significantly reduced the amount of IL-8 released by about 25%. This result is in line with a previous study showing that cinnamon extract reduced IL-8 secretion in Caco2 cells under inflammatory stimulus [7]. Indeed, increasing evi-dence links the bioactivity of polyphenols, and in particular flavonoids, to the modulation of IL-8 expression and secretion [47].

These data are particularly interesting since IL-8 concentrations in plasma of patients with chronic gastrointestinal disease are also maintained high in the remissive phases of pathologies [48]. Therefore, a decrease of IL-8 production and secretion could be an important factor to prevent a strong acute shot.

In order to deepen the anti-inflammatory properties of digested cinnamon extract, in parallel, the ciclooxigenase-2 (COX-2) protein levels were analyzed. This enzyme, among the NF-κB target, represents one of the most studied key markers of inflammation, because it catalyzes the arachidonic acid conversion in the inflammatory lipid mediators prostaglandins (PGE2).

According to previous evidence indicating that COX-2 protein expression augmented in several diseases characterized by inflammatory processes and oxidative stress, cytokines administration in Caco-2 cells induced an increment of the protein levels by about 2-fold with respect to untreated cells (200%). Interestingly, the pre-treatment with the higher digested extract concentration in polyphenols (46 μg/mL) reduced by about 35% the COX-2 levels observed under a pro-inflammatory stimulus (Figure 6B).

As a NF-κB target, COX2 expression is strictly associated to the nuclear factor regulation. Indeed, several studies reported that polyphenols, including apigenin, quercetin, kaempferol and resveratrol, inhibit COX activity both at the transcriptional levels [49]. Thus, COX2 inhibition by polyphenols might result in a PGE2 pro-inflammatory mediators reduction.

Taken together, these results showed a moderate but significant anti-inflammatory effect of digested cinnamon extract, due to all the polyphenols contained in the extract that might synergistically act to reduce acute inflammation through the inhibition of NF-κB activity.

4. Conclusions

This study suggests that even plant extracts obtained through eco-sustainable processes, for example, with aqueous solvents, can maintain their beneficial properties for human health. The antioxidant and anti-inflammatory actions are also maintained following the digestion process. Even if simulated digestion approaches with static models have some limitations, such as the inability to evaluate dynamic processes such as secretion flow rate or gastric emptying, they are widely used, due in part to their simplicity, cost-effectiveness, reproducibility and reliability. Indeed, the results obtained allowed us to hypothesize that food enriched with cinnamon extract retains the bioactive properties of the spice also after digestion.

Based on these results, two fundamental points emerge, (i) it is not important to know only which are the bioactive molecules but the analysis of the phyto-complex as a whole is the element of value that must be considered in order to be able to assign a real beneficial property to a plant extract; (ii) cellular tests show that the bioactive compounds are able to interact with cellular processes and act by preventing stress factors involving inflammation and oxidation pathways, hypothesizing an active role of cinnamon to prevent and/or mitigate non-communicable diseases, such as tumors and a wide panel of degenerative diseases. These evaluations could be of interest for the producers of cinnamon-based products to encourage the manufacturing of fresh foods, that are not subjected to cooking and/or excessive chemical-physical stress. At the same time, it would be essential to test not only the extract but the whole food product by subjecting it to digestion, chemical analysis and bioactivity study. This would make it possible to claim assumptions based on solid scientific grounds and not only on hypothetical properties related to the single bioactive compounds present in a certain extract.

Supplementary Materials: The following supporting information can be downloaded at: https://www.mdpi.com/article/10.3390/foods12030452/s1, Figure S1. Procyanidins dimer A-type and B-type. Table S1. Identification and quantification of metabolites in aqueous extract before and after digestion using mass spectrometry analysis by UHPLC-DAD-HESI-HRMS. In Table S1 are reported: peak number, probable identification, number of bonds for procyanidin species, retention time, chemical formula, theoretical and observed high-resolution m/z values, MS/MS fragments and quantification (mg/mL). (* quantified as *trans*-cinnamic acid; ** quantified as catechin).

Author Contributions: Experimental activities and analysis of the results, S.P., M.F., G.S., E.L. and F.R.; original draft preparation, I.B., S.P. and A.B.; review and editing, I.B. and A.B.; supervision of the chemical methodology and of the chemical result analysis, L.C. and P.R.; conceptualization and supervision, M.L., P.F. and P.P. All authors have read and agreed to the publication of the manuscript.

Funding: This research was funded by "FoodNET, Food Social Sensor Network", funded by Lombardy Region, Italy, grant number 2251551, POR FESR 2014–2020.

Data Availability Statement: All related data and methods are presented in this paper. Additional inquiries should be addressed to the corresponding author.

Acknowledgments: The authors wish to thank EPO srl., for the provided cinnamon samples.

Conflicts of Interest: The authors declare no conflict of interest. The funders had no role in the design of the study; in the collection, analyses, or interpretation of data; in the writing of the manuscript, or in the decision to publish the results.

References

1. Nabavi, S.F.; Di Lorenzo, A.; Izadi, M.; Sobarzo-Sánchez, E.; Daglia, M.; Nabavi, S.M. Antibacterial Effects of Cinnamon: From Farm to Food, Cosmetic and Pharmaceutical Industries. *Nutrients* **2015**, *7*, 7729–7748. [CrossRef] [PubMed]
2. Avula, B.; Smillie, T.J.; Wang, Y.H.; Zweigenbaum, J.; Khan, I.A. Authentication of True Cinnamon (*Cinnamon verum*) Utilising Direct Analysis in Real Time (DART)-QToF-MS. *Food Addit. Contam. Part A* **2015**, *32*, 1–8. [CrossRef] [PubMed]
3. Ford, P.W.; Harmon, A.D.; Tucker, A.O.; Sasser, M.; Jackoway, G.; Albornoz, G.; Grypa, R.D.; Pratt, J.L.; Cardellina, J.H. Cinnamon—Differentiation of Four Species by Linking Classical Botany to an Automated Chromatographic Authentication System. *J. AOAC Int.* **2019**, *102*, 363–368. [CrossRef] [PubMed]
4. Khasnavis, S.; Pahan, K. Sodium Benzoate, a Metabolite of Cinnamon and a Food Additive, Upregulates Neuroprotective Parkinson Disease Protein DJ-1 in Astrocytes and Neurons. *J. Neuroimmune Pharmacol.* **2012**, *7*, 424–435. [CrossRef] [PubMed]
5. Hemmati, A.A.; Alboghobeish, S.; Ahangarpour, A. Effects of Cinnamic Acid on Memory Deficits and Brain Oxidative Stress in Streptozotocin-Induced Diabetic Mice. *Korean J. Physiol. Pharmacol.* **2018**, *22*, 257–267. [CrossRef] [PubMed]
6. Liao, J.C.; Deng, J.S.; Chiu, C.S.; Hou, W.C.; Huang, S.S.; Shie, P.H.; Huang, G.J. Anti-Inflammatory Activities of Cinnamomum Cassia Constituents in Vitro and in Vivo. *Evid.-Based Complement. Altern. Med.* **2012**, *2012*, 429320. [CrossRef] [PubMed]
7. Kim, M.S.; Kim, J.Y. Intestinal Anti-Inflammatory Effects of Cinnamon Extracts in a Co-Culture Model of Intestinal Epithelial Caco-2 Cells and RAW264.7 Macrophages. *Appl. Biol. Chem.* **2017**, *60*, 553–561. [CrossRef]
8. Kim, M.S.; Kim, J.Y. Cinnamon Subcritical Water Extract Attenuates Intestinal Inflammation and Enhances Intestinal Tight Junction in a Caco-2 and RAW264.7 Co-Culture Model. *Food Funct.* **2019**, *10*, 4350–4360. [CrossRef]
9. Hagenlocher, Y.; Satzinger, S.; Civelek, M.; Feilhauer, K.; Köninger, J.; Bischoff, S.C.; Lorentz, A. Cinnamon Reduces Inflammatory Response in Intestinal Fibroblasts in Vitro and in Colitis in Vivo Leading to Decreased Fibrosis. *Mol. Nutr. Food Res.* **2017**, *61*, 1601085. [CrossRef]
10. Hariri, M.; Ghiasvand, R. Cinnamon and Chronic Diseases. *Adv. Exp. Med. Biol.* **2016**, *929*, 1–24. [CrossRef]
11. Quero, J.; Mármol, I.; Cerrada, E.; Rodríguez-Yoldi, M.J. Insight into the Potential Application of Polyphenol-Rich Dietary Intervention in Degenerative Disease Management. *Food Funct.* **2020**, *11*, 2805–2825. [CrossRef]
12. Liu, W.; Chen, X.; Li, H.; Zhang, J.; An, J.; Liu, X.; Liu, W.; Chen, X.; Li, H.; Zhang, J.; et al. Anti-Inflammatory Function of Plant-Derived Bioactive Peptides: A Review. *Foods* **2022**, *11*, 2361. [CrossRef]
13. Minekus, M.; Alminger, M.; Alvito, P.; Ballance, S.; Bohn, T.; Bourlieu, C.; Carrière, F.; Boutrou, R.; Corredig, M.; Dupont, D.; et al. A Standardised Static in Vitro Digestion Method Suitable for Food—An International Consensus. *Food Funct.* **2014**, *5*, 1113–1124. [CrossRef]
14. De Giani, A.; Pagliari, S.; Zampolli, J.; Forcella, M.; Fusi, P.; Bruni, I.; Campone, L.; Di Gennaro, P. Characterization of the Biological Activities of a New Polyphenol-Rich Extract from Cinnamon Bark on a Probiotic Consortium and Its Action after Enzymatic and Microbial Fermentation on Colorectal Cell Lines. *Foods* **2022**, *11*, 3202. [CrossRef]
15. Pineda-Vadillo, C.; Nau, F.; Dubiard, C.G.; Cheynier, V.; Meudec, E.; Sanz-Buenhombre, M.; Guadarrama, A.; Tóth, T.; Csavajda, É.; Hingyi, H.; et al. In Vitro Digestion of Dairy and Egg Products Enriched with Grape Extracts: Effect of the Food Matrix on Polyphenol Bioaccessibility and Antioxidant Activity. *Food Res. Int.* **2016**, *88*, 284–292. [CrossRef]
16. Pagliari, S.; Cannavacciuolo, C.; Celano, R.; Carabetta, S.; Russo, M.; Labra, M.; Campone, L. Valorisation, Green Extraction Development, and Metabolomic Analysis of Wild Artichoke By-Product Using Pressurised Liquid Extraction UPLC–HRMS and Multivariate Data Analysis. *Molecules* **2022**, *27*, 7157. [CrossRef]
17. Helal, A.; Tagliazucchi, D. Impact of In-Vitro Gastro-Pancreatic Digestion on Polyphenols and Cinnamaldehyde Bioaccessibility and Antioxidant Activity in Stirred Cinnamon-Fortified Yogurt. *LWT* **2018**, *89*, 164–170. [CrossRef]

18. Hidalgo, I.J.; Raub, T.J.; Borchardt, R.T. Characterization of the Human Colon Carcinoma Cell Line (Caco-2) as a Model System for Intestinal Epithelial Permeability. *Gastroenterology* **1989**, *96*, 736–749. [CrossRef]
19. Van De Walle, J.; Hendrickx, A.; Romier, B.; Larondelle, Y.; Schneider, Y.J. Inflammatory Parameters in Caco-2 Cells: Effect of Stimuli Nature, Concentration, Combination and Cell Differentiation. *Toxicol. Vitr.* **2010**, *24*, 1441–1449. [CrossRef]
20. Vance, P.G.; Keele, B.B.; Rajagopalan, K.V. Superoxide Dismutase from Streptococcus Mutans: Isolation and characterization of two forms of the enzyme. *J. Biol. Chem.* **1972**, *247*, 4782–4786. [CrossRef]
21. Oldani, M.; Manzoni, M.; Villa, A.M.; Stefanini, F.M.; Melchioretto, P.; Monti, E.; Forcella, M.; Urani, C.; Fusi, P. Cadmium Elicits Alterations in Mitochondrial Morphology and Functionality in C3H10T1/2Cl8 Mouse Embryonic Fibroblasts. *Biochim. Biophys. Acta—Gen. Subj.* **2020**, *1864*, 129568. [CrossRef] [PubMed]
22. Wang, Y.; Oberley, L.W.; Murhammer, D.W. Antioxidant Defense Systems of Two Lipidopteran Insect Cell Lines. *Free Radic. Biol. Med.* **2001**, *30*, 1254–1262. [CrossRef] [PubMed]
23. Nakamura, W.; Hosoda, S.; Hayashi, K. Purification and Properties of Rat Liver Glutathione Peroxidase. *Biochim. Biophys. Acta—Enzymol.* **1974**, *358*, 251–261. [CrossRef]
24. Habig, W.H.; Pabst, M.J.; Jakoby, W.B. Glutathione S-transferases: The first enzymatic step in mercapturic acid formation. *J. Biol. Chem.* **1974**, *249*, 7130–7139. [CrossRef] [PubMed]
25. Bradford, M.M. A Rapid and Sensitive Method for the Quantitation of Microgram Quantities of Protein Utilizing the Principle of Protein-Dye Binding. *Anal. Biochem.* **1976**, *72*, 248–254. [CrossRef]
26. Cimarosti, H.; Ashikaga, E.; Jaafari, N.; Dearden, L.; Rubin, P.; Wilkinson, K.A.; Henley, J.M. Enhanced SUMOylation and SENP-1 Protein Levels Following Oxygen and Glucose Deprivation in Neurones. *J. Cereb. Blood Flow Metab.* **2012**, *32*, 17–22. [CrossRef]
27. Wei, Z.; Chigurupati, S.; Arumugam, T.V.; Jo, D.G.; Li, H.; Chan, S.L. Notch Activation Enhances the Microglia-Mediated Inflammatory Response Associated With Focal Cerebral Ischemia. *Stroke* **2011**, *42*, 2589–2594. [CrossRef]
28. Lazarus, S.A.; Adamson, G.E.; Hammerstone, J.F.; Schmitz, H.H. High-Performance Liquid Chromatography/Mass Spectrometry Analysis of Proanthocyanidins in Foods and Beverages. *J. Agric. Food Chem.* **1999**, *47*, 3693–3701. [CrossRef]
29. Vallverdú-Queralt, A.; Regueiro, J.; Martínez-Huélamo, M.; Rinaldi Alvarenga, J.F.; Leal, L.N.; Lamuela-Raventos, R.M. A Comprehensive Study on the Phenolic Profile of Widely Used Culinary Herbs and Spices: Rosemary, Thyme, Oregano, Cinnamon, Cumin and Bay. *Food Chem.* **2014**, *154*, 299–307. [CrossRef]
30. Ali, A.; Wu, H.; Ponnampalam, E.N.; Cottrell, J.J.; Dunshea, F.R.; Suleria, H.A.R. Comprehensive Profiling of Most Widely Used Spices for Their Phenolic Compounds through Lc-Esi-Qtof-Ms2 and Their Antioxidant Potential. *Antioxidants* **2021**, *10*, 721. [CrossRef]
31. Klejdus, B.; Kováčik, J. Quantification of Phenols in Cinnamon: A Special Focus on "Total Phenols" and Phenolic Acids Including DESI-Orbitrap MS Detection. *Ind. Crops Prod.* **2016**, *83*, 774–780. [CrossRef]
32. Symma, N.; Hensel, A. Advanced Analysis of Oligomeric Proanthocyanidins: Latest Approaches in Liquid Chromatography and Mass Spectrometry Based Analysis. *Phytochem. Rev.* **2021**, *21*, 809–833. [CrossRef]
33. Quatrin, A.; Rampelotto, C.; Pauletto, R.; Maurer, L.H.; Nichelle, S.M.; Klein, B.; Rodrigues, R.F.; Maróstica Junior, M.R.; de Souza, F.B.; de Menezes, C.R.; et al. Bioaccessibility and Catabolism of Phenolic Compounds from Jaboticaba (*Myrciaria trunciflora*) Fruit Peel during in Vitro Gastrointestinal Digestion and Colonic Fermentation. *J. Funct. Foods* **2020**, *65*, 103714. [CrossRef]
34. Neilson, A.P.; Hopf, A.S.; Cooper, B.R.; Pereira, M.A.; Bomser, J.A.; Ferruzzi, M.G. Catechin Degradation with Concurrent Formation of Homo- and Heterocatechin Dimers during in Vitro Digestion. *J. Agric. Food Chem.* **2007**, *55*, 8941–8949. [CrossRef]
35. de Angelis, I.; Turco, L. Caco-2 Cells as a Model for Intestinal Absorption. *Curr. Protoc. Toxicol.* **2011**, *47*, 20.6.1–20.6.15. [CrossRef]
36. Durak, A.; Gawlik-Dziki, U.; Pecio, Ł. Coffee with Cinnamon—Impact of Phytochemicals Interactions on Antioxidant and Anti-Inflammatory in Vitro Activity. *Food Chem.* **2014**, *162*, 81–88. [CrossRef]
37. Nordberg, J.; Arnér, E.S.J. Reactive Oxygen Species, Antioxidants, and the Mammalian Thioredoxin System. *Free Radic. Biol. Med.* **2001**, *31*, 1287–1312. [CrossRef]
38. SIES, H. Strategies of Antioxidant Defense. *Eur. J. Biochem.* **1993**, *215*, 213–219. [CrossRef]
39. Primiano, T.; Sutter, T.R.; Kensler, T.W. Redox Regulation of Genes That Protect Against Carcinogens. *Comp. Biochem. Physiol. Part B Biochem. Mol. Biol.* **1997**, *118*, 487–497. [CrossRef]
40. Bernardi, S.; Del Bo, C.; Marino, M.; Gargari, G.; Cherubini, A.; Andrés-Lacueva, C.; Hidalgo-Liberona, N.; Peron, G.; González-Dominguez, R.; Kroon, P.; et al. Polyphenols and Intestinal Permeability: Rationale and Future Perspectives. *J. Agric. Food Chem.* **2020**, *68*, 1816–1829. [CrossRef]
41. Valdez, J.C.; Cho, J.; Bolling, B.W. Aronia Berry Inhibits Disruption of Caco-2 Intestinal Barrier Function. *Arch. Biochem. Biophys.* **2020**, *688*, 108409. [CrossRef] [PubMed]
42. Serra, G.; Incani, A.; Serreli, G.; Porru, L.; Melis, M.P.; Tuberoso, C.I.G.; Rossin, D.; Biasi, F.; Deiana, M. Olive Oil Polyphenols Reduce Oxysterols -Induced Redox Imbalance and pro-Inflammatory Response in Intestinal Cells. *Redox Biol.* **2018**, *17*, 348–354. [CrossRef] [PubMed]
43. Choy, K.W.; Murugan, D.; Leong, X.F.; Abas, R.; Alias, A.; Mustafa, M.R. Flavonoids as Natural Anti-Inflammatory Agents Targeting Nuclear Factor-Kappa B (NFκB) Signaling in Cardiovascular Diseases: A Mini Review. *Front. Pharmacol.* **2019**, *10*, 1295. [CrossRef] [PubMed]
44. Di Stasi, L.C.; Matos, M.J. Coumarin Derivatives in Inflammatory Bowel Disease. *Molecules* **2021**, *26*, 422. [CrossRef]

45. Schink, A.; Naumoska, K.; Kitanovski, Z.; Kampf, C.J.; Fröhlich-Nowoisky, J.; Thines, E.; Pöschl, U.; Schuppan, D.; Lucas, K. Anti-Inflammatory Effects of Cinnamon Extract and Identification of Active Compounds Influencing the TLR2 and TLR4 Signaling Pathways. *Food Funct.* **2018**, *9*, 5950–5964. [CrossRef]
46. Chen, Y.; Patel, A.; Meier, P.P.; Fantuzzi, G. Digested Early Preterm Human Milk Suppresses Tumor Necrosis Factor-Induced Inflammation and Cytotoxicity in Intestinal Epithelial Cells. *J. Pediatr. Gastroenterol. Nutr.* **2018**, *66*, e153–e157. [CrossRef]
47. Al-Khayri, J.M.; Sahana, G.R.; Nagella, P.; Joseph, B.V.; Alessa, F.M.; Al-Mssallem, M.Q. Flavonoids as Potential Anti-Inflammatory Molecules: A Review. *Molecules* **2022**, *27*, 2901. [CrossRef]
48. Bourgonje, A.R.; Hu, S.; Spekhorst, L.M.; Zhernakova, D.V.; Vila, A.V.; Li, Y.; Voskuil, M.D.; Van Berkel, L.A.; Folly, B.B.; Charrout, M.; et al. The Effect of Phenotype and Genotype on the Plasma Proteome in Patients with Inflammatory Bowel Disease. *J. Crohn's Colitis* **2022**, *16*, 414–429. [CrossRef]
49. Salaritabar, A.; Darvishi, B.; HadjiakhoonDi, F.; Manayi, A.; Sureda, A.; Nabavi, S.F.; Fitzpatrick, L.R.; Nabavi, S.M.; Bishayee, A. Therapeutic Potential of Flavonoids in Inflammatory Bowel Disease: A Comprehensive Review. *World J. Gastroenterol.* **2017**, *23*, 5097–5114. [CrossRef]

Disclaimer/Publisher's Note: The statements, opinions and data contained in all publications are solely those of the individual author(s) and contributor(s) and not of MDPI and/or the editor(s). MDPI and/or the editor(s) disclaim responsibility for any injury to people or property resulting from any ideas, methods, instructions or products referred to in the content.

foods MDPI

Communication

Aqueous Extract from *Cuminum cyminum* L. Seed Alleviates Ovalbumin-Induced Allergic Rhinitis in Mouse via Balancing of Helper T Cells

Momoko Ishida [1], Fuka Miyagawa [2], Kosuke Nishi [1,2] and Takuya Sugahara [1,2,*]

[1] Department of Bioscience, Graduate School of Agriculture, Ehime University, Matsuyama 790-8566, Ehime, Japan
[2] Food and Health Sciences Research Center, Ehime University, Matsuyama 790-8566, Ehime, Japan
* Correspondence: mars95@agr.ehime-u.ac.jp; Tel.: +81-89-946-9863

Abstract: *Cuminum cyminum* L. (cumin) seeds are widely used as a spice. Although we previously reported that the aqueous extract of cumin seeds suppresses the degranulation of rat basophilic RBL-2H3 cells, it has not been clarified whether the extract alleviates actual allergy symptoms in vivo. Therefore, in this study, we investigated the effect of oral administration of cumin seed aqueous extract (CAE) in ovalbumin (OVA)-induced allergic rhinitis. BALB/c mice were randomly divided into the following three groups: control group (five mice), OVA group (five mice), and OVA + CAE group (five mice). Allergic rhinitis was induced by sensitization (intraperitoneal, 25 μg OVA and 1.98 mg aluminum hydroxide gel) followed by challenge (intranasal, 400 μg OVA). The oral administration of CAE (25 mg/kg) reduced the sneezing frequency of OVA-induced allergic rhinitis model mice. In addition to reducing the serum immunoglobulin E and IL-4 levels, the oral administration of CAE reduced the production of T-helper type-2 (Th2) cytokines (IL-4, IL-5, IL-10, and IL-13) in the splenocytes of the model mice. Furthermore, a significant increase in the ratio of Th1 to Th2 cells was observed in the CAE-administered group. Our findings suggest that the ingestion of CAE improves T cell balance, the dominant state of Th2, and alleviates allergic rhinitis symptoms.

Keywords: *Cuminum cyminum* L.; ovalbumin; allergic rhinitis; T-helper cells; T cell balance

1. Introduction

Allergies are caused by an excessive reaction of the immune system to harmless substances (for example, pollen and food molecules) and are classified into four types (types I to IV) according to the mechanism of action. Type-I allergies, including hayfever, allergic rhinitis, and food allergies, involve a reaction called degranulation that occurs in mast cells and basophils. In response to cross-linking between the Fc receptor on the cell surface and immunoglobulin E (IgE)-antigen, the intracellular granules are released from the cells [1,2]. Chemical mediators, including histamine and leukotrienes, in granules cause immediate allergy symptoms by stimulating nerves and dilating blood vessels [3]. In addition, cytokines such as interleukin (IL)-33, IL-25 and thymic stromal lymphopoietin (TSLP) are produced from mucosal epithelial cells when stimulated by antigens; additionally, these cytokines act on activated T-helper type-2 (Th2) cells to secrete Th2-type cytokines such as IL-4, IL-5, and IL-13, thereby increasing the allergic response [4–8]. Recent studies have shown that activation of the phosphatidylinositol-3- kinase (PI3K), mitogen-activated protein kinase (MAPK), and nuclear transcription factor κB (NF-κB) pathways contributes to T cell activation, function, and differentiation, leading to eosinophilic recruitment and T-helper type-1 (Th1)/Th2 imbalance [9–13]. In addition, the treatment of inhibitors showed a reduction in Th2 cytokine production and the degree of allergic symptoms [11–13]. Therefore, the inhibition of Th2 cytokine secretion and improvement of T cell immune imbalance, as well as inhibition of histamine release and its action, could lead to preventive and therapeutic strategies for treating allergic diseases.

Cuminum cyminum L. (cumin) belongs to the *Apiaceae* (*Umbelliferae*) family; cumin seeds have a unique aroma and are widely used as a spice. The essential oil obtained from the seeds contains volatile components such as cuminaldehyde (4-isopropylbenzaldehyde), γ-terpinene, and β-pinene [14]. The essential oils and cuminaldehyde have been reported to have many biological functions, including anti-inflammatory, anti-diabetic, anti-oxidative, and anti-bacterial activities [14–19]. However, there are few reports concerning the functions of the water-soluble components in cumin seeds. We previously reported that an aqueous extract from cumin seed suppressed the antigen-induced degranulation of rat basophilic leukemia RBL-2H3 cells and passive cutaneous anaphylaxis (PCA) reaction in mice [20]. These findings suggest that the extract may alleviate allergic symptoms caused by histamine. In allergic reactions, not only chemical mediators released by degranulation but also allergy-related cytokines produced by immune cells, such as Th1 and Th2 cells, are involved [21,22]. However, it has not been clarified whether the aqueous extract from cumin seeds affects the secretion of cytokines involved in allergic reactions and actual allergy symptoms. Since cumin seed extract has been shown to downregulate PI3K phosphorylation [20], it may also exert inhibitory effects on T cell activation and differentiation by inhibiting the PI3K pathway. Hence, in this study, we investigated whether ingestion of cumin seed extract alleviates allergic symptoms and improves the secretion of T cell cytokines and T cell imbalance in an ovalbumin (OVA)-induced allergic rhinitis mouse.

2. Materials and Methods

2.1. Reagents

Albumin from chicken egg white (OVA), fetal bovine serum (FBS), penicillin, Roswell Park Memorial Institute 1640 (RPMI 1640) medium, and streptomycin were purchased from Sigma-Aldrich (St. Louis, MO, USA). Purified rat anti-mouse IgE, biotin rat anti-mouse IgE, goat affinity purified antibody to mouse IgG, and biotin-XX goat anti-mouse IgG_1 were purchased from BD Biosciences (Franklin Lakes, NJ, USA), MP Biomedicals (Santa Ana, CA, USA), and Life Technologies (Carlsbad, CA, USA), respectively.

2.2. Sample Preparation

Cumin seed powder imported from Turkey was purchased by S&B Foods Inc, (Tokyo, Japan). A sodium phosphate buffer (NaPB; pH 7.4) was used as an extraction solvent to prevent the pH from changing significantly depending on the environment and sample concentration during sample storage or use. The powder was suspended in 10 mM NaPB at 0.05 g/mL, and the suspension was stirred at 30 rpm for 24 h using an RT-30mini (TAITEC Corporation, Saitama, Japan) in a low-temperature room at 12 °C. After centrifugation at 15,000$\times$ g and 4 °C for 20 min, the supernatant was collected, adjusted to a pH of 7.4, filtered using a 0.2 μm membrane for sterilization, and used as cumin seed aqueous extract (CAE).

2.3. Animals

BALB/c mice were acquired from Clea Japan (Tokyo, Japan) and housed in an animal room in the following conditions: a temperature of 24 $\pm$ 1 °C and 12 h alternating light and dark. All mice were fed with a standard diet and water ad libitum and acclimatized for 1 week before experiments. Animal experiments were approved by the Animal Experiment Committee of Ehime University and were performed in accordance with the Guidelines of Animal Experiments of Ehime University (approval number: 08U24-1).

2.4. Mouse Model of OVA-Induced Allergic Rhinitis

A mouse model of OVA-induced allergic rhinitis was established, as shown in Figure 1a. Six-week-old female BALB/c mice (15 mice) were randomly divided into the following three groups: control group (5 mice), OVA group (5 mice), and OVA + CAE group (5 mice).

The mice of the OVA and OVA + CAE groups were sensitized with intraperitoneal injections of 200 μL phosphate buffered saline (PBS) containing 25 μg OVA and 1.98 mg aluminum hydroxide gel (Fujifilm Wako Pure Chemical, Osaka, Japan) on days 0, 7, and 14. After 1 week (day 21), the mice of the OVA and OVA + CAE groups were challenged with intranasal administration of 10 μL of PBS containing 400 μg OVA for 8 consecutive days from day 21 to 28. The mice of the control group were neither sensitized nor challenged with OVA but were injected intranasally with PBS only on day 28. The mice of the OVA + CAE group were orally administered CAE (25 mg/kg) for 8 consecutive days from day 21 to 28, while the mice of the control and OVA groups were orally administered 10 mM NaPB as a vehicle. The dosage of CAE was based on a previous study [20]. On day 28, the mice were counted for the frequency of sneezing and rubbing for 20 min after the challenge with OVA or PBS. On day 29, the mice were sacrificed, and blood and spleen were collected for further analysis.

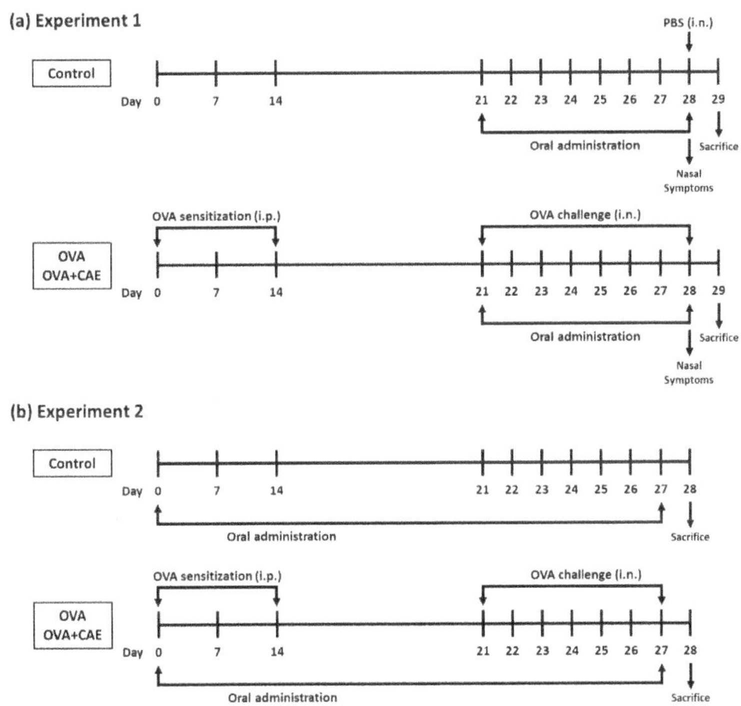

Figure 1. Experimental design used for inducing allergic rhinitis in mice. Abbreviations used: OVA: ovalbumin; PBS: phosphate buffered saline; CAE: cumin seed aqueous extract; i.p.: intraperitoneally; i.n.: intranasally. (**a**) Experiment 1 was conducted to examine the effect of CAE on OVA challenge-induced allergic rhinitis symptoms and biological parameters, and (**b**) Experiment 2 to examine the effect of CAE on T cell balance in OVA-induced allergic rhinitis mice.

2.5. *Biochemical Measurement in the Serum of OVA-Induced Allergic Rhinitis Mice*

On day 29, all mice were anesthetized with isoflurane, and blood was collected through cardiac puncture in a MiniCollect II tube (Greiner Bio-One, Kremsmünster, Austria). The tube was centrifuged at 3000× *g* and 15 °C for 10 min to obtain the serum. The samples were stored at −80 °C until further analysis. The concentrations of total IgE and total IgG_1 were evaluated using an in-house-developed ELISA as described previously [23]. The concentrations of IL-4 and interferon gamma (IFN-γ) were evaluated using respective ELISA quantitation kits (eBioscience, San Diego, CA, USA) as per the manufacturer's protocols.

The absorbance values at 415 nm for total IgE and IgG_1 and at 450 nm for the other studied factors were measured using an iMark microplate reader (Bio-Rad Laboratories, Hercules, CA, USA), and used to calculate the final concentrations.

2.6. Preparation of Splenocytes

After blood collection, the spleen was harvested and kept in sterile PBS until preparation. The spleen was passed through a 40 μm cell strainer to obtain a single cell suspension, and then the suspension underwent red blood cell lysis. The obtained splenocytes were seeded at 1.25×10^6 cells/well to a 48-well culture plate and cultured in 5% FBS-RPMI 1640 containing 100 μg/mL OVA for 72 h at 37 °C. The concentrations of total IgE and total IgG_1 in the cultured media were detected using an in-house-developed ELISA, as described previously [23]. IL-4, IL-5, IL-10, IL-13, and IFN-γ in culture media were detected by the respective ELISA quantitation kits as per the manufacturer's protocols. The absorbance values at 415 nm for total IgE and IgG_1 and at 450 nm for the other studied factors were measured using an iMark microplate reader and were used to calculate the final concentrations.

2.7. Flow Cytometric Analysis

The effect of CAE on T cell balance in the spleen of OVA-induced allergic rhinitis mice was examined in another experiment in order to obtain a greater difference. The OVA-induced model mice were established as described above, and CAE was orally administered from days 0 to 27 (Figure 1b). On day 28, the splenocytes were prepared as described above and stained using the Mouse Th1/Th2/Th17 Phenotyping Kit (BD Biosciences, Franklin Lakes, NJ, USA). Briefly, the splenocytes were seeded at 2.0×10^6 cells/well to a 24-well culture plate and cultured in 5% FBS-RPMI 1640 containing 50 ng/mL phorbol 12-myristate 13-acetate (PMA) and 1 μg/mL ionomycin for 5 h at 37 °C. After incubation, the cells were fixed, permeabilized, and stained according to the protocol using each reagent provided in the kit. The stained cells were suspended with 2% FBS-PBS prior to flow cytometric analysis. To detect Th1 and Th2 cells, the splenocytes were stained with fluorescein isothiocyanate (FITC)-labeled anti-IFN-γ, allophycocyanin (APC)-labeled anti-IL-4, and peridinin chlorophyll protein-cyanine5.5 (PerCP-Cy5.5)-labeled anti-CD4. The distribution of Th1 cells ($IFN\text{-}\gamma^+/CD4^+$) and Th2 cells ($IL\text{-}4^+/CD4^+$) was analyzed using a FACSVerse flow cytometer (Becton Dickinson, Franklin Lakes, NJ, USA). The $CD4^+$ cell population was gated among 10,000 cells, and the proportion of $IFN\text{-}\gamma^+$ or $IL\text{-}4^+$ cells was detected among them.

2.8. Statistical Analysis

Data are presented as the mean ± standard error of the mean (SEM). Statistical analysis was performed using GraphPad Prism version 8.4.3 (GraphPad Software, San Diego, CA, USA). Data were analyzed using one-way analysis of variance followed by post-hoc Dunnett's test. A value of $p < 0.05$ was considered statistically significant.

3. Results

3.1. Effect of Oral Administration of CAE on Allergy Symptoms in OVA-Induced Allergic Rhinitis Mice

Body weight loss due to allergy induction was observed in the OVA and OVA + CAE groups, whereas there was no significant difference between the groups (Table 1). There was a significant difference in the sneezing frequency between the control and OVA groups ($p < 0.01$), and the OVA + CAE group showed a decreasing tendency ($p = 0.076$) compared to the OVA group. In contrast, the rubbing frequency increased in the OVA group, but the differences between the OVA and OVA + CAE groups were not significant ($p = 0.834$).

Table 1. Effect of CAE administration on body weight and OVA challenge-induced nasal sneezing and rubbing in allergic rhinitis mice.

Parameters	Control	OVA	OVA + CAE	OVA vs. Control	OVA vs. OVA + CAE
	Mean ± SEM (n = 5)			p-Value	p-Value
Body weight (g) on day 29	23.6 ± 0.3	21.1 ± 0.7	21.0 ± 0.5	0.010	0.995
		OVA challenge on day 28			
Sneezing (number)	6.8 ± 1.6	50.0 ± 10.8	27.0 ± 6.0	0.002	0.076
Rubbing (number)	21.6 ± 7.9	44.2 ± 5.2	38.8 ± 9.0	0.098	0.834

Abbreviations used: OVA, ovalbumin; CAE, cumin seed aqueous extract; SEM, standard error of the mean. Data were analyzed by one-way analysis of variance followed by post-hoc Dunnett's test.

3.2. *Effect of CAE on Serum Levels of Immunoglobulins and Cytokines of OVA-Induced Allergic Rhinitis Mice*

Serum IgE, IgG_1, and IL-4 levels in the OVA group were significantly increased by OVA injection (Table 2). In contrast, the administration of CAE significantly reduced the serum levels of IgE ($p < 0.01$) and IL-4 ($p < 0.05$). There was no difference in the serum level of IgG_1 between the OVA and OVA + CAE groups. Although the serum level of IFN-γ was not significantly different between these groups, it increased in the OVA + CAE group compared to the OVA group ($p = 0.325$).

Table 2. Effect of CAE administration on OVA-induced alterations in serum levels of immunoglobulins and cytokines in allergic rhinitis mice.

Serum Levels	Control	OVA	OVA + CAE	OVA vs. Control	OVA vs. OVA + CAE
	Mean ± SEM (n = 5)			p-Value	p-Value
IgE (μg/mL)	0.9 ± 0.3	41.9 ± 3.7	29.3 ± 1.4	<0.001	0.004
IgG_1 (mg/mL)	0.8 ± 0.2	5.9 ± 0.2	5.6 ± 0.3	<0.001	0.438
IL-4 (pg/mL)	1.3 ± 0.2	9.5 ± 2.4	3.7 ± 0.6	0.003	0.023
IFN-γ (pg/mL)	769.5 ± 344.9	62.0 ± 37.3	716.1 ± 474.2	0.276	0.325

Abbreviations used: OVA, ovalbumin; CAE, cumin seed aqueous extract; SEM, standard error of the mean; IgE, immunoglobulin E; IgG_1, immunoglobulin G_1; IL-4, interleukin-4; IFN-γ, interferon-γ. Data were analyzed by one-way analysis of variance followed by post-hoc Dunnett's test.

3.3. *Effect of CAE on Production of Immunoglobulins and Cytokines in Splenocytes of OVA-Induced Allergic Rhinitis Mice*

The production of IgE and IgG_1 in the OVA group significantly increased following OVA injection (Table 3). In the OVA + CAE group, the IgG_1 production was not significantly affected. In addition, there was a slight decrease in IgE production; however, no significant difference ($p = 0.207$) was detected from the OVA group. IL-10 production was high in the OVA group, whereas it was significantly reduced in the OVA + CAE group ($p < 0.05$). IFN-γ production increased in the OVA + CAE group compared to that in the OVA group, and it was similar to that in the control group, though not significantly ($p = 0.650$). Conversely, the production of IL-4 ($p < 0.05$), IL-5 ($p < 0.01$), and IL-13 ($p = 0.066$) reduced in the OVA + CAE group.

Table 3. Effect of CAE administration on OVA-induced alterations in the production of immunoglobulins and cytokines in splenocytes of allergic rhinitis mice.

Serum Levels	Control	OVA	OVA + CAE	OVA vs. Control	OVA vs. OVA + CAE
	Mean ± SEM (n = 5)			p-Value	p-Value
IgE (ng/mL)	6.5 ± 2.1	37.7 ± 6.7	24.4 ± 6.9	0.004	0.207
IgG_1 (ng/mL)	7.9 ± 4.5	131.1 ± 5.0	124.9 ± 10.3	<0.001	0.767
IL-4 (pg/mL)	7.6 ± 0.6	875.6 ± 178.3	381.2 ± 121.1	<0.001	0.029
IL-5 (pg/mL)	8.7 ± 0.3	744.9 ± 73.7	331.0 ± 96.0	<0.001	0.002
IL-10 (pg/mL)	31.7 ± 4.0	829.9 ± 47.8	556.3 ± 118.9	<0.001	0.041
IL-13 (ng/mL)	0.1 ± 0.0	3.3 ± 0.4	2.0 ± 0.6	<0.001	0.066
IFN-γ (ng/mL)	1.7 ± 0.4	1.0 ± 0.2	1.5 ± 0.6	0.429	0.650

Abbreviations used: OVA: ovalbumin; CAE: cumin seed aqueous extract; SEM: standard error of the mean; IgE: immunoglobulin E; IgG_1: immunoglobulin G_1; IL-4: interleukin-4; IL-5: interleukin-5; IL-10: interleukin-10; IL-13: interleukin-13; IFN-γ: interferon-γ. Data were analyzed using one-way analysis of variance followed by post-hoc Dunnett's test.

3.4. *Effect of CAE on Th1/Th2 Balance*

The proportion of Th1/Th2 cells in the spleen was analyzed by performing flow cytometry. In this experiment, the oral administration period of CAE was changed from that used at the start of the experiment to obtain a greater difference (Figure 1b). The proportion of Th1 cells (IFN-γ^+/$CD4^+$ cells) was slightly increased in the OVA + CAE group compared to that in the OVA group, but there was no significant difference (p = 0.1767) (Figure 2). On the other hand, the proportion of Th2 cells (IL-4^+/$CD4^+$ cells) did not decrease significantly (p = 0.4867). A significant increase in the Th1/Th2 ratio was confirmed in the OVA + CAE group compared to the OVA group ($p < 0.05$).

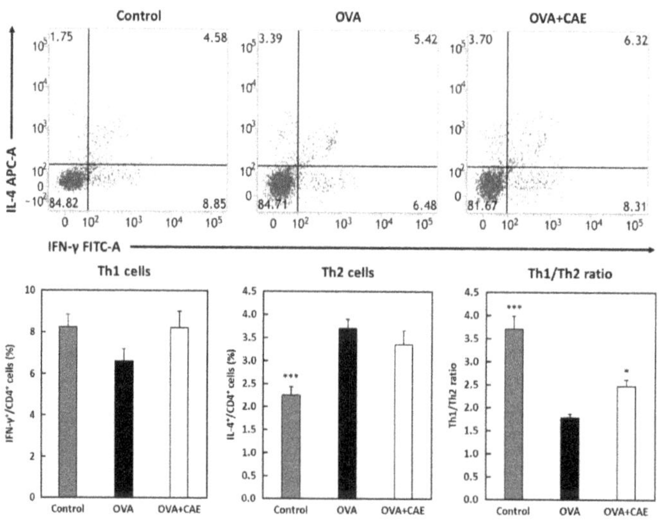

Figure 2. The percentage of Th1 and Th2 cells and Th1/Th2 ratio in spleen. Abbreviations used: OVA: ovalbumin; CAE: cumin seed aqueous extract; SEM: standard error of the mean. Splenocytes were cultured with phorbol 12-myristate 13-acetate (PMA) and ionomycin for 5 h at 37 °C. The distribution of Th1 cells (IFN-γ^+/$CD4^+$) and Th2 cells (IL-4^+/$CD4^+$) were analyzed using a flow cytometer. The Th1/Th2 ratio was observed by the ratio of the percentage of Th1 cells to the percentage of Th2 cells. Gray bar, closed bar, and open bar represent control, OVA, and OVA + CAE groups, respectively. Data are expressed as means ± SEM (n = 7). * $p < 0.05$ and *** $p < 0.001$ against OVA group by post hoc Dunnett's test.

4. Discussion

The in vivo experiments showed that the oral administration of CAE reduced the sneezing frequency in mice with allergic rhinitis induced by OVA (Table 1). In addition, the serum IgE, serum IL-4, and Th2 cytokines (IL-4, IL-5, IL-10, and IL-13) produced by splenocytes were significantly reduced in the OVA + CAE group (Tables 2 and 3). IL-4 and IL-13 are anti-inflammatory cytokines produced by activated Th2 cells. These cytokines are associated with the differentiation and maturation of B cells, which promote the expression of CD23 and MHC class II molecules and class switching to IgE [24]. IL-5 is a hematopoietic cytokine that plays an important role in the differentiation, maturation, recruitment, and activation of eosinophils at allergic inflammation sites [25]. The suppression of Th2 cytokine production from Th2 cells by the oral administration of CAE likely regulates the class switch to IgE and the infiltration of immune cells into inflammation sites, leading to a decrease in serum IgE levels and the reduction in allergy symptoms. IL-10 production in splenocytes was also reduced in the OVA + CAE group (Table 3). IL-10 is an anti-inflammatory cytokine produced by Th2 cells that regulates abnormal immune responses by acting on the activation and antigen-presenting ability of monocytes and T cells exposed to antigens. However, studies have also reported that IL-10 is essential for inducing allergy symptoms [26,27]. In our study, OVA-induced allergic rhinitis symptoms were alleviated. Therefore, it is unlikely that the suppression of IL-10 production had a negative effect. Despite the many studies conducted on this topic, the action and regulation of IL-10 in allergic reactions has not yet been elucidated; however, our results lead us to speculate that the suppression of IL-10 production by the oral administration of CAE may be involved in the alleviation of OVA-induced allergy symptoms.

The level of IFN-γ in both serum and splenocytes was slightly increased in the OVA + CAE group, suggesting that the oral administration of CAE may regulate the dominant state of Th2 cells. A significant increase in the Th1/Th2 ratio was observed in the OVA + CAE group (Figure 2). IFN-γ and IL-12 play important roles in the differentiation of Th1 cells; IL-12 is a heterodimeric pro-inflammatory cytokine that induces IFN-γ production and Th1 cell differentiation [28]. The differentiation of Th1 cells by IL-12 occurs via the activation of the transcription factor STAT4, which induces IFN-γ production [29]. The increase in the IFN-γ production in the splenocytes appeared to be slightly correlated with an increased proportion of Th1 cells in the spleen. In contrast, the production of Th2 cytokines in the splenocytes decreased, whereas the proportion of Th2 cells did not decrease. These results suggest that CAE suppresses the cytokine secretion ability of Th2 cells but does not affect the proportion of Th2 cells in the spleen; furthermore, it improves the Th1/Th2 balance by increasing the proportion of Th1 cells, resulting in the alleviation of allergic symptoms. PI3K/AKT, MAPK, and NF-κB pathways are known to be involved in T cell activation and differentiation, contributing to allergic inflammation by promoting Th1/Th2 imbalance [9–13]. Our previous study suggested that cumin seed aqueous extract contains substances downregulating PI3K activation [20]. Thus, the active substances in CAE may suppress T cell activation and differentiation by directly or indirectly inhibiting the PI3K pathway. Further research is needed to identify the active substances in CAE regulating T cell balance and to elucidate their mechanism of action.

The anti-allergic effect of CAE was also exerted in vivo, suggesting that the active substance is efficiently digested and absorbed when ingested orally. In our previous study, we identified the active substances contained in cumin seeds that inhibit degranulation in RBL-2H3 cells, i.e., apigenin, luteolin, and umbelliferose [30]. Apigenin and luteolin have also been reported to attenuate OVA-induced allergic rhinitis [31–34]. However, since apigenin and luteolin are practically insoluble in water, they are hardly extracted in CAE. Umbelliferose, a characteristic trisaccharide present in plants of the *Apiaceae* (*Umbelliferae*) family, has been generally recognized as an indigestible oligosaccharide. However, some studies have demonstrated that dietary raffinose, which is an isomer of umbelliferose, is absorbed intact in the gastrointestinal tract and exerts an anti-allergic effect [35,36]. Therefore, it is possible that umbelliferose is also absorbed from the gastrointestinal tract and exerts an anti-allergic effect in vivo. On the other hand, CAE dialyzed with a membrane of a molecular weight cut off of 500 showed a partial decrease in anti-degranulation activity compared to non-dialyzed CAE; however, the activity was still detected (Figure S1). This result suggests that CAE contains not only active substances with a molecular weight of 500 or less, but also active substances of a high molecular weight. In addition, the active substances identified in cumin seeds mentioned above are present only in small quantities, suggesting that interactions between water-soluble substances, including substances that have yet to be identified, are possibly responsible for the observed strong activities of CAE. The water-soluble cumin seed ingredients need to be further analyzed to obtain further insight into their anti-allergic properties.

In conclusion, our findings suggest that the ingestion of CAE improves the T-cell balance and alleviates allergy symptoms of allergic rhinitis in vivo. This study, which focused on the water-soluble ingredients in cumin seeds that have not previously received much attention, can contribute to the understanding of the value of cumin seeds and to the library of food-derived functional ingredients, thereby developing functional food science. In addition, cumin seeds are widely used as a spice, and the volatile non-polar ingredients in the seeds are extracted as essential oils for a range of uses. However, many promising water-soluble ingredients that remain in the seed are wasted after extraction. We believe that these findings provide insights into the biological functions of cumin seed ingredients in the field of functional foods and contribute to the effective utilization of cumin seed residue after oil extraction.

Supplementary Materials: The following supporting information can be downloaded at: https://www.mdpi.com/article/10.3390/foods11203224/s1, Figure S1: Effect of dialysis treatment on the anti-degranulation activity of cumin seed aqueous extract (CAE).

Author Contributions: Conceptualization, K.N. and T.S.; methodology, M.I., F.M., K.N. and T.S.; formal analysis, M.I. and F.M.; investigation, M.I. and F.M.; data curation, M.I. and F.M.; writing—original draft preparation, M.I.; writing—review and editing, K.N. and T.S.; supervision, T.S.; project administration, K.N. and T.S.; funding acquisition; K.N. and T.S. All authors have read and agreed to the published version of the manuscript.

Funding: This work was supported in part by a Grants-in-Aid for Scientific Research (KAKENHI Grant Number JP20K05891) from the Japan Society for the Promotion of Science.

Data Availability Statement: The data that support the findings in this study are available from the corresponding author upon reasonable request.

Acknowledgments: Animal experiments were carried out at the Division of Genetic Research of the Advanced Research Support Center (ADRES), Ehime University.

Conflicts of Interest: The authors declare no conflict of interest.

References

1. Turner, H.; Kinet, J.-P. Signalling through the high-affinity IgE receptor FcεRI. *Nature* **1999**, *402*, 24–30. [CrossRef]
2. Metcalfe, D.D.; Peavy, R.D.; Gilfillan, A.M. Mechanisms of mast cell signaling in anaphylaxis. *J. Allergy Clin. Immunol.* **2009**, *124*, 639–646. [CrossRef] [PubMed]
3. White, M. Mediators of inflammation and the inflammatory process. *J. Allergy Clin. Immunol.* **1999**, *103*, S378–S381. [CrossRef]
4. Ohno, T.; Morita, H.; Arae, K.; Matsumoto, K.; Nakae, S. Interleukin-33 in allergy. *Allergy* **2012**, *67*, 1203–1214. [CrossRef]
5. Saluja, R.; Khan, M.; Church, M.K.; Maurer, M. The role of IL-33 and mast cells in allergy and inflammation. *Clin. Transl. Allergy* **2015**, *5*, 33. [CrossRef]
6. Kubo, M. Innate and adaptive type 2 immunity in lung allergic inflammation. *Immunol. Rev.* **2017**, *278*, 162–172. [CrossRef] [PubMed]
7. Cayrol, C.; Girard, J.-P. Interleukin-33 (IL-33): A nuclear cytokine from the IL-1 family. *Immunol. Rev.* **2017**, *281*, 154–168. [CrossRef] [PubMed]
8. Murdaca, G.; Greco, M.; Tonacci, A.; Negrini, S.; Borro, M.; Puppo, F.; Gangemi, S. IL-33/IL-31 Axis in Immune-Mediated and Allergic Diseases. *Int. J. Mol. Sci.* **2019**, *20*, 5856. [CrossRef] [PubMed]
9. Zhang, W.; Nie, Y.; Chong, L.; Cai, X.; Zhang, H.; Lin, B.; Liang, Y.; Li, C. PI3K and Notch signal pathways coordinately regulate the activation and proliferation of T lymphocytes in asthma. *Life Sci.* **2013**, *92*, 890–895. [CrossRef] [PubMed]

10. Liu, L.-L.; Li, F.-H.; Zhang, Y.; Zhang, X.-F.; Yang, J. Tangeretin has anti-asthmatic effects via regulating PI3K and Notch signaling and modulating Th1/Th2/Th17 cytokine balance in neonatal asthmatic mice. *Braz. J. Med. Biol. Res.* **2017**, *50*, e5991. [CrossRef]
11. Huang, L.; Wang, M.; Yan, Y.; Gu, W.; Zhang, X.; Tan, J.; Sun, H.; Ji, W.; Chen, Z. OX40L induces helper T cell differentiation during cell immunity of asthma through PI3K/AKT and P38 MAPK signaling pathway. *J. Transl. Med.* **2018**, *16*, 74. [CrossRef] [PubMed]
12. Ma, B.; Athari, S.S.; Nasab, E.M.; Zhao, L. PI3K/AKT/mTOR and TLR4/MyD88/NF-κB signaling inhibitors attenuate pathological mechanisms of allergic asthma. *Inflammation* **2021**, *44*, 1895–1907. [CrossRef] [PubMed]
13. Yuan, J.; Liu, Y.; Yu, J.; Dai, M.; Zhu, Y.; Bao, Y.; Peng, H.; Liu, K.; Zhu, X. Gene knockdown of CCR3 reduces eosinophilic inflammation and the Th2 immune response by inhibiting the PI3K/AKT pathway in allergic rhinitis mice. *Sci. Rep.* **2022**, *12*, 5411. [CrossRef] [PubMed]
14. Korinek, M.; Handoussa, H.; Tsai, Y.-H.; Chen, Y.-Y.; Chen, M.-H.; Chiou, Z.-W.; Fang, Y.; Chang, F.-R.; Yen, C.-H.; Hsieh, C.-F.; et al. Anti-Inflammatory and Antimicrobial Volatile Oils: Fennel and Cumin Inhibit Neutrophilic Inflammation via Regulating Calcium and MAPKs. *Front. Pharmacol.* **2021**, *12*, 674095. [CrossRef]
15. Allahghadri, T.; Rasooli, I.; Owlia, P.; Nadooshan, M.J.; Ghazanfari, T.; Taghizadeh, M.; Astaneh, S.D.A. Antimicrobial Property, Antioxidant Capacity, and Cytotoxicity of Essential Oil from Cumin Produced in Iran. *J. Food Sci.* **2010**, *75*, H54–H61. [CrossRef]
16. Patil, S.B.; Takalikar, S.S.; Joglekar, M.M.; Haldavnekar, V.S.; Arvindekar, A.U. Insulinotropic and β-cell protective action of cuminaldehyde, cuminol and an inhibitor isolated from Cuminum cyminum in streptozotocin-induced diabetic rats. *Br. J. Nutr.* **2013**, *110*, 1434–1443. [CrossRef]
17. Wei, J.; Zhang, X.; Bi, Y.; Miao, R.; Zhang, Z.; Su, H. Anti-Inflammatory Effects of Cumin Essential Oil by Blocking JNK, ERK, and NF-κB Signaling Pathways in LPS-Stimulated RAW 264.7 Cells. *Evid. Based Complement. Alternat. Med.* **2015**, *2015*, 474509. [CrossRef]
18. Haque, M.R.; Ansari, S.H. Aromatic aldehyde compound cuminaldehyde protects nonalcoholic fatty liver disease in rats feeding high fat diet. *Hum. Exp. Toxicol.* **2019**, *38*, 823–832. [CrossRef]
19. Monteiro-Neto, V.; De Souza, C.D.; Gonzaga, L.F.; Da Silveira, B.C.; Sousa, N.C.F.; Pontes, J.P.; Santos, D.M.; Martins, W.C.; Pessoa, J.F.V.; Junior, A.R.C.; et al. Cuminaldehyde potentiates the antimicrobial actions of ciprofloxacin against Staphylococcus aureus and Escherichia coli. *PLoS ONE* **2020**, *15*, e0232987. [CrossRef]
20. Hada, M.; Nishi, K.; Ishida, M.; Onda, H.; Nishimoto, S.; Sugahara, T. Inhibitory effect of aqueous extract of Cuminum cyminum L. seed on degranulation of RBL-2H3 cells and passive cutaneous anaphylaxis reaction in mice. *Cytotechnology* **2019**, *71*, 599–609. [CrossRef]
21. Nakae, S.; Morita, H.; Ohno, T.; Arae, K.; Matsumoto, K.; Saito, H. Role of Interleukin-33 in Innate-Type Immune Cells in Allergy. *Allergol. Int.* **2013**, *62*, 13–20. [CrossRef]
22. Yoshimoto, T.; Matsushita, K. Innate-Type and Acquired-Type Allergy Regulated by IL-33. *Allergol. Int.* **2014**, *63*, 3–11. [CrossRef] [PubMed]
23. Mizusaki, A.; Nishi, K.; Nishiwaki, H.; Ishida, M.; Tamamoto, T.; Sugahara, T. Suppressive effect of ethanol extract from passion fruit seeds on IgE production. *J. Funct. Foods* **2017**, *32*, 176–184. [CrossRef]
24. Romagnani, S. The role of lymphocytes in allergic disease. *J. Allergy Clin. Immunol.* **2000**, *105*, 399–408. [CrossRef]
25. Nagase, H.; Ueki, S.; Fujieda, S. The roles of IL-5 and anti-IL-5 treatment in eosinophilic diseases: Asthma, eosinophilic granulomatosis with polyangiitis, and eosinophilic chronic rhinosinusitis. *Allergol. Int.* **2020**, *69*, 178–186. [CrossRef] [PubMed]
26. Mäkelä, M.J.; Kanehiro, A.; Borish, L.; Dakhama, A.; Loader, J.; Joetham, A.; Xing, Z.; Jordana, M.; Larsen, G.L.; Gelfand, E.W. IL-10 is necessary for the expression of airway hyperresponsiveness but not pulmonary inflammation after allergic sensitization. *Proc. Natl. Acad. Sci. USA* **2000**, *97*, 6007–6012. [CrossRef]
27. Polukort, S.H.; Rovatti, J.; Carlson, L.; Thompson, C.; Ser-Dolansky, J.; Kinney, S.R.M.; Schneider, S.S.; Mathias, C.B. IL-10 enhances IgE-mediated mast cell responses and is essential for the development of experimental food allergy in IL-10-deficient mice. *J. Immunol.* **2016**, *196*, 4865–4876. [CrossRef] [PubMed]
28. Trinchieri, G. Interleukin-12 and the regulation of innate resistance and adaptive immunity. *Nat. Rev. Immunol.* **2003**, *3*, 133–146. [CrossRef]
29. Zhang, Y.; Zhang, Y.; Gu, W.; Sun, B. TH1/TH2 cell differentiation and molecular signals. *Adv. Exp. Med. Biol.* **2014**, *841*, 15–44. [PubMed]
30. Ishida, M.; Ohara, R.; Miyagawa, F.; Kikuzaki, H.; Nishi, K.; Onda, H.; Yoshino, N.; Sugahara, T. Umbelliferose Isolated from *Cuminum cyminum* L. Seeds Inhibits Antigen-Induced Degranulation in Rat Basophilic Leukemia RBL-2H3 Cells. *Molecules* **2022**, *27*, 4101. [CrossRef] [PubMed]
31. Li, J.; Zhang, B. Apigenin protects ovalbumin-induced asthma through the regulation of Th17 cells. *Fitoterapia* **2013**, *91*, 298–304. [CrossRef] [PubMed]
32. Chen, F.; He, D.; Yan, B. Apigenin attenuates allergic responses of ovalbumin-induced allergic rhinitis through modulation of Th1/Th2 responses in experimental mice. *Dose Response* **2020**, *18*, 1559325820904799. [CrossRef] [PubMed]
33. Wang, S.; Wuniqiemu, T.; Tang, W.; Teng, F.; Bian, Q.; Yi, L.; Qin, J.; Zhu, X.; Wei, Y.; Dong, J. Luteolin inhibits autophagy in allergic asthma by activating PI3K/Akt/mTOR signaling and inhibiting Beclin-1-PI3KC3 complex. *Int. Immunopharmacol.* **2021**, *94*, 107460. [CrossRef] [PubMed]

34. Dong, J.; Xu, O.; Wang, J.; Shan, C.; Ren, X. Luteolin ameliorates inflammation and Th1/Th2 imbalance via regulating the TLR4/NF-κB pathway in allergic rhinitis rats. *Immunopharmacol. Immunotoxicol.* **2021**, *43*, 319–327. [CrossRef] [PubMed]
35. Watanabe, H.; Sonoyama, K.; Watanabe, J.; Yamaguchi, N.; Kikuchi, H.; Nagura, T.; Aritsuka, T.; Fukumoto, K.; Kasai, T. Reduction of allergic airway eosinophilia by dietary raffinose in Brown Norway rats. *Br. J. Nutr.* **2004**, *92*, 247–255. [CrossRef] [PubMed]
36. Sonoyama, K.; Watanabe, H.; Watanabe, J.; Yamaguchi, N.; Yamashita, A.; Hashimoto, H.; Kishino, E.; Fujita, K.; Okada, M.; Mori, S.; et al. Allergic Airway Eosinophilia Is Suppressed in Ovalbumin-Sensitized Brown Norway Rats Fed Raffinose and α-Linked Galactooligosaccharide. *J. Nutr.* **2005**, *135*, 538–543. [CrossRef]

foods

MDPI

Article

Effect of Pomegranate Extract Consumption on Satiety Parameters in Healthy Volunteers: A Preliminary Randomized Study

Angela Stockton [1] and Emad A. S. Al-Dujaili [2,*]

[1] Dietetics, Nutrition and Biological Sciences, Queen Margaret University, Edinburgh EH16 4TJ, UK
[2] Centre for Cardiovascular Science, Queen's Medical Research Institute, University of Edinburgh, Edinburgh EH16 4TJ, UK
* Correspondence: ealduja1@exseed.ed.ac.uk; Tel.: +44-131-2426777

Abstract: There has been an increasing interest in nutraceuticals and functional foods in reducing appetite and to lose weight. We assessed the effect of oral pomegranate extract (PE) and PE juice (PJ) intake vs. placebo on satiety parameters in healthy volunteers. Twenty-eight subjects (mean age 34.5 ± 13.7 years, body mass index [BMI] 25.05 ± 3.91 kg/m^2) were randomized to 3-week priming supplementation with PE (Pomanox®) or placebo. On week 3, satiety parameters were determined on 1 testing day after participants ingested a breakfast and a lunch meal with PJ juice, using 100-mm visual acuity scales (VAS) for hunger, desire to eat, fullness and satisfaction. Meal quality and palatability were also tested. The desire to eat was less at all time points in the PJ juice with PE priming group and participants were also less hungry (p = 0.044) than those who consumed placebo. There was an overall significant difference between the groups (p < 0.001). Participants in the PJ juice with PE priming group experienced significantly greater satisfaction (p = 0.036) and feeling of fullness (p = 0.02) than those in the placebo group. These findings suggest that consumption of PE could have the potential to modulate satiety indicators.

Keywords: pomegranate extract; polyphenols; obesity; satiety; appetite; visual analogue scale

1. Introduction

The global impact of obesity on morbidity, mortality and healthcare costs is enormous, with both lifestyle and dietary habits as the key determinants in the prevalence of obesity [1]. The continuous rise in the obesity epidemic worldwide [2], coupled with the lack of success of current approaches to weight management and the uncertainty around potential drug side effects, have led to hopeful consumer interest in nutraceuticals and functional foods [3]. Yeoman and Chambers [4] suggested that "the worldwide increase in the prevalence of obesity and overweight has resulted in an urgent need to better understand the nature of satiety". Appetite control has been highlighted as one of the most important factors involved in the success of dietary treatments of obesity. It is known that appetite sensations play an essential role in the control of energy intake [5], and, therefore, in the control of body weight [6,7].

Satiety occurs at the end of a meal and prevents further eating. Satiation takes place during a meal and determines the duration and size of the meal as well as how quickly it is consumed [8]. Appetite sensations, including hunger, fullness, prospective consumption, and desire to eat, are the appetite-related dimensions most consistently used in research [3,9–11]. De Graaf et al. [11] concluded from their study into biomarkers of satiation and satiety that the next important challenge in the field is to identify food ingredients which may have an impact on either satiety or satiation, with the hope that such foods may assist consumers in maintaining a healthy body weight. The capacity of specific foods and food characteristics to affect appetite (i.e., feelings of hunger and satiety) continue to be a focus of interest [12], particularly in relation to food consumption and weight management [13]. A variety of neuropeptides such as cholecystokinin and ghrelin, plus hormone secretions such as leptin, mediate these processes and influence food intake and therefore satiety [14,15].

Polyphenols have received special attention as potential complementary approaches in the management of obesity [16]. In a recent in-depth review of managing obesity through natural polyphenols [17], appetite suppressing effects of natural polyphenolic compounds have been attributed to the interplay of different mechanisms, including slowing-down secretion of appetite-stimulating hormones, inactivation of appetite sensors, modulation of melanin-concentration hormone receptors, inhibition of ghrelin secretion, increase of adiponectin levels, reduction of glucagon-like peptide 1, increase of serotonin, and modulating adipohormones and the expression of gut peptides, all involved in transferring satiety signals to the brain [18].

The consumption of pomegranate juice has grown tremendously due to its reported health benefits, which are presumably due to its high content of antioxidant polyphenols especially tannins, anthocyanins and ellagic acid derivatives [19,20]. Animal and in vitro studies have shown that pomegranate may reduce inflammation [21], lipid peroxidation, oxidative stress, blood pressure [22–24] and insulin resistance [25]. Pomegranate may therefore have a protective role against atherosclerosis, hypertension, cancer, diabetes type II, and obesity [24,25]. It has also been shown that the intake of pomegranate juice [26,27] and pomegranate extract reduced blood pressure, insulin resistance and stress hormones levels [28]. Pomegranate juice and pomegranate extract intake have also been reported to decrease food consumption and reduce body weight in animals [29,30]; a human study by González-Ortiz et al. [31] showed that the administration of 120 mL pomegranate juice daily for one month significantly decreased fat mass in adults. It has been suggested that one of the mechanisms by which pomegranate can participate in the management of obesity is through regulating appetite [24]. However, research investigating this association remains relatively limited. The aim of this study was to assess the effect of pomegranate extract and juice intake on satiety parameters in healthy volunteers. It was hypothesized that consumption of a pomegranate extract may reduce appetite, which could be evaluated by means of satiety parameters.

2. Materials and Methods

2.1. Study Design

This small-scale exploratory study on satiety was conducted over 1 day (morning and lunch time) as a randomized placebo-controlled trial, and in the framework of a previous randomized, double-blind controlled trial that investigated the effect of consumption of a pomegranate extract (PE) on blood pressure, insulin resistance, stress hormones and quality of life [28]. The protocol of the present satiety study was approved by the Institutional Review Board of Queen Margaret University, Edinburgh, UK. Written informed consent was obtained from all participants. The overall trial was registered in the ClinicalTrials.gov (accessed on 17 February 2019) (NCT02005939).

2.2. Participants

Volunteers were recruited through the university's email research digest and by word-of-mouth recommendations. The study was open to both men and women from 18–65 years of age, with a body mass index (BMI) between 18 and 34.9 kg/m^2 and no history of diabetes or cardiovascular diseases. Exclusion criteria were systemic disease, including heart disease, diabetes, liver or kidney dysfunction and immunological disorders; heavy smoking; abuse of drugs or alcohol; pregnancy and breastfeeding; allergy to either pomegranate (PE) or test foods; and history of weight control management within the last 2 months. Also, as recommended by Smeets et al. [32] restrained eaters recording a score of >7 in the three-factor eating questionnaires were ineligible as these subjects may control their intake of food and would therefore be unlikely to eat freely.

2.3. Pomegranate Extract and Placebo Capsules

The study consisted of a daily oral intake of PE capsules or placebo for 3 weeks and 1 testing day (satiety session) in the kitchen/food laboratory on week 3. Participants were randomized to the PE or the placebo group by an independent technical staff using a research randomizer website [33]. Each PE capsule contained a standardized extract of whole pomegranate fruit obtained by an eco-friendly process (Pomanox® P30, Euromed S.A., Barcelona, Spain), with the following composition: 210 mg punicalagin (the recommended daily intake to provide the beneficial effects of these antioxidants) and 328 mg total polyphenols, including other pomegranate polyphenols (e.g., flavonoids, ellagic acid and 0.37 mg anthocyanins). Placebo capsules contained maltodextrin, and both PE and placebo capsules were identical in appearance, weighing 1.08 g and providing 6.52 kcal per capsule.

2.4. Satiety Session: Breakfast and Lunch Meals

Participants came to the kitchen/food laboratory between 8:30 and 9:30 a.m. in at least 8-h fasting conditions (water was allowed). They had previously received instructions by e-mail being advised to follow their usual diets with no restrictions but requested to refrain from alcohol or extra physical activity on the previous day. On the testing day, they received a breakfast and a lunch meal. The breakfast meal (432 kcal) consisted of 60 g of Kellogg's Crunchy Nut Cornflakes cereal (243 kcal), 150 mL of semi-skimmed milk (69 kcal) and 150 mL of a pomegranate juice (PJ) or a placebo juice drink. The PJ (PomeGreat® Pure supplied by RJA Foods, UK) contained 126 mg total polyphenols as gallic acid equivalent and 72 kcal as carbohydrates in 150 mL (sugars 12 g per 100 mL). The placebo juice was composed of diluted orange juice with very low polyphenols containing the same amount of energy as the PJ by adding sucrose. The drinks were equivalent for colour and energy (Table 1) and were served in black containers to disguise a slightly different appearance and viscosity as the PJ was thicker and slightly cloudy.

Table 1. Nutritional composition of the lunch meal and drinks.

Nutritional Composition	Pasta (500 g Cooked)	Pasta Sauce (500 g)	Total Meal Composition	Pomegranate Juice (PJ)	Placebo Juice
Energy (kcal)	765	220	985	72	72
Protein (g)	28	9	37	0.6	Trace
Carbohydrates (g)	138.5	39.5	178	18	18
Sugars (g)	3.5	35	38.5		
Fat (g)	8.5	3	11.5	Trace	Trace
Saturated fat (g)	2.5	0.5	3		
Fibre (g)	12	9	21	Trace	Trace
Sodium (g)	Trace	2	2		
Polyphenols (mg GAE)				126	Negligible

GAE: gallic acid equivalent.

The lunch meal consisted of 1 kg of pasta with sauce (served hot) and another 150 mL of PJ to be consumed 30 min before lunch. The total energy of the lunch provided was 1057 kcal. The lunch composition is shown in Table 1. The pasta was comprised of 500 g cooked weight of pasta [Morrison's Italian Pasta Twists (100% durum wheat semolina)], and the sauce contained 500 g of a pre-prepared pasta sauce (Tesco Bolognese Sauce, a vegetarian tomato sauce with garlic and herbs). The choice of this food quantity was based on a feasibility trial carried out in four participants, who were provided with the test meal to eat. The pre-weighed lunch portion was served after completion of the VAS at 30 min after the preload PJ as shown in Table 2 [4]. A maximum of 1 kg of pasta and sauce could be consumed [33]. A smaller food serving size was not adopted because the cue of an empty dish could prompt meal termination [34]. The homogeneity of the test meal facilitated assessment of macronutrient and energy intake [35].

Table 2. Satiety study intervention protocol.

Time	Procedures
8:30 a.m.	Breakfast served and eaten. 60 g crunchy nut cornflakes; 150 mL semi-skim milk; 150 mL pomegranate juice (PJ) or placebo drink (PLD).
3-h interval	Participants fast and then returned to the kitchen/food laboratory for the lunch session.
11:30 a.m. (0–15 min)	Starting to complete VAS for satiety (VAS 1 baseline; zero point) 150 mL of PJ juice preload for all participants (both groups) (VAS 2 satiety and palatability variables).
12:00 noon (30 min)	Lunch presented to each participant (VAS 3 satiety variables before eating anything)
12:05 p.m. (35 min)	Participants consumed 1 spoon of lunch (VAS for palatability and meal quality variables). Afterwards, participants continued to eat the lunch.
12:15 p.m. (45 min)	Participants completed VAS 4 for satiety and meal palatability variables. A timer prompts the participants.
12:30 p.m. (60 min)	VAS 5; satiety and meal palatability (repeat VAS 4)
12:45 p.m. (75 min)	VAS 6; satiety and meal palatability (repeat VAS 4)
1:00 p.m. (90 min)	VAS 7; satiety and meal palatability (repeat VAS 4)
1:15 p.m. (105 min)	VAS 8; satiety and meal palatability (repeat VAS 4)
1:30 p.m. (120 min)	VAS 9; satiety and meal palatability (repeat VAS 4)

2.5. Study Procedures

The satiety study intervention protocol is summarized in Table 2. Briefly, between 8:30 a.m. and 1:30 p.m. subjects stayed in the laboratory in controlled conditions where the breakfast and a lunch meal were served, and visual analogue scales (VAS) were recorded to assess general personal feelings for appetite and to examine the acceptability and palatability of both the juice and the test meal qualities. A 100-mm VAS with words anchored at each end expressing the most positive and negative ratings regarding satiety-related characteristics (0 = not at all, 100 = extremely) was used. Questions for general personal feelings of appetite were the following: "How hungry do you feel?", "How strongly do you feel a desire to eat?", "How satisfied do you feel?" and "How full do you feel right now?". The characteristics of the quality of meal and juice were also measured using VAS and included the following five questions: "How visually appearing is the meal/juice?", "How much do you like the smell of the meal/juice?", "How much do you like the taste of the meal/juice?", "How much do you like the texture of the meal/juice?" and "How palatable is the meal/juice overall?".

Participants were kept in this controlled kitchen/food laboratory environment while completing all VAS scores. As shown in Table 2, participants recorded their VAS scores at baseline (VAS 1, 0 min) from the start of the intervention in the absence of food or juice, and then at 15-min intervals for 120 min (VAS 2–9). Consumption of the allocated preload juice took place immediately after baseline VAS was recorded. Changes from baseline were calculated by subtracting the baseline score (at 0 min) from the score at a certain time point (starting at 15 min and continuing through VAS time points until 120 min) as suggested by Veldhorst et al. [36]. A composite score of the VAS responses was used to report the overall subjective satiety scores for each treatment. Also, total meal quality was a composite score derived from the five quality-related variables (visual appeal, liked smell, like taste, like texture, and perceived overall palatability).

2.6. Food and Drink Diaries

A food and drink diary were used to account for dietary habits. The 3-day diet diaries were completed as previously described [23] on two occasions, once before the study began, and then on each day before, during, and after the lunch-day intervention. To monitor whether increased energy before a meal result in reduced intake at that meal, the participants' 3-day diet diary encompassing the study day was compared to the pre-intervention consecutive 3-day diet diary [37]. Energy intake (kcal) and macronutrient intakes (i.e., protein, fat, and carbohydrates) were expressed in g and calculated from diet diary data input using the method of Astbury et al. [38].

2.7. Statistical Analysis

Data were analyzed using SPSS for Windows® version 21.0 (Microsoft, IBM Corp., Armonk, NY, USA, 2012). Continuous normally distributed data are expressed as mean ± SD. Normality of all data was examined using the Shapiro-Wilk test. Differences in baseline characteristics between groups were explored using independent t-tests or the Mann-Whitney *U* test when the data were non-parametric. Data are generally presented as mean ± standard deviation (SD) unless otherwise stated. VAS for subjective ratings of hunger, desire to eat, satisfaction, and fullness during the 120 min postprandial period were analyzed using a mixed two-way model analysis of variance (ANOVA), with time (15, 30, 45, 60, 75, 90, 105, 120 min) as the within-subject factor, and treatment type (PE/placebo) as the between-subject factor. Food intake at the test meal was calculated from the amount of food (g) eaten at the ad libitum meal. It was hypothesized that PE priming would reduce the amount of food consumption at meal. Energy and macronutrient intakes were analyzed using NetWISP 4 (Tinuviel Software 2006, 7th Edition to the UK Composition of Foods, Anglesey, UK) [39]. Differences in energy and macronutrients were assessed using paired t-tests. Significant changes were set at $p \leq 0.05$.

3. Results

3.1. Study Population

Twenty-nine volunteers were recruited, but one participant did not attend the satiety session for personal reasons. The study population included 28 subjects, 14 in the PE group and 14 in the placebo group. There were no significant differences in age ($p = 0.51$) and body mass index (BMI) ($p = 0.72$) between the two groups. No side effects of PE or adverse events on the screening assessment were reported by all participants. The baseline characteristics are shown in Table 3.

Table 3. Baseline characteristics of subjects randomized to the study groups.

Characteristic	Pomegranate Extract Capsule Group Mean (SD)	Placebo Capsule Group Mean (SD)
Age (years)	34.5 (13.7)	32.6 (12.9)
Gender (M; F)	4; 10	4; 10
Height (m)	1.66 (0.06)	1.70 (0.09)
Weight (kg)	70.0 (13.7)	72.1 (12.0)
BMI (kg/m^2)	25.1 (4.5)	24.94 (3.4)

3.2. Satiety-Related Variables

VAS scores showed that there were lower levels of hunger (Figure 1) and a desire to eat (Figure 2), as well as higher levels satisfaction (Figure 3) and fullness (Figure 4), in the PE capsule group drinking PJ, compared to those who had placebo priming. At 30–90 min, participants were less hungry after consuming PE priming for 3 weeks and PJ on the test day than those who consumed placebo capsules with PJ. The desire to eat was less at all time points in participants who had taken PE capsules with PJ before lunch. These results suggest that the PE group participants with the PJ preload were generally more satisfied than the placebo group with the PJ preload, specifically 30 min after consuming lunch.

Participants in the PE group consuming PJ felt less full before lunch than those in the placebo group. However, the PE group experienced more fullness than those in the PL group as lunch progressed from 30 to 120 min.

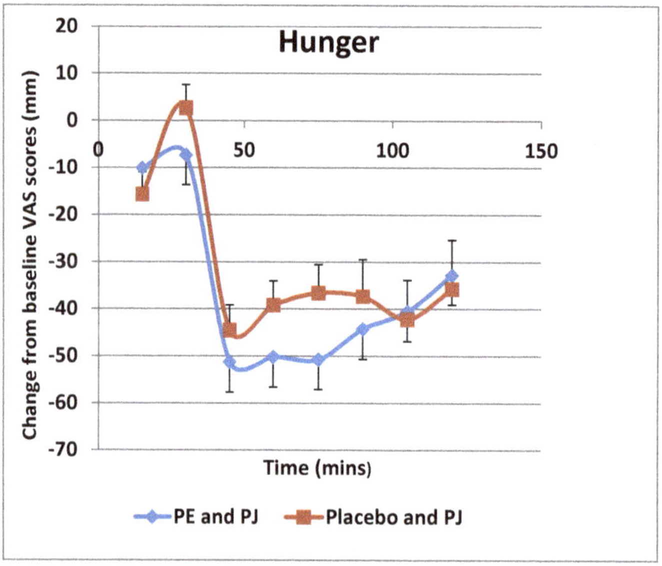

Figure 1. Change in hunger from baseline (mm VAS) over 120-min lunch meal period after 3 weeks of PE or placebo capsule consumption with pre-meal PJ intake. Results are expressed as mean (±SD). PE: pomegranate extract; PJ: pomegranate juice. There was an overall statistically significant difference between the two groups ($p = 0.044$) where the hunger feeling was significantly lower in the PE extract group.

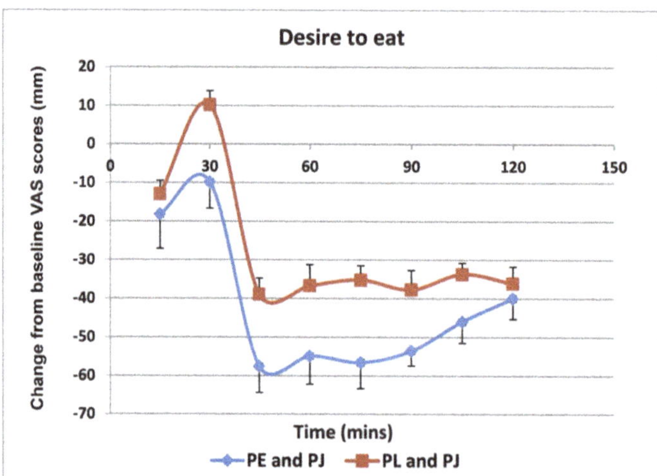

Figure 2. Change in hunger from baseline (mm VAS) over 120-min lunch meal period after 3 weeks of PE or placebo capsule consumption with pre-meal PJ intake. Results are expressed as mean (±SD). PE: pomegranate extract; PJ: pomegranate juice; PL: placebo capsules. There was an overall statistically significant difference between the two groups ($p < 0.001$) where the desire to eat was significantly lower in the PE extract group.

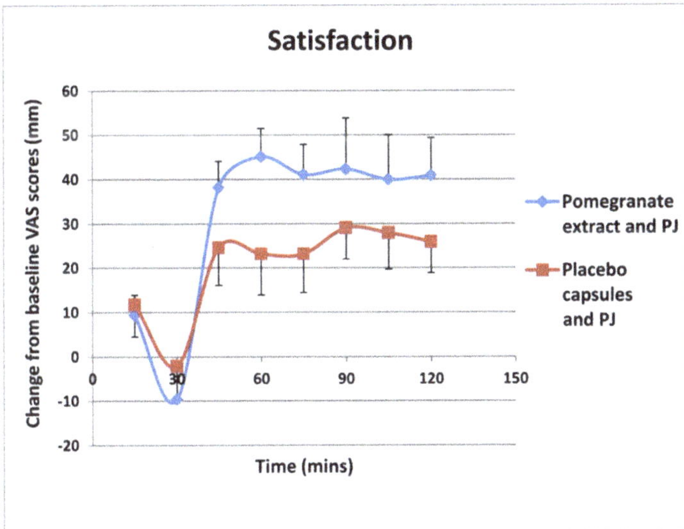

Figure 3. Change in hunger from baseline (mm VAS) over 120-min lunch meal period after 3 weeks of PE or placebo capsule consumption with pre-meal PJ intake. Results are expressed as mean (±SD). PJ: pomegranate juice. There was an overall statistically significant difference between the two groups ($p = 0.036$) with subjects in the PE group being significantly more satisfied.

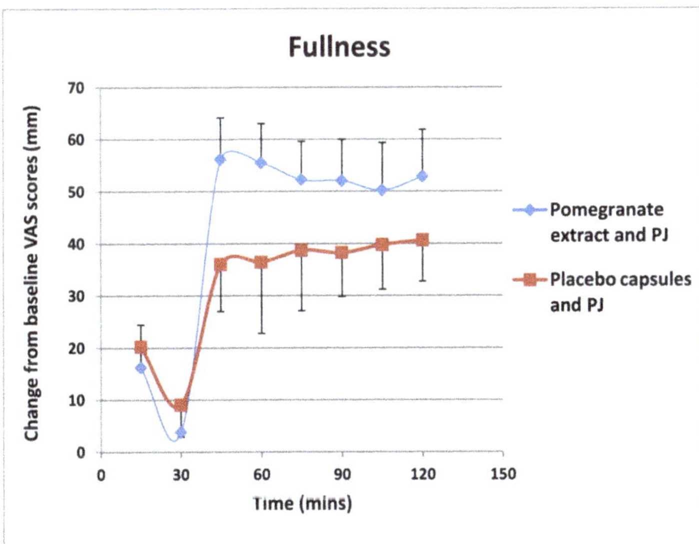

Figure 4. Change in fullness scores from baseline (mm VAS) over 120-min lunch meal period after 3 weeks of PE or placebo capsule consumption with pre-meal PJ intake. Results are expressed as mean (±SD). PJ: pomegranate juice. Subjects in the PE group experienced a greater feeling of fullness than those in the placebo group with PJ preload, which was statistically significant ($p = 0.02$).

3.3. Meal Quality

At the 35-min period, the five variables of meal quality were recorded, with VAS scores greater than 60 in the responses to each individual question (Figure 5). There was a statistically significant difference in the overall VAS score for meal quality between the study groups, with higher scores in subjects treated with PE priming (mean 72.9 ± 12.7) than in those treated with placebo (63.8 ± 5.5) (p = 0.001).

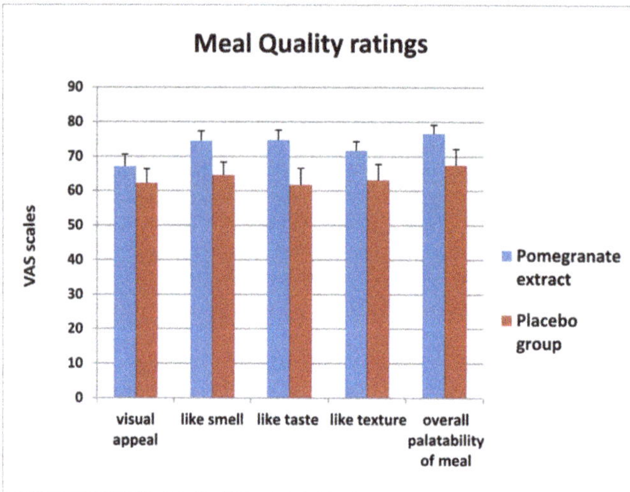

Figure 5. Meal quality ratings for the pomegranate (PE) and placebo priming groups. All VAS scores for individual characteristics were higher in the PE group as compared with placebo. The PE group liked the smell of the meal significantly more than the placebo group (p = 0.02) (Data as mean ± SD).

3.4. Meal Palatability

VAS scores taken over the entire lunch period of 120 min for meal palatability showed a significance between-group difference, with higher ratings in the PE priming group with PJ preload (mean 24.0 ± 8.9 mm) than in the placebo priming group (15.9 ± 12.1) ($p < 0.05$) (Figure 6).

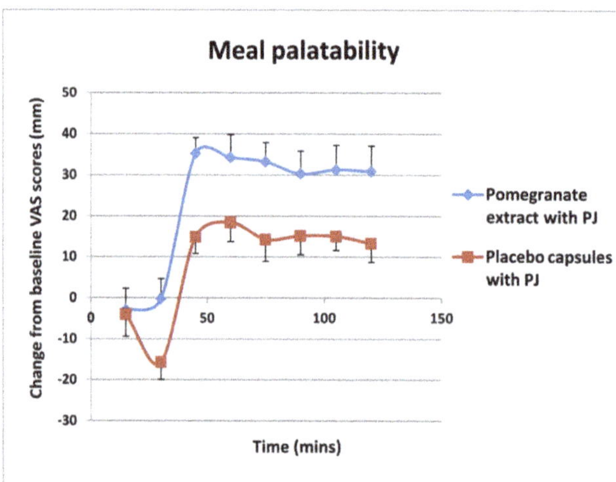

Figure 6. Meal palatability ratings over the 120-min lunch period for the pomegranate extract (PE) and placebo capsules (PL) group with pomegranate juice (PJ) preload. The meal palatability was rated significantly more highly (p = 0.05) by the PE group with the PJ preload compared to those in the placebo group (data as mean ± SD).

3.5. Meal Food Consumption

There was a significant difference in the mean weight of food consumed between the PE priming group (471 ± 100 g, range 447–636) and the placebo priming group (574 ± 169 g, range 569–1000) ($p < 0.05$), with lower amounts in the PE group.

3.6. Food and Drink Diaries

Table 4 shows the comparison of data from the 3-day diet diaries after starting PE or placebo priming and the 3-day diet diaries encompassing the satiety session day. In both study groups, there were increases in the consumption of carbohydrates, sugars, and starch, with significant differences as compared with baseline data. Increases were of a higher magnitude in the PE priming group for carbohydrates and sugars than in the placebo group, whereas changes in starch were greater for the placebo group. These increases, however, did not affect the total energy intake in either the PE group or the placebo group, with a mean difference from baseline of 196 ± 389 kcal ($p = 0.07$) and 297 ± 629 kcal ($p = 0.12$) for the PE and placebo groups, respectively.

Table 4. Satiety study energy expenditure and macronutrient intake at baseline and 3-week post study values for PE and PL groups with differences (baseline—3 weeks). Results are expressed in means with SD and SEM and p values. There were significant increases in carbohydrates, sugars, and starch in both groups.

Variable		Pomegranate Extract (PE) Group (n = 14)				Placebo (PL) Group (n = 14)			
		Mean	SD	SEM	p Value	Mean	SD	SEM	p Value
Energy (kcal)	Baseline	1913.20	415.97	107.40		1840.64	411.93	113.19	
	4th week	2109.53	512.60	132.35		2119.79	542.34	143.30	
	Change	−196.33	388.90	100.41	0.071	−279.14	629.49	168.24	0.121
Energy (kJ)	Baseline	8037.53	1744.32	450.38		7738.14	322.55	86.21	
	4th week	8893.00	2148.51	554.74		8931.14	556.36	148.69	
	Change	−855.47	1611.12	415.99	0.069	1193.00	2639.54	705.45	0.115
Protein (g)	Baseline	75.52	16.42	4.24		78.44	1345.88	359.70	
	4th week	75.23	12.69	3.28		91.34	2338.91	625.10	
	Change	0.29	17.48	4.51	0.950	−12.89	24.31	6.50	0.069
Carbohydrate (g)	Baseline	232.75	59.84	15.45		210.76	38.45	10.28	
	4th week	297.53	74.52	19.24		269.29	54.24	14.50	
	Change	−64.78	30.69	7.92	0.021	−58.53	80.59	21.54	0.018
Sugars (g)	Baseline	101.47	42.59	11.00		90.91	34.67	9.27	
	4th week	129.14	47.02	12.14		111.00	67.11	17.94	
	Change	−27.67	30.17	7.79	0.033	−20.09	29.12	7.78	0.023
Starch (g)	Baseline	127.12	44.23	11.42		112.13	27.58	7.37	
	4th week	162.75	44.19	11.41		154.53	26.64	7.12	
	Change	−35.63	31.52	8.14	0.021	−42.40	60.00	16.04	0.026
Total fat (g)	Baseline	76.73	30.22	7.80		80.28	30.08	8.04	
	4th week	71.60	26.46	6.83		77.90	44.82	11.98	
	Change	5.13	32.90	8.50	0.556	2.38	33.40	8.93	0.794

PE: pomegranate extract; PL: placebo capsules; SD: standard deviation; SEM: standard error of the mean.

4. Discussion

The purpose of this study was to conduct a preliminary testing of the potential role of PE (Pomanox®) in promoting satiety and reducing food intake. The results suggest that subjects in the PE group with the PJ preload were generally more satisfied than those treated with placebo. Participants were less hungry after PE capsules intake with PJ during the meal than those who consumed placebo capsules. Results of paired t-tests for VAS appetite measurements showed some significance between-group differences and a definite trend towards lower levels of hunger and a desire to eat, as well as higher levels of fullness and satisfaction, thus greater levels of satiety, in participants consuming PE with PJ, compared to placebo. However, the fact that these trends were noted only in the group consuming PE and juice warrants further investigation on a larger cohort.

4.1. Methodology and Compliance

The assessment of meal and juice palatability was carried out to eliminate any potential change in food and juice intake resulting from low scores in either the meal or juice palatability by the two groups. The mean scores for both groups were sufficiently high to confirm that the participants found the meal and the PE and PJ to be palatable. However, the meal quality was rated significantly more highly ($p = 0.03$) by participants who had Pomanox® PE priming and drank the PJ than those taking PL (24.0 ± 8.9 mm versus 15.9 ± 12.1 mm, respectively). VAS tests have previously been viewed critically, and in some well-controlled studies, changes in VAS failed to predict subsequent reductions in food intake [3,9,38,40]. Another criticism of VAS is the reluctance of subjects to make full use of the scale, preferring either to avoid extreme responses or to record only their responses, thereby affecting the results [41]. Other problems associated with the use of standard VAS measurements that have been discussed by Sadoul et al. [13] included non-linearity, the inherently abstract nature of overall 'hunger' feelings, and the lack of quantitative interpretation of when or how much the volunteer might wish to eat. Single-item meals can also be monotonous, and thus likely to limit consumption within the study, irrespective of the enhanced satiating potential of one of the preloads. Dietary monotony may curb excessive intake but is unlikely to be commercially viable as a solution to weight management [3].

4.2. Satiety Mechanisms

The mechanisms underlying the potential effect of PE extract on satiety remain unclear. It has been suggested that the role of polyphenols contained in pomegranate in suppressing appetite can be similar to that of sibutramine [29]. In a study by Lei et al. [30], high-dose (800 mg/kg) pomegranate leaf extract (PLE) had an inhibitory effect on the energy intake of mice and was associated with a marked decrease in serum glucose, triglycerides (TG) and total cholesterol (TC) concentrations, along with a significant decrease in the TC/high-density lipoprotein cholesterol (HDL-C) ratio, compared to the mice fed a high-fat diet. The PLE mice also showed a significant decrease in body weight and energy intake compared to the high-fat diet control group. Sibutramine, the anti-obesity drug, was used in another arm of the same study, and produced significant effects, not only in reducing body weight and adiposity, but also in lowering circulating TG and glucose concentrations, as well as the TC/HDL-C ratio. The researchers suggested that, since the effect of PLE on energy intake was like that of sibutramine, suppressing energy intake and inhibiting the intestinal absorption of dietary fat via inhibition of pancreatic lipase activity might be potential mechanisms for the anti-obesity effect of the PE extract. We have shown previously that cacao bean extract exhibited an inhibitory effect on pancreatic lipase in vitro [42]. It could also be possible that the intense flavor generated by polyphenols in the juice [43] is responsible for inducing satiety. Therefore, designing a study using VAS to measure the effect of PE extract and PJ on sensory feelings may be useful.

4.3. Food Consumption

In the present study, PE priming was associated with a significantly lower amount of food consumption during the satiety session as compared with placebo priming. Although small-scale studies are not powered to detect minimal clinically important differences, and do not provide results that are clearly statistically significant [43], our results are encouraging, and provide a good foundation for any future investigations testing the direct effect of PE on satiety and amount of food intake per meal. The analysis of food and alcohol diaries showed an increase in carbohydrate and sugar intake in both groups compared to pre-intervention levels. This could be explained by the additional energy that is provided following the ingestion of a meal and juice (high in carbohydrates and sugars), which may have affected the mean intake. In fact, Rolls et al. [34] showed that people who were offered a large portion of food ate more than those offered a small portion. Whether a different preload time, inter-meal interval, higher dose of PE priming, or different meals could have produced more significant eating effects are some aspects that can be considered in future studies. Because satiation depends on both gastric and intestinal nutrient stimulation and interactions, Zaremba et al. [44] noted that an optimal dose and timing between preload and an ad libitum test meal would likely be crucial in detecting any eating-inhibitory effects.

4.4. Limitations and Future Directions

In future studies, a crossover trial design, where VAS could be repeated during the test meal with the alternative PE/placebo treatment, might provide more accurate results [45]. For assessing inter-individual differences in eating patterns and completing the VAS, a 'within subject' study design, rather than separate groups of control and test subjects, is more commonly used, with participants acting as their own controls [15]. The effect of the PJ may also have been confounded by the priming capsules consumed at the same time. This deserves further investigation. However, effectively blinding participants to the type of extract and juice consumed was beneficial for avoiding any potentially significant effects resulting from expectation bias. Future researchers should carefully consider the randomization process for more complex studies.

Compensatory eating and responses to sensory quality vary across individuals, depending on their body size, physical activity, age, and level of restraint [4]. Our study primarily recruited participants with normal weight and unrestrained eating habits. Whether the results would have been the same with an overweight/obese participant group remains to be explored. Due to the length of lunch in our study, participants had access to 500 mL of water, and should have been able to clearly identify whether they were hungry, rather than thirsty. Conversely, this may have led to erroneous feelings of satiety (due to the water consumed, rather than the food). However, the volume of water consumed should have been recorded, and considered in the interpretation of the results. The Cochrane Collaboration Systematic Review of studies concerning the dietary supplement Chitosan and weight loss [46] recommended that investigations involving BMI and appetite should have a medium-to-long term duration (8 weeks was the average period for the trials accepted for their review), and this should be considered in future studies.

Benelam [15] cautioned that, despite satiation and satiety processes having the potential to control energy intake, many individuals habitually override the signals their bodies generate. He argued that to help people control their energy intake, thus preventing weight gain, knowledge about foods, ingredients, and the dietary patterns that can enhance satiation and satiety is still potentially useful in controlling body weight and avoiding shifts towards obesity. Our study was exploratory, and not powered to detect any statistically significant effects on satiety measures, making it more difficult to identify any meaningful effects. Livingstone et al. [41] have noted that the measurement of satiety is complex, since many physical factors, such as body weight, age, gender, and menstrual cycles, can all influence appetite and energy intake [11]. Eating behavior can also be affected by other factors, such as diet or the influence of other people who may be present. Examples include the completion of a food diary, which is prone to bias, particularly towards under-reporting of energy intake and differential misreporting of the macronutrients [47]. In the current study, food diary analysis was completed using the NetWISP software programme, which required a high degree of data processing. Some input errors might have occurred, particularly those involving inaccurate nutritional substitutions of food categories that were not available within the programme [48].

4.5. Critical Issues

Blundell [8] emphasized that care is also required when considering the legislative framework around functional food product claims. Examples of this have been 'freedom from hunger' and 'feel fuller for longer'. The European Food Safety Authority states that claims must be substantiated by scientific evidence and should be clearly understood by consumers. Since ingested foods influence appetite by a system of physiological satiety signals, functional foods could act by increasing the potency and/or duration of these signals. PE extract and PJ might be useful for public health, are readily available commercially, and appear to have no side effects. PE products could also provide a prevention and therapy alternative to pharmaceutical drugs that produce side effects or are otherwise poorly tolerated by some people. One example of this is orlistat, a drug used for weight loss, which is a highly potent pharmaceutical, and has well-recognized side effects such as diarrhoea and fatty stools. Orlistat can also interfere with the absorption of vitamins A, D, E, K and beta-carotene [49]. Moyad [50] has stated that, while companies should be given recognition for their research efforts into the effects of PE, more attention should be paid to the efficacy of experimental designs using suitable control groups.

5. Conclusions

Although this study indicates that PE and PJ could have potential in modulating satiety indicators, limitations of the study design, the small number of participants, and the outcomes obtained do not support drawing strong conclusions. However, the results obtained seem to be promising. Further trials with a larger number of volunteers and assessment of the effect of the PE extract and juice on satiety parameters are required.

Author Contributions: Conceptualization, E.A.S.A.-D. and A.S.; methodology, E.A.S.A.-D. and A.S.; validation, E.A.S.A.-D. and A.S.; formal analysis, A.S.; investigation, A.S.; resources, E.A.S.A.-D. and A.S.; data curation, A.S.; writing—original draft preparation, E.A.S.A.-D. and A.S; writing—review and editing, E.A.S.A.-D.; supervision, E.A.S.A.-D.; project administration, E.A.S.A.-D.; funding acquisition, E.A.S.A.-D. and A.S. All authors have read and agreed to the published version of the manuscript.

Funding: This research received no external funding. However, AS PhD Scholarship was initially supported by RJ, UK for a short period of time which subsequently funded by the Postgraduate Committee, Queen Margaret University, Edinburgh.

Institutional Review Board Statement: The study was conducted in accordance with the Declaration of Helsinki and approved by the Institutional Review Board of Queen Margaret University (code: PhD/APS/Pomanox/PhD-2019;NUT/DNBS/QMU).

Informed Consent Statement: Informed consent was obtained from all subjects involved in the study with no objection to publication provided that they will not be identified.

Data Availability Statement: Study data are available from the authors upon request.

Acknowledgments: We gratefully thank all participants who took part in the study as well as Euromed, Spain for supplying the Pomanox® and placebo capsules and RJ, UK for supplying the pure pomegranate juice. We thank Marta Pulido, for editorial assistance.

Conflicts of Interest: The authors declare no conflict of interest.

References

1. Gho, J.M.; Postema, P.G.; Conijn, M.; Bruinsma, N.; de Jong, J.S.; Bezzina, C.R.; Wilde, A.A.; Asselbergs, F.W. Heart failure following STEMI: A contemporary cohort study of incidence and prognostic factors. *Open Heart* **2017**, *4*, e000551. [CrossRef] [PubMed]
2. WHO. *Obesity and Overweight*; The World Health Organization: Geneva, Switzerland, 2021. Available online: http://www.who.int/mediacentre/factsheets/fs311/en/ (accessed on 26 August 2017).
3. Halford, J.C.; Harrold, J.A. Satiety-enhancing products for appetite control: Science and regulation of functional foods for weight management. *Proc. Nutr. Soc.* **2012**, *71*, 350–362. [CrossRef] [PubMed]
4. Yeomans, M.R.; Chambers, L. Satiety-relevant sensory qualities enhance the satiating effects of mixed carbohydrate-protein preloads. *Am. J. Clin. Nutr.* **2011**, *94*, 1410–1417. [CrossRef] [PubMed]
5. Murphy, K.G.; Bloom, S.R. Gut hormones and the regulation of energy homeostasis. *Nature* **2006**, *444*, 854–859. [CrossRef]
6. Drapeau, V.; King, N.; Hetherington, M.; Doucet, E.; Blundell, J.; Tremblay, A. Appetite sensations and satiety quotient: Predictors of energy intake and weight loss. *Appetite* **2007**, *48*, 159–166. [CrossRef]

7. Parra, D.; Ramelb, A.; Bandarra, N.; Kielyd, M.; Martínez, J.A.; Thorsdottirb, I. A diet rich in long chain omega-3 fatty acids modulates satiety in overweight and obese volunteers during weight loss. *Appetite* **2008**, *51*, 676–680. [CrossRef]
8. Blundell, J. Making claims: Functional foods for managing appetite and weight. *Nat. Rev. Endocrinol.* **2010**, *6*, 53–56. [CrossRef]
9. Stubbs, J.; Whybrow, S.; Lavin, J. Dietary and Lifestyle measures to enhance satiety and weight control. *Nutr. Bull.* **2010**, *35*, 113–125. [CrossRef]
10. Clegg, M.; Shafat, A. Energy and macronutrient composition of breakfast affect gastric emptying of lunch and subsequent food intake, satiety and satiation. *Appetite* **2010**, *54*, 517–523. [CrossRef]
11. de Graaf, C.; Blom, W.A.; Smeets, P.A.; Stafleu, A.; Hendriks, H.F. Biomarkers of satiation and satiety. *Am. J. Clin. Nutr.* **2004**, *79*, 946–961. [CrossRef]
12. Dansinger, M.L.; Gleason, J.A.; Griffith, J.L.; Selker, H.P.; Schaefer, E.J. Comparison of the Atkins, Ornish, Weight Watchers, and Zone diets for weight loss and heart disease risk reduction: A randomized trial. *JAMA* **2005**, *293*, 43–53. [CrossRef] [PubMed]
13. Sadoul, B.C.; Schuring, E.A.; Symersky, T.; Mela, D.J.; Masclee, A.A.; Peters, H.P. Measuring satiety with pictures compared to visual analogue scales. An exploratory study. *Appetite* **2012**, *58*, 414–417. [CrossRef] [PubMed]
14. Ahima, R.S.; Antwi, D.A. Brain regulation of appetite and satiety. *Endocrinol. Metab. Clin. N. Am.* **2008**, *37*, 811–823. [CrossRef] [PubMed]
15. Benelam, B. Satiation, satiety and their effects on eating behaviour. *Nutr. Bull.* **2009**, *34*, 126–173. [CrossRef]
16. Meydani, M.; Syeda, T.H. Dietary polyphenols and obesity. *Nutrients* **2010**, *2*, 737–751. [CrossRef]
17. Singh, M.; Thrimawithana, T.; Shukla, R.; Adhikari, B. Managing obesity through natural polyphenols: A review. *Future Foods* **2020**, *1–2*, 100002. [CrossRef]
18. Lai, C.S.; Tsai, M.L.; Badmaev, V.; Jimenez, M.; Ho, C.T.; Pan, M.H. Xanthigen suppresses preadipocyte differentiation and adipogenesis through down-regulation of PPARγ and C/EBPs and modulation of SIRT-1, AMPK, and FoxO pathways. *J. Agric. Food Chem.* **2012**, *60*, 1094–1101. [CrossRef]
19. Gil, M.I.; Tomás-Barberán, F.A.; Hess-Pierce, B.; Holcroft, D.M.; Kader, A.A. Antioxidant activity of pomegranate juice and its relationship with phenolic composition and processing. *J. Agric. Food Chem.* **2000**, *48*, 4581–4589. [CrossRef]
20. Danesi, F.; Ferguson, L.R. Could Pomegranate Juice Help in the Control of Inflammatory Diseases? *Nutrients* **2017**, *9*, 958. [CrossRef]
21. Kaplan, M.; Hayek, T.; Raz, A.; Coleman, R.; Dornfeld, L.; Vaya, J.; Aviram, M. Pomegranate juice supplementation to atherosclerotic mice reduces macrophage lipid peroxidation, cellular cholesterol accumulation and development of atherosclerosis. *J. Nutr.* **2001**, *131*, 2082–2089. [CrossRef]
22. Aviram, M.; Volkova, N.; Coleman, R.; Dreher, M.; Reddy, M.K.; Ferreira, D.; Rosenblat, M. Pomegranate phenolics from the peels, arils, and flowers are antiatherogenic: Studies in vivo in atherosclerotic apolipoprotein E-deficient (E0) mice and in vitro in cultured macrophages and lipoproteins. *J. Agric. Food Chem.* **2008**, *56*, 1148–1157. [CrossRef] [PubMed]
23. Dujaili, A.I.; Good, G.; Tsang, C. Consumption of Pomegranate Juice Attenuates Exercise - Induced Oxidative Stress, Blood Pressure and Urinary Cortisol/Cortisone Ratio in Human Adults. *EC Nutr.* **2016**, *4.6*, 982–995.
24. McFarlin, B.K.; Strohacker, K.A.; Kueht, M.L. Pomegranate seed oil consumption during a period of high-fat feeding reduces weight gain and reduces type 2 diabetes risk in CD-1 mice. *Br. J. Nutr.* **2009**, *102*, 54–59. [CrossRef] [PubMed]
25. Al-Muammar, M.N.; Khan, F. Obesity: The preventive role of the pomegranate (*Punica granatum*). *Nutrition* **2012**, *28*, 595–604. [CrossRef] [PubMed]
26. Tsang, C.; Smail, N.F.; Almoosawi, S.; Davidson, I.; Al-Dujaili, E.A. Intake of polyphenol-rich pomegranate pure juice influences urinary glucocorticoids, blood pressure and homeostasis model assessment of insulin resistance in human volunteers. *J. Nutr. Sci.* **2012**, *1*, e9. [CrossRef]
27. Al-Dujaili, E.A.S. Functional Foods need more Mechanistic Studies. *EC Nutr.* **2016**, *4.1*, 770–771.
28. Stockton, A.; Al-Dujaili, E.; McDougall, G.; Davidson, I.; Drummond, S.; Wyness, L. Effect of Pomegranate Extract Consumption on Cardiovascular Disease Risk Factors, Stress Hormones, and Quality of Life in Human Volunteers: An Exploratory Randomised, Double-Blind, Placebo-Controlled Trial. *EC Nutr.* **2015**, *2*, 396–411.
29. Cerdá, B.; Cerón, J.J.; Tomás-Barberán, F.A.; Espín, J.C. Repeated oral administration of high doses of the pomegranate ellagitannin punicalagin to rats for 37 days is not toxic. *J. Agric. Food Chem.* **2003**, *51*, 3493–3501. [CrossRef]
30. Lei, F.; Zhang, X.N.; Wang, W.; Xing, D.M.; Xie, W.D.; Su, H.; Du, L.J. Evidence of anti-obesity effects of the pomegranate leaf extract in high-fat diet induced obese mice. *Int. J. Obes.* **2007**, *31*, 1023–1029. [CrossRef]
31. González-Ortiz, M.; Martínez-Abundis, E.; Espinel-Bermúdez, M.C.; Pérez-Rubio, K.G. Effect of pomegranate juice on insulin secretion and sensitivity in patients with obesity. *Ann. Nutr. Metab.* **2011**, *58*, 220–223. [CrossRef]
32. Smeets, A.J.; Lejeune, M.P.; Westerterp-Plantenga, M.S. Effects of oral fat perception by modified sham feeding on energy expenditure, hormones and appetite profile in the postprandial state. *Br. J. Nutr.* **2009**, *101*, 1360–1368. [CrossRef] [PubMed]
33. Research Randomizer. Available online: https://www.randomizer.org/ (accessed on 10 January 2016).
34. Rolls, B.J.; Morris, E.L.; Roe, L.S. Portion size of food affects energy intake in normal-weight and overweight men and women. *Am. J. Clin. Nutr.* **2002**, *76*, 1207–1213. [CrossRef] [PubMed]
35. Burger, K.S.; Fisher, J.O.; Johnson, S.L. Mechanisms behind the portion size effect: Visibility and bite size. *Obesity* **2011**, *19*, 546–551. [CrossRef] [PubMed]

36. Veldhorst, M.A.; Nieuwenhuizen, A.G.; Hochstenbach-Waelen, A.; van Vught, A.J.; Westerterp, K.R.; Engelen, M.P.; Brummer, R.J.; Deutz, N.E.; Westerterp-Plantenga, M.S. Dose-dependent satiating effect of whey relative to casein or soy. *Physiol. Behav.* **2009**, *96*, 675–682. [CrossRef] [PubMed]
37. Thabane, L.; Ma, J.; Chu, R.; Cheng, J.; Ismaila, A.; Rios, L.P.; Robson, R.; Thabane, M.; Giangregorio, L.; Goldsmith, C.H. A tutorial on pilot studies: What, why and how. *BMC Med. Res. Methodol.* **2010**, *10*, 1. [CrossRef]
38. Astbury, N.M.; Stevenson, E.J.; Morris, P.; Taylor, M.A.; Macdonald, I.A. Dose-response effect of a whey protein preload on within-day energy intake in lean subjects. *Br. J. Nutr.* **2010**, *104*, 1858–1867. [CrossRef] [PubMed]
39. Tinuviel Software. *NetWISP (Version 3.0 and 4.0)*; Tinuviel Software: Anglesey, UK, 2006.
40. Beck, E.J.; Tosh, S.M.; Batterham, M.J.; Tapsell, L.C.; Huang, X.F. Oat beta-glucan increases postprandial cholecystokinin levels, decreases insulin response and extends subjective satiety in overweight subjects. *Mol. Nutr. Food Res.* **2009**, *53*, 1343–1351. [CrossRef]
41. Livingstone, B.E.; Robson, P.J.; Welch, R.W.; Burns, A.A.; Burrows, M.S.; Mccormack, C. Methodological issues in the assessment of satiety. *Näringsforskning* **2000**, *44*, 98–103. [CrossRef]
42. Almoosawi, S.; McDougall, J.G.; Fyfe, L.; Al-Dujaili, E.A.S. Investigating the inhibitory activity of green coffee and cacao bean extracts on pancreatic lipase. *Nutr. Bull.* **2010**, *35*, 207–212. [CrossRef]
43. Lesschaeve, I.; Noble, A.C. Polyphenols: Factors influencing their sensory properties and their effects on food and beverage preferences. *Am. J. Clin. Nutr.* **2005**, *81*, 330S–335S. [CrossRef]
44. Zaremba, S.M.M.; Drummond, S.; Steinert, R.E. The need to standardize ad libitum eating protocols in dietary fibre appetite research. *Eur. J. Clin. Nutr.* **2017**, *71*, 570–572. [CrossRef]
45. Josic, J.; Olsson, A.T.; Wickeberg, J.; Lindstedt, S.; Hlebowicz, J. Does green tea affect postprandial glucose, insulin and satiety in healthy subjects: A randomized controlled trial. *Nutr. J.* **2010**, *9*, 63. [CrossRef]
46. Jull, A.B.; Ni Mhurchu, C.; Bennett, D.A.; Dunshea-Mooij, C.A.; Rodgers, A. Chitosan for overweight or obesity. *Cochrane Database Syst. Rev.* **2008**, CD003892. [CrossRef]
47. Gibson, R.; Eriksen, R.; Lamb, K.; McMeel, Y.; Vergnaud, A.C.; Spear, J.; Aresu, M.; Chan, Q.; Elliott, P.; Frost, G. Dietary assessment of British police force employees: A description of diet record coding procedures and cross-sectional evaluation of dietary energy intake reporting (The Airwave Health Monitoring Study). *BMJ Open* **2017**, *7*, e012927. [CrossRef]
48. Beasley, J.M.; Riley, W.T.; Davis, A.; Singh, J. Evaluation of a PDA-based dietary assessment and intervention program: A randomized controlled trial. *J. Am. Coll. Nutr.* **2008**, *27*, 280–286. [CrossRef]
49. Opala, T.; Rzymski, P.; Pischel, I.; Wilczak, M.; Wozniak, J. Efficacy of 12 weeks supplementation of a botanical extract-based weight loss formula on body weight, body composition and blood chemistry in healthy, overweight subjects—A randomised double-blind placebo-controlled clinical trial. *Eur. J. Med. Res.* **2006**, *11*, 343–350.
50. Moyad, M.A. The optimal male health diet and dietary supplement program. *Urol Clin. N. Am.* **2012**, *39*, 89–107. [CrossRef]

foods

MDPI

Review

Konjac Glucomannan: An Emerging Specialty Medical Food to Aid in the Treatment of Type 2 Diabetes Mellitus

Yimeng Fang [1,†], Jiahui Ma [1,†], Pengyu Lei [1], Lei Wang [1], Junying Qu [1], Jing Zhao [1], Fan Liu [1], Xiaoqing Yan [2], Wei Wu [3], Libo Jin [1,4], Hao Ji [1] and Da Sun [1,5,*]

1 Institute of Life Sciences & Biomedical Collaborative Innovation Center of Zhejiang Province, Wenzhou University, Wenzhou 325000, China
2 Chinese–American Research Institute for Diabetic Complications, School of Pharmaceutical Sciences, Wenzhou Medical University, Wenzhou 325000, China
3 Key Laboratory for Biorheological Science and Technology of Ministry of Education, State and Local Joint Engineering Laboratory for Vascular Implants, Bioengineering College of Chongqing University, Chongqing 400044, China
4 Wenzhou City and Wenzhou OuTai Medical Laboratory Co., Ltd. Joint Doctoral Innovation Station, Wenzhou Association for Science and Technology, Wenzhou 325000, China
5 Wenzhou City and Kunlong Technology Co., Ltd. Joint Doctoral Innovation Station, Wenzhou Association for Science and Technology, Wenzhou 325000, China
* Correspondence: sunday@wzu.edu.cn
† These authors contributed equally to this work.

Abstract: There are many factors causing T2DM; thus, it is difficult to prevent and cure it with conventional treatment. In order to realize the continuous intervention of T2DM, the treatment strategy of combining diet therapy and traditional medication came into being. As a natural product with the concept of being healthy, konjac flour and its derivatives are popular with the public. Its main component, Konjac glucomannan (KGM), can not only be applied as a food additive, which greatly improves the taste and flavor of food and extends the shelf life of food but also occupies an important role in T2DM. KGM can extend gastric emptying time, increase satiety, and promote liver glycogen synthesis, and also has the potential to improve intestinal flora and the metabolic system through a variety of molecular pathways in order to positively regulate oxidative stress and immune inflammation, and protect the liver and kidneys. In order to establish the theoretical justification for the adjunctive treatment of T2DM, we have outlined the physicochemical features of KGM in this article, emphasizing the advantages of KGM as a meal for special medical purposes of T2DM.

Keywords: konjac glucomannan; type 2 diabetes mellitus; FSMP; diet therapy; biomedical function

1. Introduction

Type 2 diabetes mellitus (T2DM) is a metabolic syndrome characterized by decreased insulin sensitivity and insufficient insulin secretion, which is related to the unhealthy metabolism of sugar, fat, amino acids, water, and electrolytes. Genetics [1,2], insulin abnormalities [3], and metabolic disturbance are key factors contributing to the disease. Nowadays, studies have shown that the gut microbiome may also influence the development of T2DM [4,5]. More than 422 million people worldwide have been diagnosed with diabetes mellitus, and the disease claims more than 1.6 million lives annually, according to a recent World Health Organization report [6]. T2DM will affect 439 million people worldwide by the year 2030, making up about 90% to 95% of all cases [7,8]. However, traditional treatment cannot cure it.

T2DM is treated with drugs (such as biguanides, glucosidase inhibitors, and thiazolidinediones) that manage blood pressure, cholesterol, and blood glucose levels [9]. However, long-term use of pharmaceuticals can lead to adverse effects such as bloating and diarrhea [10], and most individuals with T2DM will eventually require insulin therapy to maintain normal blood sugar levels. In this way, the idea of homology between medicine and food is gradually rising, such as control diets to alleviate the progressive worsening of diabetes [11]. The combination of diet and medicine can control the levels of blood sugar and urine sugar without or with only minimal medicine, especially for patients with mild illness, while those with moderate or severe illnesses even can reduce the amount of medication they require. Among the many therapeutic options studied, there is a growing interest in exploring the therapeutic effects of dietary fiber as a special medical food [12]. Consuming dietary fiber has a multitude of metabolic benefits that are unrelated to changes in weight, including impacts on various metabolic and inflammatory diseases, enhanced insulin sensitivity, and regulation of certain gut hormones. Konjac glucomannan (KGM) is a kind of dietary fiber derived from konjac, that is not easily hydrolyzed by human digestive tract enzymes, and can directly enter the colon and be used by intestinal microorganisms [13]. Long-term consumption can also reduce calories, balance body salt, and improve obesity and diabetes, to improve the sub-health of the human body [14].

Studies have shown that KGM as a sticky dietary fiber can improve subjective satiety and reduce appetite, which is very beneficial for patients with T2DM [15]. Adding KGM as a dietary additive to noodles can not only slow down the aging rate of noodles, but also improve the storage stability of cooked noodles [16]. This diet may have potential long-term health outcomes, such as lower blood lipids, postprandial blood glucose and insulin levels, and the benefits of this blood sugar response are highly dependent on the type of substrate and the dose of KGM [17,18]. Compared with other dietary fibers (cereal fiber and vegetable fiber), the purified KGM has a single component and is safer for the human body. The amount of dietary fiber per 100 g of other diets (such as soybeans, wheat, and corn) is much lower than that of KGM. In addition, KGM has a wide range of sources (Figure 1), simple extraction process, high efficiency, and considerable price advantages. The most important thing is that in addition to regulating blood sugar, it can also regulate blood lipids, intestinal flora, oxidative stress, immune suppression and so on. However, it should be noted that dietary fiber itself does not have high nutritional value and cannot be used as a long-term staple food. Although KGM cannot be used as a drug in the treatment of disease, it can be used as a special medical food to assist the treatment of type 2 diabetes mellitus. Since increasing evidence linking the consumption of highly processed foods with an increased risk of noncommunicable diseases poses a public health challenge [19–21], here we review the structure and physicochemical properties of naturally extracted KGM, discuss its benefits on various physiological and biochemical indexes of type 2 diabetes mellitus, and explain the effects of related mechanisms and pathways, in order to reveal its great potential in food for a special medical purpose (FSMP).

Figure 1. Glucomannan from different sources.

2. Extraction and Purification of KGM

Amorphophallus is a perennial herb of Araceae. It has up to 170 varieties, mainly distributed in Southeast Asia and Africa. These varieties are perennial plants with an underground stem in the form of a corm and a highly dissected umbrella-shaped leaf blade [22]. Konjac has a history of more than 2000 years and is a very popular food in China and Japan. Although konjac is poisonous and has a special smell, it is often called "fishy smell". However, further processing and processing can solve this problem perfectly well. Moreover, the smaller the fishy smell, the higher the purity and the better the quality. The specific processing process is as follows.

KGM is usually obtained from natural high-quality konjac without pesticides and fertilizers (mainly 3 years old, weighing 300–1500 g), extracted by a series of broken walls and refined by multi-stage purification [23,24]. The KGM content in fresh konjac tubers is about 30%, and the highest content of refined konjac flour is 96.9% [25]. From the structure of konjac corm, KGM is mainly distributed in the storage parenchyma below the epidermis of konjac, where a large number of idioblast (also known as cystic cells) and ordinary cells are uniformly dispersed [26]. Ordinary cells contain starch, cellulose, and other components, and the texture is more brittle and very easy to break into dust. However, KGM only exists in idioblast with high hardness, good toughness, and is not easy to break [26]. Although there is a certain amount of crude protein, cellulose, and mineral elements in idioblast, the purity of KGM is enough to meet its medical and food applications.

At present, there are mainly two ways of dry processing and wet processing in industry. The steps of dry processing include washing, peeling, slicing, fixing, drying, grinding and screening of konjac (Figure 2) [27]. In principle, according to the differences in composition, toughness, and hardness between idioblast and ordinary cells, ordinary cells are broken first by mechanical crushing, in which starch, cellulose, and other impurities are gradually crushed into konjac fly powder, while idioblast will not be broken under general crushing conditions, still maintaining the integrity of the particles. Through high-intensity repeated collision and friction, the impurities on the surface of the particles will continue to detach, and then be removed by sieving or wind separation, leaving translucent konjac flour particles. Usually, the KGM treated in this way has a slight smell, but when added to food or water, the smell can also be well masked. Therefore, most people can accept it.

The wet processing principle of KGM is similar to that of the dry process, except that the wet process uses a liquid medium when it is subjected to various mechanical forces such as shear, impact, and extrusion. Its advantage is that in the process of contact between konjac and liquid medium, the soluble impurities in idioblast will gradually dissolve out and then be removed by solid–liquid separation, leaving only glucomannan particles with higher purity. However, KGM is easy to swell and agglomerate in the presence of water, so some blocking solvents (such as ethanol and isopropanol) need to be added. For example, ethanol precipitation is the most commonly used method to obtain refined KGM in the laboratory [28]. Compared with dry processing, wet processing has the advantages of effective impurity removal, higher yield, and better viscosity. It has absolute advantages in safety and environmental protection, and is more favored by researchers. However, wet processing has seasonal requirements for konjac, and more importantly, has a high cost, which limits its application in the industrial processing of konjac flour [29,30].

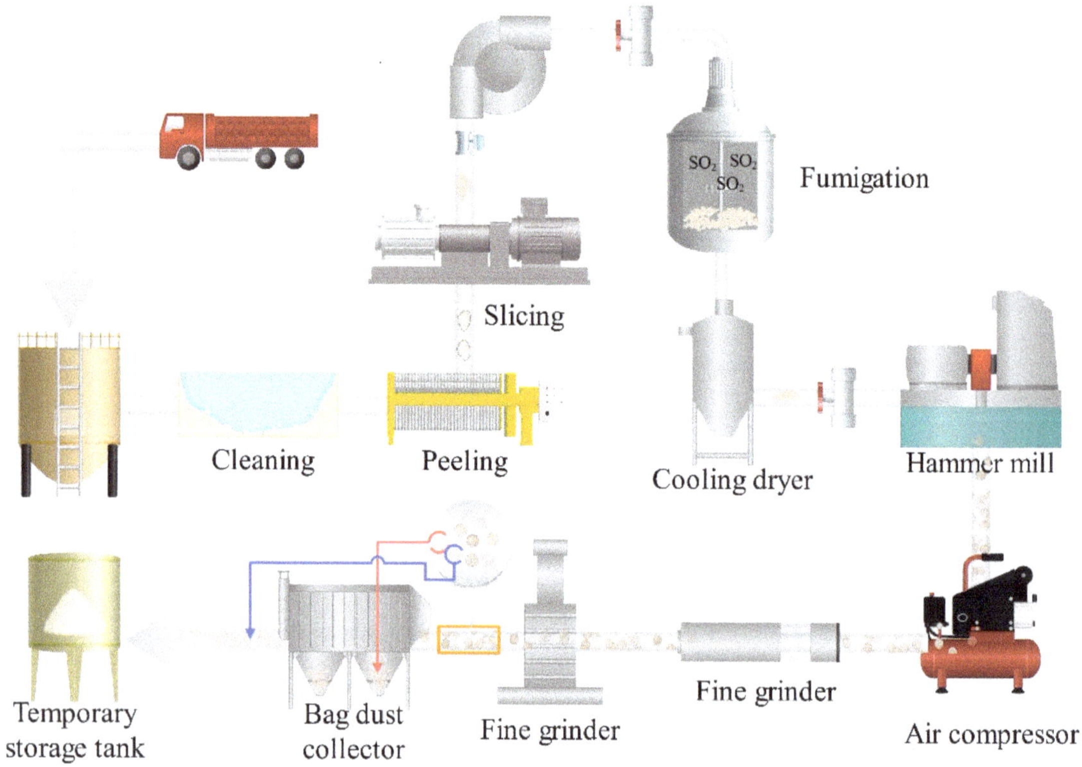

Figure 2. Flow chart of dry preparation of KGM in industry. After washing, peeling, and slicing, the konjac chips will be mixed with SO_2 and hot air in a special device, and the color will be determined by fumigation, then it will be put into the drying equipment immediately, and the konjac chips will be crushed to the powder by hammer grinder and fine grinder. Finally, under the action of grinding and sorting machine and bag dust collector, the finer flying powder such as starch and cellulose will be removed, and only KGM will be retained.

3. Chemical Structure and Physicochemical Properties of KGM

3.1. Chemical Structure of KGM

The chemical structure of polysaccharides has a strong influence on their functional and nutritional properties or biological activity [31]. The basic framework of KGM is composed of D-glucose residues and D-mannose residues (molar ratio 1:1.6~1.7) and is polymerized by β-1,4-pyranoside bonds [32]. Part of the side chain can be formed at the C-3 position of the main chain mannose or the C-6 position of the sugar unit, by the linkage between acetyl groups (Figure 3) [33]. KGM has two native conformations, alpha (amorphous) and beta (crystalline) [34]. The molecular weight distribution of KGM is approximately normal since it is a relatively homogeneous polysaccharide. However, KGM has different molecular weight distributions that are affected by their origin, processing, and storage times [24]. It is reported that KGM has a mean molecular weight of 5.83×10^5 g/mol [35].

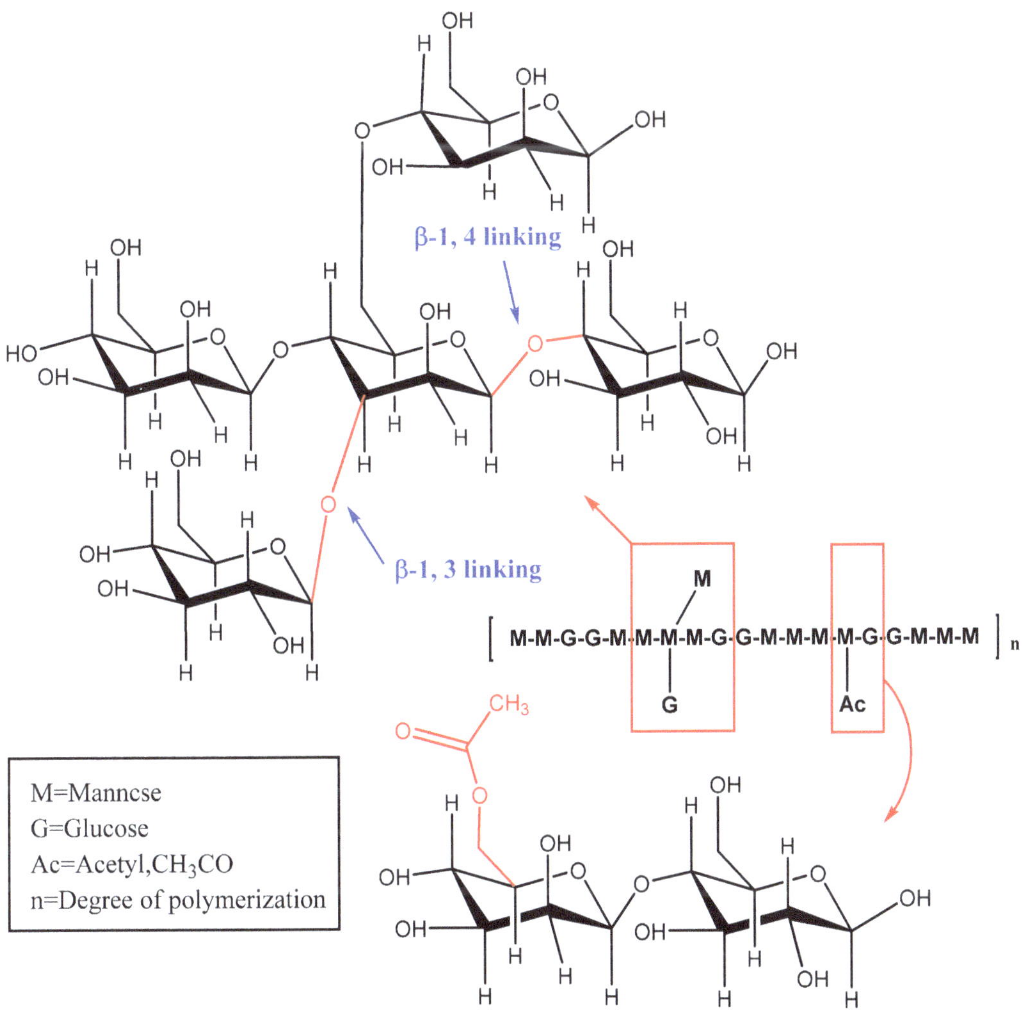

Figure 3. Schematic diagram of KGM structure.

3.2. Physical and Chemical Properties and Function of KGM in Food and Diabetes

When KGM is dissolved in water, it can maintain high performance in an acidic environment, but it is easily unstable in an alkaline environment. For example, these gels remain essentially unchanged in strength, even after repeated heating at 100 °C [36]. A possible explanation for this may be that the acetyl group of KGM is lost under alkaline conditions, which causes aggregation and entanglement of split self-body substrates, which eventually results in the formation of a localized, continuous gel reticular structure (Figure 4) [37]. These properties may also explain the reduced diffusion of glucose in the gut, such as the complex network of gel films that prevent glucose transfer through dialysis.

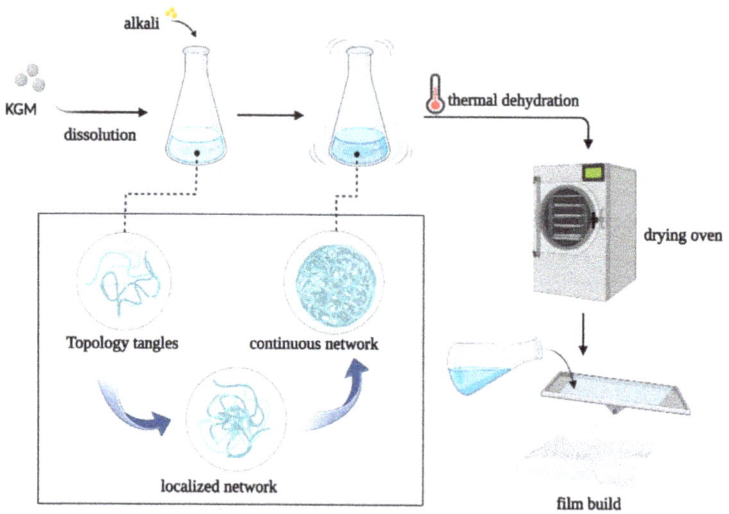

Figure 4. Alkali treatment causes the molecules of KGM to gather and form gels, which are then heated and dehydrated to obtain thin films.

Furthermore, the role of KGM is also related to viscosity and rheological properties. Firstly, KGM can form a high viscosity solution because of its high molecular weight, strong adsorption to water, and electrical neutrality [31]. The apparent viscosity of 1% (*w*/*w*) KGM is about 30,000 cps, making it the most viscosity natural polysaccharide [38,39]. A moderate increase in dietary viscosity by KGM will help T2DM to meet the challenge of hunger in weight management. Secondly, fiber can lessen the rise in postprandial blood glucose and insulin concentration in both normal and diabetic individuals due to its rheological properties [40]. KGM solution is a typical pseudoplastic fluid with shear-thinning properties. The rheological curve can be fitted by the power law equation $\tau = KD^n$, where τ is shear stress (Pa), K is viscosity index ($Pa{\cdot}s^n$), and D is shear rate (s^{-1}) [41]. The mechanism by which KGM improves metabolic control may be related to its rheological properties, such as increasing the viscosity of digestive juices and slowing down the absorption of food in the small intestine, thus reducing postprandial blood glucose and prolonging satiety.

The physical and chemical properties of KGM not only bring many benefits to diabetics, but they also provide more choices for the diabetic diet. Many konjac foods can be produced by using the gelling property of KGM, including konjac powder, konjac knot, konjac cake, konjac tofu, and konjac jelly. It not only meets people's psychological and taste needs, but also maintains the effects of konjac weight loss and fullness, promotes oral health, enhances the growth and vitality of beneficial organisms in the colon, and has a combination of nutrients (such as cholesterol). KGM can also be used as a preservative and fat substitute. The diet of diabetics is affected by diseases, so they should reasonably distribute their food intake among three meals a day, KGM can be added to meat products as a fat substitute, which can effectively improve meat texture, thicken, reduce fat, and improve water retention [42]. Low-glycemic index (GI) diets are thought to reduce postprandial glycemia, resulting in more stable blood glucose concentrations [43,44]. For example, the GI of most fruits and vegetables is relatively low, so it is suitable for consumption by those with T2DM [45]. KGM can be used as a fresh-keeping agent for fruits and vegetables to give patients a better experience in vision and taste [46,47].

4. Biomedical Function of KGM to T2DM

As a kind of natural polysaccharide with unique structure and physicochemical properties, KGM has shown brilliant medical value and prospects. Pharmacological studies on KGM have shown that it has the potential to regulate blood lipids, blood sugar, inflammation, oxidative stress, and intestinal microorganisms (Table 1) [48]. It can relieve the deterioration of T2DM by regulating physiological and biochemical indexes of various organs (Figure 5).

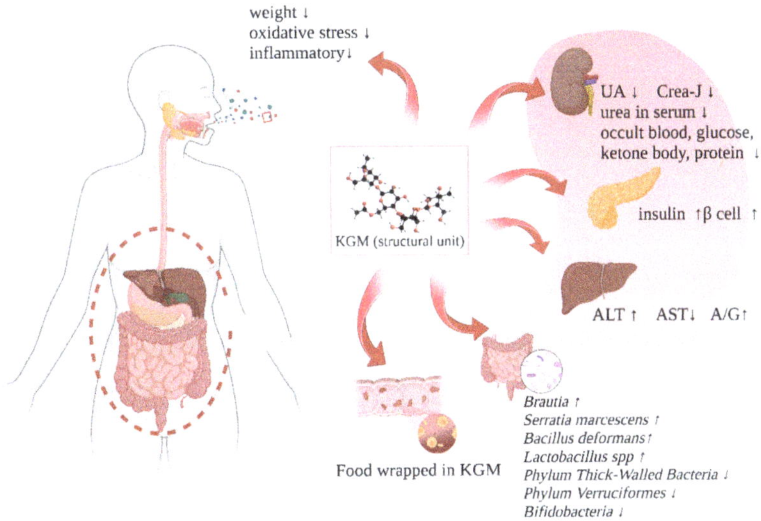

Figure 5. The benefits of the intake of KGM to the human body.

4.1. Regulate Blood Lipid

Dyslipidemia often occurs in the latent and developmental stages of T2DM and is the major cause of complications of diabetics. Therefore, the regulation of blood lipids is one of the therapeutic methods for T2DM [49,50]. Plenty of studies have observed that KGM regulates the metabolic parameters of lipid and cholesterol levels [51]. The long-term consumption of low-dose KGM can improve LDL-C and cholesterol, and reduce blood lipid levels [51]. According to clinical studies, diabetic patients with high cholesterol levels can reduce their cholesterol levels by adding 0.7 g of KGM per 100 calories to their diet [52]. Although the mechanism of the hypolipidemic effect has not been fully understood, it is presumed to be achieved by increasing the excretion of sterols or bile acids, and this presumption has been confirmed by experimental studies in animals [53]. Compared with other dietary supplements, KGM lowers the primary endpoint of metabolic control and various indicators of diabetes such as plasma cholesterol, LDL-C, total/HDL-C ratio, and ApoB [22]. KGM supplementation increased the expression of lipid metabolism-related genes (PPARα, CPT1, Hs1) and lipid transport-related genes (FABP1, apoB100, and CD36), and decreased the expression of lipid synthesis-related genes (srebp1, PPARγ, Fas) [54]. It is apparent that KGM is capable of improving the weight and blood lipids of most obese people, and has the potential to prevent and improve obesity in a limited sense.

It can improve obesity through the following mechanisms: (I) KGM is difficult to be decomposed by human enzymes (mainly endo-1,4-beta-mannanase) [17]; (II) KGM is a soluble fiber that forms a viscous, gel-like mass when hydrated in the stomach, which contributes to a reduction in the rate of gastric emptying and can induce satiety [55]; (III) KGM promotes intestinal peristalsis [56]. (IV) KGM metabolizes energy by excreting feces; (V) the hydrolysate of KGM can stimulate the growth of lactic acid bacteria and play a probiotic role [57]. KGM can also be used in combination with drugs, which is quite effective in the treatment of obese teenagers [58]. Orlistat has become the only drug approved for the treatment of adolescent obesity in the European Union. KGM can reduce the side effects of orlistat, such as transient gastrointestinal adverse reactions [58].

4.2. Maintenance of Glucose Homeostasis

The uptake, utilization, storage, regeneration, and metabolism of glucose are important for maintaining glucose homeostasis in patients with diabetes. An imbalance in glucose homeostasis occurs for individuals with T2DM [59]. Insulin is the most important component of blood sugar control. Therefore, impaired signal transmission will give rise to a disturbance of blood sugar levels [60]. It was found that insulin signaling genes and proteins reduced in diabetic rats, and KGM consumption could increase gene up-regulation and insulin pathway expression, resulting in normalized insulin secretion and lowered blood sugar levels [61]. In addition, insulin receptor substrate 1 (IRS1) and phosphatidylinositol 3-kinase (PI3K) play crucial roles in mediating insulin metabolism. Reduced IRS1 activity can lead to mutations in the tyrosine kinase hinsr gene in diabetics, so as to disrupt insulin signaling [62]. An experiment found that KGM can reduce the mice's body weight and blood sugar level significantly, restore the normal proliferation of pancreatic β-cells, and increase IRS1 and PI3K expression levels (Figure 6e) [63]. This suggests that the reasons why KGM can prevent insulin resistance in diabetic rats may be due to the following two points: (I) inducing the expression of activating proteins related to pancreatic β-cell repair and (II) regeneration of pancreatic β-cells through receptor tyrosine kinase pathway.

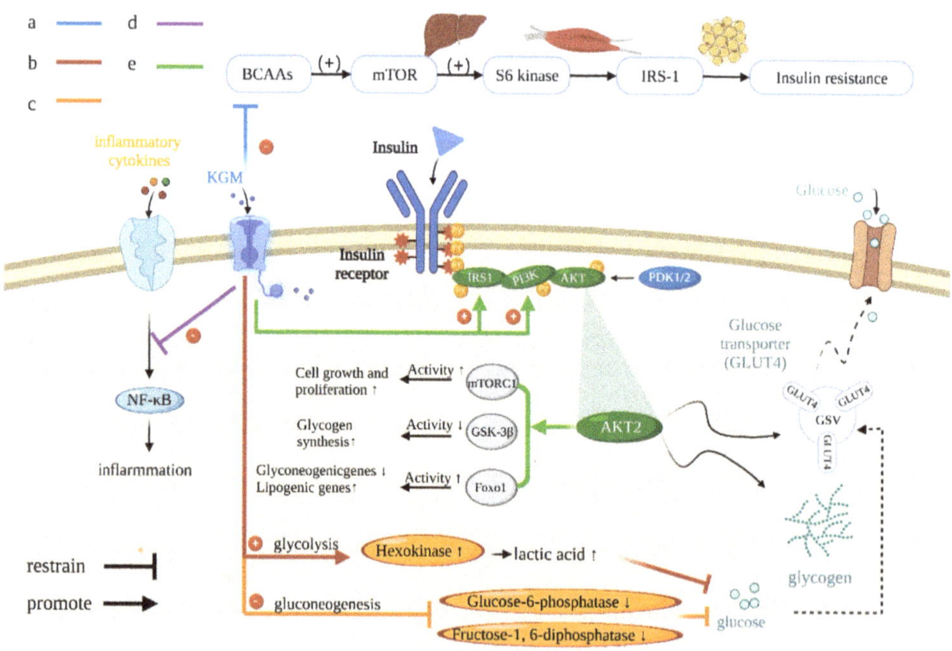

Figure 6. The related mechanism of maintaining sugar homeostasis by KGM. (**a**) KGM reduces insulin resistance through the BCAA pathway. (**b**) KGM increases hexokinase activity and glycolysis. (**c**) KGM regulates the activity of fructose-1 and glucose-6-bisphosphatase and reduces gluconeogenesis. (**d**) KGM inhibits the signal pathway of related inflammatory factors and reduces the immune inflammatory response associated with diabetes. (**e**) KGM up-regulates the expression of IRS1 and PI3K to improve the insulin signaling pathway.

Reducing glucose intake is one of the treatments to lower postprandial glucose in patients with T2DM [64]. There are two main mechanisms associated with this therapeutic target. The first is the reduction of the diffusion rate of glucose. This is related to the ability of different molecules to thicken and swell the hydration network and the subsequent size of the resulting viscosity [65]. The second mechanism is the inhibition of carbohydrate hydrolases (α-amylase and α-glucosidase) in the small intestine. KGM exerts its hypoglycemic effect mainly through the first modality. The dissolved KGM is in the state of gel, which can wrap nutrients, slow down the flow of food in the digestive tract, prolong the residence time of food paste in the gastric cavity, form a protective membrane barrier, and effectively inhibit the value of postprandial blood glucose. Although there is no direct evidence that KGM has the second mechanism of regulation, there is evidence that konjac can lower blood glucose by inhibiting α-amylase and α-glucosidase activity [66].

The increase in circulating blood glucose concentration plays a key role in the pathogenesis of chronic diabetic complications [67,68]. The combination of high sugar and high fat will often lead to a series of human health problems, including eye, kidney, heart, liver, nerve, and other dysfunctions. One element influencing the onset and progression of diabetic nephropathy is a long-term aberrant rise in glomerular microcirculatory filtration pressure. Animal experiments have shown that KGM can treat diabetic nephropathy, normalize rat glomerular structure, lower serum uric acid, creatinine-J, urine occult blood, blood glucose, a ketone body, and protein levels [69]. The ability of the liver to store glycogen is an important indicator of T2DM. KGM can promote the synthesis of glycogen in the liver, reducing blood sugar and alleviating the progression of diabetes, which is beneficial to the protection of the liver. Elevated serum alanine aminotransferase (ALT) and aspartate aminotransferase (AST) reflect the degree of hepatocellular damage, and serum alkaline phosphatase (ALP) activity represents the severity of cholestasis [70]. An elevated ALT level is closely related to T2DM, causing abnormal metabolism of leucine, lysine, and glutamic acid. A study was conducted on mice with liver injury prepared by subcutaneous injection of CCl_4 as a model [71], it was proved that the administration of KGM could down-regulate the abnormal increase of ALT and AST in mice serum, up-regulate the A/G ratio in serum, and increase the abnormally low superoxide dismutase (SOD) content and decrease the abnormally high malondialdehyde content in the liver homogenate of mice caused by liver injury. This demonstrates the protective effect of KGM on experimental liver injury caused by CCl_4 in mice.

In glucose metabolism, KGM greatly affects the enzymatic activity of glycolytic and gluconeogenic enzymes. As a rate-limiting enzyme of glycolysis, hexokinase regulates glucose uptake and energy production by converting glucose into glucose 6-phosphate. When blood glucose levels fall below normal requirements, glucose-6-phosphate acts as a gluconeogenesis enzyme in the liver, releasing glucose through glucose-6-phosphatase [72]. Researchers found that in diabetic rats, hexokinase, glucose-6-phosphate dehydrogenase, and glycogen levels were significantly reduced ($p < 0.05$), while gluconeogenesis (glucose-6-phosphatase) was significantly increased. In diabetic rats administered 80 mg/kg of KGM, the levels of hexokinase, glucose-6-phosphate dehydrogenase, and glycogen were increased, but the levels of gluconeogenesis enzymes were decreased [73,74]. The intensity of gluconeogenic enzymes was significantly reduced, which led to the hypothesis that KGM treatment could improve the activity of enzymes involved in glucose metabolism to improve insulin sensitivity and up-regulate more glucose for energy production (Figure 6b,c) [61].

4.3. *Regulation of Oxidative Stress and Inflammation*

As a systemic chronic disease, diabetes is bound to be affected by oxidative stress and inflammation [75,76]. Some inflammatory markers such as TNF-α, IL-6, and monocyte chemoattractant protein-1 (MCP-1) have been shown to be associated with adipose tissue and are the primary causes of insulin resistance [77,78]. Detecting and regulating changes in oxidative stress and inflammation may prevent the progression of diabetes. Studies have found that KGM can increase aromatic amino acid (AAA) metabolism-related intestinal bacteria (*Lactobacillus*, *Ruminococcus-1*, and *Bifidobacterium*) and AAA metabolites (kynurenic acid, desaminotyrosine, 3-hydroxy-3-(3-hydroxyphenyl) propanoic acid-O-sulphate and hippuric acid), which may also contribute to the reduction of systemic inflammation and thus reduce symptoms of oxidative stress and T2DM [79]. KGM appears to regulate oxidative stress and inflammation positively, which indicates that it may be effective in preventing or treating diabetes, although further research and data are necessary to confirm this.

4.4. Effects on Gut Microbes

The gut microbiota play a crucial role in the body's physiological processes [15,80,81]. Intestinal microflora can regulate the Mammalian target rapamycin (mTORC) signal pathway, thus inhibit tyrosine phosphorylation, reduce protein level, down-regulate insulin signal transduction, and exerting the effect of intestinal microorganisms on diabetes [82]. Diabetic patients have different intestinal flora species as well as different numbers than healthy individuals due to metabolic dysregulation. The levels of Enterococci in patients with diabetes will increase, while those of *Bifidobacterium vulgatus* and other *Bifidobacterium* species decrease when their blood glucose rises, leading to the aggravation of diabetes. Conditional pathogenic bacteria, such as *Bacteroides*, *Escherichia coli*, and *Desulfovibrio*, have been observed to increase in diabetic patients [83]. Meanwhile, most of the metabolites and products of these microflorae are mediated by mTOR pathway [84]. For example, BCAA is synthesized by intestinal microorganisms. The high content of BCAA can activate mTOR and its downstream effector S6K1 in the liver, muscle, and adipose tissue, and sustained activation leads to serine phosphorylation of IRS-1, which inhibits IRS-1 and leads to insulin resistance (Figure 6a) [85]; mTORC activates anabolism (protein and fat biosynthesis) and inhibits autophagy. Its overactivation may lead to diabetes [84]. The reason why KGM reduces insulin resistance may be achieved by down-regulating the abundance of BCAA-synthesizing-associated flora [86], including *Clostridium* spp., *Bacteroides* spp., *Prevotella* spp., *Klebsiella* spp., *E. coli*, *Streptococcus* spp., and *S. aureus* [86].

Table 1. Summary of biomedical functions of diabetes mellitus by KGM.

KGM Activity	Model	Study Design and Period (DUS)	Dosage Form (DSF)	Results	Reference
Regulation of blood lipids	Eight in adults and four in children	3 g/day; three weeks	Unlimited	Reduced LDL cholesterol and non-HDL cholesterol of 10% and 7%.	[51]
	Twenty-two diabetic subjects	3.6 g/day; 28 days	Gelatin capsules	Alleviated the elevated cholesterol, LDL-cholesterol, apo B, and ratios of total/HDL-cholesterol.	[52]
	Seven hundred and twenty healthy fish	With 0, 0.5, 1 and 2% KGM twice a day; 60 days	Fodder	Improved growth performance, antioxidant activity, immune response, and lipid metabolism in juvenile pompano.	[54]
Regulation of blood sugar	Six rats in each group	I-control rats fed standard pellet diet alone II-KGM control (120 mg/kg body weight (b.w.)) III-T2DM (HFD + STZ-40 mg/kg b.w.) IV-T2DM + KGM (80 mg/kg b.w.) V-T2DM + RSG (4 mg/kg b.w.); 28 days	Fodder	Improved cell glucose-tolerance; regulated glycolytic, gluconeogenesis enzymes; reduced the stored glycogen in the liver, and restored liver enzymes.	[61]
	Three-month-old male	0.06 g/mg/kg b.w. and 0.12 g/mg/kg b.w.; one month	Fodder	Significantly increased IRS-1 level expression, proliferated properly and consistently increased the PI3-K expression level.	[63]
	Twelve-week-old male Wistar rats ($n = 8$)	102 mg/kg b.w.; at0,30,60, and 120 min	Konjac solution	Konjac has hypoglycemic and antioxidant activities in vitro and in vivo.	[66]
	Seven-week-old male Wistar rats (180–220 g)	160 mg/kg b.w. of glucomannans; treatment for once every day; gavage for 4 weeks	Fodder	Lower levels of fasting blood glucose, serum insulin, and glycated serum protein; improve urea cycle, metabolism of lipid, glucose, and amino acids.	[69]
	Rats (weight 22~28 g)	50,100,200 mg/kg b.w.; once a day; six days	Konjac solution	The levels of ALT and AST decreased, the ratio of serum albumin to globulin increased, and hepatocytes were protected.	[71]
Regulation of oxidative stress and inflammation	Male C57BL/6J mice (body weight 20 ± 2 g)	500 mg/kg b.w.; once a day; 4 weeks	Konjac solution	Lactobacillus, Ruminococcus_1 and Bifidobacterium and AAA metabolites such as kynurenic acid, desaminotyrosine, 3hydroxy-3-(3hydroxyphenyl) propanoic acid-O-sulphate, and hippuric acid were added.	[79]
Effects on gut microbes	Wistar rats (180–220 g)	2 mL/200 g of BW; once daily for 4 weeks	fodder	Decreased the abundance of microbial BCAA biosynthesis-related genes and ameliorated the host BCAA metabolism.	[86]

5. Conclusions and Future Prospect

There are several advantages of KGM and there are also some problems. First, high doses cause minor side effects with high doses, such as hiccups, bloating, and diarrhea [87,88]. Fortunately, there have only been a handful of reports on side effects associated with KGM in a normal diet. It has been demonstrated that a short-term intake of 3.9 g of KGM per day does not cause changes in bowel function [89]. Currently, there is no unified regulatory standard on a global scale within KGM, which needs to rely on the improvement and perfection of the food management industry as it advances. The KGM also provides an opportunity to formulate or revise FSMP-related policies to promote the rapid and orderly development of the FSMP industry (Figure 7a) [90–92].

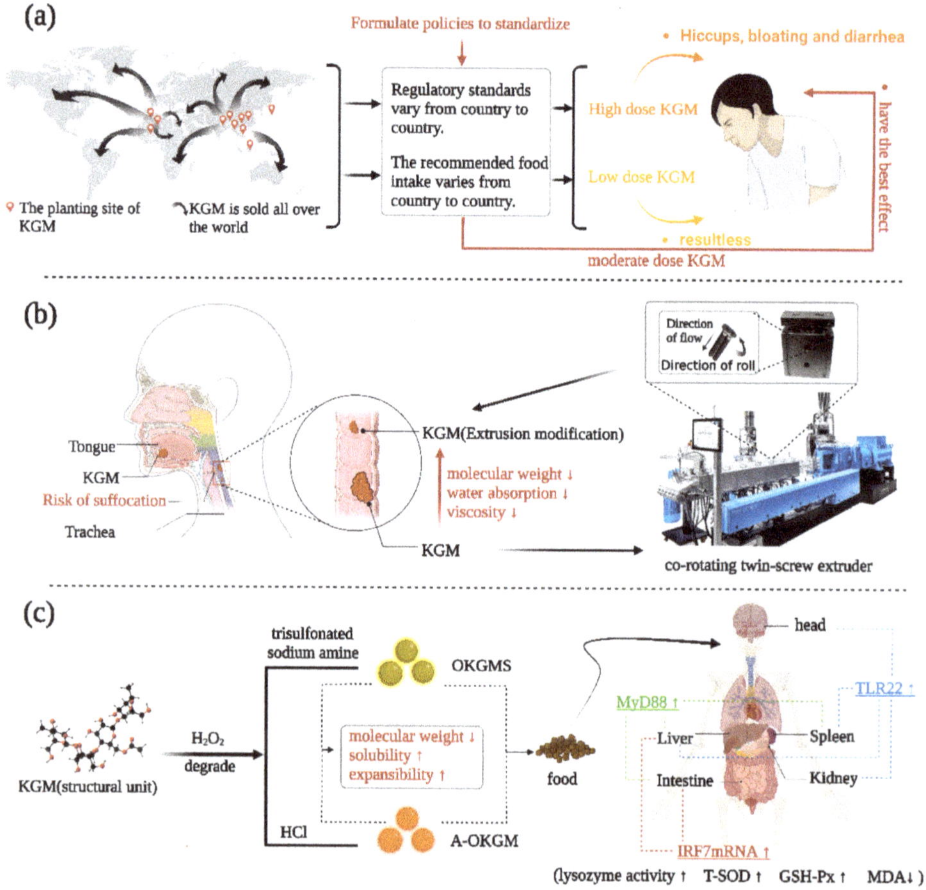

Figure 7. Some solutions to KGM's problems. (**a**) The establishment of a unified regulatory system can better regulate the average daily intake of KGM and reduce the side effects. (**b**) KGM that has undergone physical modification can lessen the asphyxia risk. (**c**) Chemically altered KGM derivatives can improve the body's immunological indicators.

Secondly, the highwater absorption of KGM is associated with the risk of asphyxiation. Viscosity plays a crucial role in the swallowing process [93]. The thicker liquid gives the muscles more time to react by slowing down the flow of food in the mouth and throat stages [82], and the likelihood of swallowing incidents is decreased as a result [94]. After, KGM repeated machine extrusion and high temperature grinding the water absorption and viscosity will significantly reduce, which can reduce the risk of asphyxiation of products containing KGM (Figure 7b) [95].

Thirdly, the solubility of natural KGM in water decreases with increasing molecular weight, and it has several shortcomings as a feed additive, such as low solubility, high expansibility, and ease to cause abdominal distension. The molecular weight of KGM degraded by H_2O_2 is significantly lower than that of raw materials, and it has peculiar solubility and does not easily cause abdominal distension. In addition, studies have shown that when the molecular weight of polysaccharides decreases to 100–200 kDa, the biological activity will be significantly increased [96]. After *S. prenanti* ate feed containing oxidized konjac glucomannan sulfates (OKGMS) and acidolysis-oxidized konjac glucomannan (A-OKGM), its growth and immune function were improved (Figure 7c) [97,98].

Fourthly, whiteness is one of the important indicators to measure the quality of konjac and its products, and the market value of konjac powder will decrease after browning. Therefore, in the initial processing of konjac, SO_2 fumigation or sulfite soaking is usually used to protect the color of konjac powder, which may lead to sulfur dioxide residue in KGM [25]. At present, it can be reduced by improving drying methods (vacuum freeze drying, microwave drying) or using natural browning inhibitors (such as citric acid, mercaptan, ascorbic acid, and oxalic acid) [99].

As the challenges faced by KGM in production applications are gradually overcome, its demonstrated advantages as a biomedical function of FSMP are attracting attention. KGM can improve the taste, flavor, and visual enjoyment of food as an additive. It can also play a powerful role as a dietary supplement and in combination with drugs [100]. Moreover, the hypoglycemic effect of different purities of KGM is also different, and one study found that 90% of the hypoglycemic effect is the best [101–104]. Thus, KGM may also replace the traditional sugars in drugs, bringing a blessing to patients with obesity, hyperglycemia, hypertension, and metabolic syndrome. The integration of such dietary fiber into conventional medical practice may reduce the required drug dose, thus improving the overall economic benefits and therapeutic efficacy [39]. Furthermore, continuous innovation of product dosage forms, continuous refinement of disease coverage, gradual precision of nutritional interventions, and accelerated exploration of clinical treatment concepts will open up industrial development paths for FSMP based on KGM.

Author Contributions: D.S., H.J. and L.J. contributed to the conception of this review; Y.F. and J.M. analyzed the literature and wrote the manuscript; D.S., F.L., P.L. and Y.F. completed the figure drawings; D.S., H.J., X.Y., J.Z., L.W., W.W. and J.Q. revised the manuscript. All authors have read and agreed to the published version of the manuscript.

Funding: This work was funded by the National Natural Science Foundation of China (51901160).

Data Availability Statement: Not applicable.

Conflicts of Interest: The authors declare no conflict of interest.

Abbreviations

T2DM	type 2 diabetes mellitus
KGM	Konjac glucomannan
FSMP	food for special medical purpose
GI	glycemic index
IRS1	insulin receptor substrate 1
PI3K	phosphatidylinositol 3-kinase
ALT	alanine aminotransferase
AST	aspartate aminotransferase
ALP	serum alkaline phosphatase
SOD	superoxide dismutase
OKGM	oxidized konjac glucomannan sulfates
A-OKGM	acidolysis-oxidized konjac glucomannan
BCAA	Branched Chain Amino Acid

References

1. Fletcher, B.; Gulanick, M.; Lamendola, C. Risk Factors for Type 2 Diabetes Mellitus. *J. Cardiovasc. Nurs.* **2002**, *16*, 17–23. [CrossRef]
2. American Diabetes Association. Diagnosis and Classification of Diabetes Mellitus. *Diabetes Care* **2013**, *36*, S67–S74. [CrossRef] [PubMed]
3. Griffin, S. Diabetes precision medicine: Plenty of potential, pitfalls and perils but not yet ready for prime time. *Diabetologia* **2022**, *65*, 1913–1921. [CrossRef]
4. Maskarinec, G.; Raquinio, P.; Kristal, B.; Wilkens, L.; Franke, A.; Lim, U.; Marchand, L.L.; Lampe, J.; Hullar, M. The Gut Microbiome and Diabetes Status in the Multiethnic Cohort. *Curr. Dev. Nutr.* **2020**, *4*, 1450. [CrossRef]
5. Jiang, H.; Cai, M.; Shen, B.; Wang, Q.; Zhang, T.; Zhou, X. Synbiotics and Gut Microbiota: New Perspectives in the Treatment of Type 2 Diabetes Mellitus. *Foods* **2022**, *11*, 2438. [CrossRef]
6. Sun, J.; Ren, J.; Hu, X.; Hou, Y.; Yang, Y. Therapeutic effects of Chinese herbal medicines and their extracts on diabetes. *Biomed. Pharmacother.* **2021**, *142*, 111977. [CrossRef] [PubMed]
7. Chen, L.; Magliano, D.J.; Zimmet, P.Z. The worldwide epidemiology of type 2 diabetes mellitus—Present and future perspectives. *Nat. Rev. Endocrinol.* **2012**, *8*, 228–236. [CrossRef] [PubMed]
8. Wu, Y.; Ding, Y.; Tanaka, Y.; Zhang, W. Risk Factors Contributing to Type 2 Diabetes and Recent Advances in the Treatment and Prevention. *Int. J. Med. Sci.* **2014**, *11*, 1185–1200. [CrossRef]
9. Zhou, K.; Pedersen, H.K.; Dawed, A.Y.; Pearson, E.R. Pharmacogenomics in diabetes mellitus: Insights into drug action and drug discovery. *Nat. Rev. Endocrinol.* **2016**, *12*, 337–346. [CrossRef]
10. Lee, S.-H.; Yoon, K.-H. A Century of Progress in Diabetes Care with Insulin: A History of Innovations and Foundation for the Future. *Diabetes Metab. J.* **2021**, *45*, 629–640. [CrossRef]
11. Ojo, O. Dietary Intake and Type 2 Diabetes. *Nutrients* **2019**, *11*, 2177. [CrossRef] [PubMed]
12. Mao, T.; Huang, F.; Zhu, X.; Wei, D.; Chen, L. Effects of dietary fiber on glycemic control and insulin sensitivity in patients with type 2 diabetes: A systematic review and meta-analysis. *J. Funct. Foods* **2021**, *82*, 104500. [CrossRef]
13. Connolly, M.L.; Lovegrove, J.A.; Tuohy, K.M. Konjac glucomannan hydrolysate beneficially modulates bacterial composition and activity within the faecal microbiota. *J. Funct. Foods* **2010**, *2*, 219–224. [CrossRef]
14. Tester, R.F.; Al-Ghazzewi, F.H. Beneficial health characteristics of native and hydrolysed konjac (*Amorphophallus konjac*) glucomannan: Health characteristics of native and hydrolysed konjac glucomannan. *J. Sci. Food Agric.* **2016**, *96*, 3283–3291. [CrossRef] [PubMed]
15. Wu, X.; Song, M.; Qiu, P.; Li, F.; Wang, M.; Zheng, J.; Wang, Q.; Xu, F.; Xiao, H. A metabolite of nobiletin, 4′-demethylnobiletin and atorvastatin synergistically inhibits human colon cancer cell growth by inducing G0/G1 cell cycle arrest and apoptosis. *Food Funct.* **2018**, *9*, 87–95. [CrossRef] [PubMed]
16. Zhou, Y.; Qin, J.; Wang, Y.; Wang, Y.; Cheng, Y. Gastrointestinal and metabolic effects of noodles-based konjac glucomannan in rats. *Food Nutr. Res.* **2019**, *63*. [CrossRef]
17. Nagasawa, T.; Kimura, T.; Yoshida, A.; Tsunekawa, K.; Araki, O.; Ushiki, K.; Ishigaki, H.; Shoho, Y.; Suda, I.; Hiramoto, S.; et al. Konjac Glucomannan Attenuated Triglyceride Metabolism during Rice Gruel Tolerance Test. *Nutrients* **2021**, *13*, 2191. [CrossRef]
18. Yoshida, A.; Kimura, T.; Tsunekawa, K.; Araki, O.; Ushiki, K.; Ishigaki, H.; Shoho, Y.; Suda, I.; Hiramoto, S.; Murakami, M. Glucomannan Inhibits Rice Gruel-Induced Increases in Plasma Glucose and Insulin Levels. *Ann. Nutr. Metab.* **2020**, *76*, 259–267. [CrossRef]
19. Adams, J.; Hofman, K.; Moubarac, J.-C.; Thow, A.M. Public health response to ultra-processed food and drinks. *BMJ* **2020**, *369*, m2391. [CrossRef]
20. Gibney, M.J.; Forde, C.G.; Mullally, D.; Gibney, E.R. Ultra-processed foods in human health: A critical appraisal. *Am. J. Clin. Nutr.* **2017**, *106*, 717–724. [CrossRef]
21. Monteiro, C.A.; Cannon, G.; Levy, R.B.; Moubarac, J.-C.; Louzada, M.L.; Rauber, F.; Khandpur, N.; Cediel, G.; Neri, D.; Martinez-Steele, E.; et al. Ultra-processed foods: What they are and how to identify them. *Public Health Nutr.* **2019**, *22*, 936–941. [CrossRef] [PubMed]
22. Devaraj, R.D.; Reddy, C.K.; Xu, B. Health-promoting effects of konjac glucomannan and its practical applications: A critical review. *Int. J. Biol. Macromol.* **2019**, *126*, 273–281. [CrossRef] [PubMed]
23. Parry, J.-M. Konjac Glucomannan. In *Food Stabilisers, Thickeners and Gelling Agents*; Imeson, A., Ed.; Wiley-Blackwell: Oxford, UK, 2009; pp. 198–217, ISBN 978-1-4443-1472-4.
24. Khan, H. Marya Konjac (*Amorphophallus konjac*). In *Nonvitamin and Nonmineral Nutritional Supplements*; Elsevier: Amsterdam, The Netherlands, 2019; pp. 307–312. ISBN 978-0-12-812491-8.
25. Ye, S.; Zongo, A.W.-S.; Shah, B.R.; Li, J.; Li, B. Konjac Glucomannan (KGM), Deacetylated KGM (Da-KGM), and Degraded KGM Derivatives: A Special Focus on Colloidal Nutrition. *J. Agric. Food Chem.* **2021**, *69*, 12921–12932. [CrossRef] [PubMed]
26. Shi, Y.; Tao, Y.; Lu, Y.; Fei, X. Morphology of starch and manna granules in corms of Amorphophallus COn/ac. *Guihaia* **1998**, *18*, 75–78.
27. NIE, C.; GAO, Q. Research Progress on Deep Processing and Application of Konjak. *Food Sci. Technol.* **2022**, *47*. [CrossRef]
28. Xu, W.; Wang, S.; Ye, T.; Jin, W.; Liu, J.; Lei, J.; Li, B.; Wang, C. A simple and feasible approach to purify konjac glucomannan from konjac flour—Temperature effect. *Food Chem.* **2014**, *158*, 171–176. [CrossRef]

29. Ye, T.; Wang, L.; Xu, W.; Liu, J.; Wang, Y.; Zhu, K.; Wang, S.; Li, B.; Wang, C. An approach for prominent enhancement of the quality of konjac flour: Dimethyl sulfoxide as medium. *Carbohydr. Polym.* **2014**, *99*, 173–179. [CrossRef]
30. Yuan, Z.; Wu, D.; Wu, H.; Li, X. China journal of Chinese Materia. *Medica* **2003**, *28*, 324–327.
31. Tester, R.; Al-Ghazzewi, F. Glucomannans and nutrition. *Food Hydrocoll.* **2017**, *68*, 246–254. [CrossRef]
32. Cui, T.; Liu, R.; Wu, T.; Sui, W.; Zhang, M. Influence of Konjac Glucomannan and Frozen Storage on Rheological and Tensile Properties of Frozen Dough. *Polymers* **2019**, *11*, 794. [CrossRef]
33. Jiang, M.; Li, H.; Shi, J.; Xu, Z. Depolymerized konjac glucomannan: Preparation and application in health care. *J. Zhejiang Univ. Sci. B* **2018**, *19*, 505–514. [CrossRef] [PubMed]
34. Meng, F.; Liu, D.; LI, Y. Research progress of the structures, properties and modifications of Konjacglucomannan. *Sci. Technol. Food Ind.* **2016**, *37*, 394–400. [CrossRef]
35. Katsuraya, K.; Okuyama, K.; Hatanaka, K.; Oshima, R.; Sato, T.; Matsuzaki, K. Constitution of konjac glucomannan: Chemical analysis and 13C NMR spectroscopy. *Carbohydr. Polym.* **2003**, *53*, 183–189. [CrossRef]
36. Yang, D.; Yuan, Y.; Wang, L.; Wang, X.; Mu, R.; Pang, J.; Xiao, J.; Zheng, Y. A Review on Konjac Glucomannan Gels: Microstructure and Application. *Int. J. Mol. Sci.* **2017**, *18*, 2250. [CrossRef] [PubMed]
37. Hou, Y.; Nie, J. Shaoping Study and application on the deacetylated gelation of konjac glucomannan. *J. Shaanxi Norm. Univ.* **2022**, *50*, 1–14. [CrossRef]
38. Du, X.; Li, J.; Chen, J.; Li, B. Effect of degree of deacetylation on physicochemical and gelation properties of konjac glucomannan. *Food Res. Int.* **2012**, *46*, 270–278. [CrossRef]
39. Chua, M.; Baldwin, T.C.; Hocking, T.J.; Chan, K. Traditional uses and potential health benefits of Amorphophallus konjac K. Koch ex N.E.Br. *J. Ethnopharmacol.* **2010**, *128*, 268–278. [CrossRef]
40. InterAct Consortium. Dietary fibre and incidence of type 2 diabetes in eight European countries: The EPIC-InterAct Study and a meta-analysis of prospective studies. *Diabetologia* **2015**, *58*, 1394–1408. [CrossRef]
41. Yoshimura, M.; Nishinari, K. Dynamic viscoelastic study on the gelation of konjac glucomannan with different molecular weights. *Food Hydrocoll.* **1999**, *13*, 227–233. [CrossRef]
42. Chen, J.; Zhao, J.; Li, X.; Liu, Q.; Kong, B. Composite Gel Fabricated with Konjac Glucomannan and Carrageenan Could Be Used as a Cube Fat Substitute to Partially Replace Pork Fat in Harbin Dry Sausages. *Foods* **2021**, *10*, 1460. [CrossRef]
43. Vlachos, D.; Malisova, S.; Lindberg, F.A.; Karaniki, G. Glycemic Index (GI) or Glycemic Load (GL) and Dietary Interventions for Optimizing Postprandial Hyperglycemia in Patients with T2 Diabetes: A Review. *Nutrients* **2020**, *12*, 1561. [CrossRef] [PubMed]
44. Zafar, M.I.; Mills, K.E.; Zheng, J.; Regmi, A.; Hu, S.Q.; Gou, L.; Chen, L.-L. Low-glycemic index diets as an intervention for diabetes: A systematic review and meta-analysis. *Am. J. Clin. Nutr.* **2019**, *110*, 891–902. [CrossRef] [PubMed]
45. Atkinson, F.S.; Brand-Miller, J.C.; Foster-Powell, K.; Buyken, A.E.; Goletzke, J. International tables of glycemic index and glycemic load values 2021: A systematic review. *Am. J. Clin. Nutr.* **2021**, *5*, 1625–1632. [CrossRef]
46. Li, X.; Xiao, N.; Xiao, G.; Bai, W.; Zhang, X.; Zhao, W. Lemon essential oil/vermiculite encapsulated in electrospun konjac glucomannan-grafted-poly (acrylic acid)/polyvinyl alcohol bacteriostatic pad: Sustained control release and its application in food preservation. *Food Chem.* **2021**, *348*, 129021. [CrossRef]
47. Hashemi, S.M.B.; Jafarpour, D. The efficacy of edible film from Konjac glucomannan and saffron petal extract to improve shelf life of fresh-cut cucumber. *Food Sci. Nutr.* **2020**, *8*, 3128–3137. [CrossRef] [PubMed]
48. Zhao, Y.; Jayachandran, M.; Xu, B. In vivo antioxidant and anti-inflammatory effects of soluble dietary fiber Konjac glucomannan in type-2 diabetic rats. *Int. J. Biol. Macromol.* **2020**, *159*, 1186–1196. [CrossRef]
49. Zhang, C.; Chen, J.; Yang, F. Konjac Glucomannan, a Promising Polysaccharide for OCDDS. *Carbohydr. Polym.* **2014**, *104*, 175–181. [CrossRef]
50. Lu, Y.; Zhang, J.; Zhang, Z.; Liang, X.; Liu, T.; Yi, H.; Gong, P.; Wang, L.; Yang, W.; Zhang, X.; et al. Konjac Glucomannan with Probiotics Acts as a Combination Laxative to Relieve Constipation in Mice by Increasing Short-Chain Fatty Acid Metabolism and 5-Hydroxytryptamine Hormone Release. *Nutrition* **2021**, *84*, 111112. [CrossRef]
51. Hayeeawaema, F.; Wichienchot, S.; Khuituan, P. Amelioration of Gut Dysbiosis and Gastrointestinal Motility by Konjac Oligo-Glucomannan on Loperamide-Induced Constipation in Mice. *Nutrition* **2020**, *73*, 110715. [CrossRef]
52. Patel, J. Diabetes: Managing dyslipidaemia. *BMJ Clin. Evid.* **2008**, *2008*, 610.
53. Szalat, A.; Durst, R.; Leitersdorf, E. Managing dyslipidaemia in type 2 diabetes mellitus. *Best Pract. Res. Clin. Endocrinol. Metab.* **2016**, *30*, 431–444. [CrossRef]
54. Ho, H.V.T.; Jovanovski, E.; Zurbau, A.; Blanco Mejia, S.; Sievenpiper, J.L.; Au-Yeung, F.; Jenkins, A.L.; Duvnjak, L.; Leiter, L.; Vuksan, V. A systematic review and meta-analysis of randomized controlled trials of the effect of konjac glucomannan, a viscous soluble fiber, on LDL cholesterol and the new lipid targets non-HDL cholesterol and apolipoprotein B. *Am. J. Clin. Nutr.* **2017**, *105*, 1239–1247. [CrossRef]
55. Chen, H.-L.; Sheu, W.H.-H.; Tai, T.-S.; Liaw, Y.-P.; Chen, Y.-C. Konjac Supplement Alleviated Hypercholesterolemia and Hyperglycemia in Type 2 Diabetic Subjects—A Randomized Double-Blind Trial. *J. Am. Coll. Nutr.* **2003**, *22*, 36–42. [CrossRef] [PubMed]
56. Dziadek, K.; Kopeć, A.; Piątkowska, E.; Leszczyńska, T. High-Fructose Diet-Induced Metabolic Disorders Were Counteracted by the Intake of Fruit and Leaves of Sweet Cherry in Wistar Rats. *Nutrients* **2019**, *11*, 2638. [CrossRef] [PubMed]

57. Li, Y.; Liang, S.; Shao, Y.; Li, Y.; Chen, C.; You, C.; Monroig, Ó.; Rahimnejad, S.; Tocher, D.R.; Wang, S. Impacts of dietary konjac glucomannan supplementation on growth, antioxidant capacity, hepatic lipid metabolism and inflammatory response in golden pompano (*Trachinotus ovatus*) fed a high fat diet. *Aquaculture* **2021**, *545*, 737113. [CrossRef]
58. Keleszade, E.; Willner, T.; Patterson, M.; Trangmar, S.; Kolida, S.; Costabile, A. A pilot study to assess the effect of a fibre and mineral formulation on satiety and satiation when taken as part of a calorie restriction diet in overweight and obese women. *J. Funct. Foods* **2020**, *74*, 104157. [CrossRef]
59. Guo, L.; Yokoyama, W.; Chen, M.; Zhong, F. Konjac glucomannan molecular and rheological properties that delay gastric emptying and improve the regulation of appetite. *Food Hydrocoll.* **2021**, *120*, 106894. [CrossRef]
60. Al-Ghazzewi, F.H.; Tester, R.F. Efficacy of cellulase and mannanase hydrolysates of konjac glucomannan to promote the growth of lactic acid bacteria. *J. Sci. Food Agric.* **2012**, *92*, 2394–2396. [CrossRef]
61. Rogovik, A.L.; Goldman, R.D. Should weight-loss supplements be used for pediatric obesity? *Can. Fam. Physician* **2009**, *55*, 257–259.
62. Röder, P.V.; Wu, B.; Liu, Y.; Han, W. Pancreatic regulation of glucose homeostasis. *Exp. Mol. Med.* **2016**, *48*, e219. [CrossRef]
63. Bettedi, L.; Yan, A.; Schuster, E.; Alic, N.; Foukas, L.C. Increased mitochondrial and lipid metabolism is a conserved effect of Insulin/PI3K pathway downregulation in adipose tissue. *Sci. Rep.* **2020**, *10*, 3418. [CrossRef] [PubMed]
64. Li, X.; Jayachandran, M.; Xu, B. Antidiabetic effect of konjac glucomannan via insulin signaling pathway regulation in high-fat diet and streptozotocin-induced diabetic rats. *Food Res. Int.* **2021**, *149*, 110664. [CrossRef] [PubMed]
65. Fatchiyah, F.; Christian, N.; Soeatmadji, D. Reducing IRS-1 Activation Cause Mutation of Tyrosine Kinase Domain hINSR Gene on Type-2 Diabetes Mellitus Patients. *Bioinformation* **2013**, *9*, 853–857. [CrossRef]
66. Fatchiyah, F.; Nurmasari, D.A.; Masruro, N.; Rohmah, N.R.; Triprisila, L.F.; Mulyati, M.; Yamada, T.; Ohta, T. Level of mRNA Insulin Gene and Blood Glucose STZ-Induced Diabetic Rat are Improved by Glucomannan of Amorphophallus muelleri Blume from East Java Forest Indonesia. *J. Trop. Life Sci.* **2019**, *9*, 163–169. [CrossRef]
67. Rasouli, H.; Hosseini-Ghazvini, S.M.-B.; Adibi, H.; Khodarahmi, R. Differential α-amylase/α-glucosidase inhibitory activities of plant-derived phenolic compounds: A virtual screening perspective for the treatment of obesity and diabetes. *Food Funct.* **2017**, *8*, 1942–1954. [CrossRef] [PubMed]
68. Fabek, H.; Messerschmidt, S.; Brulport, V.; Goff, H.D. The effect of in vitro digestive processes on the viscosity of dietary fibres and their influence on glucose diffusion. *Food Hydrocoll.* **2014**, *35*, 718–726. [CrossRef]
69. Gamboa-Gómez, C.I.; Guerrero-Romero, F.; Sánchez-Meraz, M.A.; Simental-Mendía, L.E. Hypoglycemic and antioxidant properties of konjac (*Amorphophallus konjac*) in vitro and in vivo. *J. Food Biochem.* **2020**, *44*, e13503. [CrossRef] [PubMed]
70. Spanakis, E.K.; Singh, L.G.; Siddiqui, T.; Sorkin, J.D.; Notas, G.; Magee, M.F.; Fink, J.C.; Zhan, M.; Umpierrez, G.E. Association of glucose variability at the last day of hospitalization with 30-day readmission in adults with diabetes. *BMJ Open Diab. Res. Care* **2020**, *8*, e000990. [CrossRef]
71. Tong, L.; Chi, C.; Zhang, Z. Association of various glycemic variability indices and vascular outcomes in type-2 diabetes patients: A retrospective study. *Medicine* **2018**, *97*, e10860. [CrossRef]
72. Chen, H.; Nie, Q.; Hu, J.; Huang, X.; Zhang, K.; Nie, S. Glucomannans Alleviated the Progression of Diabetic Kidney Disease by Improving Kidney Metabolic Disturbance. *Mol. Nutr. Food Res.* **2019**, *63*, 1801008. [CrossRef]
73. Iluz-Freundlich, D.; Zhang, M.; Uhanova, J.; Minuk, G.Y. The relative expression of hepatocellular and cholestatic liver enzymes in adult patients with liver disease. *Ann. Hepatol.* **2020**, *19*, 204–208. [CrossRef]
74. Lin, H.-M.; Pang, J.; Deng, S.-G. Protective Effect of Konjac Glucomannan on Carbon Tetrachloride Induced Liver Injury in Mice. *J. Zhejiang Ocean. Univ.* **2009**, *28*, 465–467.
75. Tang, B.L. Glucose, glycolysis, and neurodegenerative diseases. *J. Cell. Physiol.* **2020**, *235*, 7653–7662. [CrossRef]
76. Ahmed, D.; Khan, M.I.; Sharma, M.; Khan, M.F. Novel pentacyclic triterpene isolated from seeds of *Euryale Ferox Salisb.* ameliorates diabetes in streptozotocin induced diabetic rats. *Interdiscip. Toxicol.* **2018**, *11*, 275–288. [CrossRef]
77. Subash-Babu, P.; Ignacimuthu, S.; Alshatwi, A.A. Nymphayol increases glucose-stimulated insulin secretion by RIN-5F cells and GLUT4-mediated insulin sensitization in type 2 diabetic rat liver. *Chem.-Biol. Interact.* **2015**, *226*, 72–81. [CrossRef]
78. Wellen, K.E. Inflammation, stress, and diabetes. *J. Clin. Investig.* **2005**, *115*, 1111–1119. [CrossRef]
79. Asmat, U.; Abad, K.; Ismail, K. Diabetes mellitus and oxidative stress—A concise review. *Saudi Pharm. J.* **2016**, *24*, 547–553. [CrossRef]
80. Anto Michel, N.; Colberg, C.; Buscher, K.; Sommer, B.; Pramod, A.B.; Ehinger, E.; Dufner, B.; Hoppe, N.; Pfeiffer, K.; Marchini, T.; et al. Inflammatory Pathways Regulated by Tumor Necrosis Receptor–Associated Factor 1 Protect From Metabolic Consequences in Diet-Induced Obesity. *Circ. Res.* **2018**, *122*, 693–700. [CrossRef]
81. Bourebaba, L.; Bedjou, F.; Röcken, M.; Marycz, K. Nortropane alkaloids as pharmacological chaperones in the rescue of equine adipose-derived mesenchymal stromal stem cells affected by metabolic syndrome through mitochondrial potentiation, endoplasmic reticulum stress mitigation and insulin resistance alleviation. *Stem. Cell Res. Ther.* **2019**, *10*, 178. [CrossRef]
82. Jeon, J.; Jang, J.; Park, K. Effects of Consuming Calcium-Rich Foods on the Incidence of Type 2 Diabetes Mellitus. *Nutrients* **2018**, *11*, 31. [CrossRef]
83. Song, S.; Lee, J.E. Dietary Patterns Related to Triglyceride and High-Density Lipoprotein Cholesterol and the Incidence of Type 2 Diabetes in Korean Men and Women. *Nutrients* **2018**, *11*, 8. [CrossRef]

84. Chen, H.; Ji, H.; Kong, X.; Lei, P.; Yang, Q.; Wu, W.; Jin, L.; Sun, D. Bacterial Ghosts-Based Vaccine and Drug Delivery Systems. *Pharmaceutics* **2021**, *13*, 1892. [CrossRef]
85. Jung, M.-J.; Lee, J.; Shin, N.-R.; Kim, M.-S.; Hyun, D.-W.; Yun, J.-H.; Kim, P.S.; Whon, T.W.; Bae, J.-W. Chronic Repression of mTOR Complex 2 Induces Changes in the Gut Microbiota of Diet-induced Obese Mice. *Sci. Rep.* **2016**, *6*, 30887. [CrossRef]
86. Eckburg, P.B.; Bik, E.M.; Bernstein, C.N.; Purdom, E.; Dethlefsen, L.; Sargent, M.; Gill, S.R.; Nelson, K.E.; Relman, D.A. Diversity of the Human Intestinal Microbial Flora. *Science* **2005**, *308*, 1635–1638. [CrossRef]
87. Krebs, M.; Brunmair, B.; Brehm, A.; Artwohl, M.; Szendroedi, J.; Nowotny, P.; Roth, E.; Fürnsinn, C.; Promintzer, M.; Anderwald, C.; et al. The Mammalian Target of Rapamycin Pathway Regulates Nutrient-Sensitive Glucose Uptake in Man. *Diabetes* **2007**, *56*, 1600–1607. [CrossRef]
88. Laplante, M.; Sabatini, D.M. mTOR Signaling in Growth Control and Disease. *Cell* **2012**, *149*, 274–293. [CrossRef]
89. Chen, H.; Nie, Q.; Hu, J.; Huang, X.; Yin, J.; Nie, S. Multiomics Approach to Explore the Amelioration Mechanisms of Glucomannans on the Metabolic Disorder of Type 2 Diabetic Rats. *J. Agric. Food Chem.* **2021**, *69*, 2632–2645. [CrossRef]
90. Zalewski, B.M.; Chmielewska, A.; Szajewska, H. The effect of glucomannan on body weight in overweight or obese children and adults: A systematic review of randomized controlled trials. *Nutrition* **2015**, *31*, 437–442.e2. [CrossRef]
91. Keithley, J.K.; Swanson, B.; Mikolaitis, S.L.; DeMeo, M.; Zeller, J.M.; Fogg, L.; Adamji, J. Safety and Efficacy of Glucomannan for Weight Loss in Overweight and Moderately Obese Adults. *J. Obes.* **2013**, *2013*, 610908. [CrossRef]
92. Arvill, A.; Bodin, L. Effect of short-term ingestion of konjac glucomannan on serum cholesterol in healthy men. *Am. J. Clin. Nutr.* **1995**, *61*, 585–589. [CrossRef]
93. Lei, P.; Chen, H.; Ma, J.; Fang, Y.; Qu, L.; Yang, Q.; Peng, B.; Zhang, X.; Jin, L.; Sun, D. Research progress on extraction technology and biomedical function of natural sugar substitutes. *Front. Nutr.* **2022**, *9*, 952147. [CrossRef]
94. Bonelli, A.; Menna, P.; Minotti, G.; Angeletti, S.; Comandini, A.; Picollo, R.; Quarchioni, E.; Russo, V.; Salvatori, E.; Ferravante, F.; et al. Safety and tolerability of a novel oral nutritional supplement in healthy volunteers. *Clin. Nutr.* **2021**, *40*, 946–955. [CrossRef]
95. Folwarski, M.; Kłęk, S.; Zoubek-Wójcik, A.; Szafrański, W.; Bartoszewska, L.; Figuła, K.; Jakubczyk, M.; Jurczuk, A.; Kamocki, Z.; Kowalczyk, T.; et al. Foods for Special Medical Purposes in Home Enteral Nutrition-Clinical Practice Experience. Multicenter Study. *Front. Nutr.* **2022**, *9*, 906186. [CrossRef] [PubMed]
96. Zhang, W.; Ren, X.; Zhang, L.; Chen, J. Preparation and Performance of Thickened Liquids for Patients with Konjac Glucomannan-Mediated Dysphagia. *Molecules* **2022**, *27*, 2194. [CrossRef]
97. Cichero, J.A.Y.; Steele, C.; Duivestein, J.; Clavé, P.; Chen, J.; Kayashita, J.; Dantas, R.; Lecko, C.; Speyer, R.; Lam, P.; et al. The Need for International Terminology and Definitions for Texture-Modified Foods and Thickened Liquids Used in Dysphagia Management: Foundations of a Global Initiative. *Curr. Phys. Med. Rehabil. Rep.* **2013**, *1*, 280–291. [CrossRef]
98. Tatirat, O. Physicochemical properties of extrusion-modified konjac glucomannan. *Carbohydr. Polym.* **2012**, *7*, 1545–1551. [CrossRef]
99. Chen, M.; Wang, H.; Yan, Q.; Zheng, Q.; Yang, M.; Lv, Z.; He, M.; Feng, L.; Zhao, J.; Tang, T.; et al. Effects of dietary oxidized konjac glucomannan sulfates (OKGMS) and acidolysis-oxidized konjac glucomannan (A-OKGM) on the immunity and expression of immune-related genes of Schizothorax prenanti. *Fish Shellfish Immunol.* **2016**, *56*, 96–105. [CrossRef]
100. Zhang, L. Effects of oxidized konjac glucomannan on the intestinal microbial flora and intestinal morphology of Schizothorax prenanti. *Aquacult. Int.* **2017**, *18*, 233–250. [CrossRef]
101. Zheng, Q.; Wu, Y.; Xu, H. Effect of dietary oxidized konjac glucomannan on Schizothorax prenanti growth performance, body composition, intestinal morphology and intestinal microflora. *Fish Physiol. Biochem.* **2015**, *41*, 733–743. [CrossRef]
102. Liu, M.; Zou, T.; Li, H.; Zhou, M.; Cheng, Z. Research Status of the Technology for Reducing Sulfur Dioxide Residue in Konjac Flour. *Farm Prod. Process.* **2021**, *2021*, 59–62.
103. Vuksan, V.; Sievenpiper, J.L.; Xu, Z.; Wong, E.Y.Y.; Jenkins, A.L.; Beljan-Zdravkovic, U.; Leiter, L.A.; Josse, R.G.; Stavro, M.P. Konjac-Mannan and American Ginsing: Emerging Alternative Therapies for Type 2 Diabetes Mellitus. *J. Am. Coll. Nutr.* **2001**, *20*, S370–S380. [CrossRef] [PubMed]
104. Deng, J.; Zhong, J.; Long, J.; Zou, X.; Wang, D.; Song, Y.; Zhou, K.; Liang, Y.; Huang, R.; Wei, X.; et al. Hypoglycemic effects and mechanism of different molecular weights of konjac glucomannans in type 2 diabetic rats. *Int. J. Biol. Macromol.* **2020**, *165*, 2231–2243. [CrossRef] [PubMed]

Disclaimer/Publisher's Note: The statements, opinions and data contained in all publications are solely those of the individual author(s) and contributor(s) and not of MDPI and/or the editor(s). MDPI and/or the editor(s) disclaim responsibility for any injury to people or property resulting from any ideas, methods, instructions or products referred to in the content.

MDPI AG
Grosspeteranlage 5
4052 Basel
Switzerland
Tel.: +41 61 683 77 34

Foods Editorial Office
E-mail: foods@mdpi.com
www.mdpi.com/journal/foods

Disclaimer/Publisher's Note: The title and front matter of this reprint are at the discretion of the Guest Editors. The publisher is not responsible for their content or any associated concerns. The statements, opinions and data contained in all individual articles are solely those of the individual Editors and contributors and not of MDPI. MDPI disclaims responsibility for any injury to people or property resulting from any ideas, methods, instructions or products referred to in the content.

www.ingramcontent.com/pod-product-compliance
Lightning Source LLC
Chambersburg PA
CBHW041944260926
51850CB00018B/685
* 9 7 8 3 7 2 5 8 7 7 4 9 2 *